Indoor Positioning and Navigation

Indoor Positioning and Navigation

Editor

Simon Tomažič

MDPI • Basel • Beijing • Wuhan • Barcelona • Belgrade • Manchester • Tokyo • Cluj • Tianjin

Editor
Simon Tomažič
Laboratory of Control Systems
and Cybernetics, Faculty of
electrical engineering
University of Ljubljana
Ljubljana
Slovenia

Editorial Office
MDPI
St. Alban-Anlage 66
4052 Basel, Switzerland

This is a reprint of articles from the Special Issue published online in the open access journal *Sensors* (ISSN 1424-8220) (available at: www.mdpi.com/journal/sensors/special_issues/position_navigation).

For citation purposes, cite each article independently as indicated on the article page online and as indicated below:

LastName, A.A.; LastName, B.B.; LastName, C.C. Article Title. *Journal Name* **Year**, *Volume Number*, Page Range.

ISBN 978-3-0365-1913-5 (Hbk)
ISBN 978-3-0365-1912-8 (PDF)

Contents

sensors

MDPI

Editorial

Indoor Positioning and Navigation

Simon Tomažič

Faculty of Electrical Engineering, University of Ljubljana, 1000 Ljubljana, Slovenia; simon.tomazic@fe.uni-lj.si; Tel.: +386-1-4768-760

check for updates

Citation: Tomažič, S. Indoor Positioning and Navigation. *Sensors* **2021**, *21*, 4793. https://doi.org/10.3390/s21144793

Received: 8 July 2021
Accepted: 12 July 2021
Published: 14 July 2021

Recently, the social and commercial interest in location-based services (LBS) has been increasing significantly. The scientific field of indoor localization and navigation has undergone rapid development due to many studies considering advanced mobile and communication technologies [1]. The purpose of developing new localization algorithms and navigation systems is to enable autonomous mobile systems [2] to use these solutions in performing a specific task or to assist people who have lost their ability to navigate [3] (e.g., blind people, people with Alzheimer's disease). A secure, user-friendly, and accurate indoor positioning system (IPS) that can run on a smartphone could open the door to many innovative applications and create new business opportunities. The global indoor location market is expected to reach a worth of $40.99 billion by 2022 [1]. A low-cost real-time locating systems (RTLS) are very useful as they can guide people through airports, shopping malls, museums, etc. [1], or improve production control and logistics [4]. Modern smartphones are equipped with numerous sensors (inertial sensors, camera, barometer) and communication modules (WiFi, Bluetooth, NFC, LTE/5G, ultra-wideband), which enable the implementation of different localization algorithms, namely visual localization, inertial navigation system and radio localization [1]. For the mapping of indoor environment and localization of autonomous mobile systems, LIDAR sensor is also commonly used besides smartphone sensors [2,5,6]. Since visual localization and the inertial navigation systems (INS) are sensitive to external disturbances, the sensor fusion approaches based on Kalman filters and (deep) neural networks can be used to implement robust localization algorithms, as proposed in [3,7]. The localization algorithms need to be optimized in order to be computationally efficient, which is essential for real-time processing and low energy consumption on a smartphone or robot. A practical indoor positioning system (IPS) should have characteristics such as ease of implementation, acceptable localization accuracy, scalable system, feasible system cost and minimal computational complexity.

In general, there are two main approaches to indoor localization, namely infrastructure-based and infrastructureless approaches [1]. The latter generally uses fingerprints of environmental features such as sound, light, magnetic field, or is based on smartphone sensors (e.g., accelerometer, gyroscope, etc.). Infrastructure-based methods can use pre-installed visual sensors or wireless technologies such as ZigBee, WiFi, Ultra-Wideband (UWB) [2,6], Radio Frequency Identification (RFID) [5], and Bluetooth Low Energy (BLE) [1]. An infrastructure-based indoor positioning system can be expensive, either because of the methods required or because of the expensive hardware components. Installing the necessary infrastructure is often time-consuming and labor-intensive, which drives up the price. To solve this problem, authors S. Tomažič and I. Škrjanc [8] developed an automated indoor localization system that combines all the necessary components to realize low-cost Bluetooth localization with the least data acquisition and network configuration overhead. The proposed system incorporates a sophisticated visual-inertial localization algorithm for a fully automated collection of Bluetooth signal strength data. The visual-inertial SLAM algorithm, which is part of the ARCore library, successfully fuses information from the camera and inertial sensors to provide accurate localization in large spaces over a long period of time. A suitable collection of measurements can be made quickly and easily, clearly defining which part of the space is not yet well covered by measurements. An automated system has many advantages in various approaches to data collection, e.g., quick and easy construction of datasets for model development or fingerprinting, beacon parameter studies, in-depth studies of beacon location and density in a given environment, and quick maintenance of the database as the environment changes [8].

Radio-based indoor positioning can generally be divided into three groups: Proximity, Time of Flight (TOF) measurements [9], and Received Signal Strength Indicator (RSSI)-based methods [10]. In some cases, these were also supplemented with an Angle of Arrival (AOA) localization approach [11]. Depending on the localization approach chosen to compute the current position, the following methods are most commonly used: triangulation, trilateration, and fingerprinting. Among these, fingerprinting is perhaps the most popular due to its simplicity: it is based on signal strength (RSSI) and its procedure is essentially to collect the signal from the transmitters and assign it to a specific position [1]. Since the RSSI suffers from low stability due to interference with objects and environmental effects, the authors Shi et al. [10] proposed a tri-partition RSSI classification and its tracing algorithm as an RSSI filter, which enables lower variance range. Using the filter improves the accuracy of trilateration-based positioning by 20.5%. In recent years, many commercial WiFi devices support the collection of physical layer channel state information (CSI). CSI is an index that can characterize signal characteristics with finer granularity than RSSI. Compared to RSSI, CSI can avoid the effects of multipath and noise by analyzing the properties of multichannel subcarriers. To improve the indoor location accuracy and the efficiency of the algorithm, Wang et al. [12] proposed a hybrid fingerprint location technology based on RSSI and CSI. The accurate hybrid fingerprint database was constructed after the dimensionality reduction of the obtained high-dimensional data values. Weighted k-nearest neighbor (WKNN) algorithm was applied to reduce the complexity of the algorithm during the online positioning phase.

BLE is one of the most widely used technologies in ubiquitous computing and many Internet-of-Things (IoT) applications because it offers many advantages such as low power consumption and low cost. As 5G technology is rising across the world and UWB chip is available in the latest smartphones, IPS can leverage and integrate these technologies in the future to develop a better IPS [1]. Authors Blaszkiewicz et al. [13] presented an interesting positioning solution, which utilizes existing radiating cables in tunnels or corridors without the need to deploy a dense network of reference nodes. Radiating cables are mostly used to provide radio communication in tunnels or corridors, but can also be used to estimate the position of a mobile terminal along the cable.

To improve indoor localization based on the time-of-arrival (TOA) principle, Deng et al. [9] proposed a clock synchronization solution for dynamic networks called Multi–Gaussian Variational Message Passing (M-VMP) method. The latter improved the positioning accuracy and convergence speed for mixed Line-of-Sight (LOS) and Non-Line-of-Sight (NLOS) environments. To optimize the anchor node density in TOA positioning approach the authors Deng et al. [14] developed a location source optimization algorithm based on fuzzy comprehensive evaluation. Authors Groth et al. [11] went one step further and developed a calibration-free single-anchor indoor localization method, including a dedicated algorithm and all necessary hardware modules. They showed that a single base station equipped with an ESPAR antenna to perform the measurements could be used to find the position of an unknown BLE tag without calibration or recalibration, since low-cost reference modules installed on walls inside the test area provided enough reference information for the positioning algorithm.

Global navigation satellite systems (GNSS) have long been employed for LBS to navigate and provide accurate and reliable location information in outdoor environments. However, GNSS signals are too weak to penetrate buildings and unable to provide reliable indoor LBS. In order to overcome this problem, authors Uzun et al. [15] proposed an indoor positioning system using Global Positioning System (GPS) signals in the 433 MHz Industrial Scientific Medical (ISM) band. The proposed method is based on down-converting (DC) repeaters and an up-converting (UC) receiver. The repeaters receive outdoor GPS signals at

1575.42 MHz (L1 band), down-convert them to the 433 MHz ISM band, then amplify and retransmit them to the indoor environment. When GPS signals at 433 MHz are received by the up-converting receiver, it amplifies these signals and up-converts them back to the L1 frequency. Then, the commercially available GPS receiver calculates the pseudo-ranges.

In the field of indoor localization, visible light positioning (VLP) systems are also very promising. Authors Amsters et al. [16] presented an innovative solution for calibrating VLP systems using a mobile robot to facilitate data acquisition. The new approach significantly improved performance compared to previous studies, almost doubling the accuracy of LED localization. The authors showed that the ambient illumination had little impact on the proposed method. Authors Jaenal et al. [17] developed an appearance-based robot localization in 2D with a sparse, lightweight map of the environment consisting of descriptor–pose image pairs. The authors proposed a piecewise approximation of the geometry of descriptor manifold through a tessellation of the so-called Patches of Smooth Appearance Change (PSACs), which defines their appearance map. The authors' proposal is based on the assumption that the global image descriptors form a manifold articulated by the camera pose that adequately approximates the Image Manifold. On the map, the presented robot localization method applies both a Gaussian Process Particle Filter (GPPF) to perform camera tracking and a place recognition technique for re-localization within the most likely PSACs according to the observed descriptor.

Due to the benefits of indoor positioning technology, numerous indoor navigation applications have been deployed in large buildings, such as hospitals, airports, and train stations, to guide visitors to their destinations. A commonly used user interface displayed on smartphones is a 2D floor map with a route to the destination. Navigation instructions, such as turn left, turn right, and go straight on, are displayed on the screen when the user arrives at an intersection. However, due to the limitations of a 2D navigation map, users may be under mental pressure and become confused while making a connection between the real environment and the 2D navigation map before proceeding. For this reason, authors Huang et al. [18] developed ARBIN, an augmented reality-based navigation system that displays navigation instructions on the real environment screen for ease of use. The positions are determined using BLE beacons and RSSI models.

Several indoor localization solutions are already being used in practice or are ripe for implementation in real-world environments. Nevertheless, there is still much room for development of new approaches and standards in the field of indoor localization. Hopefully, these new standards will make new solutions as useful as GNSS is for outdoor localization and navigation.

Funding: This research received no external funding.

Institutional Review Board Statement: Not applicable.

Informed Consent Statement: Not applicable.

Conflicts of Interest: The author declares no conflict of interest.

References

1. Subedi, S.; Pyun, J.Y. A Survey of Smartphone-Based Indoor Positioning System Using RF-Based Wireless Technologies. *Sensors* **2020**, *20*, 7230. [CrossRef] [PubMed]
2. Orjales, F.; Losada-Pita, J.; Paz-Lopez, A.; Deibe, A. Towards Precise Positioning and Movement of UAVs for Near-Wall Tasks in GNSS-Denied Environments. *Sensors* **2021**, *21*, 2194. [CrossRef] [PubMed]
3. Mahida, P.; Shahrestani, S.; Cheung, H. Deep Learning-Based Positioning of Visually Impaired People in Indoor Environments. *Sensors* **2020**, *20*, 6238. [CrossRef] [PubMed]
4. Rácz-Szabó, A.; Ruppert, T.; Bántay, L.; Löcklin, A.; Jakab, L.; Abonyi, J. Real-Time Locating System in Production Management. *Sensors* **2020**, *20*, 6766. [CrossRef] [PubMed]
5. Fu, W.; Liu, R.; Wang, H.; Ali, R.; He, Y.; Cao, Z.; Qin, Z. A Method of Multiple Dynamic Objects Identification and Localization Based on Laser and RFID. *Sensors* **2020**, *20*, 3948. [CrossRef] [PubMed]
6. Vazquez-Lopez, R.; Herrera-Lozada, J.C.; Sandoval-Gutierrez, J.; von Bülow, P.; Martinez-Vazquez, D.L. Sensor Information Sharing Using a Producer-Consumer Algorithm on Small Vehicles. *Sensors* **2021**, *21*, 3022. [CrossRef] [PubMed]

7. Jamil, F.; Iqbal, N.; Ahmad, S.; Kim, D.H. Toward Accurate Position Estimation Using Learning to Prediction Algorithm in Indoor Navigation. *Sensors* **2020**, *20*, 4410. [CrossRef] [PubMed]
8. Tomažič, S.; Škrjanc, I. An Automated Indoor Localization System for Online Bluetooth Signal Strength Modeling Using Visual-Inertial SLAM. *Sensors* **2021**, *21*, 2857. [CrossRef] [PubMed]
9. Deng, Z.; Tang, S.; Jia, B.; Wang, H.; Deng, X.; Zheng, X. Cooperative Localization and Time Synchronization Based on M-VMP Method. *Sensors* **2020**, *20*, 6315. [CrossRef] [PubMed]
10. Shi, Y.; Shi, W.; Liu, X.; Xiao, X. An RSSI Classification and Tracing Algorithm to Improve Trilateration-Based Positioning. *Sensors* **2020**, *20*, 4244. [CrossRef] [PubMed]
11. Groth, M.; Nyka, K.; Kulas, L. Calibration-Free Single-Anchor Indoor Localization Using an ESPAR Antenna. *Sensors* **2021**, *21*, 3431. [CrossRef] [PubMed]
12. Wang, J.; Park, J. An Enhanced Indoor Positioning Algorithm Based on Fingerprint Using Fine-Grained CSI and RSSI Measurements of IEEE 802.11n WLAN. *Sensors* **2021**, *21*, 2769. [CrossRef] [PubMed]
13. Blaszkiewicz, O.; Sadowski, J.; Stefanski, J. Position Estimation in Corridors Along the Coupled Mode of Radiating Cables. *Sensors* **2020**, *20*, 5064. [CrossRef] [PubMed]
14. Deng, Z.; Tang, S.; Deng, X.; Yin, L.; Liu, J. A Novel Location Source Optimization Algorithm for Low Anchor Node Density Wireless Sensor Networks. *Sensors* **2021**, *21*, 1890. [CrossRef] [PubMed]
15. Uzun, A.; Ghani, F.A.; Ahmadi Najafabadi, A.M.; Yenigün, H.; Tekin, I. Indoor Positioning System Based on Global Positioning System Signals with Down- and Up-Converters in 433 MHz ISM Band. *Sensors* **2021**, *21*, 4338. [CrossRef] [PubMed]
16. Amsters, R.; Demeester, E.; Stevens, N.; Slaets, P. Calibration of Visible Light Positioning Systems with a Mobile Robot. *Sensors* **2021**, *21*, 2394. [CrossRef] [PubMed]
17. Jaenal, A.; Moreno, F.A.; Gonzalez-Jimenez, J. Appearance-Based Sequential Robot Localization Using a Patchwise Approximation of a Descriptor Manifold. *Sensors* **2021**, *21*, 2483. [CrossRef] [PubMed]
18. Huang, B.C.; Hsu, J.; Chu, E.T.H.; Wu, H.M. ARBIN: Augmented Reality Based Indoor Navigation System. *Sensors* **2020**, *20*, 5890. [CrossRef] [PubMed]

sensors

Article

An Automated Indoor Localization System for Online Bluetooth Signal Strength Modeling Using Visual-Inertial SLAM

Simon Tomažič * and Igor Škrjanc

Faculty of Electrical Engineering, University of Ljubljana, 1000 Ljubljana, Slovenia; igor.skrjanc@fe.uni-lj.si
* Correspondence: simon.tomazic@fe.uni-lj.si; Tel.: +386-1-4768-760

Abstract: Indoor localization is becoming increasingly important but is not yet widespread because installing the necessary infrastructure is often time-consuming and labor-intensive, which drives up the price. This paper presents an automated indoor localization system that combines all the necessary components to realize low-cost Bluetooth localization with the least data acquisition and network configuration overhead. The proposed system incorporates a sophisticated visual-inertial localization algorithm for a fully automated collection of Bluetooth signal strength data. A suitable collection of measurements can be quickly and easily performed, clearly defining which part of the space is not yet well covered by measurements. The obtained measurements, which can also be collected via the crowdsourcing approach, are used within a constrained nonlinear optimization algorithm. The latter is implemented on a smartphone and allows the online determination of the beacons' locations and the construction of path loss models, which are validated in real-time using the particle swarm localization algorithm. The proposed system represents an advanced innovation as the application user can quickly find out when there are enough data collected for the expected radiolocation accuracy. In this way, radiolocation becomes much less time-consuming and labor-intensive as the configuration time is reduced by more than half. The experiment results show that the proposed system achieves a good trade-off in terms of network setup complexity and localization accuracy. The developed system for automated data acquisition and online modeling on a smartphone has proved to be very useful, as it can significantly simplify and speed up the installation of the Bluetooth network, especially in wide-area facilities.

Keywords: indoor localization; visual-inertial SLAM; constrained optimization; path loss model; particle swarm optimization; Bluetooth low energy; beacon

Citation: Tomažič, S.; Škrjanc, I. An Automated Indoor Localization System for Online Bluetooth Signal Strength Modeling Using Visual-Inertial SLAM. *Sensors* **2021**, *21*, 2857. https://doi.org/10.3390/s21082857

Academic Editor: David Plets

Received: 18 March 2021
Accepted: 16 April 2021
Published: 19 April 2021

1. Introduction

Recently, the social and commercial interest in Location-Based Services (LBS) is increasing significantly. The scientific field of indoor localization and navigation has experienced rapid development through many studies that consider advanced mobile and communication technologies [1]. The purpose of developing new localization algorithms and navigation systems is to enable autonomous mobile systems to use these solutions in performing a specific task or to assist people who have lost their ability to navigate [2] (e.g., blind people, people with Alzheimer's disease). A secure, user-friendly, and accurate localization method that can run on a smartphone could open the door to many innovative applications and create new businesses opportunities. The global indoor location market is expected to reach a worth of $40.99 billion by 2022 [3]. A low-cost indoor positioning system (IPS) is very helpful as it can guide people through airports, shopping malls, museums etc. [1,4]. In these environments, it can be very difficult to figure out where to go without spending a lot of time looking for directions. Accurate indoor localization represents a major challenge, mainly due to the fact that GNSS (Global Navigation Satellite System) signals are not available indoors [5]. The modern smartphone is increasingly used as a Personal Navigation System (PNS) because it contains many sensors (accelerometer, gyroscope, magnetometer, altimeter, camera) and communication modules (Bluetooth, WiFi, NFC, 5G, ultra-wideband), and, last but not least, almost everyone has it in their pocket [1,6]. Moreover, smartphones contain increasingly powerful multicore processors that allow the implementation of computationally intensive localization algorithms. A practical Indoor Positioning System (IPS) should have characteristics such as ease of implementation, acceptable localization accuracy, scalable system, feasible system cost and minimal computational complexity [7].

There are two main approaches to indoor localization, namely infrastructure-based and infrastructure-less approaches [8]. The latter generally uses fingerprints from environmental features such as sound, light, magnetic field or smartphone sensors. Infrastructure-based methods may use pre-installed visual sensors or wireless technologies such as ZigBee [9], WiFi, Ultra-Wideband (UWB) [3], Radio Frequency Identification (RFID) [10] and Bluetooth Low Energy (BLE) [1]. An infrastructure-based indoor positioning system can be expensive, either because of the methods required or because of the expensive hardware components. Among all these technologies, BLE is one of the most widely used in ubiquitous computing and many Internet-of-Things (IoT) applications because it has many advantages such as low power consumption and low cost [7,11]. BLE transmitters or beacons are portable and easy to deploy, they have the potential to provide high positioning accuracy, and they can provide advanced services to users. ZigBee consumes less energy than BLE, but it is not as widely supported on smartphones. WiFi is widely supported but it has a higher energy requirement than BLE [1]. Regarding Ultra-Wideband (UWB), which is an emerging technology (still not supported on smartphones) in the field of indoor positioning, Dardari et al. [12] provided a detailed comparative analysis of UWB positioning technologies.

Bluetooth low-energy transmitters are used in combination with various standards or protocols, such as iBeacon [13] and Eddystone [14], which define how data packets are transmitted. For localization purposes, the advertising mode, in which data packets are sent periodically, is most established. In this mode, beacon messages are advertised on the three primary channels 37, 38 and 39, to reduce interference with other wireless technologies (e.g., with WiFi channels 1, 6 and 11) and increase redundancy. In addition to broadcast transmissions, the advertising channels allow a device to be discovered and securely connected.

Radio-based indoor positioning can generally be divided into three groups: Proximity, Time of Flight (TOF) measurements, and Received Signal Strength Indicator (RSSI)-based methods [15]. In some cases, these have also been supplemented with an Angle of Arrival (AOA) localization approach [16]. Depending on the localization approach chosen to calculate the current position, the following methods are most commonly used: Triangulation, trilateration and fingerprinting. Among these, fingerprinting is perhaps the most popular due to its simplicity: It is based on signal strength (RSSI) and its procedure is basically to collect the signal from the transmitters and assign it to a specific position. It consists of two phases: The offline (the calibration or training) phase [11] and the online (the positioning) phase [17]. The most common fingerprinting matching algorithms can be classified as: (a) probabilistic; (b) deterministic, such as k-nearest neighbor (or weighted k-nearest neighbor); and (c) machine learning- and sparse sampling-based [18,19]. Positioning using BLE fingerprinting has the potential to achieve high accuracy if sufficiently dense training data are available. However, this process is time- and labor-intensive, which is its main drawback [20]. Another problem associated with the fingerprinting method is that the time complexity of the execution phase increases with the size of the localization area. Moreover, the instability of RSSI in the indoor environment forces frequent updating of the radio map database [21]. Therefore, a better alternative to the fingerprinting approach could be model-based methods, among which trilateration is the most popular [15]. Localization solutions based on trilateration mostly use path loss models to estimate the range from the RSSI of the available beacons [11]. In addition to the trilateration localization method, weighted centroid localization also relies on the signal propagation model to estimate the distance from the RSSI [22]. The advantages of using models over the fingerprinting approach are a more effortless adjustment of the radio maps as the environment changes and better extrapolation of RSSI signals for areas where measurements have not been made.

The main challenge in fingerprinting- and model-based IPS is that accuracy is affected by several factors, such as signal-related and environment-related factors [15]. In terms of signal-related factors, the following can often occur: Large fluctuations, reflection, path loss, non-line-of-sight conditions and multipath fading, and many of these factors also depend on the type of material present in the environment [23]. For environment-related factors, the main ones are changes in hardware or furniture, the presence of people, or ambient humidity conditions [24].

Regardless of the localization technology used, there are several more or less labor-intensive approaches to collect localization data in the modeling or fingerprinting calibration phase, which can be classified as follows [11,18,20]:

- A fully manual approach consisting of a calibration phase, e.g., the traditional manual survey, where the user collects the signal at discrete and uniformly distributed survey points.
- A semi-automated approach that attempts to reduce the time and effort of the calibration phase, e.g., with the use of interpolation-based methods, the user attempts to construct a signal map from a sparse set of fingerprints collected while walking through a space.
- A fully automated approach that does not require any calibration phase. It uses only online RSSI measurements or those that in some way merge the calibration phase and the positioning phase, e.g., implicit or explicit crowdsourcing that involves users in data collection.

An automated system has many advantages in various approaches to data collection, e.g., quick and easy construction of datasets for model development or fingerprinting, beacon parameter studies, in-depth studies of beacon location and density in a particular environment, and rapid maintenance of the database as the environment changes [8,20].

Related Work

In this section, the works related to the study presented in this article are analyzed. First, an overview of the most commonly used technologies in indoor positioning is given, with a later focus on the technologies that use radio frequency signals, with a greater emphasis on BLE technology.

Self-localization of Unmanned Ground Vehicles (UGV) in indoor environments is already a well-developed field, as they can be equipped with more powerful hardware and additional sensors, e.g., depth camera, stereo camera, LIDAR, ultrasonic sensors, etc. [25]. In conjunction with the sensor LIDAR, the method of simultaneous localization and mapping (SLAM) [26] in particular has become established, which can also process the information obtained from the camera [27] (visual SLAM). Within the visual localization approach, several algorithms have been developed to determine an agent's motion (person, vehicle, robot) to which the camera is attached. The established methods include: Simultaneous Localization and Mapping (SLAM) [27], Visual Odometry (VO) [28,29], Structure From Motion (SFM) and image-to-map matching [30]. The SLAM and SFM methods are fairly computationally demanding since they construct a 3D map of the environment in addition to the motion estimation.

For indoor localization purposes, the Inertial Navigation System (INS) [16,31] based on the Inertial Measurement Unit (IMU) is often used as a complementary system to radio- or visual-based localization. The pedestrian inertial navigation system, which uses a dead reckoning approach, usually consists of a pedometer (or step counter) and a digital compass, which enable to calculate the current pose according to the starting point [3,32].

According to the currently existing solutions and studies [33], the most suitable smartphone indoor localization approach is based on the measurement of WiFi [2], Bluetooth [34] and geomagnetic field signals [35]. However, localization methods based on only one technology or sensor often cannot provide the required positioning accuracy, so new approaches based on sensor fusion [3,36] are needed. Many studies show that a combination of inertial sensors and camera-based techniques has great potential since they provide better robustness and higher localization accuracy [37]. The fusion of data from a camera and inertial sensors can be performed using Kalman filters as proposed by Sirtkaya et al. [38]. Since image processing places a heavy load on the processor and therefore the battery, visual localization is not always the best choice for indoor localization. Using visual localization within a smartphone application can also be inconvenient as the user has to constantly provide the correct direction of the camera according to the environment [39]. This is not necessary with the radiolocation, which is a significant advantage.

The disadvantage of indoor radiolocation over visual localization is that it requires some infrastructure. From this point of view, WiFi localization takes precedence over Bluetooth-based localization because WiFi networks are already present in most buildings. However, WiFi localization does not enable high localization accuracy in existing networks since WiFi access points are usually rather sparsely located [2]. Due to this disadvantage in WiFi localization, Bluetooth Low Energy (BLE) [5,40] has been established for the purpose of indoor localization. BLE has many advantages over WiFi: from the low price of transmitters, low power consumption and better robustness to the smaller size of integrated circuits [41,42]. Li and Ma [43] took a hybrid approach using BLE and WiFi. They used BLE tags and BLE/WiFi repeaters. Their system performs position calculations based on two values: RSSI fingerprint and cell of origin. They conducted experiments in a rest area and an office area and obtained errors of 1.2 and 1.37 m for these areas, respectively.

In the field of radiolocation, three approaches based on an analysis of Bluetooth signal strengths have been established: Methods that consider the strongest base station; methods that require the construction of path loss models, i.e., models of signal strengths (and use trilateration) [1,11,24]; and methods based on the principle of "fingerprints" [21,42]. Tosi et al. [44] reviewed in their study the main methodologies adopted to investigate BLE performance: they provided an analysis of the maximum number of connectable sensors, throughput, latency, power consumption and maximum achievable range, with the aim of identifying the current limitations of BLE technology. Zue et al. [45] proposed a graph optimization-based approach that combines fingerprinting-based methods and range-based methods. The authors performed a test in an area of 90 m $\times$ 37 m with two different numbers of beacons, namely 24 beacons (sparse) and 48 beacons (dense), to see the effect of beacon density on the position estimates. They obtained errors of 2.26 and 1.27 m with the sparse beacon environment and dense beacon environment, respectively. In [46], the authors implemented a fingerprinting algorithm with fuzzy logic type-2 that is suitable for use as an indoor positioning method with BLE beacons with an average localization error of 0.43 m. Since this solution is based on fingerprinting, it requires a time-consuming offline phase and needs several additional algorithms for support. Thus, the computational resource consumption and algorithmic program complexity are comparatively high.

In previous studies, researchers have already focused on developing automated or semi-automated data acquisition systems [47]. Peng et al., developed an efficient method to create and update a fingerprint database using an Unmanned Ground Vehicle (UGV) platform. To collect BLE and WiFi fingerprints, they used a smartphone installed on the UGV. According to the obtained results, the root mean square error of the positioning results was reduced by 20% compared to the traditional fingerprint collection methods. Gao et al. [48] proposed the so-called path survey technique, in which radio maps are created from a sparse set of BLE and WiFi fingerprints. These were collected in such a way that a person walked through a room. With the obtained results, they showed that with a path-survey technique (and using Gaussian processes), the created maps are of the same quality as in the case of manual collection. The authors of [18] proposed a semi-automated fingerprint collection system (static rotating platform) to simplify and speed up the data collection process for 50%. Similarly, Nastac et al. [49] discussed the problem of automated data acquisition in fingerprint-based indoor positioning. As a platform for collecting data, they used a robot with simple wheel encoder sensors for odometry-based positioning. The benefit of using a robot was most evident in the time savings, as the data acquisition time could be reduced from 16 to 2 h for 3000 observations. In the process, the measurement accuracy could also be increased compared to the manual approach.

When setting up a BLE network, the main challenge is to mark the BLE transmitters' locations on the map of the space and then collect the RSSI measurements for constructing the path loss model. These tasks are rather time-consuming since the network operator spends at least half a minute to measure the locations of each beacon (e.g., using a tape measure) or to collect the signal strengths at each measurement point. Therefore, this paper presents a fully automated approach for data collection and online model development on a smartphone, which is a significant advance over state-of-the-art solutions. The latter have one of the drawbacks, such as the requirement for manual measurement and inserting the beacon's locations, offline data processing, or the use of expensive accessories, such as a wheeled robot. The proposed system allows real-time determination of in which part of the space there is a lack of measurements or which transmitter model is not yet good enough. With the implemented fusion of visual SLAM and inertial navigation on a smartphone, the proposed system enables accurate localization during BLE data collection, which is a prerequisite for good models of signal strengths. These are used for parallel online localization with the Particle Swarm Optimization (PSO)-based method, which enables reinitialization of the SLAM algorithm and real-time validation of the models. In the online model development phase, the beacon's locations and the model parameters are calculated using the constrained nonlinear optimization method. Assume that the model construction completion criterion is met and the BLE localization accuracy is satisfactory. In this case, the visual SLAM algorithm can be turned off and the position is calculated using only the PSO method, which results in a lower CPU and battery load. In contrast to fingerprinting methods, the proposed system is quickly adaptable to changes in the room (e.g., a new beacon or furniture layout) and supports a crowdsourcing approach to data collection.

The paper is organized as follows. Section 2 describes a fully automated data collection approach based on visual-inertial SLAM. Section 3 provides the background to the online path loss model construction on a smartphone. In Section 4 the particle swarm optimization (PSO) method is proposed and described in detail. Section 5 presents the results of path loss model construction and Section 6 deals with pedestrian localization using the proposed algorithm. A discussion and concluding remarks are presented in Sections 7 and 8, respectively.

2. SLAM-Based Fully Automated Approach for Data Collection

When gathering measurements of BLE signal strengths, real-time localization that enables accurate determination of reference positions in space is essential. In previous studies, we have already shown that an algorithm combining an inertial and visual navigation system is best suited for this purpose [37,50]. We have developed a visual odometry that can very accurately determine the position of the smartphone in the short term by tracking incremental movements. The problem with visual odometry is that the position drifts over time and the error accumulates [51]. To improve the localization accuracy over time using a camera and the smartphone's inertial sensors, we used Google's ARCore library [52] in our fully automated system for data acquisition and model development. The ARCore algorithm is based on the state-of-the-art visual-inertial SLAM [53,54] or Concurrent Odometry and Mapping (COM) [55], which allows determining the pose of a smartphone with a monocular camera in an unknown environment, such as ORB-SLAM [56], MIMC-VINS [57] or LSD-SLAM [27]. Feature detection helps in calculating the relative change in position by searching for the same unique feature between frames. Using motion tracking capabilities, ARCore tracks the smartphone's position relative to the world Coordinate System (C. S.) by identifying key feature points (see Figure 1). This visual information is combined with inertial measurements (i.e., linear acceleration and angular velocity) from IMU to estimate the relative 6-DOF pose (position and orientation) of the smartphone with respect to the world coordinate system. ARCore has the ability to track up to 20 images simultaneously, with tracking occurring on the smartphone. In addition, ARCore can store up to 1000 reference images in an image database.

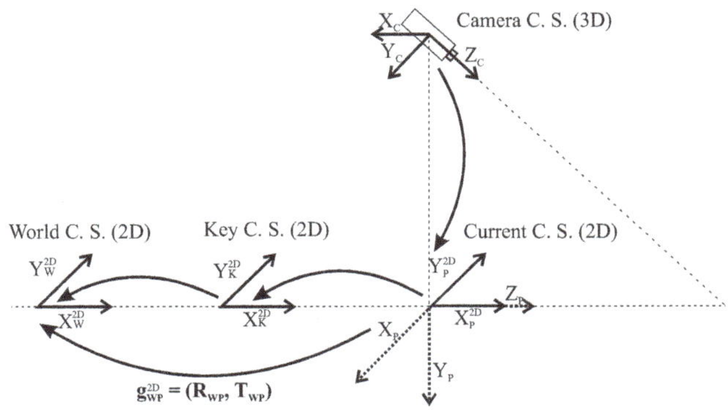

Figure 1. The ARCore tracking algorithm calculates transformation from camera C. S. to world C. S.

ARCore's tracking algorithm computes the 3D transformation from the world C. S. to the camera C. S. (Figure 1). To simplify pose representation, a 3D-to-2D transformation is required. First, to obtain the current 2D pose of a smartphone in the world C. S., the camera C. S. must be transformed to the current C. S. (from 3D to 2D), and second, the transformation g_{WP}^{2D} from the current C. S. to the world C. S. must be determined. The rotation matrix $\boldsymbol{R_{WP}}$ determines the heading and the translation vector $\boldsymbol{T_{WP}}$ determines the position of the smartphone in 2D space [37]. The world C. S., which is determined by the first frame, is not movable and it represents a reference C. S. in which the final result of the visual localization is expressed.

The concept of the ARCore COM algorithm is shown in Figure 2. The algorithm is divided into three main modules:

- Motion tracking module;
- Mapping module;
- Localization module.

ARCore takes visual data (captured images from smartphone camera) and inertial data (from IMU) as the main input sources. Each of the modules has a clear responsibility for the continuous COM process and they are highly dependent on each other.

The motion tracking module, also known as the front-end module, receives input data to compute a locally accurate pose estimate. From the visual data, it identifies good feature points and provides the captured feature descriptors. These descriptors also define orientation, gravity direction, scale and other aspects. The estimated pose is based on the feature descriptor correspondences from the previous frame. This module maintains only a limited history of tracked motion and treats any previously generated estimated feature point pose as fixed. Inertial sensors help correlate spatial features observed in one frame with spatial features observed in a subsequent frame to efficiently determine the change in pose. Therefore, the estimated pose can be performed with high frequency (due to the limited matching time) with a locally accurate pose. On the other hand, the estimated pose has almost no control over the drift or no means to check for loop closure since only a very limited history is stored.

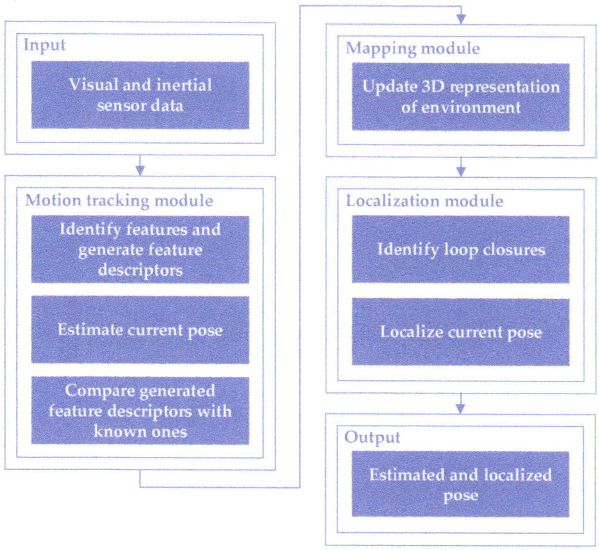

Figure 2. The concept of the ARCore concurrent odometry and mapping algorithm [55].

The mapping module, also called the back-end module, takes as input the previously defined feature descriptors from the current frame. In this module, a 3D representation of the environment is created based on a stored variety of maps, feature descriptors, and estimated device location. It has an extensive history of 3D feature positions in the environment and poses of the smartphone. Matching the newly generated features with the fixed features in the multitude of maps is an extensive process. Therefore, the updating is done with a relatively low frequency. The 3D representation is sent to the localization module. Within the mapping module, ARCore provides spatial understanding based on plane detection in the depth map. Using the same feature points used for motion tracking, ARCore looks for clusters of points that appear to lie on common surfaces. These can be anything, such as floors, tables, or walls. These surfaces are stored as planes with a specific boundary.

The localization module identifies discrepancies between stored and observed feature descriptors. It performs loop closure by minimizing these discrepancies (e.g., with a least squares adjustment or bundle adjustment [58]) to output a localized pose. If the error (difference between estimated pose and matched pose) is greater than a certain threshold, the localized pose snaps to the pose from the descriptor matching. This can happen for a variety of reasons (e.g., significant drift). Difficulties occur for real-world surfaces without texture (e.g., white wall) because ARCore uses feature points to detect surfaces. No useful features can be found in textureless regions. In case the SLAM algorithm fails, inertial navigation based on a pedometer and a digital compass becomes crucial. Inertial navigation is combined with visual localization using an extended Kalman filter (as described in our previous work [37]).

In summary, the ARCore algorithm performs two types of outputs from the input data (see Figure 2). First, it provides a locally accurate and high frequency estimated pose with no control segment such as loop closure or drift correction. Second, it provides a localized pose with a lower update frequency due to extensive steps such as descriptor matching (with the full history) or an adaptation process.

Suppose path loss models for the current space do not already exist on the server. In that case, the visual-inertial localization method described above allows any smartphone application user to begin collecting BLE signal strength data for online path loss model construction and subsequent radiolocation. The smartphone user only needs to walk through the room while the visual-inertial localization system simultaneously determines their current pose and creates a visual and radio map of the environment. While using the smartphone application, it collects the following data into the database: BLE signal strengths with time and pose (position and orientation of the smartphone) where individual RSSI of a beacon with a unique ID is captured. The local data stored on the smartphone are used for online model construction, as described in the next section. The collected data, along with the path loss model parameters, are also sent to the server, where they are used for further post-processing and realization of the crowdsourcing approach.

3. Online Path Loss Model Construction

Measurements of BLE signal strengths can be described using a path loss (or path attenuation) model that determines a reduction in signal strength as the receiver moves away from the transmitter. The model has the general form of a nonlinear equation with three parameters:

$$d = K_1 \cdot ratio^{K_2} + K_3, \quad (1)$$

where $ratio = R/TXP = R/-59$ ($TXP = -59$ dBm represents a signal strength at distance of $d = 1$ m). From Equation (1) the signal strength R can be expressed:

$$R = TXP \cdot \left(\frac{d - K_3}{K_1}\right)^{K_2^{-1}}, \quad (2)$$

where the distance d is equal to:

$$d = \sqrt{(x - x_0)^2 + (y - y_0)^2 + h^2}. \quad (3)$$

Equation (3) contains the position (x, y) at which the signal strength R is measured, the location of the beacon in space (x_0, y_0) and the distance h, which represents the height difference between the smartphone and the known height at which the beacon is fixed. Constrained nonlinear optimization (in which a trust-region method [59,60] is used) can be used to determine the model parameters K_1, K_2, K_3 and the position of the beacon in space (x_0, y_0) in a way that model (2) fits the measurements well. Since the optimization problem has multiple solutions, it is very important to strictly set the boundaries of model parameters and limit the space of beacons' locations.

Constrained Nonlinear Optimization

Constrained nonlinear minimization [61–63] involves finding the vector x, representing the local minimum of the scalar function $f(x)$, taking into account the given constraints on the vector x. These constraints can be given as: A linear constraint $(A \cdot x \leq b, Aeq \cdot x = beq)$, a nonlinear constraint ($c(x) \leq 0$ or $ceq(x) = 0$) or a constraint with given bounds (box constraints) $l \leq x \leq u$. In the following, the emphasis will be on nonlinear optimization with box constraints l (vector of lower bounds) and u (vector of upper bounds). In this case, we solve the following problem:

$$\min\{f(x) \text{ such that } l \leq x \leq u\}. \quad (4)$$

In the vector of lower and upper bounds, some of the components may be unconstrained (equal to $\pm\infty$). In the optimization process, the method returns a sequence of strictly feasible points. With the goal of preserving feasibility while achieving robust convergence behavior, the following techniques are used. In the first technique, the unconstrained Newton step is substituted with a scaled modified Newton step. In this way, the two-dimensional subspace S is defined using the preconditioned conjugate gradient method [64]. The latter ensures global convergence. In the second technique, reflections [61] are used to increase the step size. The scaled modified Newton step is obtained by considering the Kuhn–Tucker necessary conditions for Equation (4):

$$(D(x))^{-2}g = 0, \tag{5}$$

where $D(x)$ is a diagonal scaling matrix:

$$D(x) = diag(|r_k|^{-1/2}). \tag{6}$$

The vector $r(x)$ is defined below according to the gradient $g = \nabla f(x)$ and bounds, for each $1 \leq i \leq n$:

- If $u_i < \infty$ and $g_i < 0$ then $r_i = x_i - u_i$;
- If $l_i > -\infty$ and $g_i \geqslant 0$ then $r_i = x_i - l_i$;
- If $u_i = \infty$ and $g_i < 0$ then $r_i = -1$;
- If $l_i = -\infty$ and $g_i \geqslant 0$ then $r_i = 1$.

It is important to point out that the nonlinear system in Equation (5) is not differentiable when $r_i = 0$. However, by taking into account the restriction $l < x < u$ and maintaining strict feasibility, such points can be avoided. According to the nonlinear system of Equation (5), the scaled modified Newton step s_k is computed as the solution of the linear system:

$$\widehat{M}_k D_k s_k^N = -\widehat{g}_k, \tag{7}$$

at the kth iteration, where:

$$\widehat{g}_k = D_k^{-1} g_k = \text{diag}(|r_k|^{1/2}) g_k \tag{8}$$

and:

$$\widehat{M}_k = D_k^{-1} H_k D_k^{-1} + \text{diag}(g_k) J_k^r. \tag{9}$$

In Equation (9), $J^r(x)$ represents the Jacobian matrix of $|r(x)|$ and H is the Hessian matrix. The Jacobian matrix J^r is diagonal, where each element is equal to 0, -1 or 1. For the case where all elements of l and u are finite, the Jacobian matrix is equal to $J^r = \text{diag}(\text{sign}(g))$. If there is a point at which $g_i = 0$, then r_i might not be differentiable. However, at such a point the Jacobian $J_{ii}^r = 0$ is determined. Such non-differentiability is not a problem, since it is not important which value r_i takes. Indeed, the function $|r_i| \cdot g_i$ is continuous, although $|r_i|$ is discontinuous at this point.

Equation (7) considers the use of the affine transformation $\widehat{x} = D_k x$ (the matrix D_k is a symmetric matrix: $D_k = D_k^T$). This transformation converts the constrained nonlinear optimization problem into an unconstrained nonlinear optimization problem, minimizing the function in the new $\widehat{x}$ coordinates. This means that in this case the following minimization problem is solved:

$$min\{\widehat{\psi}_k(\widehat{s}) = \frac{1}{2}\widehat{s}^T \widehat{M}_k \widehat{s} + \widehat{g}_k^T \widehat{s} \quad \text{such that} \quad ||\widehat{s}|| \leq \Delta_k\} \tag{10}$$

or, if $s = D_k^{-1}\widehat{s}$ is considered, then the following minimization problem can be solved in the original space:

$$min\{\psi_k(s) = \frac{1}{2}s^T M_k s + s^T g_k \quad \text{such that} \quad ||D_k s|| \leq \Delta_k\}, \tag{11}$$

where:

$$M_k = H_k + C_k \tag{12}$$

and:

$$C_k = D_k \text{diag}(g_k) J_k^r D_k. \tag{13}$$

In Equation (10) Δ_k is a positive scalar representing confidence interval width. During the optimization, it is adjusted according to the rules of the trust-region method [59,60].

The second technique of the constrained nonlinear optimization provides feasibility with the following condition $l \leq x \leq u$ to be satisfied. Within this technique, so-called reflections are used, which have the purpose of increasing the step size. A single reflection step can be described as follows. If p represents a step that exceeds the first bound constraint (the ith upper or ith lower bound), then the reflection step is defined as $p^R = p$ and $p_i^R = -p_i$ for the ith element for which a constraint is required.

4. Localization with Particle Swarm Optimization

Particle Swarm Optimization (PSO) is an optimization method that was initially described by Kennedy, Eberhart and Shi [65,66]. In a practical application [67], they showed that the PSO method could be used to simulate social behavior within a school of fish or a flock of birds. PSO is a stochastic optimization method that robustly solves a given problem in successive iterations according to the selected objective function. This determines whether the candidate solutions of the problem are appropriately improved at each step. The candidate solutions are represented by particles forming a particle swarm. The solution to the optimization problem can be achieved by moving these particles within a constrained region using simple mathematical formulas. These determine a new position and velocity of the particles in each iteration. In the case where all particles converge to a common position, the global optimum of the objective function is found.

Let the position of the particle be denoted by $\boldsymbol{x}_i \in \mathbb{R}^n$ and its velocity by $\boldsymbol{v}_i \in \mathbb{R}$. In the swarm with S elements, the best position of each particle is denoted by $\boldsymbol{p}_i$ and the best position of the whole swarm is denoted by $\boldsymbol{g}$. The number of all required particles S can be inferred from the experimental results or calculated using the suggested formulas (e.g., $S = 2\sqrt{n} + 10$, where n is the search area's dimension).

In the PSO method [66] a new position of particle $\boldsymbol{x}_i^{k+1}$ is determined by the model of particle movement that takes into account its current current velocity $\boldsymbol{v}_i^k$, position $\boldsymbol{x}_i^k$, the distance $\boldsymbol{d}_{p_i}^k$ between the particle's best known position $\boldsymbol{p}_i^k$ and the particle's current position $\boldsymbol{x}_i^k$ and the distance $\boldsymbol{d}_{g_i}^k$ between the best position of the whole swarm $\boldsymbol{g}^k$ at a given time and the particle's current position $\boldsymbol{x}_i^k$:

$$\boldsymbol{x}_i^{k+1} = \boldsymbol{x}_i^k + \boldsymbol{v}_i^{k+1}, \quad (14)$$

where:

$$\boldsymbol{v}_i^{k+1} = \omega \boldsymbol{v}_i^k + \Delta \boldsymbol{v}_i^k, \quad (15)$$

$$\Delta \boldsymbol{v}_i^k = \varphi_g r_g (\boldsymbol{g}^k - \boldsymbol{x}_i^k) + \varphi_p r_p (\boldsymbol{p}_i^k - \boldsymbol{x}_i^k) = \varphi_g r_g \boldsymbol{d}_{g_i}^k + \varphi_p r_p \boldsymbol{d}_{p_i}^k. \quad (16)$$

The parameter ω can be used to determine whether accurate local exploration (a low value) or faster convergence (a high value) is preferred. The parameter represents the remembering rate or inertia weight [66], which determines to what extent the previous direction of the particle is preserved. The parameter φ_g is a swarm confidence factor (or a social learning factor) and the parameter φ_p is a self confidence factor (or cognitive learning factor). The parameters r_g and r_p, which change randomly in the interval $[0, 1]$, represent random accelerations in the directions of the best position of the particle $\boldsymbol{p}_i$ and the best position of the whole swarm $\boldsymbol{g}$. In the process of the PSO configuration, according to a given optimization problem, great emphasis should be put on the choice of parameters ω, φ_p and φ_g, which have the most significant impact on the performance of the method [68–71].

The goal of PSO-based localization is to find a particle (i.e., the position of the smartphone) (x, y) in two-dimensional space for which the corresponding vector of signal strengths $\boldsymbol{R_P} = [f_{R_1}(d_1), f_{R_2}(d_2), ..., f_{R_m}(d_m)]$, obtained by the constructed path loss models $R_i = f_{R_i}(d_i)$ (for $i = 1, ..., m$), is the most similar to the vector of current measurements of signal strengths $\boldsymbol{R_M}$ according to the objective function:

$$f_c = \sum_{i=1}^{m} |R_{M_i} - R_{P_i}|, \tag{17}$$

where m is the number of BLE transmitters considered for the calculation of the current position. The distance d_i (for $i = 1, ..., m$) represents the Euclidean distance between the current position of the smartphone (x, y) and the position of the ith BLE transmitter (x_{0_i}, y_{0_i}).

Using the path loss models, the PSO localization method generates fingerprints for all particles in the swarm that represent the local radio map. The particles are uniformly distributed in the grid of the chosen size at the beginning of each optimization.

5. Online Path Loss Model Construction Results

In order to evaluate the proposed system for automated localization data collection, online modeling and real-time PSO localization on a smartphone (Samsung Galaxy S9), an experiment was performed in am ~80 m^2 laboratory, where ten Bluetooth beacons from Kontakt.io [72] were installed. The beacons were distributed around the room (taking into account certain constraints due to the layout of walls, doors, pillars and windows), as shown in Figure 3 (green squares). To reduce the influence of the presence of human bodies on the measurements of the signal strengths, the transmitters were mounted at the height of 2 m above the floor.

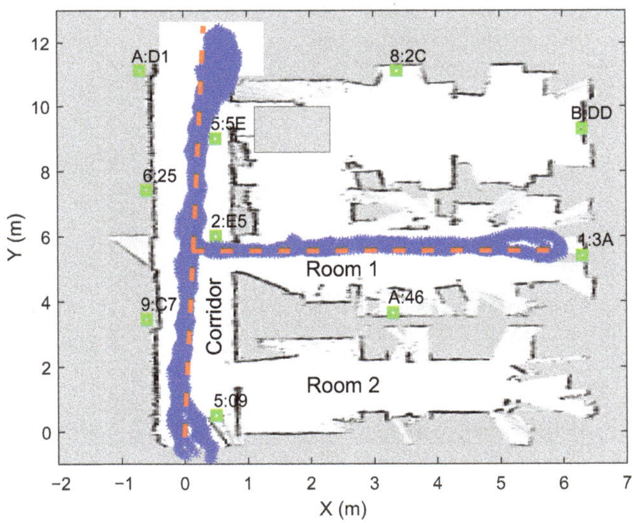

Figure 3. Locations of BLE beacons (green squares) and walked path (determined with the visual-inertial SLAM) during automated measurement capture. The ground truth path is marked with the red dashed line. The white pixels represent empty space and grey pixels represent occupied space.

The measurements of the signal strengths from all ten beacons were collected in a way that a pedestrian repeatedly walked the path that is shown in Figure 3 while the smartphone recorded the data. The current position of the smartphone (calculated every 50 ms) was tracked by the algorithm combining the SLAM and the inertial navigation system (see Section 2). All of the positions where the measurements of the signal strengths were performed can be seen in Figure 3. This figure shows that the obtained position points deviate slightly from the intended route (the lines were marked on the ground). Part of this error is due to the visual-inertial localization algorithm (strong magnetic interference affects the digital compass) determining the position relative to the starting point, and part is due to the inaccurate walking of pedestrians along the line. Our results on visual localization include that under optimal environmental conditions, the errors are at most of a few tens of centimeters in a typical room-sized environment (without significant drift errors). However, as the speed of movement increases and in low light conditions, the overall performance may decrease.

Using the constrained nonlinear optimization and the signal strengths measurements from all Bluetooth beacons distributed in the room, the path loss models are constructed online on the smartphone. Figure 4 shows the fitting of the models to the measurements of the signal strengths after 30, 70 and 150 s of walking (for the BLE beacon with MAC address DA:57:30:EC:6A:D1 and position $(-0.7, 11.10)$) to demonstrate the influence of the amount of the collected measurements. After 30 s of walking, the pedestrian covered a distance of about 10 m along the corridor. After 70 s, half a T-shaped path was covered, and after 150 s, the pedestrian covered the entire T-shaped path (i.e., 36 m) from the start and back. The online construction of the path loss models and the determination of the beacons' locations are performed in iterations every 10 s when enough new data are collected. The constrained nonlinear optimization runs in parallel to the real-time PSO localization on the smartphone, which means that they do not affect each other when individual CPU cores are fully utilized.

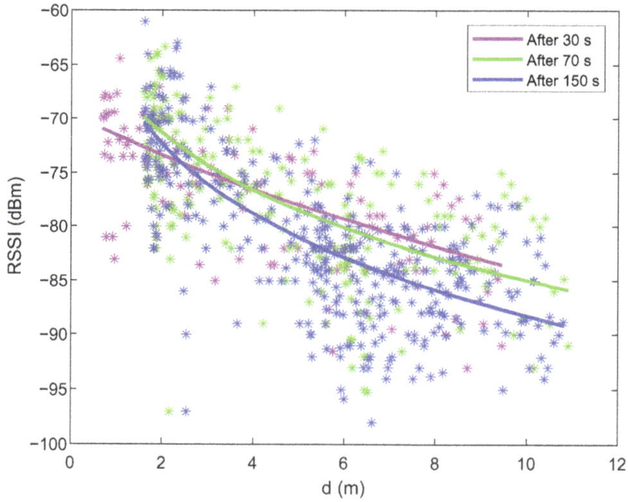

Figure 4. Fitting the online constructed path loss models (2) to the measurements of the signal strengths after 30, 70 and 150 s of walking (for the BLE beacon with MAC address DA:57:30:EC:6A:D1). Since the data contain a lot of noise, the coefficient of determination for all models is quite low: $R^2 = 0.5$.

In the laboratory where the experiment was conducted, there were many obstacles (walls, wooden barriers, pillars) between the transmitters and the receiver (smartphone), while the pedestrian walked along the path shown in Figure 3. Due to absorption and reflection from the obstacles, the signal strength measurements were very noisy and scattered at the same distance from the beacon (see Figure 4).

During data collection and online model development, the measurements and the path loss models shift left and right according to the distance d (see Figure 4) since the positions of the beacons are estimated with the constrained nonlinear optimization algorithm in addition to the model parameters. With more available data, the positions of the beacons converge towards the correct position (see the upper part of Figure 5). Still, after a certain number of iterations, they do not improve any more because the measurements contain a lot of noise. For model improvement, it is crucial to collect measurements at different distances (from 1 m to 10 m) from the beacon because, in this way, the whole RSSI signal range is covered (e.g., from −59 to −100 dBm). The lower part of Figure 5 shows the BLE signal range (the difference between maximum and minimum signal strength) as a function of the travelled distance. As the signal range increased, the error of the beacon's location decreased. When the signal range reached values close to 40 dBm, the error entered the steady-state. After 30 s of data acquisition, the amount of data was still too small or the measurements are not well distributed throughout the room, which is reflected in the transmitter's inaccurate position and a worse model 1 in Figure 4. After 70 s, the signal range and the beacon's location were satisfactory and the path loss model was also close to the optimum. With the additional data, the models and beacons' locations can be improved only slightly.

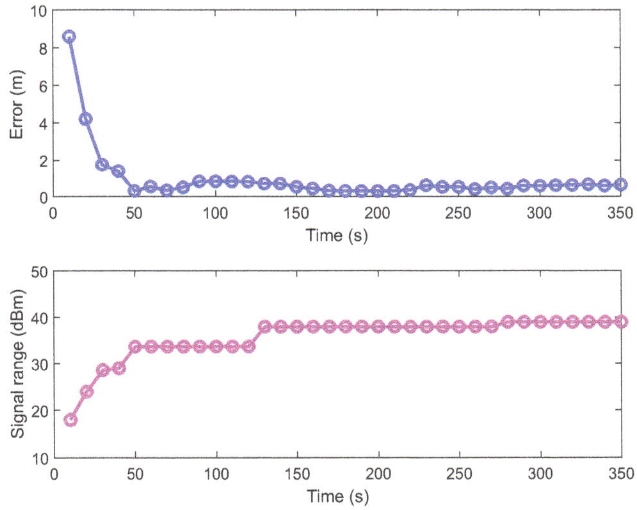

Figure 5. The upper part of the figure shows the error between the beacon's actual location (with MAC address DA:57:30:EC:6A:D1) and the position calculated by the constrained nonlinear optimization method. The error depends on the time or the travelled distance. The lower part of the figure shows the BLE signal range (the difference between maximum and minimum signal strength) as a function of time or the travelled distance.

In the online data collection process and model development, it is essential to know when there is enough data for a given beacon. For this purpose, two (stopping) criteria can be used, first, when the beacon's location changes less than the selected threshold during two consecutive iterations (in our case 0.1 m) and second, when the signal range is large enough (in our case 35 dBm). The duration of collecting the measurements and the path length are worse for estimating whether there are enough measurements. Namely, if the measurements are collected only in the same position (for a long time) or at a long distance from the transmitter, they are not useful for the model construction.

Once the positions of all beacons and the path loss models are computed, the goal of the proposed system is achieved. However, if there are any changes to the space (e.g., change in the position of the beacons), the whole process can be repeated very quickly. The data and models can be created by one user or by multiple users of the smartphone application. With the crowdsourcing approach, large amounts of data can be collected on the server, which means that the models created online can definitely be improved offline using more sophisticated and computationally intensive methods (e.g., using the SUHICLUST algorithm that creates fuzzy path loss models with confidence intervals [50]), if necessary.

6. Particle Swarm Localization Results

In order to evaluate the online developed path loss models and the PSO-based localization algorithm, the measurements of signal strengths were collected with a smartphone along the same path used to construct the models and determine the beacons' locations (see Figure 3). As the pedestrian walked along the path, each time the smartphone receiver measured at least one new signal strength that was greater than −95 dBm, a vector of signal strengths $\boldsymbol{R_M}$ was created from which the current position was determined. In our experiment, this happened on average every 30 ms, with each beacon sending a packet every 100 ms. For each beacon, the measurements of the signal strengths were independently filtered using the Savitzky–Golay filter (with frame length 99) [73] to reduce variability. Each vector of current measurements (or fingerprint) consists of at least three signal strengths from different beacons. In our experiment, the fingerprints contained five elements, as this is optimal according to our previous study [50]. To determine the accuracy of the PSO localization algorithm in real time, the visual-inertial SLAM algorithm was used to record the positions where the vectors $\boldsymbol{R_M}$ were created.

During the experiment, the pedestrian walked for 5 minutes in the laboratory. Their current location (see Figure 6) was determined in real time using the PSO localization method on a smartphone (Samsung Galaxy S9), where the latest path loss models were used. These were created online from the latest available data and they were further improved with new data if the stopping criteria were not yet met (see previous subsection).

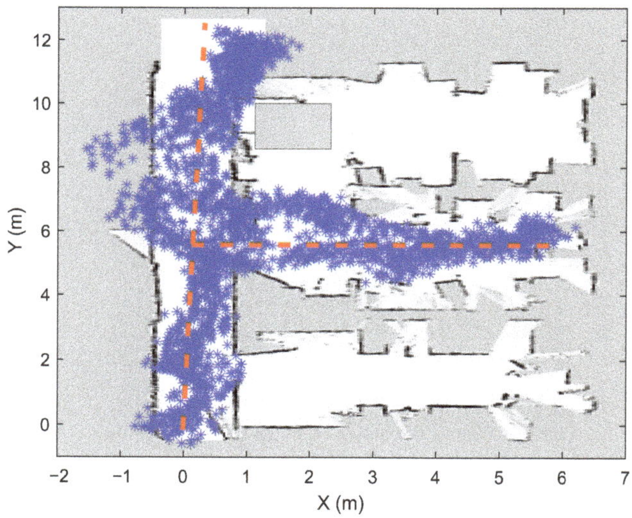

Figure 6. The indoor positioning results (after 5 min of walking) obtained using the online constructed path loss models and the PSO localization method.

When selecting the PSO search area, the current position of the receiver (i.e., the smartphone) must be approximately known. During PSO initialization, the position can be calculated using the location of the nearest beacon (with the highest signal strength) or using the fast linear trilateration method. In this case, the room-level accuracy is acceptable. Thereafter, when PSO provides the first positioning results, the current position of the receiver can be used in the next iteration to determine the PSO search area (all particles are arranged around the current position).

With the aim of obtaining the optimal parameters of the PSO method, the offline simulation experiments were performed in the Matlab environment. In this way, the following parameters were obtained: $\omega = 0.06$, $\varphi_p = 0.1$ and $\varphi_g = 0.1$. The optimal size of the search area with a square shape is 0.8 m × 0.8 m. By choosing the boundaries of the search area, a filtering effect is achieved in addition to reducing the computational complexity.

Since the PSO algorithm must operate in real-time on a smartphone, the parameters were chosen as a trade-off between computational speed and localization accuracy. By running simulations, the number of particles and iterations needed was also determined. The use of eight particles and twenty iterations within the PSO optimization proved to be optimal, as a larger number of particles or iterations only increases the computational complexity and does not improve the result.

As path loss models and beacons' locations were more accurately determined with more available data, the localization error decreased. The graph of the cumulative distribution function (Figure 7) shows that after 310 s of walking in 70 % of the position estimates, the error was less than 1 m and the average error was less than 0.8 m. For comparison, when incomplete models built with a smaller dataset were used, the results show that after 30 s of walking in 44 % of the position estimates the error was smaller than 1 m and after 70 s there were 52 % of the position estimates where the error was smaller than 1 m.

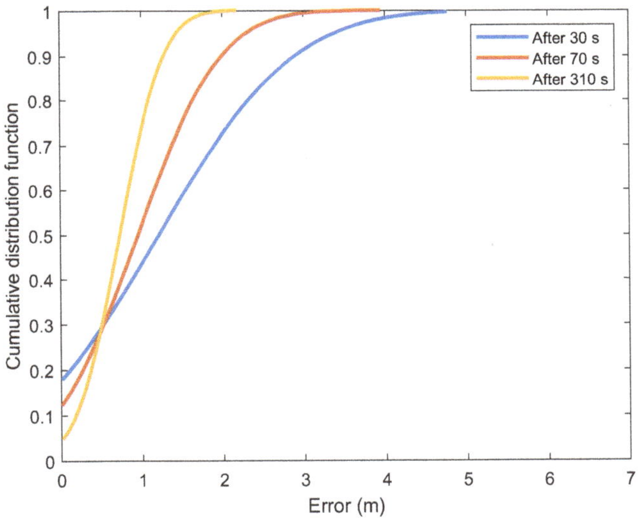

Figure 7. The cumulative distribution functions for the positioning errors in determining the positions by the PSO method after 30, 70 and 310 s of walking.

The results of the online construction of the path loss model show that in the first 80 s after data collection began, the models and the beacons' locations changed drastically. Consequently, the positioning errors obtained with the PSO localization method (see Figure 8) were substantially larger in the first 80 s (or for the first eight model series). Thereafter, the mean positioning error decreased more slowly. The decrease of error can be described by an exponential model $e = 0.35\,exp(-0.04t) + 0.98\,exp(-0.0006t)$ (where t is time) with the coefficient of determination $R^2 = 0.8$. In the first evaluation of the mean error, about 150 positions and in the last evaluation, about 5500 positions were used for error estimation. The number of calculated positions (with PSO method) used for estimating the mean error increased practically linearly with time. Since path loss model construction and PSO localization can run simultaneously on a smartphone, the comparison of the current radio and visual-inertial localization results can be used as a real-time criterion to determine in which part of space additional signal strength data need to be collected to improve the latest online computed models and consequently the localization accuracy.

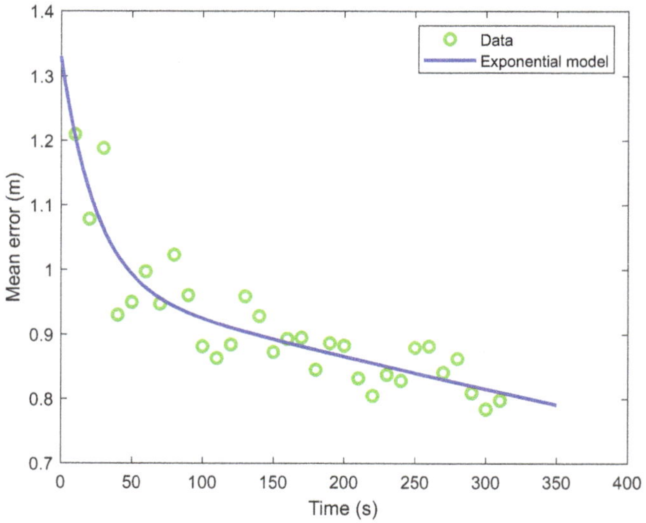

Figure 8. The mean localization error as a function of time or the travelled distance. The decrease of error can be described by an exponential model $e = 0.35\, exp(-0.04\, t) + 0.98\, exp(-0.0006\, t)$ with the coefficient of determination $R^2 = 0.8$.

When comparing the computational complexity, the localization algorithm based on the PSO method turns out to be better than the one based on the nonlinear trilateration [74], as it is 20 times faster for the same localization accuracy (on the smartphone, it spends about 5 ms computing a position). The experiment was performed on a Samsung Galaxy S9 smartphone (with a Samsung Exynos 9810 processor) using offline signal strength measurements and prebuilt path loss models. Average execution time for both localization methods was calculated from 5000 repetitions of the different position calculations.

7. Discussion

With the implementation of an automated system for BLE data collection, online path loss model construction, and real-time PSO localization on a smartphone, radiolocation becomes significantly less time- and labor-intensive by reducing configuration time (i.e., the time from the start of data collection to the moment when Bluetooth localization becomes operational) by more than 90%. With the manual collecting of signal strength data in the $\sim$80 m^2 laboratory, where at least 80 measurement points are needed (previous studies showed that at least one measurement point per square meter is required), the network operator would spend about 40 min (half a minute to manually measure each location in the room and collect the signal strengths). To measure ten beacons' locations in our laboratory, the network operator would spend an additional 5 min. A non-automated localization system also requires some time to transfer the data to the computer and compute the path loss models or create the map of fingerprints. For comparison, the proposed automated system required 5 min for collecting all signal strength data (more than 5000 measurement points), calculating path loss models, and determining all beacons' locations.

Compared to our previous study [37,50,74], in which a visual odometry algorithm was developed to determine the reference positions during the collection of signal strengths, the new proposed automated system has many advantages. The visual-inertial SLAM does not require an initial camera calibration with a checkerboard to obtain intrinsic and extrinsic parameters describing the transformation between the camera and base coordinate systems. Therefore, the smartphone does not need to be at the same height above the ground during the walk. The visual-inertial SLAM also includes loop closure, which is very important to reduce position drift and error accumulation. Therefore, the proposed system can be used for data collection in wide area facilities such as airports, hospitals, museums, etc. The visual localization system based on SLAM is very accurate, but it can also fail, especially in dark places and places where the surfaces are without texture. In such a case, the fusion with inertial sensors and methods becomes crucial. The visual-based localization system has proven to be less suitable to be used as a primary localization system, as the image processing puts a heavy load on the processor, which drains the battery quickly. During the experiment, the two big cores of the Samsung Exynos 9810 processor (with four big cores operating at 2.9 GHz and four small cores operating at 1.9 GHz) were loaded more than 90% on average while using visual localization. By contrast, PSO localization is computationally much less demanding and can run on a small core (loaded less than 60% on average), saving energy.

In addition to the visual-inertial SLAM, another important component of the proposed system is the constrained nonlinear optimization, which allows constructing path loss models and determining the beacons' locations online on a smartphone. The results of the experiment show that the beacons' locations can be determined fairly accurately, as the final average errors are less than half a meter for all the beacons. These errors are not problematic since the obtained beacons' locations can also be represented as virtual beacons' locations that best fit the corresponding path loss models with the constrained parameters. In this way the deviations (due to the obstacles) in the path loss model can be partially compensated for with the virtual beacon's location and consequently the model can better fit the measurements. It has been found that the constraints on the parameters of the path loss model can be quite stringent, since the same beacons located in open space have very similar path loss models. The constraints on the beacons' locations do not impose an additional burden since the dimensions of the area where the transmitters are located are usually known. When implementing algorithms on a smartphone, the main concern is that they can run in real time or in limited time. Therefore, a trade-off between the accuracy and speed of the algorithms is required. Since the proposed system is designed to support the crowdsourcing approach, all the collected data are transferred to the server database where the models already created online can be additionally improved if needed. The proposed solution is suitable for both BLE network operators and end users. Anyone who enables visual localization on a smartphone can start simultaneous localization and data collection. Based on the proposed criteria (signal range, beacons' location stability and accuracy of radiolocation compared to visual-inertial localization), the following information is available to the user collecting data: For which beacon in the room there are not enough measurements yet or in which part of the room the radiolocation is still poor and additional measurements are needed.

For real-time radiolocation on a smartphone, the PSO-based method was chosen instead of nonlinear trilateration. The reason is that PSO localization has been shown to be much faster. However, in order to achieve good localization accuracy, the parameters of the PSO optimization method must be carefully chosen, otherwise the algorithm may become slow and inefficient. The PSO localization is more flexible in the sense of a small temporary radio map generation, which consumes very little memory for the same accuracy as the fingerprinting method. The obtained localization results are slightly worse (with an average error of 0.8 m) than in the case where the locations of the beacons are known and the path loss models are constructed offline using more sophisticated methods [50] (with an average error of 0.5 m). Since most location applications (for large facilities such as airports) do not require high localization accuracy, the obtained results represent a good compromise according to the complexity of the network deployment and the localization accuracy.

The proposed system is not limited to collecting BLE signal strengths and PSO radiolocation. Depending on the needs, it can be modified to collect data about the WiFi, UWB or geomagnetic field signals and perform localization based on these signals.

8. Conclusions

This paper presents a sophisticated indoor localization system that combines visual-inertial positioning for accurate data collection and Bluetooth positioning for real-time smartphone tracking with low power consumption. The visual-inertial SLAM algorithm, which is part of the ARCore library, successfully fuses information from the camera and inertial sensors to provide accurate localization in large spaces over a long period of time. In this way, a suitable collection of measurements of signal strengths can be made quickly and easily, clearly defining which part of the space is not yet well covered by measurements. The path loss models and the beacons' locations are determined online on a smartphone using the constrained nonlinear optimization, which is the main contribution. The constrained nonlinear optimization considers all the necessary bounds on the model parameters and the map of the space when constructing the path loss models. The obtained beacons' locations are determined fairly accurately (with an average error of 0.5 m) and the models fit the measurements well, although they contain a lot of noise. From the obtained radiolocation results, it can be seen that the proposed PSO-based localization algorithm (with an average error of 0.8 m) combined with online constructed models can meet most localization requirements in smartphone applications. The developed system for automated data collection and online modeling on a smartphone has proven to be very useful as it can greatly simplify and speed up the installation of the Bluetooth network, especially in wide-area facilities such as airports, hospitals, museums etc., where the configuration time can be reduced by more than 90%. Further development of the system will focus on the more sophisticated fusion of the radiolocation and inertial system.

Author Contributions: Conceptualization, S.T. and I.Š.; methodology, S.T. and I.Š.; software, S.T. and I.Š.; validation, S.T. and I.Š.; formal analysis, S.T. and I.Š.; investigation, S.T. and I.Š.; resources, S.T. and I.Š.; data curation, S.T. and I.Š.; writing—original draft preparation, S.T. and I.Š.; writing—review and editing, S.T. and I.Š.; visualization, S.T. and I.Š.; supervision, S.T. and I.Š.; project administration, S.T. and I.Š. All authors have read and agreed to the published version of the manuscript.

Funding: This research received no external funding.

Institutional Review Board Statement: Not applicable.

Informed Consent Statement: Not applicable.

Conflicts of Interest: The authors declare no conflict of interest.

References

1. Subedi, S.; Pyun, J.Y. A Survey of Smartphone-Based Indoor Positioning System Using RF-Based Wireless Technologies. *Sensors* **2020**, *20*. [CrossRef]
2. Montoliu, R.; Sansano, E.; Gascó, A.; Belmonte, O.; Caballer, A. Indoor Positioning for Monitoring Older Adults at Home: Wi-Fi and BLE Technologies in Real Scenarios. *Electronics* **2020**, *9*. [CrossRef]
3. Dinh, T.M.T.; Duong, N.S.; Sandrasegaran, K. Smartphone-Based Indoor Positioning Using BLE iBeacon and Reliable Lightweight Fingerprint Map. *IEEE Sens. J.* **2020**, *20*, 10283–10294. [CrossRef]
4. Ramos, J.; Costa, A.; Novais, P.; Neves, J. Interactive Guiding and Localization Platform. *Int. J. Artif. Intell.* **2014**, *12*, 63–78.
5. Essa, E.; Abdullah, B.A.; Wahba, A. Improve Performance of Indoor Positioning System using BLE. In Proceedings of the 2019 14th International Conference on Computer Engineering and Systems (ICCES), Cairo, Egypt, 17 December 2019; pp. 234–237. [CrossRef]
6. Honkavirta, V.; Perala, T.; Ali-Loytty, S.; Piche, R. A comparative survey of WLAN location fingerprinting methods. In Proceedings of the 2009 6th Workshop on Positioning, Navigation and Communication, Hannover, Germany, 19 March 2009; pp. 243–251.
7. Misal, S.R.; Prajwal, S.R.; Niveditha, H.M.; Vinayaka, H.M.; Veena, S. Indoor Positioning System (IPS) Using ESP32, MQTT and Bluetooth. In Proceedings of the 2020 Fourth International Conference on Computing Methodologies and Communication (ICCMC), Erode, India, 11–13 March 2020; pp. 79–82. [CrossRef]
8. Taşkan, A.K.; Alemdar, H. Obstruction-Aware Signal-Loss-Tolerant Indoor Positioning Using Bluetooth Low Energy. *Sensors* **2021**, *21*. [CrossRef] [PubMed]
9. Fang, S.; Wang, C.; Huang, T.; Yang, C.; Chen, Y. An Enhanced ZigBee Indoor Positioning System With an Ensemble Approach. *IEEE Commun. Lett.* **2012**, *16*, 564–567. [CrossRef]

10. Zhang, D.; Yang, L.T.; Chen, M.; Zhao, S.; Guo, M.; Zhang, Y. Real-Time Locating Systems Using Active RFID for Internet of Things. *IEEE Syst. J.* **2016**, *10*, 1226–1235. [CrossRef]
11. Kuxdorf-Alkirata, N.; Maus, G.; Brückmann, D. Efficient calibration for robust indoor localization based on low-cost BLE sensors. In Proceedings of the 2019 IEEE 62nd International Midwest Symposium on Circuits and Systems (MWSCAS), Dallas, TX, USA, 4–7 August 2019; pp. 702–705. [CrossRef]
12. Dardari, D.; Conti, A.; Ferner, U.; Giorgetti, A.; Win, M.Z. Ranging With Ultrawide Bandwidth Signals in Multipath Environments. *Proc. IEEE* **2009**, *97*, 404–426. [CrossRef]
13. Apple. iBeacon. Available online https://developer.apple.com/ibeacon/ (accessed on 1 March 2021).
14. Google. Eddystone. Available online: https://developers.google.com/beacons/ (accessed on 1 March 2021).
15. Naghdi, S.; O'Keefe, K. Trilateration with BLE RSSI Accounting for Pathloss Due to Human Obstacles. In Proceedings of the 2019 International Conference on Indoor Positioning and Indoor Navigation (IPIN), Pisa, Italy, 30 September–3 October 2019; pp. 1–8. [CrossRef]
16. Ye, F.; Chen, R.; Guo, G.; Peng, X.; Liu, Z.; Huang, L. A Low-Cost Single-Anchor Solution for Indoor Positioning Using BLE and Inertial Sensor Data. *IEEE Access* **2019**, *7*, 162439–162453. [CrossRef]
17. Yadav, R.K.; Bhattarai, B.; Gang, H.; Pyun, J. Trusted K Nearest Bayesian Estimation for Indoor Positioning System. *IEEE Access* **2019**, *7*, 51484–51498. [CrossRef]
18. de Blasio, G.S.; Rodríguez-Rodríguez, J.C.; García, C.R.; Quesada-Arencibia, A. Beacon-Related Parameters of Bluetooth Low Energy: Development of a Semi-Automatic System to Study Their Impact on Indoor Positioning Systems. *Sensors* **2019**, *19*. [CrossRef]
19. Echizennya, K.; Kondo, K. Estimation of indoor position and motion direction for smartphones using DNN to BLE beacon signal strength. In Proceedings of the 2020 IEEE International Conference on Consumer Electronics—Taiwan (ICCE-Taiwan), Taoyuan, Taiwan, 28–30 September 2020; pp. 1–2. [CrossRef]
20. Sato, A.; Nakajima, M.; Kohtake, N. Rapid BLE Beacon Localization with Range-Only EKF-SLAM Using Beacon Interval Constraint. In Proceedings of the 2019 International Conference on Indoor Positioning and Indoor Navigation (IPIN), Pisa, Italy, 30 September–3 October 2019; pp. 1–8. [CrossRef]
21. Nakano, M.; Komuro, N.; Kawamoto, K. Indoor Positioning Method based on BLE Location Fingerprint with Statistics Approach. In Proceedings of the 2019 IEEE 8th Global Conference on Consumer Electronics (GCCE), Osaka, Japan, 15–18 October 2019; pp. 1160–1163. [CrossRef]
22. Nagah Amr, M.; ELAttar, H.M.; Abd El Azeem, M.H.; El Badawy, H. An Enhanced Indoor Positioning Technique Based on a Novel Received Signal Strength Indicator Distance Prediction and Correction Model. *Sensors* **2021**, *21*. [CrossRef] [PubMed]
23. Phutcharoen, K.; Chamchoy, M.; Supanakoon, P. Accuracy Study of Indoor Positioning with Bluetooth Low Energy Beacons. In Proceedings of the 2020 Joint International Conference on Digital Arts, Media and Technology with ECTI Northern Section Conference on Electrical, Electronics, Computer and Telecommunications Engineering (ECTI DAMT NCON), Pattaya, Thailand, 11–14 March 2020; pp. 24–27. [CrossRef]
24. Naghdi, S.; O'Keefe, K. Detecting and Correcting for Human Obstacles in BLE Trilateration Using Artificial Intelligence. *Sensors* **2020**, *20*. [CrossRef] [PubMed]
25. Weichselbaum, J.; Zinner, C.; Gebauer, O.; Pree, W. Accurate 3D-vision-based obstacle detection for an autonomous train. *Comput. Ind.* **2013**, *64*, 1209–1220. [CrossRef]
26. Durrant-Whyte, H.; Bailey, T. Simultaneous localization and mapping: Part I. *IEEE Robot. Autom. Mag.* **2006**, *13*, 99–110. [CrossRef]
27. Engel, J.; Schöps, T.; Cremers, D. LSD-SLAM: Large-Scale Direct Monocular SLAM. In *Computer Vision—ECCV 2014, Proceedings of the 13th European Conference, Zurich, Switzerland, 6–12 September 2014*; Part II; Springer International Publishing: Cham, Switzerland, 2014; pp. 834–849.
28. Nister, D.; Naroditsky, O.; Bergen, J. Visual odometry. In Proceedings of the 2004 IEEE Computer Society Conference on Computer Vision and Pattern Recognition (CVPR 2004), Washington, DC, USA, 27 June–2 July 2004; Volume 1, pp. 652–659.
29. Silva, B.; Burlamaqui, A.; Gonçalves, L. On Monocular Visual Odometry for Indoor Ground Vehicles. In Proceedings of the 2012 Brazilian Robotics Symposium and Latin American Robotics Symposium, Fortaleza, Brazil, 16–19 October 2012; pp. 220–225.
30. Straub, J.; Hilsenbeck, S.; Schroth, G.; Huitl, R.; Möller, A.; Steinbach, E. Fast relocalization for visual odometry using binary features. In Proceedings of the 2013 IEEE International Conference on Image Processing, Melbourne, VIC, Australia, 15–18 September 2013; pp. 2548–2552.
31. Hili, K.; Sammut, A. Pedestrian tracking through inertial measurements. In Proceedings of the 2016 International Conference on Indoor Positioning and Indoor Navigation (IPIN), Alcala de Henares, Spain, 4–7 October 2016; pp. 1–8.
32. Dehghanian, V.; Lowe, M. RSS-INS integration for cooperative indoor positioning. In Proceedings of the 2016 International Conference on Indoor Positioning and Indoor Navigation (IPIN), Alcala de Henares, Spain, 4–7 October 2016; pp. 1–7.
33. Dwiyasa, F.; Lim, M.H. A survey of problems and approaches in wireless-based indoor positioning. In Proceedings of the 2016 International Conference on Indoor Positioning and Indoor Navigation (IPIN), Alcala de Henares, Spain, 4–7 October 2016; pp. 1–7.

34. Kawecki, R.; Korbel, P.; Hausman, S. Influence of User Mobility on the Accuracy of Indoor Positioning with the use of RSSI and Particle Filter Algorithm. In Proceedings of the 2019 Signal Processing Symposium (SPSympo), Krakow, Poland, 17–19 September 2019; pp. 105–108. [CrossRef]
35. Davidson, P.; Piché, R. A Survey of Selected Indoor Positioning Methods for Smartphones. *IEEE Commun. Surv. Tutorials* **2017**, *19*, 1347–1370. [CrossRef]
36. Guo, X.; Shao, W.; Zhao, F.; Wang, Q.; Li, D.; Luo, H. WiMag: Multimode Fusion Localization System based on Magnetic/WiFi/PDR. In Proceedings of the 2016 International Conference on Indoor Positioning and Indoor Navigation (IPIN), Alcala de Henares, Spain, 4–7 October 2016; pp. 1–8.
37. Tomažič, S.; Škrjanc, I. Fusion of visual odometry and inertial navigation system on a smartphone. *Comput. Ind.* **2015**, *74*, 119–134. [CrossRef]
38. Sirtkaya, S.; Seymen, B.; Alatan, A.A. Loosely coupled Kalman filtering for fusion of Visual Odometry and inertial navigation. In Proceedings of the 16th International Conference on Information Fusion, Istanbul, Turkey, 9–12 July 2013; pp. 219–226.
39. Marouane, C.; Maier, M.; Leupold, A.; Linnhoff-Popien, C. Visual odometry using motion vectors from visual feature points. In Proceedings of the 2016 International Conference on Indoor Positioning and Indoor Navigation (IPIN), Alcala de Henares, Spain, 4–7 October 2016; pp. 1–8.
40. Bluetooth SIG. Specification of the Bluetooth System. 2010. Available online: https://www.bluetooth.org/docman/handlers/downloaddoc.ashx?doc_id=229737 (accessed on 1 February 2017).
41. Zhuang, Y.; Syed, Z.; Georgy, J.; El-Sheimy, N. Autonomous Smartphone-based WiFi Positioning System by Using Access Points Localization and Crowdsourcing. *Pervasive Mob. Comput.* **2015**, *18*, 118–136. [CrossRef]
42. Faragher, R.; Harle, R. Location Fingerprinting With Bluetooth Low Energy Beacons. *IEEE J. Sel. Areas Commun.* **2015**, *33*, 2418–2428. [CrossRef]
43. Li, H.; Ma, H. A Low Complexity Low Power Indoor Positioning System Based on Wireless Received Signal Strength. In Proceedings of the 2018 IEEE 20th International Conference on e-Health Networking, Applications and Services (Healthcom), Ostrava, Czech Republic, 17–20 September 2018; pp. 1–6. [CrossRef]
44. Tosi, J.; Taffoni, F.; Santacatterina, M.; Sannino, R.; Formica, D. Performance Evaluation of Bluetooth Low Energy: A Systematic Review. *Sensors* **2017**, *17*. [CrossRef] [PubMed]
45. Zuo, Z.; Liu, L.; Zhang, L.; Fang, Y. Indoor Positioning Based on Bluetooth Low-Energy Beacons Adopting Graph Optimization. *Sensors* **2018**, *18*. [CrossRef] [PubMed]
46. Al-Madani, B.; Orujov, F.; Maskeliūnas, R.; Damaševičius, R.; Venčkauskas, A. Fuzzy Logic Type-2 Based Wireless Indoor Localization System for Navigation of Visually Impaired People in Buildings. *Sensors* **2019**, *19*. [CrossRef]
47. Peng, Y.; Niu, X.; Tang, J.; Mao, D.; Qian, C. Fast Signals of Opportunity Fingerprint Database Maintenance with Autonomous Unmanned Ground Vehicle for Indoor Positioning. *Sensors* **2018**, *18*. [CrossRef] [PubMed]
48. Gao, C.; Harle, R. Easing the survey burden: Quantitative assessment of low-cost signal surveys for indoor positioning. In Proceedings of the 2016 International Conference on Indoor Positioning and Indoor Navigation (IPIN), Alcala de Henares, Spain, 4–7 October 2016; pp. 1–8. [CrossRef]
49. Nastac, D.; Lehan, E.; Iftimie, F.A.; Arsene, O.; Cramariuc, B. Automatic Data Acquisition with Robots for Indoor Fingerprinting. In Proceedings of the 2018 International Conference on Communications (COMM), Bucharest, Romania, 14–16 June 2018; pp. 321–326. [CrossRef]
50. Tomažič, S.; Dovžan, D.; Škrjanc, I. Confidence-Interval-Fuzzy-Model-Based Indoor Localization. *IEEE Trans. Ind. Electron.* **2019**, *66*, 2015–2024. [CrossRef]
51. Engel, J.; Koltun, V.; Cremers, D. Direct Sparse Odometry. *IEEE Trans. Pattern Anal. Mach. Intell.* **2018**, *40*, 611–625. [CrossRef]
52. Google. ARCore. Available online: https://developers.google.com/ar/ (accessed on 6 March 2021).
53. Piao, J.C.; Kim, S.D. Adaptive Monocular Visual–Inertial SLAM for Real-Time Augmented Reality Applications in Mobile Devices. *Sensors* **2017**, *17*. [CrossRef] [PubMed]
54. Morar, A.; Băluţoiu, M.A.; Moldoveanu, A.; Moldoveanu, F.; Butean, A.; Asavei, V. Evaluation of the ARCore Indoor Localization Technology. In Proceedings of the 2020 19th RoEduNet Conference: Networking in Education and Research (RoEduNet), Bucharest, Romania, 11–12 December 2020; pp. 1–5. [CrossRef]
55. Nerurkar, E.; Lynen, S.; Zhao, S. System and Method for Concurrent Odometry and Mapping. US Patent US2017/0336511 A1, 23 November 2017.
56. Mur-Artal, R.; Tardós, J.D. ORB-SLAM2: An Open-Source SLAM System for Monocular, Stereo, and RGB-D Cameras. *IEEE Trans. Robot.* **2017**, *33*, 1255–1262. [CrossRef]
57. Eckenhoff, K.; Geneva, P.; Huang, G. MIMC-VINS: A Versatile and Resilient Multi-IMU Multi-Camera Visual-Inertial Navigation System. *IEEE Trans. Robot.* **2021**, 1–21. [CrossRef]
58. Triggs, B.; McLauchlan, P.F.; Hartley, R.I.; Fitzgibbon, A.W. Bundle Adjustment—A Modern Synthesis. In Proceedings of the International Workshop on Vision Algorithms: Theory and Practice (ICCV '99), Corfu, Greece, 20–25 September 1999; Springer: Berlin/Heidelberg, Germany, 1999; pp. 298–372.
59. Moré, J.J.; Sorensen, D.C. Computing a Trust Region Step. *SIAM J. Sci. Stat. Comput.* **1983**, *4*, 553–572. [CrossRef]
60. Steihaug, T. The Conjugate Gradient Method and Trust Regions in Large Scale Optimization. *SIAM J. Numer. Anal.* **1983**, *20*, 626–637. [CrossRef]

61. Coleman, T.; Li, Y. An Interior Trust Region Approach for Nonlinear Minimization Subject to Bounds. *SIAM J. Optim.* **1996**, *6*, 418–445. [CrossRef]
62. Branch, M.A.; Coleman, T.F.; Li, Y. A Subspace, Interior, and Conjugate Gradient Method for Large-Scale Bound-Constrained Minimization Problems. *SIAM J. Sci. Comput.* **1999**, *21*, 1–23. [CrossRef]
63. Coleman, T.F.; Verma, A. A Preconditioned Conjugate Gradient Approach to Linear Equality Constrained Minimization. *Comput. Optim. Appl.* **2001**, *20*, 61–72.:1011271406353. [CrossRef]
64. Nocedal, J.; Wright, S.J. *Numerical Optimization*, 2nd ed.; Springer: New York, NY, USA, 2006; pp. 102–120.
65. Kennedy, J.; Eberhart, R. Particle swarm optimization. In Proceedings of IEEE International Conference on Neural Networks, Perth, WA, Australia, 27 November–1 December 1995; IEEE: Piscataway, NJ, USA, 1995; pp. 1942–1948.
66. Shi, Y.; Eberhart, R. A modified particle swarm optimizer. In Proceedings of the 1998 IEEE International Conference on Evolutionary Computation Proceedings—IEEE World Congress on Computational Intelligence (Cat. No.98TH8360), Anchorage, AK, USA, 4–9 May 1998; pp. 69–73.
67. Kennedy, J. The particle swarm: Social adaptation of knowledge. In Proceedings of the IEEE International Conference on Evolutionary Computation, Indianapolis, IN, USA, 13–16 April 1997; pp. 303–308.
68. Taherkhani, M.; Safabakhsh, R. A novel stability-based adaptive inertia weight for particle swarm optimization. *Appl. Soft Comput.* **2016**, *38*, 281–295. [CrossRef]
69. Shi, Y.; Eberhart, R.C. Parameter Selection in Particle Swarm Optimization. In Proceedings of the 7th International Conference on Evolutionary Programming VII (EP '98), San Diego, CA, USA, 25–27 March 1998; Springer: London, UK, 1998; pp. 591–600.
70. Eberhart, R.C.; Edu, E.I. Comparing inertia weights and constriction factors in particle swarm optimization. In Proceedings of the 2000 Congress on Evolutionary Computation, La Jolla, CA, USA, 16–19 June 2000; pp. 84–88.
71. Trelea, I.C. The particle swarm optimization algorithm: Convergence analysis and parameter selection. *Inf. Process. Lett.* **2003**, *85*, 317–325. [CrossRef]
72. Kontakt.io. Bluetooth Beacons. Available online: https://kontakt.io/beacons-tags/ (accessed on 10 March 2021).
73. Schafer, R.W. What Is a Savitzky-Golay Filter? [Lecture Notes]. *IEEE Signal Process. Mag.* **2011**, *28*, 111–117. [CrossRef]
74. Tomažič, S.; Škrjanc, I. Indoor RSSI-based Localization using Fuzzy Path Loss Models. In Proceedings of the 2018 IEEE International Conference on Fuzzy Systems (FUZZ-IEEE), Rio de Janeiro, Brazil, 8–13 July 2018; pp. 1–8. [CrossRef]

sensors

MDPI

Article

Indoor Positioning System Based on Global Positioning System Signals with Down- and Up-Converters in 433 MHz ISM Band

Abdulkadir Uzun [1,2], Firas Abdul Ghani [1], Amir Mohsen Ahmadi Najafabadi [1], Hüsnü Yenigün [3] and İbrahim Tekin [1,*]

[1] Electronics Engineering, Sabanci University, Istanbul 34956, Turkey; kadiruzun@sabanciuniv.edu (A.U.); firas@sabanciuniv.edu (F.A.G.); amirahmadi@sabanciuniv.edu (A.M.A.N.)
[2] Radar Electronic Warfare and Intelligence Systems Division, ASELSAN A.Ş., Istanbul 34906, Turkey
[3] Computer Science and Engineering, Sabanci University, Istanbul 34956, Turkey; yenigun@sabanciuniv.edu
* Correspondence: tekin@sabanciuniv.edu; Tel.: +90-(216)-483-9534

Abstract: In this paper, an indoor positioning system using Global Positioning System (GPS) signals in the 433 MHz Industrial Scientific Medical (ISM) band is proposed, and an experimental demonstration of how the proposed system operates under both line-of-sight and non-line-of-sight conditions on a building floor is presented. The proposed method is based on down-converting (DC) repeaters and an up-converting (UC) receiver. The down-conversion is deployed to avoid the restrictions on the use of Global Navigation Satellite Systems (GNSS) repeaters, to achieve higher output power, and to expose the GPS signals to lower path loss. The repeaters receive outdoor GPS signals at 1575.42 MHz (L1 band), down-convert them to the 433 MHz ISM band, then amplify and retransmit them to the indoor environment. The front end up-converter is combined with an off-the-shelf GPS receiver. When GPS signals at 433 MHz are received by the up-converting receiver, it then amplifies and up-converts these signals back to the L1 frequency. Subsequently, the off-the-shelf GPS receiver calculates the pseudo-ranges. The raw data are then sent from the receiver over a 2.4 GHz Wi-Fi link to a remote computer for data processing and indoor position estimation. Each repeater also has an attenuator to adjust its amplification level so that each repeater transmits almost equal signal levels in order to prevent jamming of the off-the-shelf GPS receiver. Experimental results demonstrate that the indoor position of a receiver can be found with sub-meter accuracy under both line-of-sight and non-line-of-sight conditions. The estimated position was found to be 54 and 98 cm away from the real position, while the 50% circular error probable (CEP) of the collected samples showed a radius of 3.3 and 4 m, respectively, for line-of-sight and non-line-of-sight cases.

Keywords: down-conversion; GPS; indoor positioning; navigation; RF repeaters; up-conversion

check for updates

Citation: Uzun, A.; Ghani, F.A.; Ahmadi Najafabadi, A.M.; Yenigün, H.; Tekin, İ. Indoor Positioning System Based on Global Positioning System Signals with Down- and Up-Converters in 433 MHz ISM Band. *Sensors* **2021**, *21*, 4338. https://doi.org/10.3390/s21134338

Academic Editor: Simon Tomažič

Received: 26 May 2021
Accepted: 18 June 2021
Published: 25 June 2021

1. Introduction

Global indoor positioning is an emerging market whose size is forecast to grow from USD 6.1 billion to USD 17.0 billion by 2025 [1]. Many researchers, in the field of indoor positioning, have proposed different solutions to solve this sophisticated problem. Some of the proposed technologies for indoor positioning are based on IEEE 802.11 [2–7], Bluetooth [8–11], Zigbee [12], radio frequency identification devices (RFIDs) [13], visible light [14], acoustic [15,16], and ultrasound [17,18]. GNSS-based solutions have also proven to be a candidate for the indoor positioning problem [19–21]. Some of the GNSS-based solutions are based on high sensitivity GNSS (HS-GNSS) [22], assisted GNSS (A-GNSS) [23], pseudolites [24–31], GNSS repeaters [32,33], repealites [34,35], and peer-to-peer cooperative positioning [36]. Among the GNSS-based techniques, HS-GNSS and A-GNSS technologies require no infrastructure within the indoor environment, while pseudolite and repeater-based approaches require infrastructure.

Ideally, from a navigational point of view, a GPS-based solution to the indoor positioning problem would result in integrated outdoor and indoor applications such as asset tracking, vehicular navigation, and emergency services. However, GPS-based solutions currently experience a multitude of weaknesses due to (a) reflections and multipath errors from the indoor environment, (b) attenuation (up to 30 dB [37]) from walls and other structural components, (c) non-line-of-sight conditions caused by corners and objects, (d) changes in the indoor environment due to people and moving objects, and (e) RF devices operating in the same frequency band and their interference to the indoor positioning system.

In addition to the listed drawbacks, GPS-based systems operating in L1, L2, and L5 bands also suffer from the restrictions on the use of GNSS repeaters. Such restrictions aim to prevent repeaters from interfering with other GNSS systems in the vicinity. The Electronic Communications Committee (ECC), European Telecommunication Standards Institute (ETSI), and the US policy "Manual of Regulations and Procedures for Federal Radio Frequency Management" present the practices and restrictions on the use of GNSS repeaters [38–41]. These standards reduce the coverage of GPS repeaters by limiting the output power and the maximum allowable amplification for repeater structures in the L1, L2, and L5 frequency bands.

In [32,42], a GPS-based indoor positioning technique with three repeaters is shown; however, the repeaters operate solely in the L1 band and, therefore, are restricted to the aforementioned restrictive policies in their usage.

This paper presents a new GPS-based approach that does not contradict the restrictions in the aforementioned standards. The proposed indoor positioning system in this paper operates in the 433 MHz ISM band, hence it is not subject to the amplification restrictions in GPS frequencies. The proposed repeaters down-convert GPS signals in the 1575.42 MHz (L1 band) to the 433 MHz ISM band, allowing signal coverage to be increased, with the higher permitted power levels in the 433 MHZ ISM band. In addition to the higher power levels permitted in the 433 MHz ISM band, the free space path loss in the 433 MHz frequency is 11.22 dB less, as presented in Figure 1, while the penetration through walls is higher than at 1575.42 MHz (GPS frequency) or 2.4 GHz (Wi-Fi frequency). Therefore, in terms of coverage and compatibility with existing rules on the use of repeaters, the proposed system could outperform the existing indoor positioning systems that are operating at frequencies higher than 433 MHz.

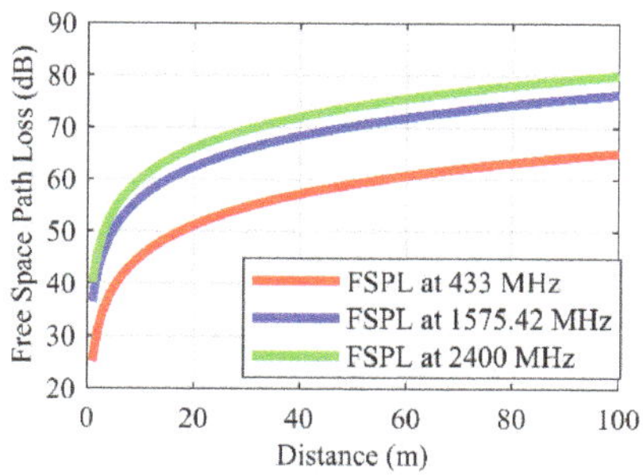

Figure 1. Free space path loss at 2.4 GHz, 1575.42 MHz, and 433 MHz.

The system we propose in this paper differs from the existing systems with its following properties. The down- and up-conversion schemes have previously been proposed in [43–45] for indoor positioning applications with GPS signals. While in [43,44], a down-converting repeater is proposed, neither indoor positioning nor GPS signal down-conversion is demonstrated. In [45], only the down-converting repeater and up-converting receiver circuits are presented, to show that GPS signal fidelity is preserved. However, in [45], it is not shown how the proposed circuits perform in an indoor environment for positioning. In [46], the indoor positioning of an up-converting receiver at a point on a line between two down-converting repeaters (1-dimensional (1D) positioning) is proposed;

however, non-line-of-sight conditions and two-dimensional (2D) positioning of the receiver are not addressed. In [47], we refer to the patent application of a system that can be used with different positioning systems (i.e., BeiDou, Galileo, GLONASS) by changing the register sets in the down- and up-converters to work in the 433 MHz frequency band or even in other ISM bands. In the latter case, the indoor antennas should also be changed to the selected frequency band.

Upon review of readily available publications and to the best of the authors' knowledge, this paper is the first experimental study that demonstrates that 2D indoor positioning can be achieved by transmitting GPS signals, at 433 MHz, from three repeaters in an indoor environment, where the position of the receiver can be estimated by calculating the distance between each repeater and the receiver, on the plane that is formed by the deployed repeaters. The indoor positioning concept that is proposed in this paper is described visually in Figure 2.

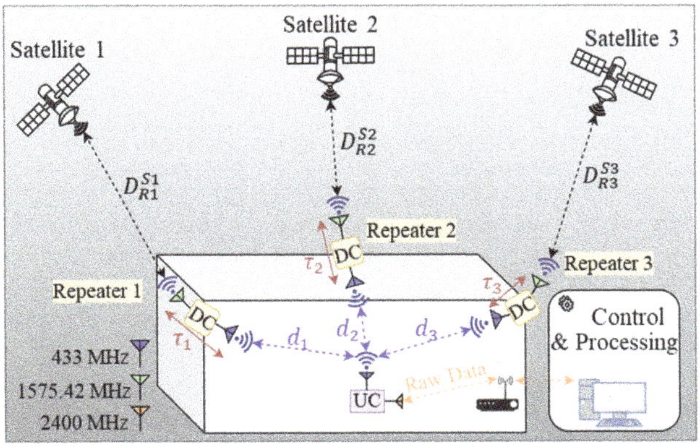

Figure 2. The proposed indoor positioning system. *i*: index of the repeaters ($i = 1, 2, 3$); *j*: index of the satellites ($j = 1, 2, \ldots, n$); *Ri*: *i*th repeater; *Sj*: *j*th satellite; d_i: indoor distance from repeater *Ri* to the receiver (with a clock bias of $t_{bias}^{receiver}$); D_{Ri}^{Sj}: distance from ith repeater (*Ri*:) to the *j*th satellite (*Sj*); τ_i: propagation delay of the *i*th repeater; t_{bias}^{Sj}: satellite clock bias of the *j*th satellite.

In this particular paper, we present two experiments for 2D indoor positioning in the 433 MHz ISM band: one experiment is under line-of-sight conditions, and another is under non-line-of-sight conditions. Note that adding a 4th repeater will allow us to achieve 3D indoor positioning.

The rest of the paper is organized as follows: Section 2 introduces the proposed indoor positioning system and describes its hardware, software and algorithms. Section 3 describes the experimental framework and the real-life environment where we tested our system. Section 4 analyzes and discusses the results we achieved when the proposed system and technique are used for 2D indoor positioning. Finally, Section 5 concludes this paper.

2. Indoor Positioning System in 433 MHz ISM Band

The proposed indoor positioning system is composed of two unique subsystems: GPS down-converting repeaters and an up-converter integrated with an off-the-shelf GPS receiver. The block diagrams of the GPS down-converting repeaters and the up-converter with the off-the-shelf GPS receiver are presented in Figure 3. The red dashed line represents the 433 MHz RF link.

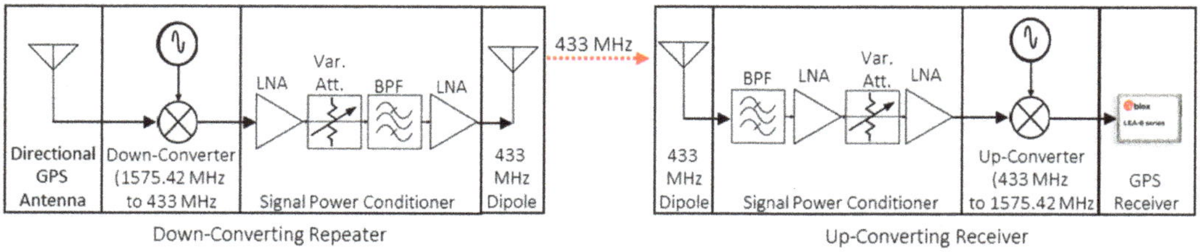

Figure 3. RF blocks in down-converting repeater (**left**) and up-converting receiver (**right**).

2.1. GPS Down-Converting Repeater

The GPS down-converting repeater subsystem comprises RF blocks and supporting blocks, which are presented in Table 1. The RF blocks refer to the components that together form the RF path, through which the GPS signal propagates, while the supporting blocks sustain the regular operation of RF blocks. The GPS down-converting repeater subsystem is implemented using the listed components and blocks in Table 1.

Table 1. Repeater subsystem and its components.

Blocks	Components	Block Properties
Directional GPS Antenna	Tango 20 off-the-shelf active GPS antenna with conic reflector	34.45 dBi gain 60 degree 3 dB beam width
Directional GPS Antenna	TB-JEBT-4R2GW + bias tee	Provides DC to GPS antenna and transmits RF to down converter
Down-Converter	ADRF6820 quadrature demodulator,	8.7 dB block loss Programmable over SPI from controller block
Down-Converter	ZX10Q-2-5-S 90 degree power combiner	Combines down-converter I/Q outputs
Signal Power Conditioner	LHA-13LN + low noise amplifier	22.43 dB gain 0.9 dB noise figure
Signal Power Conditioner	DAT-31R5A-SP + digital step attenuator (variable attenuator)	0 to 31.5 dB adjustable attenuation by 0.5 dB step size and can be adjusted from controller
Signal Power Conditioner	DBP.433.T.A.30 band pass filter at 433 MHz	1.7 dB insertion loss at 433 MHz, 19 MHz 3 dB bandwidth
Signal Power Conditioner	LHA-13LN + as the second low noise amplifier	22.43 dB gain 0.9 dB noise figure
433 MHz Dipole	433 MHz dipole on FR4	2.1 dBi gain 80 degree 3 dB beam width on both sides
Supporting Blocks	Voltage Regulator	Provides DC to system components and blocks
Supporting Blocks	Controller with Wi-Fi for programming attenuator and down-converter	Wi-Fi connection to remote PC, SPI connection to ADRF6820, and attenuator

The implemented GPS down-converting repeater is depicted in Figure 4. Fabricated and modelled 1575.42 MHz outdoor directional GPS antenna and 433 MHz dipole antenna are demonstrated in Figure 5. The RF blocks of the GPS down-converting repeater subsystem are as follows; a directional outdoor GPS antenna (proposed in [32]) and its bias-tee, a 1575.42 to 433 MHz down-converter, a signal power conditioner and filter block (where an LNA, a variable attenuator, a 433 MHz band pass filter, and a second LNA are cascaded), and an indoor 433 MHz dipole antenna. The supporting blocks are a controller over Wi-Fi that interfaces between the RF blocks and the user, and a voltage regulator that provides the required DC to RF blocks.

The active directional GPS antennas pick-up the GPS signals from satellites in the 3 dB beam width. The conic reflector reduces the beam width by 30 degrees. Therefore, the 3 dB beam width of the fabricated antenna is measured as 60 degrees. A bias-tee is placed before the down-converter to provide the DC voltage that is required by the active directional GPS antenna.

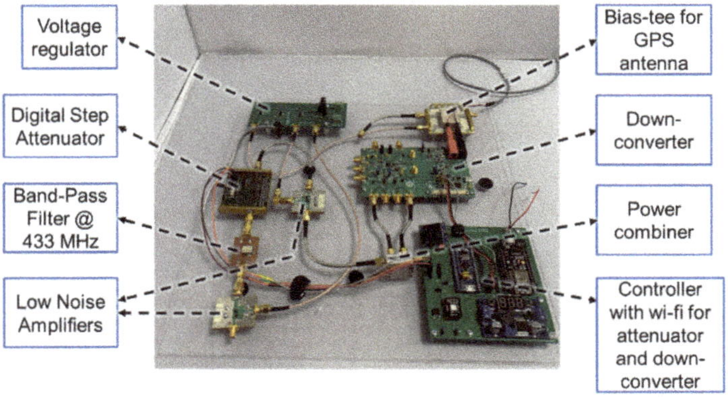

Figure 4. Implemented GPS down-converting repeater subsystem excluding the antennas.

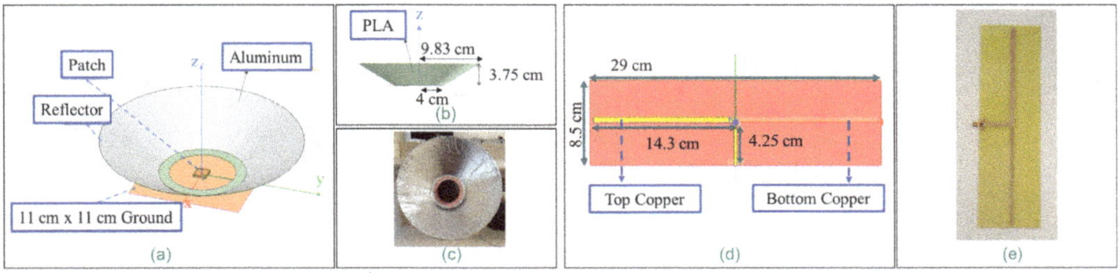

Figure 5. Modelled directional GPS antenna (**a**,**b**), fabricated directional GPS antenna (**c**), modelled 433 MHz dipole antenna (**d**), fabricated 433 MHz dipole antenna (**e**).

When the GPS signal at 1575.42 MHz is received by the antenna, the signal passes to the down-converter. The down-converter converts the signal from 1575.42 to 433 MHz. The amplifiers in the signal power conditioner and filter block then amplify the signal. The amplification level can be adjusted by changing the attenuation of the digital step attenuator, which is able to attenuate the signal from 0 to 31.5 dB. By adjusting the attenuation value of each attenuator, the gain of the overall down-converter subsystem may be set to different values. The bandpass filter at 433 MHz is also deployed to eliminate signals and harmonics out of the band. Furthermore, the signal power conditioner block has another amplifier following the band pass filter. After the second amplifier, the down-converted GPS signals are retransmitted to indoors via a 433 MHz dipole antenna at each repeater.

In terms of adjustments, the repeater board attenuation levels can be set to different values from a remote computer using a Wi-Fi connection. In this way, the gain of a repeater can be adjusted to prevent near-far effects and also keep the signal level from each repeater at a similar level. With this structure, the down-conversion of GPS signals enables users to deploy higher gain GPS receivers than those that are limited by international standards. Additionally, operating in 433 MHz allows us to further increase the gain of the proposed repeaters.

2.2. Up-Converting Receiver

The up-converting receiver subsystem is comprised of the RF blocks and supporting blocks, which are presented in Table 2.

Table 2. Receiver subsystem and its components.

Blocks	Components	Block Properties
433 MHz Dipole	433 MHz dipole on FR4	2.1 dBi gain, 80 degree 3 dB beam width on both sides
Signal Power Conditioner	DBP.433.T.A.30 band pass filter at 433 MHz	1.7 dB insertion loss at 433 MHz, 19 MHz 3 dB bandwidth
Signal Power Conditioner	LHA-13LN + low noise amplifier	22.43 dB gain 0.9 dB noise figure
Signal Power Conditioner	DAT-31R5A-SP + digital step attenuator (variable attenuator)	0 to 31.5 dB adjustable attenuation by 0.5 dB step size and can be adjusted from controller
Signal Power Conditioner	LHA-13LN + as the second low noise amplifier	22.43 dB gain 0.9 dB noise figure
Up-Converter	I/Q power divider, bias tees	Provides I/Q inputs and required DC levels to ADRF6720-27
Up-Converter	ADRF6720-27 quadrature modulator	1.2 dB block loss Programmable over SPI from controller block
Off-the-shelf GPS Receiver	LEA-6T chipset	
Supporting Blocks	Voltage Regulator	Provides DC to system components and blocks
Supporting Blocks	Controller with Wi-Fi for programming attenuator and down-converter	Wi-Fi connection to remote PC, SPI connection to ADRF6820, and attenuator

The up-converting receiver is implemented using the listed components and blocks in Table 2 and is presented in Figure 6. The aforementioned 433 MHz dipole antenna in Figure 5e is also used as the receiver indoor antenna. The RF blocks of the up-converting receiver are (a) a 433 MHz indoor dipole antenna, (b) a signal power conditioner block where a 433 MHz band pass filter, (c) an LNA, (d) a variable attenuator, (e) a second, cascaded LNA, (f) an I/Q power divider, (g) bias tees that provide the required DC to I/Q inputs of the up-converter, (h) a 433 to 1575.42 MHz up-converter, and (i) an off-the-shelf GPS receiver (u-Blox LEA-6T®). The controller over Wi-Fi provides interfaces between RF blocks and the user. Moreover, a voltage regulator is designed to provide the required DC voltage to the RF blocks. In this subsystem, the controller is also connected to the custom GPS receiver, and the raw data from the GPS receiver can be sent over Wi-Fi to a remote computer.

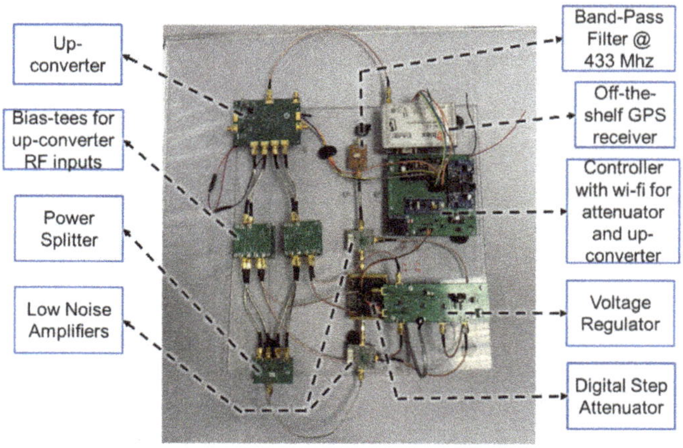

Figure 6. Implemented up-converting receiver excluding the antenna.

The retransmitted 433 MHz positioning signals are picked up by the 433 MHz indoor dipole antennas. In the receiver subsystem, the received signals are first filtered and then amplified. Similar to the repeater subsystem, the amplification levels are adjusted by changing the attenuation of the digital step attenuator, which is able to attenuate the signal from 0 to 31.5 dB. By adjusting the attenuation value of the attenuator, the gain of the overall up-converter subsystem may be set to different values. The subsystem also utilizes a second amplifier to further amplify the signal. These up-converted GPS signals then propagate to the off-the-shelf GPS receiver, which is integrated within the up-converting receiver subsystem.

As mentioned previously, for further flexibility, the up-converting repeater attenuation levels can be set to different values from a remote computer using a Wi-Fi connection. Using a remote computer with an internet connection, where the ephemeris data of GPS satellites are available, the estimation of the indoor position starts when the raw data are sent from the receiver. Therefore, one can conclude that the established Wi-Fi link between the proposed up-converting receiver, and the computer where calculations are done will provide a hot start to the system.

2.3. Algorithm for Indoor Position Estimation

In an indoor environment, although the signal loss and GPS coverage problems can be overcome with the proposed repeaters, this solution requires additional algorithms that takes the non-line-of-sight propagation and repeater delay into account. The proposed technique introduces a new path in that the distance between satellite and the receiver becomes different from that of the normal operation of an off-the-shelf receiver during which there is a line-of-sight distance between the satellite and the receiver. The GPS signals in the proposed scheme come to the repeater first and, then, reach to the receiver as seen in Figure 2.

As part of the indoor position estimation, an algorithm (run on MATLAB®) is designed and run on a remote computer that is connected to the system via Wi-Fi. The Wi-Fi link is established with a Wi-Fi module on the controller block of the receiver board. The raw data (such as pseudo-range, carrier-to-noise ratio, satellite azimuth, and elevation angles, etc.) obtained from the off-the-shelf receiver are transmitted to the remote computer through the Wi-Fi connection. The routine summarized in Figure 7 has the following four steps: raw data reception, satellite selection and satellite-to-repeater distance calculations, cleaning pseudo-range from satellite-to-repeater distances and satellite biases, and finally, running the least squares navigation (LSNAV) algorithm, which is a least squares solution that minimizes the sum of the square of the residual errors [48].

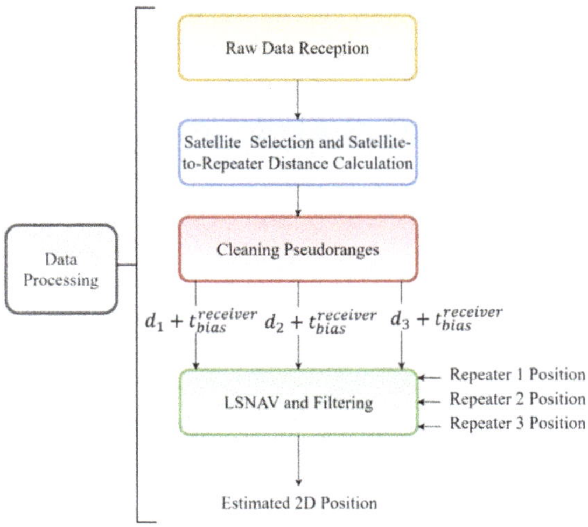

Figure 7. Flow of the algorithm.

In the first part, we remove the biases from the GPS pseudo-range measurement using the models of the troposphere, ionosphere, GPS satellite clocks, GPS satellite movement during signal propagation, and Earth rotation.

Navigation data are also collected as part of raw data from the off-the-shelf receiver, which contains GPS time of the week (ITOW) of the navigation epoch. ITOW field indicates the GPS time at which the navigation epoch occurred. Each navigation solution is triggered by the tick of the 1 kHz clock nearest to the desired navigation solution time. This tick is referred to as a navigation epoch. If the navigation solution attempt is successful, one of the results is an accurate measurement of time in the time base of the chosen GNSS system, called GNSS system time. The difference between the calculated GNSS system time and receiver local time is called the clock bias (and the clock drift is the rate at which this bias is changing). Navigation data also contain an Earth-centered Earth-fixed (ECEF) coordinates solution of the off-the-shelf receiver; however, this solution would not be correct in our system, since the satellite signals come from three different repeaters. This result in erroneous the off-the-shelf receiver position.

The data of each satellite, received by the off-the-shelf receiver on the up-converting receiver, are recorded. The recorded data include GPS satellite ID numbers, satellite azimuth and elevation angles, carrier-to-noise ratios (CNR) related to each satellite signal, and the GPS time of the week. GPS satellites transmit ephemeris through which the receiver can estimate the position of the satellites in the Earth-centered Earth-fixed (ECEF) coordinates system. In addition to a GPS satellite's location (current and predicted), ephemeris includes the orbital parameters, clock bias, date, timing, health, and an almanac (a reduced subset of ephemeris of all satellites) exist. The ephemeris data can be collected using an online server to enable hot start in the proposed system. Moreover, the ephemeris data are used to calculate the satellite's clock bias.

Each of the collected pseudo-range is a sum of the following distance terms: satellite-to-repeater distance, the receiver clock bias, satellite clock bias, repeater delay, the indoor distance from the corresponding repeater to the indoor receiver position.

The second step of the MATLAB® algorithm is designed for choosing the GPS satellites, which are seen from the repeaters. For each repeater, specific satellites are chosen based on their CNR levels and locations. Each satellite's position, clock bias, and distance from the corresponding repeater are calculated using the satellite's ephemeris data. In the third step of the algorithm, the pseudo-range (PR) from each repeater to the selected satellite for the corresponding repeater is cleaned by subtraction operation on the right-hand side of Equation (1). Thus, the indoor distance from the receiver to each repeater is calculated. The calculated indoor distances also include the receiver clock bias. The indoor distances between each repeater and the indoor receiver is calculated using Equation (1).

$$d_i + t_{bias}^{receiver} = PR_{Ri}^{Sj} - \left(D_{Ri}^{Sj} + \tau_i \times c + t_{bias}^{Sj} \times c \right) \quad (1)$$

In Equation (1), i is the index of the repeaters ($i = 1, 2, 3$), j is the index of the satellites ($j = 1, 2, \ldots, n$), Ri represents the ith repeater, Sj represents the jth satellite, d_i represents the indoor distance from repeater Ri to the receiver (with a clock bias of $t_{bias}^{receiver}$), D_{Ri}^{Sj} stands for the distance from ith repeater (Ri) to the jth satellite (Sj), τ_i is the propagation delay of the ith repeater, t_{bias}^{Sj} is the satellite clock bias of the jth satellite, and c represents the speed of the light. When more than one satellite is chosen for a repeater, the $d_i + t_{bias}^{receiver}$ value is calculated by averaging the calculations from each satellite selected for that repeater. Thus, all the distances in Figure 2 can be solved with the proposed system.

The final step of the proposed algorithm is using the LSNAV algorithm to calculate the 2D indoor position for each sample collected. The indoor distances between the receiver and each repeater with receiver clock bias are calculated using Equation (1). Then, these calculated values and the repeater positions are given as the inputs to the LSNAV algorithm. It is important to note that the calculated indoor distances still have a receiver clock bias. Since the receiver clock bias ($t_{bias}^{receiver}$) is the same in all indoor distances, it is removed by the LSNAV algorithm by subtracting them from each other, and the indoor position is calculated. This algorithm is run for each sample. The obtained results are filtered with a moving average filter, and the indoor position is estimated as an average of these filtered points. More details regarding the estimation are provided under Section 4.

3. Experimental Setups for Indoor Positioning

The indoor positioning experiments were performed in the Sabancı University Faculty of Engineering and Natural Sciences. The floorplan is depicted in Figure 8.

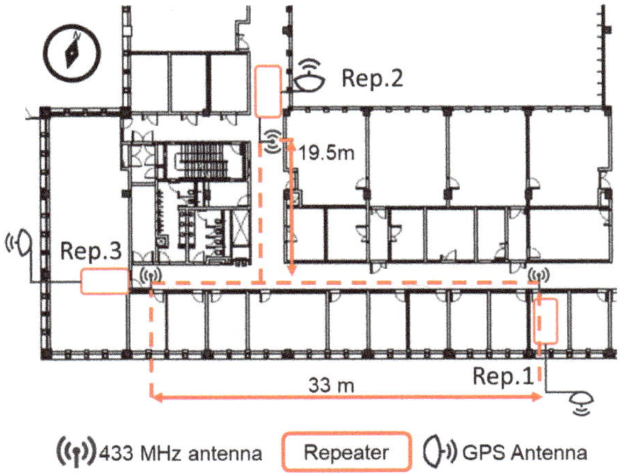

Figure 8. The location of repeaters on the floor plan of Sabancı University Faculty of Engineering and Natural Sciences (FENS) Building 2nd Floor.

The satellite positions in the sky have been observed as depicted in Figure 9, and the directional GPS antenna orientations are set accordingly.

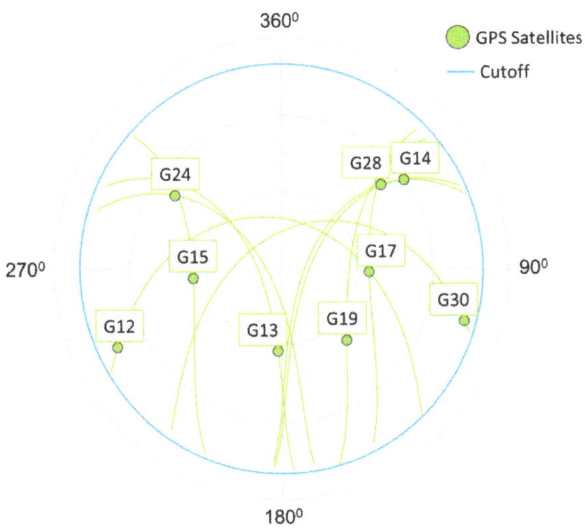

Figure 9. The satellite locations in the sky during the experiments.

The hardware delays from GPS antennas, amplifiers, attenuators, filters, converters, bias-tees, and cables are measured and presented as group delays (τ_i) in Table 3. The group delay of each repeater is measured with a 20 GHz oscilloscope. As a result, for each repeater, GPS antenna and cabling have different measured values in the order of nanoseconds. These values are used in Equation (1) to calibrate the system. Moreover, each repeater position is also provided in Table 3. As mentioned earlier, two experiments are presented in this paper. While the repeater positions are fixed in the same location in both experiments, as presented in Table 3, the receiver position is different for each experiment. In the first experiment, the receiver is located at the intersection of 2 corridors, in that it is on the line-of-sight of all 3 repeaters. However, in the second experiment, the receiver is located closer to Repeater 1 (R1), in that it is not on the line-of-sight of Repeater 2 (R2).

Table 3. Repeater configurations.

Repeater Number (Ri)	Measured Group Delay (τ_i)	Position (Latitude, Longitude)
R1	48.5 ns	40.89072, 29.37955
R2	103 ns *	40.89048, 29.37941
R3	86 ns	40.89044, 29.37966

The latitudes and longitudes are given in degrees. * Repeater 2 has a third LHA-13LN+ amplifier.

In all experiments, there has been no switching, and all three repeaters have been transmitted simultaneously.

3.1. Setup for Experiment 1

Figure 10 shows the first experimental setup and indoor distances between each receiver and repeater. The repeater configurations are kept as provided in Table 3. Although Repeater 2 has a third amplifier, its attenuator is set to 15 dB. Therefore, its gain is 7 dB higher from Repeaters 2 and 3. The outdoor GPS antenna directions are set such that the beam of an antenna does not overlap in azimuth with other GPS antennas in the setup. The directional GPS antennas in Repeaters 1, 2, and 3 are, respectively, towards the geographic east, south, and west. In this experiment, the receiver is located at the intersection of two perpendicular corridors to provide a line-of-sight condition.

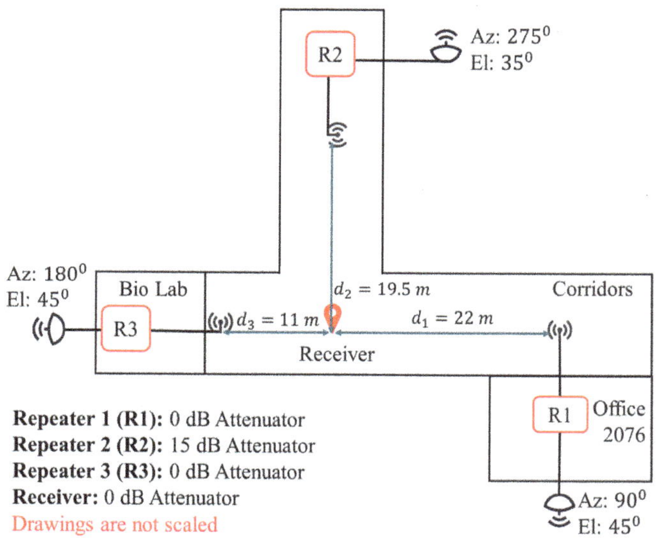

Figure 10. Distance between the receiver and each repeater, directional GPS antenna azimuth (Az) and elevation (El), and the attenuator value in each repeater in Experiment 1.

The real values of the indoor distances d_1, d_2, and d_3 are determined with a physical measurement using a laser pointer for a later comparison with the estimated values. The distance measurements are performed with respect to the known coordinates of the building.

3.2. Setup for Experiment 2

Figure 11 shows the second experimental setup and indoor distances between each receiver and repeater. In this experiment, a non-line-of-sight condition is formed by locating the receiver closer to Repeater 1. Repeater 2 does not directly see the receiver due to the corner at the intersection of two perpendicular corridors. The repeater configurations are kept as provided in Table 3. Repeater 2, which has an additional amplifier, is used with 5 dB attenuation to compensate for the non-line-of-sight condition and the scattering from the corner. Therefore, its gain is 17 dB higher than Repeaters 2 and 3.

The outdoor GPS antenna directions are the same as Experiment 1.

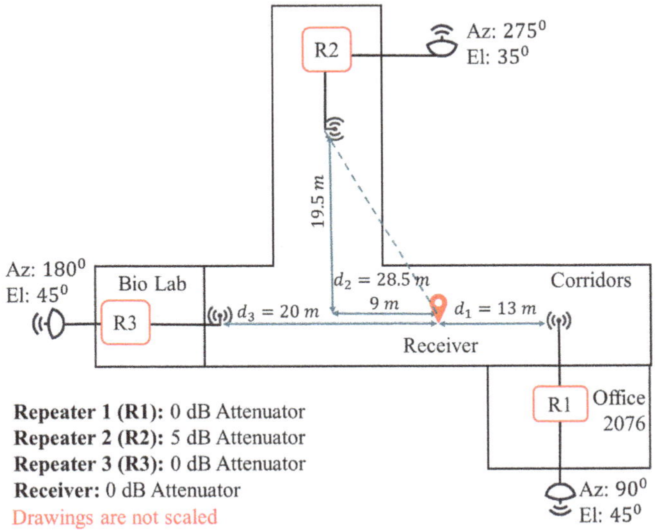

Figure 11. Distance between the receiver and each repeater, directional GPS antenna azimuth (Az) and elevation (El), and the attenuator value in each repeater in Experiment 2.

The real values of indoor distances d_1, d_2, and d_3 are determined with a physical measurement using a laser pointer for a later comparison with the estimated values. The distance measurements are performed with respect to the well-known coordinates of the building.

4. Results and Discussion

In this section, the techniques performed for the indoor position estimation with the proposed hardware, algorithm, and methods are presented along with the experimental data gathered in the experiments in a real indoor environment. Among these, the estimated indoor position, the satellites seen by each directional outdoor GPS antenna during the experiments, estimated distances from each repeater to the receiver with receiver clock bias, the CNR of each satellite signal within the angle of view of the corresponding repeater, and the 50% CEP from the estimated position are graphically visualized and presented in this section. The MATLAB® routine with LSNAV algorithm, presented in Section 2, has been utilized for the position estimation of each sample collected in the experiments.

The satellites, in the angle of view of each repeater in Experiments 1 and 2, are presented in Figure 12. It should be noted that for this approach to work properly, it is important to know which repeater transfers the signal, coming from a particular satellite, into the building. This is in fact one of the reasons for the decision to utilize directional GPS antennas.

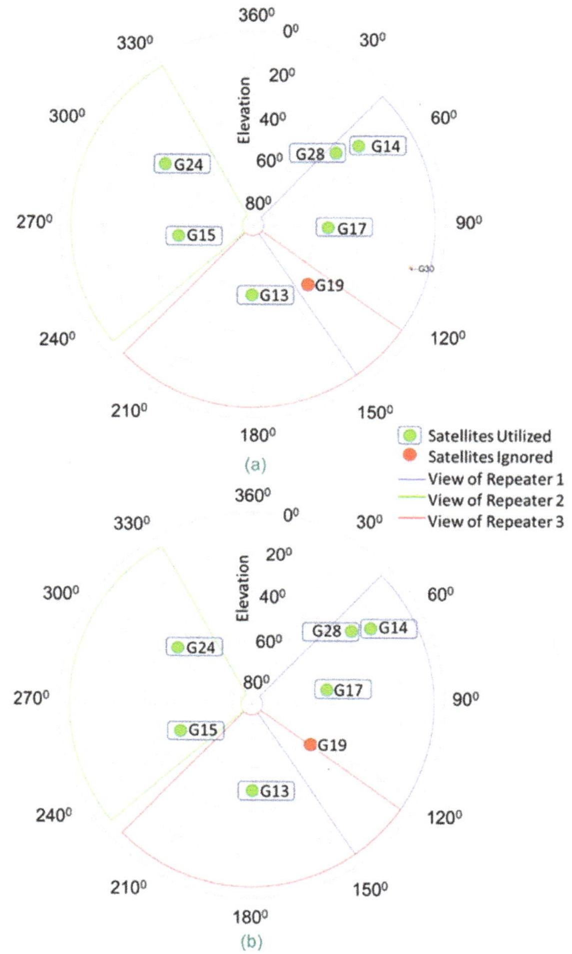

Figure 12. Satellites in each repeater's angle of view in Experiments (**a**) 1 and (**b**) 2.

The association of the repeater, used for the signal of a particular GPS satellite, is accomplished by using the position, the azimuth, and the elevation of the directional GPS antenna and considering the GPS satellites that are in the targeted region of that particular antenna. If a satellite falls into the targeted region of multiple GPS antennas, we do not use any measurement based on that satellite, since we cannot be sure which repeater transferred the signal of that satellite into the building. In these experiments, therefore, the satellite labeled as G19 in Figure 12 (red point) has been ignored, while the satellites represented with green points have been utilized in position estimation for both experiments. Additionally, although the directional antenna beam widths are 60 degrees, the experiment results have shown that the GPS antennas can receive signals from a 90-degree angle in azimuth (Figure 12).

4.1. Results of Experiment 1

The first experiment is performed in the real indoor environment, which is presented in Figure 10, to demonstrate the performance of the proposed indoor positioning system under line-of-sight conditions. The position for each sample is estimated with the LSNAV algorithm using the GPS satellite's signal, which is within the angle of view of each repeater. As mentioned previously, G19 is ignored because it is seen by more than one repeater. Under these conditions, the CNR of the received GPS signals, from the selected satellites for the indoor position estimation, is provided in Figure 13.

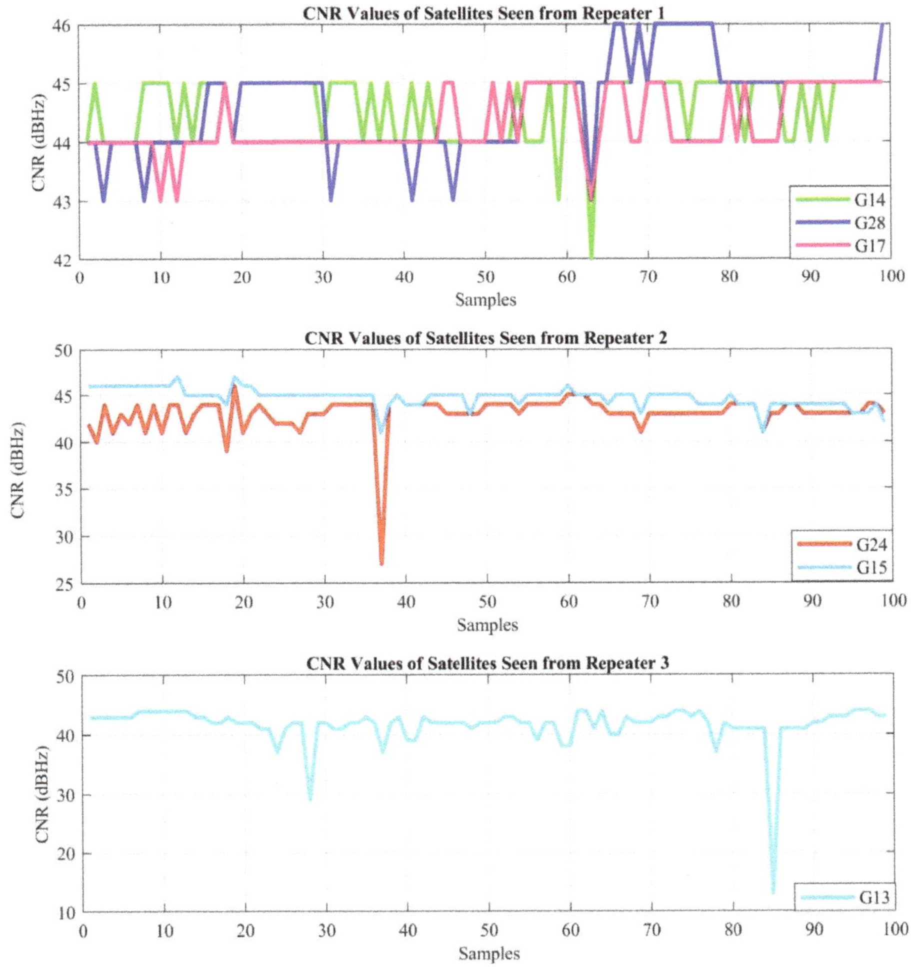

Figure 13. CNR of each GPS signal received by the repeaters.

The sum of the indoor distances (d_1, d_2, and d_3), including receiver clock bias ($t_{bias}^{receiver}$), are calculated in Equation (1) and subtracted from each to obtain the difference terms (d_2 + $t_{bias}^{receiver}$)−(d_1 + $t_{bias}^{receiver}$) and (d_3 + $t_{bias}^{receiver}$)−(d_1 + $t_{bias}^{receiver}$) for each sample. Figure 14a,b shows the indoor distance difference terms in which the receiver clock biases are removed due to the subtraction, while in Figure 14c, it demonstrates the individual distances with the receiver clock bias ($d_1 + t_{bias}^{receiver}$, $d_2 + t_{bias}^{receiver}$, $d_3 + t_{bias}^{receiver}$).

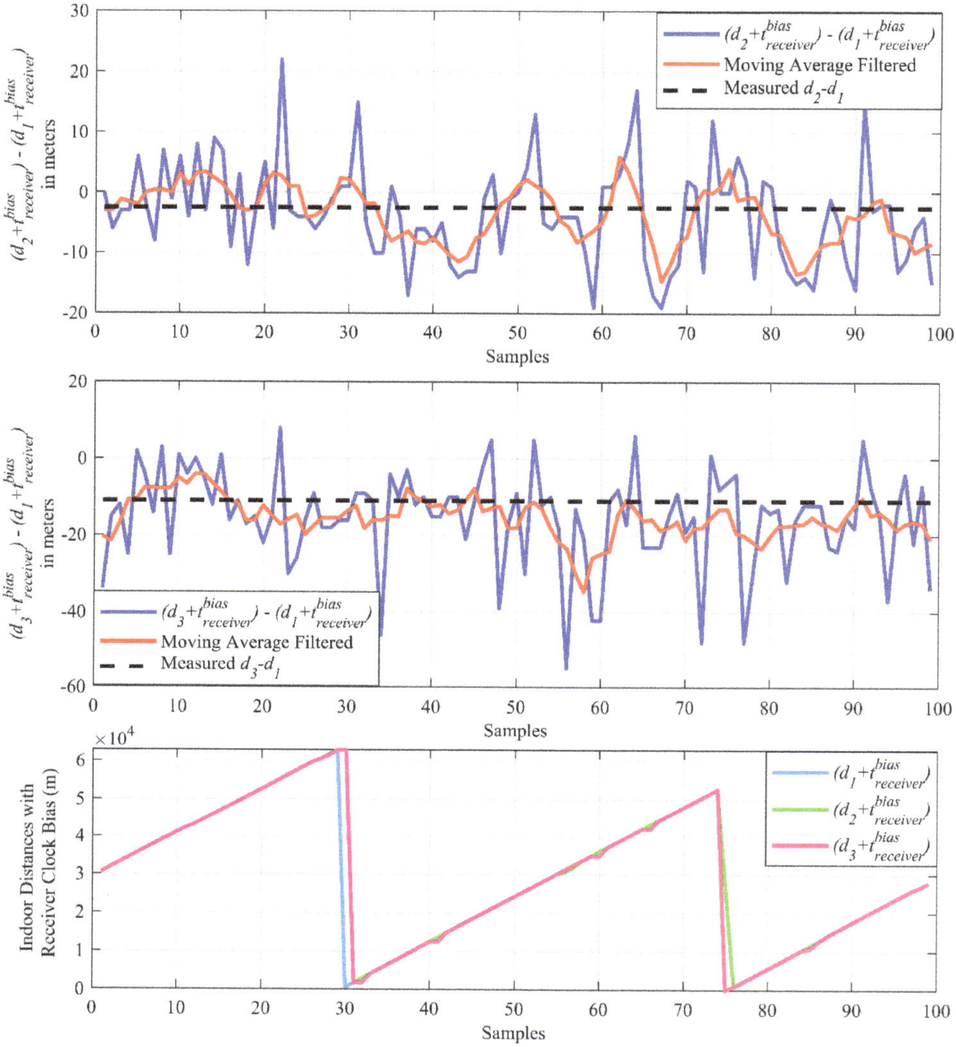

Figure 14. The indoor distance differences (**a**,**b**) and indoor distances with the receiver clock bias term (**c**).

As seen in Figure 14c, the indoor distances with receiver clock bias (d_1 + $t_{bias}^{receiver}$, d_2 + $t_{bias}^{receiver}$, d_3 + $t_{bias}^{receiver}$) can go up to thousands of meters. When the receiver clock bias is removed, with subtraction, the indoor differences can be found in the range of tens of meters. The calculated indoor distance difference terms (blue curve in Figure 14a,b) are averaged with a moving average filter of a five-sample window size. The resulting terms (red curve in Figure 14a,b) show a closer result to the measured indoor distance differences (black dashed line in Figure 14a,b), which can be calculated using the measured indoor distances d_1, d_2, and d_3 in Figure 10.

Figure 15a depicts the GPS repeater and outdoor GPS antenna positions according to their latitude and longitude in Experiment 1. These results show that the LSNAV algorithm has resulted in a solution for 74 out of 100 collected samples. The reason why the LSNAV does not result in a solution for every sample collected is addressed subsequently in the CNR provided in Figure 13. For some samples, due to environmental changes or multipath, an abrupt change in CNR occurs. For these samples, that do not have good CNR from all three repeaters, LSNAV does not result in a solution.

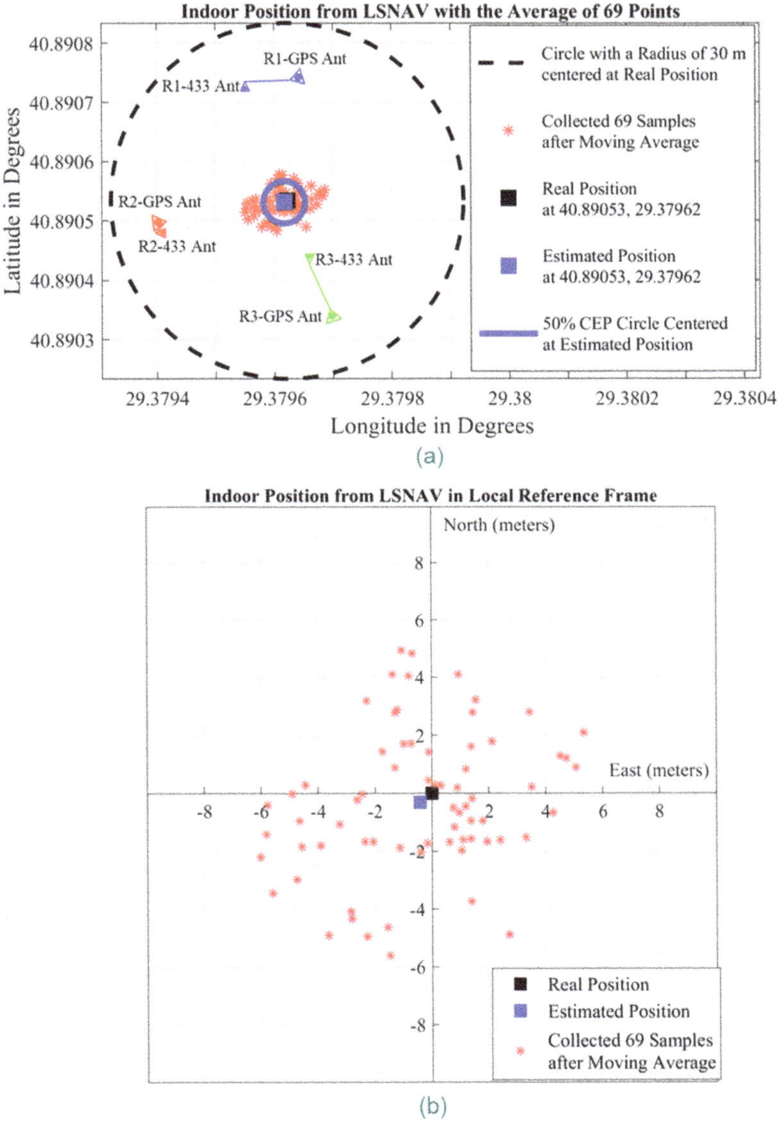

Figure 15. The results of the Experiment 1: (**a**) repeater positions, outdoor GPS antenna positions, collected samples (red dots), real position of the receiver (black square), estimated position (blue square), and CEP (blue circle) on latitude–longitude graph; (**b**) collected samples (red dots), real position of the receiver (black square), estimated position (blue square) on local reference frame whose center is the real position of the receiver.

Moreover, a moving average filter with a five-sample window size has been applied over the 74 results from LSNAV, which then give 69 location estimations. The resulting 69 locations are depicted in Figure 15a,b with red dots. Figure 15a presents the results on a latitude–longitude graph, whereas Figure 15b presents the results on a local reference frame where the center is the real position of the receiver, which was previously determined by physical measurements in the experimental setup with respect to the building whose coordinates and plan are well known. We calculated the radius of 50% CEP as 3.3 m and plotted it in Figure 15a. Moreover, the indoor position is estimated as the mean of these 69 points obtained with moving average filtering. The real position of the receiver is represented with a black square, while the estimated position is presented with a blue square in Figure 15a,b. The estimated location is obtained by averaging 69 points, which are the output of the moving average filter, and is 54 cm away from the real position, as can be seen in Figure 15b.

Because the estimated position from 69 points is 54 cm away from the real position (Figure 15), a sub-meter accuracy has been achieved for line-of-sight conditions. In addition to the 50% CEP with a radius of 3.3 m from the estimated position, we also calculated each of the 69 points distances from the real position and presented it in Figure 16. The average error of the individual samples, from the real position, is 3.37 m.

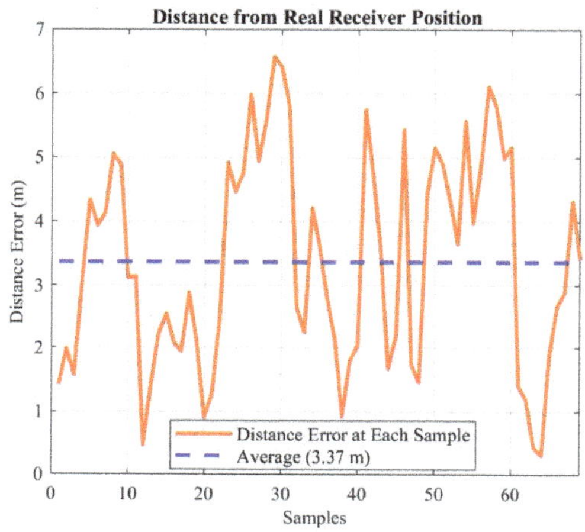

Figure 16. Distance from the real position for each of 69 points.

4.2. Results of Experiment 2

Experiment 2 setup in Figure 11 is set to demonstrate the performance of the proposed indoor positioning system under non-line-of-sight conditions in a real indoor environment. The CNR of GPS signals received from the selected GPS satellites are provided in Figure 17.

Similar to Experiment 1, the average position is estimated with the LSNAV algorithm. In this experiment also, G19 is ignored, as it is within the angle of view of more than one repeater. Additionally, although G15 is seen by Repeater 2, its CNR is much lower than G24; therefore, it is also ignored in the calculations.

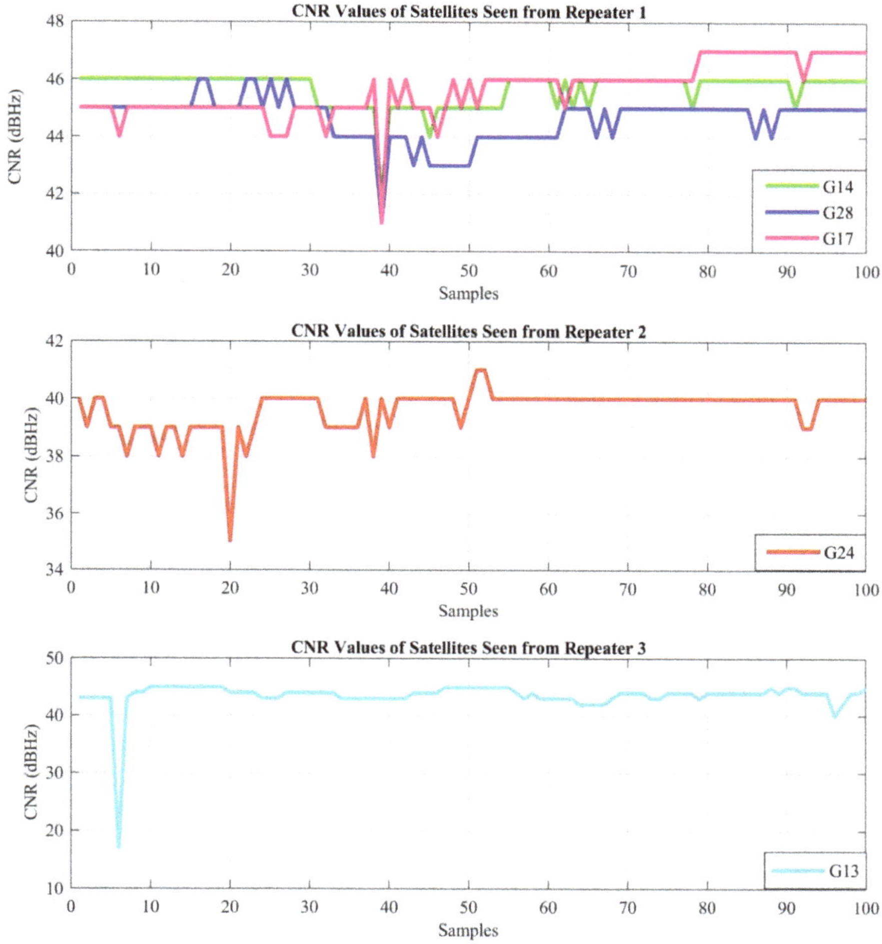

Figure 17. CNR of each GPS signal received by the repeaters.

The indoor distances (d_1, d_2, and d_3), with receiver clock bias ($t_{bias}^{receiver}$) and their difference at each sample (d_2 + $t_{bias}^{receiver}$)−(d_1 + $t_{bias}^{receiver}$) and (d_3 + $t_{bias}^{receiver}$)−(d_1 + $t_{bias}^{receiver}$) are plotted and demonstrated in Figure 18. The resulting terms (red curve in Figure 18a,b) show a closer result to the measured indoor distance differences (black dashed line in Figure 18a,b), which can be calculated using the measured indoor distances d_1, d_2, and d_3 in Figure 11.

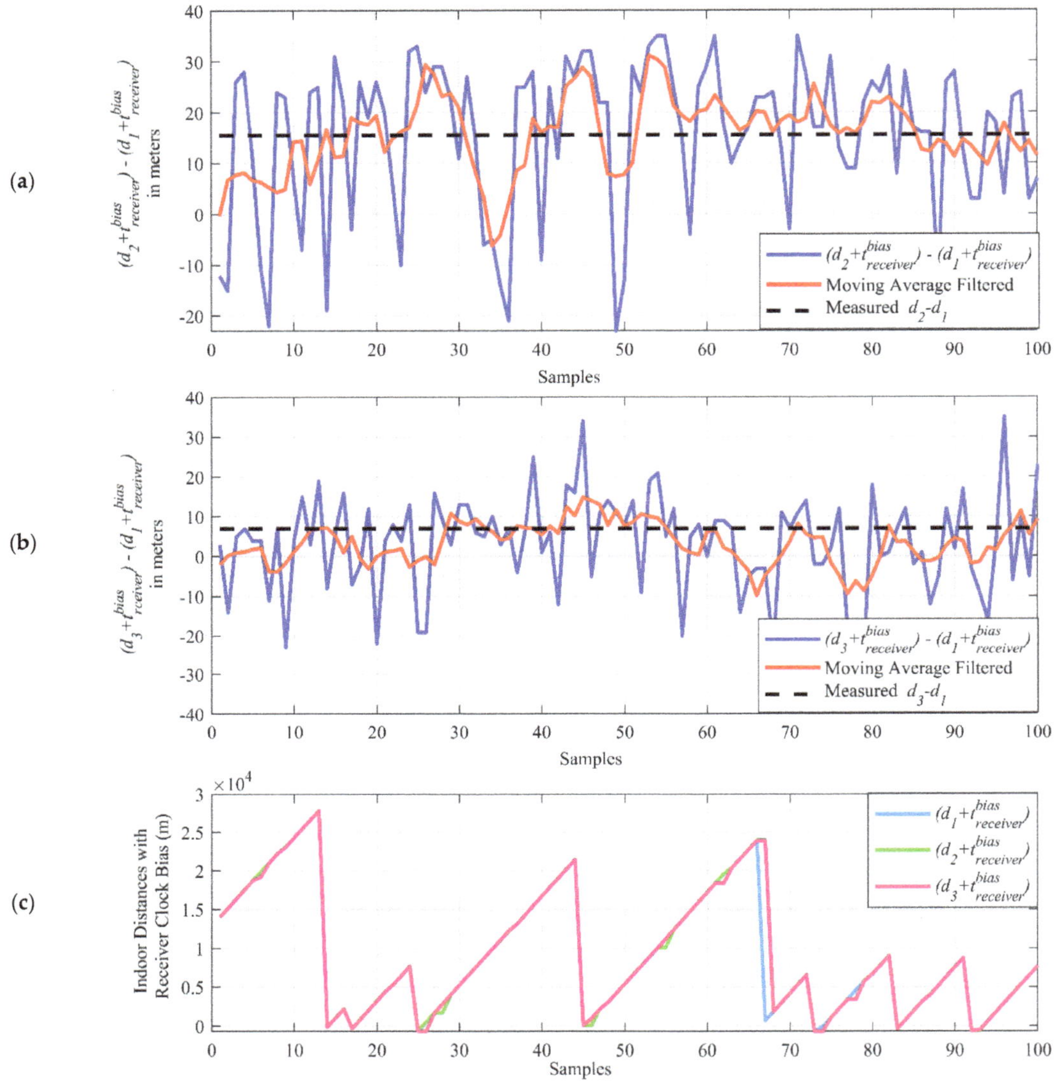

Figure 18. The indoor distance differences (**a**,**b**) and indoor distances with the receiver clock bias term (**c**).

Figure 19a depicts the GPS repeater and outdoor GPS antenna positions, according to their latitude and longitude, in Experiment 2. In this experiment, the LSNAV algorithm has resulted in solutions for 35 out of 100 collected samples. Then, a moving average filter with a five-sample window size has been applied over these 35 converged solutions, and 30 location estimations have been obtained (Figure 19a,b with red dots). Figure 19a demonstrates the results on a latitude–longitude graph, whereas Figure 19b presents the results on a local reference frame in which the center is the real position of the receiver, which was previously determined by distance measurement in the experimental setup with respect to the known coordinates of the building structure. The indoor position is estimated as the mean of these 30 points. The real position of the receiver (black square) and estimated position (blue square) have been plotted in Figure 19a,b. It is seen that the estimated position is 98 cm away from the real position in Figure 19b.

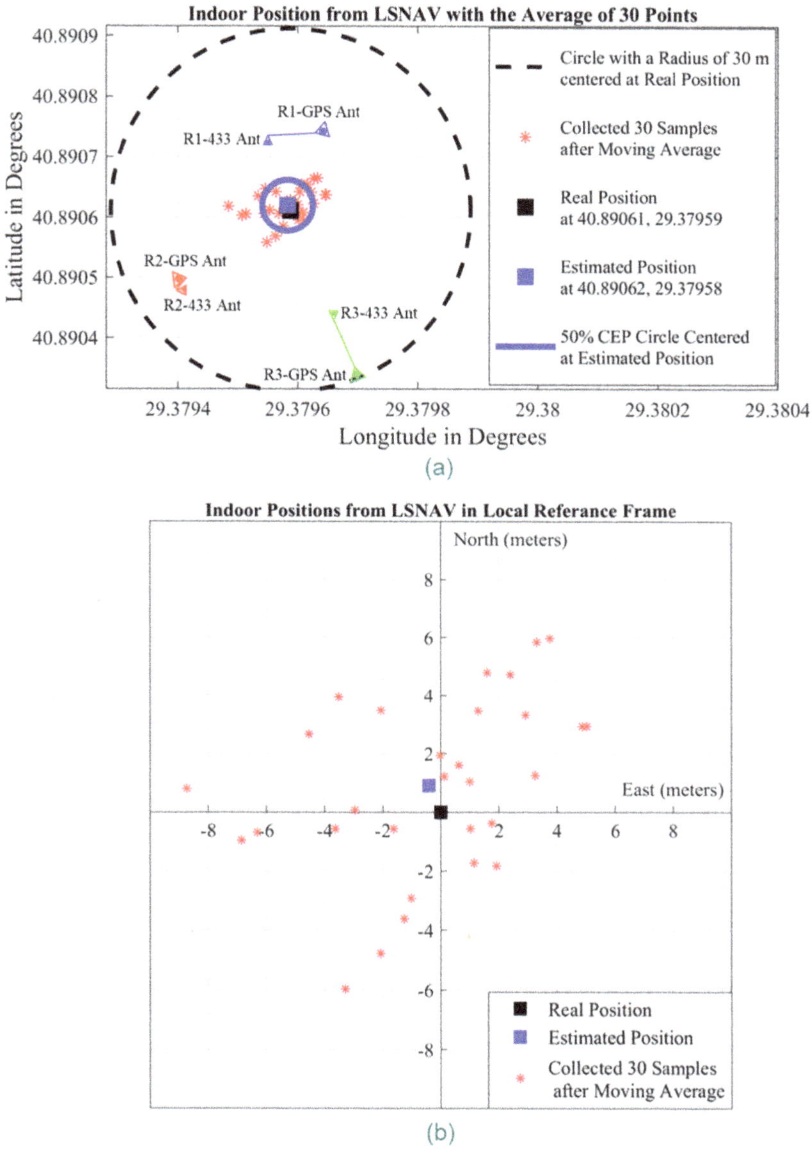

Figure 19. The results of the Experiment 2: (**a**) repeater positions, outdoor GPS antenna positions, collected samples (red dots), real position of the receiver (black square), estimated position (blue square), and CEP (blue circle) on latitude–longitude graph; (**b**) collected samples (red dots), real position of the receiver (black square), estimated position (blue square) on local reference frame whose center is the real position of the receiver.

The 50% CEP is calculated to have a radius of 4 m and drawn also in Figure 19a.

Moreover, the distance error from the real receiver position for each of the 30 points has been depicted in Figure 20. The average error of the individual samples from the real position is 4.17 m.

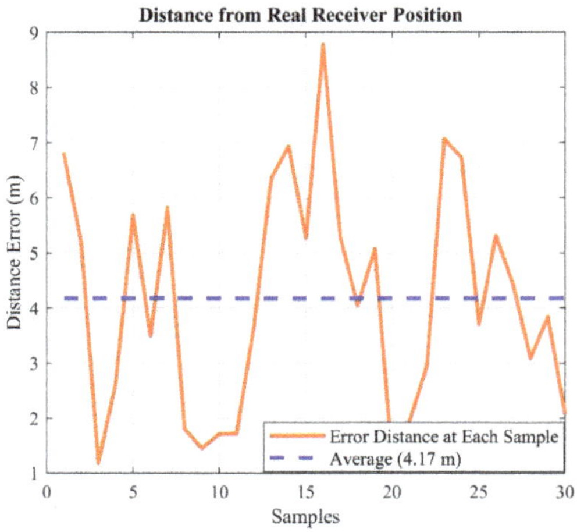

Figure 20. Deviation from the real position for each of 30 points.

Under non-line-of-sight conditions in the experiments, while estimated position error increases, it is still possible to achieve sub-meter accuracy. To quantify the increased error amount, one can compare the radius of the 50% CEP circles in Figures 15a and 19a. The radius of the 50% CEP is higher in Experiment 2. Additionally, when CNR values in Figure 17 are compared with those in Figure 13, we can conclude that the non-line-of-sight condition reduces the CNR. In the proposed system, variable attenuators are used to compensate for the non-line-of-sight conditions. To adjust, we can easily set the attenuation level, the retransmitted GPS power levels, and adjust the CNR levels so that the receivers can receive almost equal signal levels from each direction and each satellite. That is the reason why the attenuation level of Repeater 2 is set to a lower level in Experiment 2.

5. Conclusions

The implemented 433 MHz down-converting repeaters with variable attenuators and 433 MHz up-converting receiver with variable attenuator have been used in 2D positioning successfully for the first time. It is also shown that, with the help of cascaded LNA and variable attenuator, the proposed system for 2D positioning has gained immunity to the non-line-of-sight conditions. The power level of the signals that are transmitted to the receiver can be adjusted by varying the attenuation so that a receiver can pick up different satellite signals coming from different indoor paths almost equally. The GPS signals, which are down-converted and retransmitted by repeaters, are sent to the indoor environment and picked-up by an up-converting receiver. After up-conversion, the raw data are collected by an off-the-shelf receiver and then transmitted over Wi-Fi to a remote computer for processing. The estimated positions are found to be only 54 and 98 cm away from the actual receiver position, for line-of-sight and non-line-of-sight cases, respectively. Therefore, experiment results show that sub-meter accuracy can be achieved with the transmission of GPS signals in the 433 MHz ISM band in an indoor environment.

Author Contributions: Conceptualization, İ.T. and H.Y.; funding acquisition, İ.T. and H.Y.; software, F.A.G., A.U., and A.M.A.N.; investigation, A.U. and F.A.G.; testing, A.U., F.A.G., A.M.A.N., H.Y., and İ.T.; writing—original draft preparation, A.U. and F.A.G.; writing—review and editing, A.U., F.A.G., H.Y., and İ.T.; project administration, İ.T. and H.Y. All authors have read and agreed to the published version of the manuscript.

Funding: This research has been supported by the Scientific and Technological Research Council of Turkey (TÜBİTAK), Grant No: 116E752.

Institutional Review Board Statement: Not Applicable.

Informed Consent Statement: Not Applicable.

Data Availability Statement: Not Applicable.

Conflicts of Interest: The authors declare no conflict of interest. The funders had no role in the design of the study; in the collection, analyses, or interpretation of data; in the writing of the manuscript, or in the decision to publish the results.

References

1. MarketsandMarkets. Indoor Location Market by Component (Hardware, Solutions, and Services), Deployment Mode, Organization Size, Technology, Application, Vertical (Retail, Transportation and Logistics, Entertainment), and Region—Global Forecast to 2025 (Report Code: TC 2878). May 2020. Available online: https://www.marketsandmarkets.com/Market-Reports/indoor-location-market-989.html (accessed on 20 May 2021).
2. Vasisht, D.; Kumar, S.; Katabi, D. Decimeter-Level Localization with A Single Wifi Access Point. In Proceedings of the 13th USENIX Symposium on Networked Systems Design and Implementation (NSDI'16), Santa Clara, CA, USA, 16–18 March 2016; pp. 165–178.
3. Cypriani, M.; Lassabe, F.; Canalda, P.; Spies, F. Open Wireless Positioning System: A Wi-Fi-Based Indoor Positioning System. In Proceedings of the 2009 IEEE 70th Vehicular Technology Conference Fall (VTC 2009-Fall), Anchorage, AK, USA, 20–23 September 2009; pp. 1–5.
4. Zou, H.; Jiang, H.; Lu, X.; Xie, L. An Online Sequential Extreme Learning Machine Approach to Wifi Based Indoor Positioning. In Proceedings of the 2014 IEEE World Forum on Internet of Things (WF-IoT), Seoul, Korea, 6 March 2014; pp. 111–116.
5. Ciurana, M.; Cugno, S. WLAN Indoor Positioning Based on TOA with Two Reference Points. In Proceedings of the 4th Workshop on Positioning, Navigation and Communication 2007 (WPNC'07), Hannover, Germany, 22–22 March 2007; pp. 23–28.
6. Hoang, M.K.; Haeb-Umbach, R. Parameter Estimation and Classification of Censored Gaussian Data with Application to Wifi Indoor Positioning. In Proceedings of the 38th International Conference on Acoustics, Speech, and Signal Processing (ICASSP 2013), Vancouver, BC, Canada, 26–31 May 2013; pp. 3721–3725.
7. Faheem, F. Ibeacon Based Proximity and Indoor Localization System. Master's Thesis, Purdue University, West Lafayette, IN, USA, 2016.
8. Faragher, R.; Harle, R. Location Fingerprinting with Bluetooth Low Energy Beacons. *IEEE J. Sel. Areas Commun.* **2015**, *33*, 2418–2428. [CrossRef]
9. Madhavapeddy, A.; Tse, A. A Study of Bluetooth Propagation Using Accurate Indoor Location Mapping. In Proceedings of the UbiComp 2005: Ubiquitous Computing, Tokyo, Japan, 11–14 September 2005; pp. 105–122.
10. Zafari, F.; Papapanagiotou, I.; Christidis, K. Microlocation for Internet of Things Equipped Smart Buildings. *IEEE Internet Things J.* **2016**, *3*, 96–112. [CrossRef]
11. Castillo-Cara, M.; Lovón-Melgarejo, J.; Bravo-Rocca, G.; Orozco-Barbosa, L.; García-Varea, I. An Analysis of Multiple Criteria and Setups for Bluetooth Smartphone-Based Indoor Localization Mechanism. *J. Sens.* **2017**, *2017*. [CrossRef]
12. Aykac, M.; Ergun, E.; Aldin, N.B. Zigbee-Based Indoor Localization System with The Personal Dynamic Positioning Method and Modified Particle Filter Estimation. *Analog. Integr. Circuits Signal Process.* **2017**, *92*, 263–279. [CrossRef]
13. Ni, L.M.; Liu, Y.; Lau, Y.C.; Patil, A.P. LANDMARC: Indoor Location Sensing Using Active RFID. *Wirel. Netw.* **2004**, *10*, 701–710. [CrossRef]
14. Kuo, Y.-S.; Pannuto, P.; Hsiao, K.-J.; Dutta, P. Luxapose: Indoor Positioning with Mobile Phones and Visible Light. In Proceedings of the 20th Annual International Conference on Mobile Computing and Networking, Maui, HI, USA, 7–11 September 2014; pp. 447–458.
15. Liu, K.; Liu, X.; Li, X. Guoguo: Enabling Fine-Grained Smartphone Localization Via Acoustic Anchors. *IEEE Trans. Mob. Comput.* **2016**, *15*, 1144–1156. [CrossRef]
16. Huang, W.; Xiong, Y.; Li, X.Y.; Lin, H.; Mao, X.; Yang, P.; Liu, Y.; Wang, X. Swadloon: Direction Finding and Indoor Localization Using Acoustic Signal by Shaking Smartphones. *IEEE Trans. Mob. Comput.* **2015**, *14*, 2145–2157. [CrossRef]
17. Priyantha, N.B. *The Cricket Indoor Location System*; Massachusetts Institute of Technology: Cambridge, MA, USA, 2005.
18. Hazas, M.; Hopper, A. Broadband Ultrasonic Location Systems for Improved Indoor Positioning. *IEEE Trans. Mob. Comput.* **2006**, *5*, 536–547. [CrossRef]
19. Peterson, B.; Bruckner, D.; Heye, S. Measuring GPS Signals Indoors. In Proceedings of the 10th International Technical Meeting of the Satellite Division of The Institute of Navigation (ION GPS 1997), Kansas City, MO, USA, 16–19 September 1997; pp. 615–624.
20. Moeglein, M.; Krasner, N. An Introduction to SnapTrack Server-Aided GPS Technology. In Proceedings of the 11th International Technical Meeting of the Satellite Division of The Institute of Navigation (ION GPS 1998), Nashville, TN, USA, 15–18 September 1998; pp. 333–342.
21. Garin, L.J.; Chansarkar, M.; Miocinovic, S.; Norman, C.; Hilgenberg, D. Wireless Assisted GPS-SiRF Architecture and Field Test Results. In Proceedings of the 12th International Technical Meeting of the Satellite Division of The Institute of Navigation (ION GPS 1999), Nashville, TN, USA, 14–17 September 1999; pp. 489–498.
22. Andrianarison, M. New Methods and Architectures for High Sensitivity Hybrid GNSS Receivers in Challenging Environments. Ph.D. Thesis, University of Toulouse, Toulouse, France, 2018.

23. Monnerat, M.; Couty, R.; Vincent, N.; Huez, O.; Chatre, E. The Assisted GNSS, Technology and Applications. In Proceedings of the 17th International Technical Meeting of the Satellite Division of The Institute of Navigation (ION GNSS 2004), Long Beach, CA, USA, 21–24 September 2004.
24. Klein, D.; Parkinson, B.W. The Use of Pseudo-Satellites for Improving GPS Performance. *J. Institude Navig.* **1984**, *31*, 303–315. [CrossRef]
25. Rapinski, J.; Cellmer, S.; Rzepecka, Z. Modified GPS/Pseudolite Navigation Message. *J. Navig.* **2012**, *65*, 711–716. [CrossRef]
26. Rizos, C.; Roberts, G.; Barnes, J.; Gambale, N. Experimental Results of Locata: A High Accuracy Indoor Positioning System. In Proceedings of the 2010 International Conference on Indoor Positioning and Indoor Navigation (IPIN), Zurich, Switzerland, 15–17 September 2010. [CrossRef]
27. Kee, C.; Jun, H.; Yun, D. Indoor Navigation System Using Asynchronous Pseudolites. *J. Navig.* **2003**, *56*, 443–455. [CrossRef]
28. Xingli, G.; Baoguo, Y.; Xianpeng, W.; Yongqin, Y.; Ruicai, J.; Heng, Z.; Chuanzhen, S.; Lu, H.; Boyuan, W. A New Array Pseudolites Technology for High Precision Indoor Positioning. *IEEE Access* **2019**, *7*, 153269–153277. [CrossRef]
29. Petrovski, I.; Okano, K.; Kawaguchi, S.; Torimoto, H.; Suzuki, K.; Toda, M.; Akita, J. Indoor Code and Carrier Phase Positioning with Pseudolites and Multiple GPS Repeaters. In Proceedings of the 16th International Technical Meeting of the Satellite Division of The Institute of Navigation (ION GPS/GNSS 2003), Portland, OR, USA, 9–12 September 2003; pp. 1135–1143.
30. Xu, R.; Chen, W.; Xu, Y.; Ji, S. A New Indoor Positioning System Architecture Using Gps Signals. *Sensors* **2015**, *15*, 10074–10087. [CrossRef]
31. Ma, C.; Yang, J.; Chen, J.; Tang, Y. Indoor and Outdoor Positioning System Based on Navigation Signal Simulator and Pseudolites. *Adv. Space Res.* **2018**, *62*, 2509–2517. [CrossRef]
32. Ozsoy, K.; Bozkurt, A.; Tekin, I. Indoor Positioning Based on Global Positioning System Signals. *Microw. Opt. Technol. Lett.* **2013**, *55*, 1091–1097. [CrossRef]
33. Fluerasu, A.; Jardak, N.; Vervisch-picois, A.; Samama, N. GNSS Repeater Based Approach for Indoor Positioning: Current Status. In Proceedings of the European Navigational Conference ENC-GNSS 2009, Naples, Italy, 3–6 May 2009.
34. Vervisch-Picois, A.; Samama, N. First Experimental Performances of The Repealite Based Indoor Positioning System. In Proceedings of the 2012 International Symposium on Wireless Communication Systems, Paris, France, 28–31 August 2012; pp. 636–640.
35. Selmi, I.; Samama, N.; Vervisch-Picois, A. A New Approach for Decimeter Accurate GNSS Indoor Positioning Using Carrier Phase Measurements. In Proceedings of the 2013 International Conference on Indoor Positioning and Indoor Navigation, Montbeliard, France, 28–31 October 2013.
36. Morosi, S.; Martinelli, A.; DelźRe, E. Peer-to-peer cooperation for GPS positioning. *Int. J. Satell. Commun. Netw.* **2017**, *35*, 323–341. [CrossRef]
37. Mautz, R. Overview of Current Indoor Positioning Systems. *Geod. Cartogr.* **2009**, *35*, 18–22. [CrossRef]
38. Electronic Communications Committee (ECC) within the European Conference of Postal and Telecommunications Administration (CEPT). *Technical and Operational Provisions Required for the Use of GNSS Repeaters*; ECC REPORT 129; ECC: Dublin, Ireland, 2009. Available online: https://docdb.cept.org/download/f9323e3e-0577/ECCREP129.PDF (accessed on 20 May 2021).
39. Electronic Communications Committee (ECC) within the European Conference of Postal and Telecommunications Administration (CEPT). *Regulatory Framework for Global Navigation Satellite System (GNSS) Repeaters*. ECC REPORT 145. Available online: https://docdb.cept.org/download/582 (accessed on 20 May 2021).
40. European Telecommunications Standard Institute (ETSI). Electromagnetic Compatibility and Radio Spectrum Matters (ERM); Short Range Devices; Global Navigation Satellite Systems (GNSS) Repeaters; Harmonized EN Covering the Essential Requirements of Article 3.2 of the R&TTE Directive (ETSI EN 302 645). Available online: https://www.etsi.org/deliver/etsi_en/302600_302699/302645/01.01.01_60/en_302645v010101p.pdf (accessed on 20 May 2021).
41. U.S. Department of Commerce National Telecommunications and Information Administration (NTIA). *Manual of Regulations and Procedures for Federal Radio Frequency Management*; Redbook. Available online: https://www.ntia.doc.gov/page/2011/manual-regulations-and-procedures-federal-radio-frequency-management-redbook (accessed on 20 May 2021).
42. Tekin, I.; Bozkurt, A.; Ozsoy, K. Indoor Positioning System Based on GPS Signals and Pseudolites with Outdoor Directional Antennas. U.S. Patent US20120286992A1, 15 November 2012.
43. Orhan, O.; Tekin, I.; Bozkurt, A. Down-converter for GPS applications. In Proceedings of the 12th Mediterranean Microwave Symposium, Istanbul, Turkey, 3–4 September 2012.
44. Orhan, O. An Indoor Positioning System Based on Global Positioning System. Master's Thesis, Sabanci University, Istanbul, Turkey, 2013.
45. Uzun, A.; Ghani, F.A.; Yenigun, H.; Tekin, I. A Novel GNSS Repeater Architecture for Indoor Positioning Systems in ISM Band. In Proceedings of the 2020 IEEE International Symposium on Antennas and Propagation and North American Radio Science Meeting, Montreal, QC, Canada, 5–10 July 2020; pp. 1631–1632.
46. Uzun, A. A Novel Indoor Positioning System Based on GPS Repeaters in 433 MHZ ISM Band. Master's Thesis, Sabanci University, Istanbul, Turkey, 2020.

47. Tekin, I.; Yenigun, H.; Uzun, A.; Ghani, F.A. A GNSS Repeater Architecture and Location Finding Method for Indoor Positioning Systems Using Lower Frequencies than GNSS Signals. No. PCT/TR2020/050202, 13 March 2020.
48. Constell, Inc. *Constellation Toolbox User's Guide*, Version 7.00. Available online: https://docplayer.net/64176453-Constellation-toolbox.html (accessed on 20 May 2021).

sensors

MDPI

Article

Calibration-Free Single-Anchor Indoor Localization Using an ESPAR Antenna

Mateusz Groth *, Krzysztof Nyka and Lukasz Kulas

Department of Microwave and Antenna Engineering, Faculty of Electronics, Telecommunications and Informatics, Gdansk University of Technology, Narutowicza 11/12, 80-233 Gdansk, Poland; krznyka@pg.edu.pl (K.N.); lukasz.kulas@pg.edu.pl (L.K.)
* Correspondence: mateusz.groth@pg.edu.pl

Abstract: In this paper, we present a novel, low-cost approach to indoor localization that is capable of performing localization processes in real indoor environments and does not require calibration or recalibration procedures. To this end, we propose a single-anchor architecture and design based on an electronically steerable parasitic array radiator (ESPAR) antenna and Nordic Semiconductor nRF52840 utilizing Bluetooth Low Energy (BLE) protocol. The proposed algorithm relies on received signal strength (RSS) values measured by the receiver equipped with the ESPAR antenna for every considered antenna radiation pattern. The calibration-free concept is achieved by using inexpensive BLE nodes installed in known positions on the walls of the test room and acting as reference nodes for the positioning algorithm. Measurements performed in the indoor environment show that the proposed approach can successfully provide positioning results better than those previously reported for single-anchor ESPAR antenna localization systems employing the classical fingerprinting method and relying on time-consuming calibration procedures.

Keywords: Internet of Things (IoT); wireless sensor network (WSN); switched-beam antenna; electronically steerable parasitic array radiator (ESPAR) antenna; indoor positioning; received signal strength (RSS); fingerprinting

check for updates

Citation: Groth, M.; Nyka, K.; Kulas, L. Calibration-Free Single-Anchor Indoor Localization Using an ESPAR Antenna. *Sensors* **2021**, *21*, 3431. https://doi.org/10.3390/s21103431

Academic Editor: Simon Tomažič

Received: 30 March 2021
Accepted: 10 May 2021
Published: 14 May 2021

1. Introduction

Positioning and navigation systems play an important role in daily lives, since global navigation satellite system (GNSS) applications cover a number of different location-based services, such as wildlife protection, road applications, and security and safety [1]. Even though GNSS provide reliable positioning outdoors [2], the satellite signals cannot reach indoors [3], so global positioning system (GPS) receivers cannot be applied in indoor environments. Considering that people spend more than 80% of their lifetime indoors [4], indoor positioning is at least as important as outdoor positioning. Additionally, different applications may benefit from indoor positioning, such as marketing and sales [5], health [6], or security and emergency solutions [7]. Thus, a number of indoor positioning technologies have been developed and are currently the subjects of further research. Non-radio technologies such as magnetic [8], visual [9], and inertial [10] technologies can be distinguished. On the other hand, radio technologies include ultra-wideband (UWB) signals [11] and radio frequency (RF) standards, of which the most popular are Wi-Fi and Bluetooth [12], RFID [13], and NFC [14].

Considering RF methods, there are two popular approaches for indoor positioning: geometry mapping and fingerprinting [15]. Due to its popularity, most often 2.4 GHz solutions based on common systems such as 802.11, 802.15.4, or Bluetooth Low Energy (BLE) are utilized. The last technology provides the benefits of relatively inexpensive devices to create positioning systems, as well as being low maintenance due to the long battery life of such devices. Additionally, BLE devices often come with integrated inertial measurement units (IMU), so one can relatively easily implement additional data fusion algorithms relying on IMU sensors [16].

The geometry mapping approach utilizes the estimation of geometric parameters, such as the angle of arrival (AoA) of the signal and the distance of its source with respect to certain reference information [17]. The main drawback of this approach is the potential estimation error that may occur due to the multipath effect in the propagation environment.

Fingerprinting is an approach that minimizes multipath effects by using an offline map of signal features collected in certain locations of the area as a reference to find the best match for the signal received from the unknown position [18]. The advantage of the solution is the relatively low computational power needed to perform the estimation. Nevertheless, in large areas, the calibration phase might be time- and effort-consuming, since the density of the radio map impacts the accuracy of the final estimation. This can be minimized through the utilization of the channel state information (CSI) [19], which includes the amplitude and phase of each orthogonal frequency-division multiplexing (OFDM) channel subcarrier [15] to improve the estimation accuracy in locations outside of the reference points. Unfortunately, this approach requires much higher computational power to perform the signal processing, making the system more expensive, and it is not applicable in BLE.

The fingerprinting approach has two crucial downsides that limit its applicability. First is the aforementioned calibration phase, which requires a large number of precisely measured spatial samples of the signal properties to assure the best positioning results [20]. In dynamic environments, the system needs to be recalibrated regularly as the radio map becomes outdated when the propagation environment changes over time. A number of solutions have been researched to simplify the recalibration process [21–23]; nevertheless, in most cases, additional infrastructure is needed. The other drawback is the number of reference nodes that need to be installed on site to collect information about the signal. To some extent, the more reference nodes are used, the more accurate the estimation is, as the received signal strength (RSS) vectors are longer [24]. On the other hand, the number of devices that create the positioning system also affects the final cost of the installation. To solve this issue, single-anchor positioning has been proposed as one of the solutions for simplification and cost reduction in indoor positioning system installations [25–27]. In such solutions, instead of many reference modules, only a single device equipped with a reconfigurable antenna able to modify its radiation pattern is used.

In this paper, we present a novel method for calibration-free single-anchor indoor positioning. The solution is evaluated utilizing cost-effective BLE modules as auxiliary reference nodes located in fixed positions on room walls. The nodes act as beacons, transmitting reference signals to be considered during position estimation; in this way, the solution is less sensitive to changes in the environment. Additionally, no human-operated calibration or recalibration is needed, since the system is based on current signals received from the reference nodes. Furthermore, simple BLE beacons have relatively long battery life and low retail cost, thus the use of reference modules does not significantly influence the total cost of the system. As a result, such systems could be particularly valuable in scenarios where regular maintenance is difficult or impossible, including busy organizational units such as hospitals or airports. For the single-anchor localization device, the BLE module together with the energy-efficient electronically steerable parasitic array radiator (ESPAR) antenna were chosen. With the aforementioned antenna, the beam can be controlled by shortening or opening passive elements surrounding the active element in the center.

The resulting novel, calibration-less, single-anchor localization system was tested, providing better results than other systems previously reported in the literature [26,27]. Additionally, the calibration-free approach decreases the risk of accuracy loss over time and reduces the system maintenance costs. According to the authors' knowledge, there is no other calibration-free indoor positioning system using the single-anchor ESPAR antenna concept. Therefore, the main contribution of this paper is the original approach to single-anchor position estimation by utilizing the angular diversity capabilities of an ESPAR antenna and the reference signals of the reference nodes, providing zero-calibration functionality. Additionally, the proposed approach provides at least 13.7% higher accuracy than available single-anchor approaches relying on time-consuming prior calibration [26,27].

The rest of this paper is organized as follows. In Section 2, the related work is described. Section 3 includes the antenna characterization. Section 4 presents the proposed calibration-free algorithm. The test environment and results are presented in Sections 5 and 6, respectively, while concluding remarks are presented in Section 7.

2. Related Work

An analysis of the parameters of RF signals allows one to estimate the position of a transceiver based on spatial information. To this day, a number of position estimation techniques have been developed [28], however from the perspective of applicability in wireless sensor networks (WSNs), those that do not impose high computation costs and are the easiest to introduce in real scenarios are the most interesting. From this point of view, algorithms that rely on RSS values are the most attractive, as they provide relatively high accuracy while keeping the implementation complexity reasonably low. Consequently, such algorithms can easily be implemented in simple WSN nodes with integrated RSS readouts.

One of the most popular RSS-based localization methods is the trilateration method [29], which is a trigonometric approach where the distances from three access points (APs) of known locations are calculated based on the chosen propagation model. Knowing the positions of the APs and the estimated distances of the mobile nodes from each of them, the positions of the node can be estimated by solving the equation:

$$\begin{cases} R_1 = \sqrt{(x - X_1)^2 + (y - Y_1)^2 (z - Z_1)^2} \\ R_2 = \sqrt{(x - X_2)^2 + (y - Y_2)^2 (z - Z_2)^2} \\ R_3 = \sqrt{(x - X_3)^2 + (y - Y_3)^2 (z - Z_3)^2} \end{cases} \tag{1}$$

where R_j is the estimated range from AP_j and X_j, Y_j, Z_j are the coordinates of AP_j. The main drawback of this approach is the influence of the environment on the RSS values and the need to use complex propagation models to provide a proper mathematical description of the environment. Similarly, in an approach called centroid localization, N closest APs are chosen based on RSS values and the node positions are estimated by calculating the centroid formula [30]:

$$(x, y) = \left(\frac{X_1 + X_2 + \cdots + X_N}{N}, \frac{Y_1 + Y_2 + \cdots + Y_N}{N} \right) \tag{2}$$

Another approach to indoor positioning based on the strength of the received signal is called fingerprinting [31]. In this method, two phases can be distinguished, namely offline and online phases. During the offline phase, the system learns the RSS values at the number of defined reference points by collecting measurements across the scene from all of the APs. The values are stored in the database together with the coordinates of each point. During the online phase, the actual position estimation is performed and RSS values are collected and compared with the values in the database. The localization of the device is estimated by calculating the Euclidean distance between the each vector of the RSS values from the offline phase and the ones measured during the online phase, which can be described as:

$$D_j = \sqrt{\sum_{i=1}^{I} \left(RSS_{online_i} - RSS_{offline_{i,j}} \right)^2} \tag{3}$$

where I is the total number of APs for which RSS values are stored, RSS_{online_i} is the RSS value measured during the online phase by the ith AP, $RSS_{offline_{i,j}}$ is the RSS value for the ith AP stored in the database, and j is the consecutive number of fingerprints. In the simplest version, to determine the position, the distance is calculated for every reference point and the estimated position is the reference position for which the Euclidean distance has the lowest value. A number of improvements to this method have been presented, such as the arithmetic mean distance from K nearest neighbors (KNN) [32] and the weighted mean distance (WKNN).

Indoor positioning based on a single anchor allows one to achieve the localization functionalities using smaller infrastructure in terms of the number of devices. Since UWB signals are more robust to multipath effects [33], to achieve higher estimation accuracy, UWB ranging is often utilized in single anchor positioning. In [34], the authors propose a hybrid approach, which combines UWB ranging modules with data processing from inertial sensors that consist of a 3-axis accelerometer for the step counting and 3-axis magnetometer for the azimuth angle estimation. Additionally, a real-time Kalman filter is used to smooth the ranging results. Inertial measurement data can also be a support for direction of arrival (DoA) and ranging based on typical 2.4 GHz protocols. For example, in [35], the authors present a fusion algorithm for BLE transceiver signal angle and distance estimation based on Kalman filtering and simplified pedestrian dead reckoning (PDR). Angle estimation is achieved with signal-phase measurements on a uniform linear antenna array, while the range is estimated with a path loss model based on RSS. A 10-element uniform linear antenna array was also used for DoA fingerprinting in [36]. In this approach, the angle of arrival is estimated using the minimum variance distortionless response (MVDR) estimator. Position estimation was performed based on a DoA fingerprinting approach using spatial spectrum measurements, where the measured spatial spectrum is compared with the set of training data using the Pearson correlation coefficients. In [37], the authors propose a multipath-based, single-anchor positioning system, which non-coherently acquires directional measurements by exploiting specular multipath components.

Switched-beam antennas are often used for single anchor positioning systems, whereby a single base station in a given area can perform the positioning instead of a number reference nodes, which reduces the deployment cost of the system [25–27]. The benefit of such an approach is that it can be integrated with the most popular wireless communication standards such as Wi-Fi and BLE to reduce the number of base stations used as APs. The most promising switched-beam antenna concepts, which can be used in practical single-anchor localization systems, are those that are inexpensive for mass production and easy to integrate with low-cost transceivers, with simple microcontrollers performing RSS measurements and localization estimation [26,38].

Due to the RSS instability and changes in the environment, the radio map for the fingerprinting method deteriorates over time and needs to be updated regularly. Therefore, solutions that simplify or automate the recalibration process give the user the possibility to reduce the maintenance efforts and costs of the system. One such solution was presented in [39], where radio maps for the offline phase of fingerprinting were generated automatically. The authors used IoT Wi-Fi sensors as scanning nodes to analyze the environment. The obtained RSS measurements were then utilized to periodically generate new radio maps using a pre-processed path loss template. The updates were generated by estimating the RSS values for each cell of the 1 m $\times$ 1 m grid. In [23], invariant RSS statistics were introduced to eliminate the need for offline recalibration. Particle filters can also be adopted to use crowd-sourced fingerprinting maps for the recalibration [21] and to fuse PDR and positioning estimation data in order to determine and re-estimate the divergence of particle trajectories. These re-estimated trajectories can be adopted to update the radio map. In [22], APs were used to detect permanent changes in RSS values and to modify the radio map accordingly. The update was performed using a Voronoi diagram, excluding from the recalibration those APs for which RSS characteristics changed significantly. Then, the RSS delta was added to each fingerprint as a final stage of recalibration.

Often, the advantages of CSI are harnessed to achieve calibration-free localization. This can be done as described in [40], where the construction of a theoretical CSI fingerprinting database without a site survey was presented. In this approach, the constellation diagram is utilized to represent the relation between the phase difference and AoAs. Theoretical AoAs are derived with respect to existing access points (APs) for any given position and are transformed into the phase difference for a fingerprinting database.

Among other positioning methods, solutions based on product-moment [41] and least-squares [42] correlations between the RSS and the estimated signal strength have been presented. Such methods do not require offline calibration and differ only in the optimization criteria to be calculated; nevertheless, they require assumptions based on the channel propagation models. A further step was taken by the authors of [43], whereby the relation between the RSS and the distance from the APs was used to construct the Voronoi diagram of the area relative to the particular AP locations. During the positioning, the ambiguity region was estimated as the Voronoi cell of the AP of the strongest RSS and was split into two half-planes by analyzing the relative relation between the RSS of each pair of APs. The final position was estimated as the center mass of the final ambiguity region.

3. ESPAR Antenna for Single-Anchor Localization

The ESPAR antenna considered for the proposed calibration-free indoor positioning system is an interesting, less expensive, more energy-efficient alternative solution to complex antenna arrays [44,45]. Radiation patterns from ESPAR antennae can be modified by changing load impedances connected to passive elements located around the active element. For the proposed calibration-free single-anchor indoor localization method, a simplified concept, as previously presented in [45], was implemented. In this design, as depicted in Figure 1, the antenna consists of 12 passive elements located around the active element, to which the signal output of the transceiver is connected. Each of the 12 passive elements can be shorted or opened through single-pole, double-throw (SPDT) field-effect transistor (FET) switches steered individually from a microcontroller. Shorted passive elements become reflectors while opened ones are directors for the active element. As a result, a directional radiation pattern, as shown in Figure 2, can be created and rotated with a 30° discrete step, forming 12 different directional radiation patterns. The utilization of FET switches brings a significant reduction of power consumption via the antenna circuit as compared to varactor-based solutions [46] and can be successfully used in DoA estimation of battery-powered IoT modules [38,47].

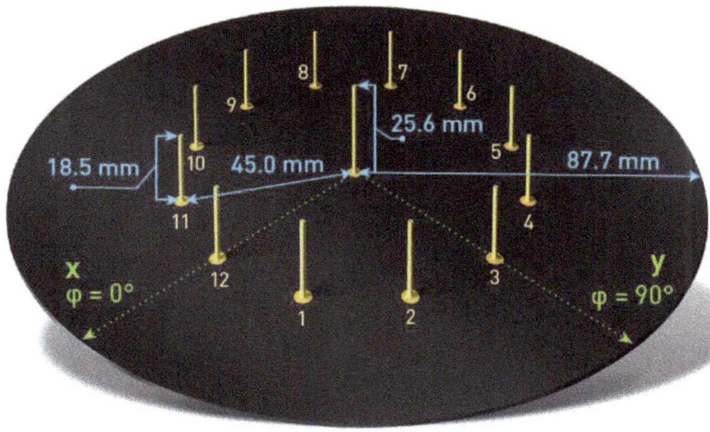

Figure 1. Electronically steerable parasitic array radiator (ESPAR) antenna design.

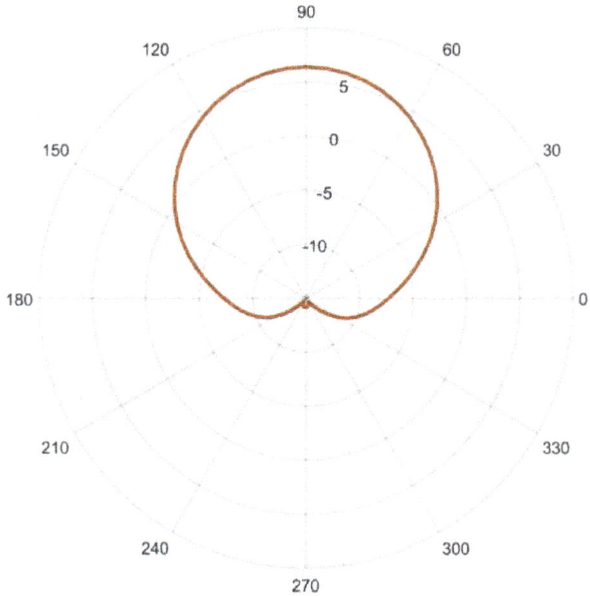

Figure 2. Simulated ESPAR antenna radiation pattern gain (in dBi) at 2.484 GHz for the steering vector $V_{max}^{1} = [111110000000]$.

The antenna designed with FEKO electromagnetic simulation software in [45] has a center frequency of 2.484 GHz and directional radiation patterns at 3 dB beamwidth of 73.2°, which can be formed by shorting five consecutive passive elements. Therefore, the antenna configurations can be denoted using the steering vector $\mathbf{V}_{\mathrm{max}}^{\mathrm{n}} = [v_1\ v_2 \cdots v_s \cdots v_{11}\ v_{12}]$, where $v_s = 0$ when the sth element is shorted and $v_s = 1$ when it is opened. All of the vectors considered following the steering vector notation and the corresponding radiation patterns are presented in Table 1. It has to be noted that from the localization system perspective, each configuration can be considered as a separate AP for indoor positioning, as each RSS from incoming RF signals for each radiation pattern has a different spatial distribution.

Table 1. The antenna's main beam directions for different steering vectors applied to the Electronically Steerable Parasitic Array Radiator (ESPAR) antenna.

n	φ_{max}^{n}	V_{max}^{n}
1	90°	111110000000
2	120°	011111000000
3	150°	001111100000
4	180°	000111110000
5	210°	000011111000
6	240°	000001111100
7	270°	000000111110
8	300°	000000011111
9	330°	100000001111
10	0°	110000000111
11	30°	111000000011
12	60°	111100000001

The antenna was constructed following the described design on 1.55 mm FR4 laminate. The passive elements were constructed with a 2-mm-diameter silver-plated copper rod, while the active element used the same 25.6-mm-long copper rod as an extension of the

SMA connector soldered directly to the laminate. To provide beam-switching capability to the antenna, NJG1681MD7 GaAs FET MMICs SPDT switches were chosen. The antenna prototype, as presented in Figure 3, was equipped with Arduino Shield headers for the convenient connection of different WSN modules compatible with the standard. Additionally, 12 LEDs were installed at the bottom of the PCB to represent the status of each passive element. A similar antenna prototype was used in [26] for indoor positioning based on the fingerprinting method.

(**a**)

(**b**)

Figure 3. ESPAR antenna: (**a**) top view; (**b**) bottom view.

4. Proposed Calibration-Free Algorithm

In this paper, we propose a novel, low-complexity RSS-based method of RF-based indoor positioning using ESPAR antennas that is dedicated to WSN applications. The method integrates aspects of the aforementioned fingerprinting and trilateration approaches and uses active auxiliary reference nodes to decrease the negative influence of indoor RF signal propagation effects, as well as over-time radio map drift. Additionally, by employing the switched-beam ESPAR antenna, the system can implement the single-anchor approach, whereby the RSS measurements and position estimation are performed with only one device, resulting in a significant reduction of the total number of APs in the system. Furthermore, RSS analysis based on beam-switching is much less complex and hardware-independent than direction findings based on phase-shift analysis, which additionally is impossible to implement in certain WSN nodes where IQ samples are not available for the system developer.

The algorithm utilizes simple transmitters as auxiliary reference nodes installed in known positions. The reference nodes transmit signals that are received by the base station (BS) equipped with the ESPAR antenna. For each antenna configuration, the RSS for each reference node is measured and stored in a database. At the same time, the RSS of the localized node is also measured and stored for analysis in the next step. For each reference node, the RSS vector of jth device can be denoted as:

$$\mathbf{V}_{ref_j} = \left[RSS_{ref_{j1}},\ RSS_{ref_{j2}},\ \ldots,\ RSS_{ref_{j12}}\right] \tag{4}$$

where $RSS_{ref_{ji}}$ is the received signal strength of jth node for the ith antenna configuration.

The RSS vector for the localized node can be defined by analogy as:

$$\mathbf{V}_{loc} = [RSS_{loc1},\ RSS_{loc2},\ \ldots,\ RSS_{loc12}] \tag{5}$$

where RSS_{loc_i} is the received signal strength for ith antenna configuration.

For the localization, two stages can be distinguished. In the first phase, the algorithm looks for the reference nodes closest to the localized node. In order to do this, the Euclidean distance D_j between $\mathbf{V}_{ref_j}$ and $\mathbf{V}_{loc}$ for each jth reference node is calculated as:

$$D_j = \|\mathbf{V}_{loc} - \mathbf{V}_{ref_j}\| = \sqrt{\sum_{i=1}^{12}\left(RSS_{loc_i} - RSS_{ref_{ji}}\right)^2} \tag{6}$$

Since each of the RSS values is measured for different antenna characteristics, the reference nodes for which the calculated distance is the lowest are considered to be in the closest vicinity of the localized node. All of the calculated distances D_j can be ordered from the lowest to the highest value:

$$D_{j=k_1} \le D_{j=k_2} \le D_{j=k_3} \le \cdots \le D_{j=k_J} \tag{7}$$

where J is the total number of reference nodes and $k_1, \ldots, k_J$ are indices of the distances arranged in growing order. Then, for the second phase, K reference nodes for which the calculated distance is the lowest are chosen, so that the actual position can be estimated based on the position of the chosen nodes with indices $\{k_1, \ldots, k_K\}$ using the weighted K nearest neighbors method. To this end, the weights for each node are calculated as:

$$w_j = \frac{D_j}{\sum_{i=1}^{K} D_{k_i}} \tag{8}$$

for $j = \{k_1, \ldots, k_K\}$, where K is the number of reference nodes chosen for the second phase. With this result, the final $(x,\ y)$ position estimation of the localized node is determined as a sum of the weighted positions of the considered K reference nodes:

$$(x, y) = \left(\sum_{j=k_1}^{k_K} w_j x_j, \sum_{j=k_1}^{k_K} w_j y_j \right) \tag{9}$$

where (x_n, y_n) are the coordinates of the jth reference node.

An exemplary localization for 8 reference nodes and K = 3 is illustrated in Figure 4. The nodes for which the distance D_j is the smallest were chosen for the second phase of the estimation process and are marked in red. Red and yellow arrows represent signals chosen for the estimation. The hatched area represents the potential position of the localized node based on the position of the chosen reference nodes. The final estimation of the localized node position is represented by the orange circle. A complete flow diagram of the position estimation procedure is shown in Figure 5.

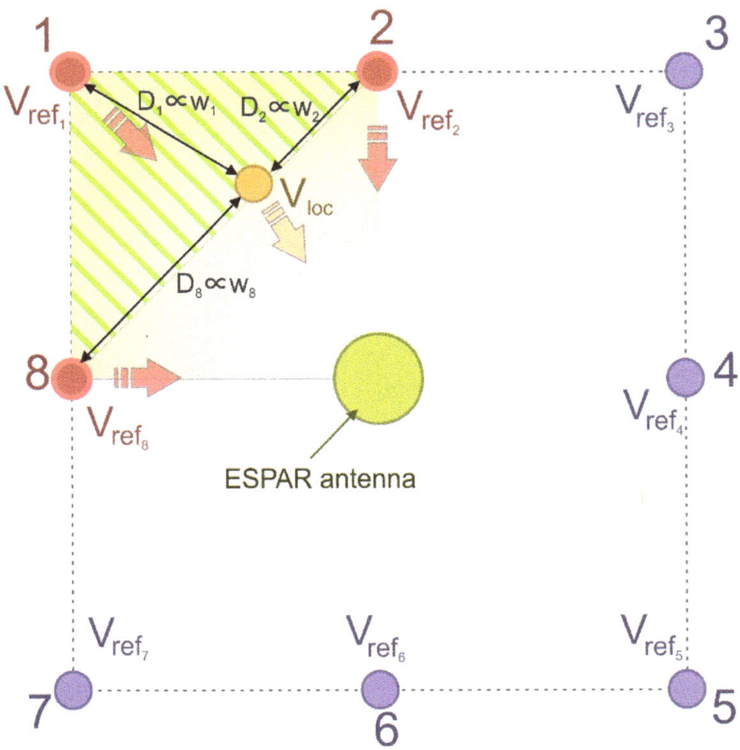

Figure 4. Proposed calibration-free algorithm overview. The position of localized node V_{loc} (marked as a yellow dot) is calculated using 3 reference nodes (K = 3, marked in red), for which the associated Euclidean distances $D_{j=k_2} < D_{j=k_1} < D_{j=k_8}$ calculated in the first phase of the estimation are the smallest (see text for explanations).

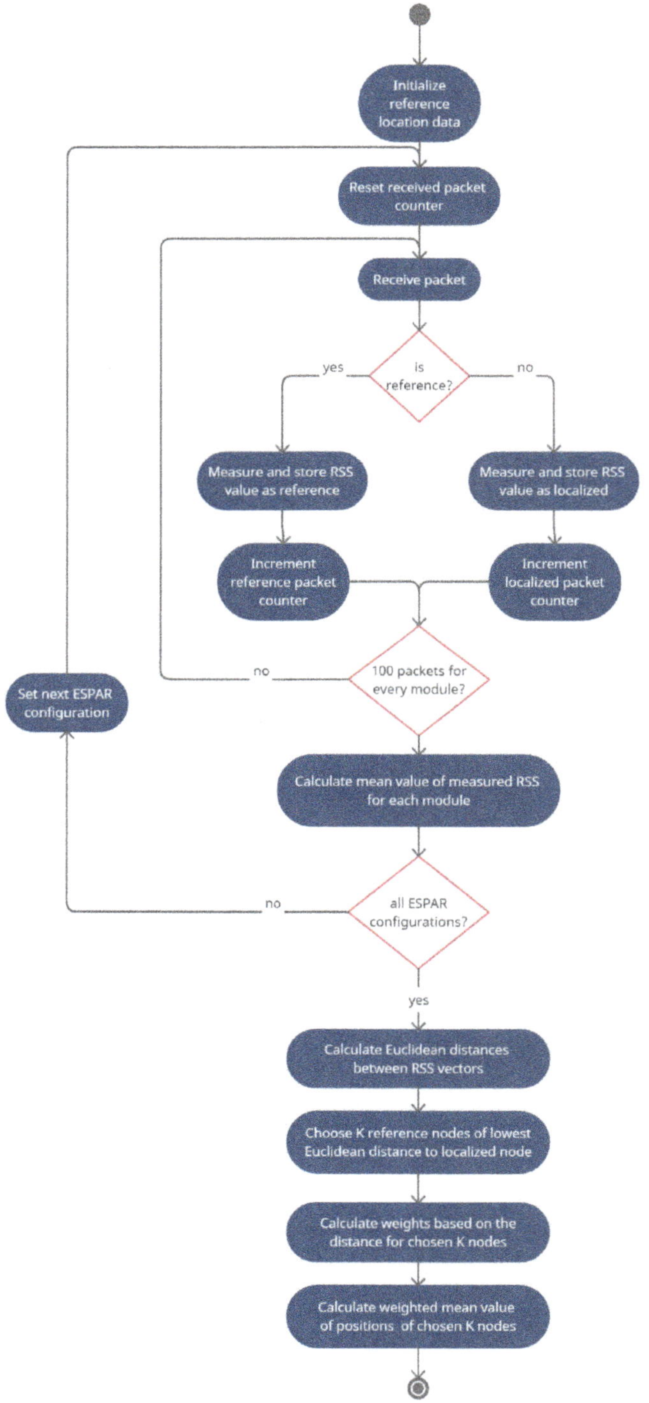

Figure 5. Activity diagram for the proposed calibration-free indoor positioning method (see text for explanations).

5. Test Environment

To verify the proposed calibration-free indoor positioning procedure using single-anchor estimation based on an ESPAR antenna, a dedicated test area and equipment were prepared. For the experiment, the ESPAR antenna was connected to a dedicated, custom-made WSN module based on the Bluetooth Low Energy (BLE)-compliant Nordic nRF52840 SoC. The antenna output of the board was connected to the ESPAR antenna through an SMA connector, while the antenna was steered using GPIO connectors, as shown in Figure 6.

Figure 6. The ESPAR antenna connected to the custom-made nRF52840 wireless sensor network (WSN) board via Arduino Shield header pins. The WSN board's radio frequency (RF) signal output is connected to the ESPAR antenna's center element via a black SMA cable.

Additionally, 25 BLE Nordic nRF52840 dongles, each with an integrated 2.4 GHz PCB antenna, were prepared as the reference and localized nodes. The nRF52840 SoC is equipped with a 32-bit ARM Cortex-M4F, has output power programmable from +8 dBm to −20 dBm, and has −96 dBm sensitivity. Even though the functionality and computation power of the nRF52840 dongles exceed the requirements for reference nodes, they were chosen to assure the integrity of the system, minimizing the potential influence of hardware diversity on the obtained positioning evaluation results. Nevertheless, in real applications, for the localized and reference devices, simple BLE beacons, which retail for 2–3 USD per item, can be utilized.

For the test environment, a 5.6 m × 6.6 m laboratory room was chosen, as the dimensions of this room were most similar to test environments in which other single-anchor positioning methods were evaluated in the literature [26,27]. The base station equipped with the ESPAR antenna was installed on the ceiling in the center of the room, while the 24 reference nodes were placed at even distances on the walls 1.5 m above the floor, which was 1.5 m below the base station, as shown in Figures 7 and 8. The positions of the devices were measured with a precise laser distance measurer and the measuring tape to assure appropriate placement accuracy. The localized node and all of the reference nodes were set into advertising mode and the transmission power was set to 0 dBm. To conduct the measurement session, a 4.5 m × 4.5 m grid was added on the floor with a 0.5 m step, which resulted in 90 test points, as illustrated in Figure 9.

Figure 7. Test environment with reference nodes installed on the walls and the ESPAR antenna integrated with the custom-made WSN board in a single housing mounted on the ceiling.

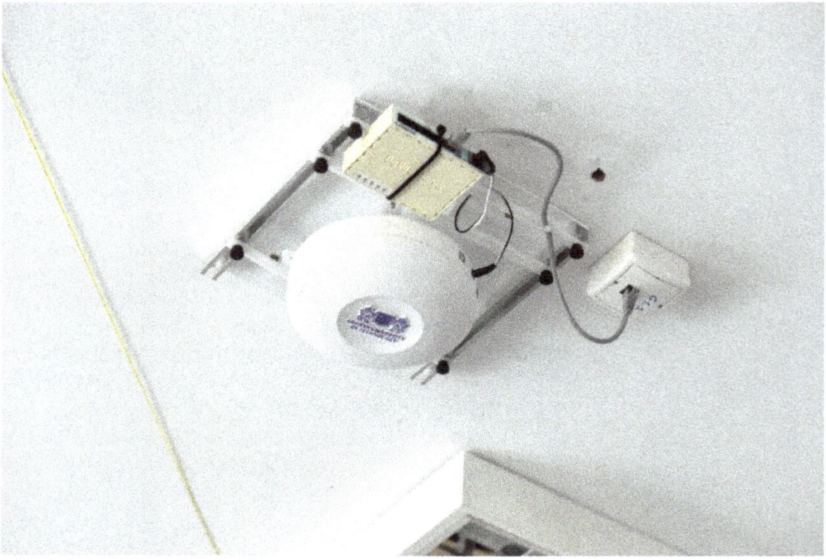

Figure 8. The ESPAR antenna integrated together with the custom-made WSN board installed within a custom-made housing and mounted on the ceiling.

Figure 10 presents the RSS values measured in the base station equipped with the ESPAR antenna from a single reference node for each of the 12 antenna configurations. One can easily notice that an approximately 8 dB spread between the maximum and minimum values is present in the measurements. The results show that in the proposed setup, there is noticeable angular diversity between the directions of ESPAR antenna radiation patterns, which is necessary for the proposed single-anchor positioning process. Similar results were obtained for all of the considered reference nodes.

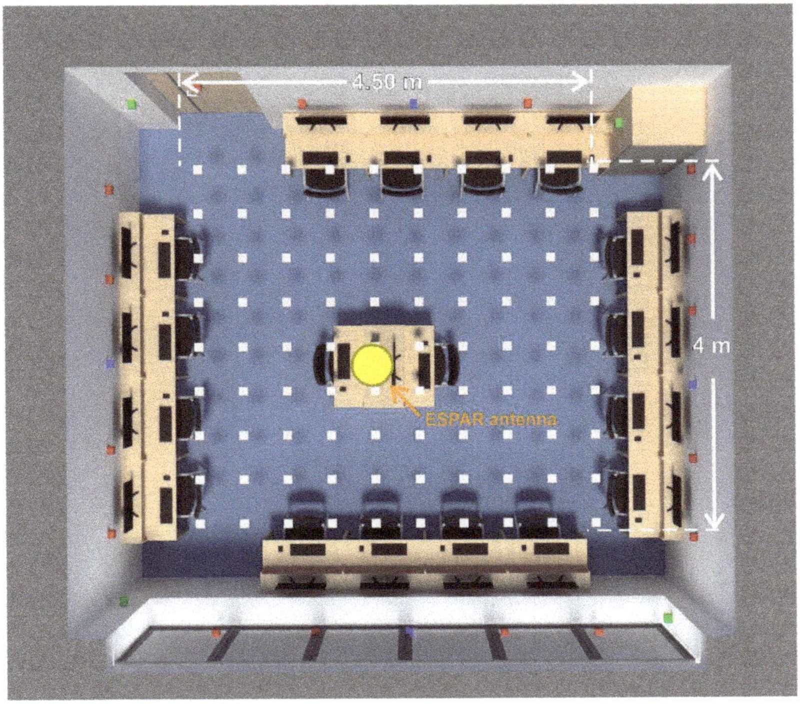

Figure 9. A 3D plan of the test environment, with the test positions marked as white squares and the reference nodes as blue, green, and red squares (see explanations in text).

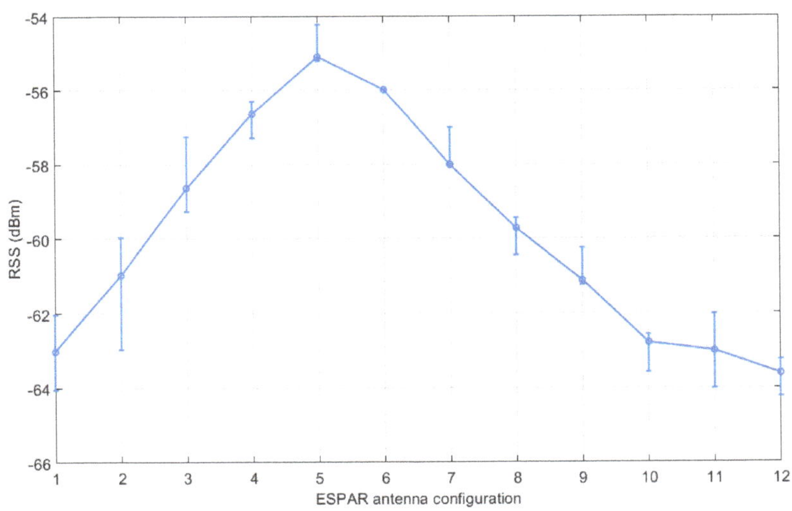

Figure 10. Received signal strength from a single reference module with respect to ESPAR antenna configurations. Error bars represent the maximal received signal strength (RSS) deviation within 100 measurement packets. The standard deviation for RSS values $\sigma = 0.316$ dB.

6. Measurement Results and Discussion

To evaluate the proposed method, for every grid point, a series of measurements in the aforementioned test environment were conducted. To this end, the localized device was placed at each point of a 4.5 m × 4 m grid shown in Figure 9. At every test point, RSS measurements of 100 transmitted packets were collected for each antenna configuration from each of the reference nodes and the localized node, giving a total of 30,000 randomly collected RSS values for one grid point. To avoid RSS fluctuation between Bluetooth channels, a single channel (number 38) was chosen for the analysis. For each test point, the mean value of the RSS measurements was calculated. Measurements took approximately 20 min for each test point.

In Table 2 and Figure 11, the resultant accuracy for different numbers of K chosen for the second phase of the estimation are presented. The results indicate that the more reference modules are considered in the algorithm in the second phase, the more accurate the estimation is. However, one can notice that for more than 4 reference modules, the increase of the accuracy is less significant.

Table 2. The estimation accuracy results (maximum error, mean value, and root mean square error (RMSE)) calculated for varying numbers of reference modules considered for the second phase of the proposed localization algorithm.

K	Max. Error (m)	Mean Error (m)	RMSE (m)
3	5.36	2.16	2.40
4	4.51	1.87	2.10
5	3.93	1.67	1.94
8	3.95	1.54	1.78

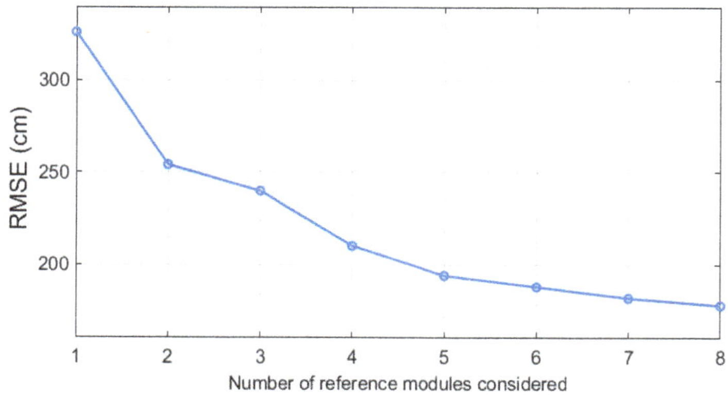

Figure 11. Root mean square error (RMSE) for different numbers of modules of the lowest Euclidean distance considered for the second phase of the proposed localization algorithm.

To examine how much the number of reference nodes could be reduced, additional tests were performed. To this end, three different layouts for reference nodes were investigated: 4 nodes in the corners of the room (only green reference nodes—see Figure 8), 4 nodes in the middle of each wall (only blue reference nodes), and a total of 8 nodes with a combination of both blue and green reference nodes. For each configuration, the algorithm was set to choose 2, 3, and 4 nodes of the lowest Euclidean distance for the position estimation. The results are presented in Table 3.

Table 3. Estimation accuracy of the proposed localization algorithm, including the maximum error, mean value, and root mean square error (RMSE) calculated for various limited layouts of reference modules.

K	Reference Modules Configuration	Max. Error (m)	Mean Error (m)	RMSE (m)
2	corners (4 modules)	4.89	2.22	2.46
	middle (4 modules)	4.89	2.02	2.24
	corners and middle (8 modules)	5.52	2.19	2.49
	all (24 modules)	5.42	2.29	2.54
3	corners (4 modules)	4.09	1.82	2.05
	middle (4 modules)	3.68	1.97	2.14
	corners and middle (8 modules)	3.97	1.98	2.18
	all (24 modules)	5.36	2.16	2.40
4	corners (4 modules)	3.98	1.72	1.85
	middle (4 modules)	3.85	1.74	1.86
	corners and middle (8 modules)	4.64	1.88	2.14
	all (24 modules)	4.51	1.87	2.10

The results show that even a significant reduction of the number of reference modules does not affect the accuracy of the estimation. In fact, for every analyzed case with a different number of modules K, the accuracy increased when compared to the configuration with 24 reference nodes. Additionally, for each value of K, the reduced configurations exhibited very similar accuracy when compared to each other. Thus, it can be concluded that 4 reference nodes per room of a similar size is enough to obtain sufficient position estimation. Nevertheless, one can easily notice that the obtained results are not satisfactory, since both the mean error and maximal error are similar or higher to those that would be calculated with a simple method that always provides estimation in the center of the room (mean error of 1.81 m).

To investigate the possibility for further accuracy improvement, the influence of the RSS normalization, as proposed in [48], on the accuracy was verified. The aim of this modification of RSS values is to mitigate the potential influence of hardware diversity between the reference nodes or uneven distance of the ESPAR antenna from the nodes. To this end, each RSS vector was normalized by rescaling the elements to fall into the range of [0, 1] using the following formula:

$$RSS_{normref_{j}i} = \frac{RSS_{ref_{j_i}} - min\left(\mathbf{V}_{ref_j}\right)}{max\left(\mathbf{V}_{ref_j}\right) - min\left(\mathbf{V}_{ref_j}\right)} \tag{10}$$

where $\mathbf{V}_{ref_j} = \left\{RSS_{ref_{j_1}}, \ldots, RSS_{ref_{j_{12}}}\right\}$ for the jth reference node, $RSS_{normref_{j}i} \in [0,1]$, and:

$$RSS_{norm_{loc_i}} = \frac{RSS_{loc_i} - min(\mathbf{V}_{loc})}{max(\mathbf{V}_{loc}) - min(\mathbf{V}_{loc})} \tag{11}$$

where $\mathbf{V}_{loc} = \{RSS_{loc_1}, \ldots, RSS_{loc_{12}}\}$ and $RSS_{norm_{loc_i}} \in [0,1]$. In consequence, the Euclidean distances rely only on the information regarding the direction of each node and can be calculated as:

$$D_{normj} = \sqrt{\sum_{i=1}^{12} \left(RSS_{norm_{loc_i}} - RSS_{normref_{ji}}\right)^2} \quad (12)$$

The accuracy results obtained for normalized RSS vectors are shown in Table 4.

Table 4. The overall estimation accuracy results (maximum error, mean value, and RMSE) calculated with received signal strength (RSS) normalization introduced in the proposed localization algorithm.

K	Reference Modules Configuration	Max. Error (m)	Mean Error (m)	RMSE (m)
2	corners (4 modules)	5.04	2.42	2.62
	middle (4 modules)	4.06	1.63	1.87
	corners and middle (8 modules)	5.09	2.36	2.59
	all (24 modules)	4.21	2.18	2.38
3	corners (4 modules)	3.95	1.63	1.82
	middle (4 modules)	3.58	1.39	1.58
	corners and middle (8 modules)	3.52	1.84	2.04
	all (24 modules)	3.90	1.99	2.18
4	corners (4 modules)	3.58	1.68	1.82
	middle (4 modules)	4.13	1.63	1.76
	corners and middle (8 modules)	3.93	1.55	1.77
	all (24 modules)	3.84	1.86	2.04

It can be noticed that for almost all of the considered configurations, the RSS normalization provides a further increase of accuracy as compared to the results in Table 3. The best accuracy is achievable for 4 reference nodes and $K = 3$ is used in the second phase of the algorithm, resulting in a mean localization error of 1.39 m, indicating that this configuration is optimal for this method. As shown in Table 5, the reference nodes installed in the middle of each wall provide 13.7% higher accuracy when compared to the fingerprinting ESPAR-antenna-based approach presented in [26], which resulted in a mean localization error of 1.61 m for measurements conducted in a room of a similar size. Additionally, the proposed system provides 17.7% higher accuracy when compared to the single-anchor DoA-based approach presented in [27] for a switched-beam antenna, where measurements were performed in an area measuring 7.2 m × 8 m with 24 test points, resulting in a mean localization error of 1.69 m.

Table 5. Comparison of accuracy results for the presented approach and other single-anchor indoor positioning methods.

Positioning Method	Mean Error (m)	Calibration Needed
Calibration-free single-anchor	1.39	NO
Single-anchor fingerprinting [26]	1.61	YES
Single-anchor DoA [27]	1.69	YES

Figures 12 and 13 present further analyses of the obtained results for individual test points. One can notice that high error values occur only incidentally, and in most cases near the walls or the furniture.

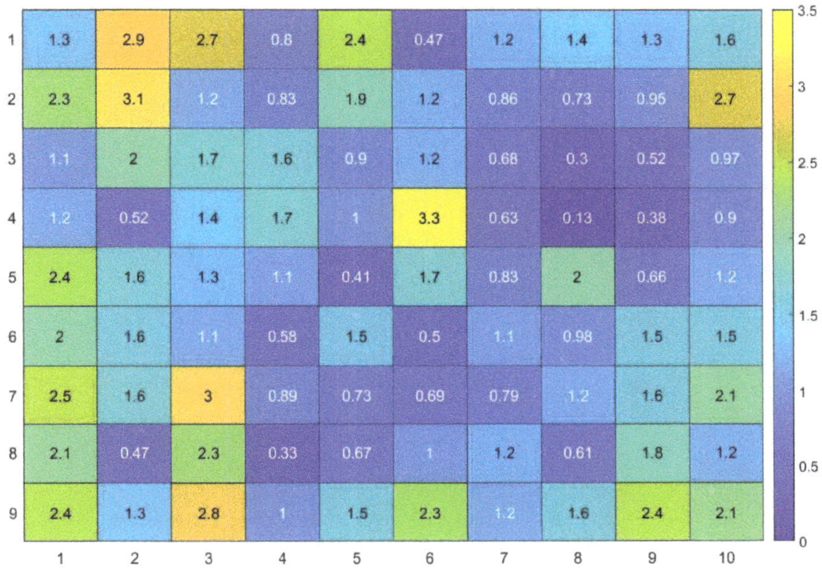

Figure 12. Estimation error (m) calculated for each test point.

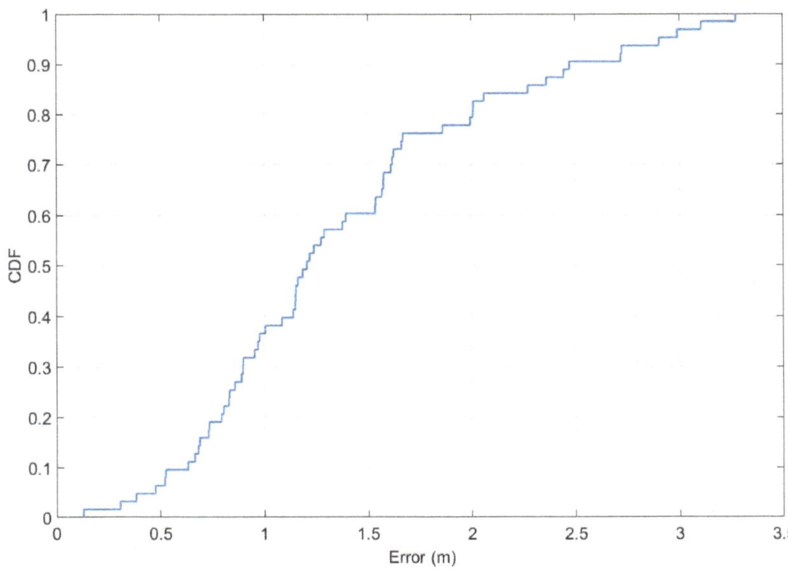

Figure 13. Localization errors in the form of a cumulative distribution function.

To verify the reproducibility of the estimation results and the stability of RSS measurements, for each test point, the collected measurements were respectively divided into 2, 3, 5, 10, 20, and 50 equal datasets of 50, 30, 20, 10, 5, and 2 RSS values measured from each node. For each dataset, position estimation was performed and the results were compared with the results achieved for the full dataset of 100 RSS values. Additionally, the approximated time difference between the first and the last estimation and the standard deviation of the mean estimation error were calculated. As shown in Table 6, even for the estimation where 2 packets per node were used, the obtained results were stable and very similar to those for the mean value calculated with the full dataset of 100 packets per node.

Table 6. Deviation of mean estimation error results calculated for a reduced number of RSS values.

Number of Packets	50	30	20	10	5	2
Number of estimations for each test point	2	3	5	10	20	50
Approximated time difference between the first and last estimation [min]	10	12	16	18	19	19.6
Maximal difference of mean estimation error when compared to full dataset [m]	0	0.01	0.01	0.01	0.03	0.07
Standard deviation of mean error [m]	0.0001	0.0004	0.001	0.002	0.008	0.01

7. Conclusions

In this paper, a calibration-free single-anchor indoor localization method, including a dedicated algorithm and all necessary hardware modules, has been proposed. It has been shown that a single WSN base station equipped with an ESPAR antenna to perform the measurements can be used to find the position of an unknown BLE tag without calibration or recalibration, since inexpensive reference modules installed on walls within the test area provide enough reference information for the positioning algorithm. An analysis of different layouts of reference modules indicated that the use of only 4 modules is sufficient for the method to provide reliable results. Moreover, optional normalization of RSS vectors allows further increases of the accuracy. Since the algorithm does not need high computational power, it can be implemented in a simple WSN sensor, which additionally contributes to the system's cost effectiveness. Measurements in a real-life 5.6 m $\times$ 6.6 m environment have shown that the proposed single-anchor approach employing an ESPAR antenna in a base station mounted on the ceiling is applicable in real-life scenarios, achieving a mean accuracy of 1.39 m, which is approximately 13.7% better than the accuracy of the fingerprinting method (1.61 m) reported previously with a similar setup and antenna [26] and approximately 17.7% better than the single-anchor approach presented in [27], both of which additionally require much a more time-consuming deployment calibration and regular recalibrations to adjust to changing propagation environments. Furthermore, the obtained results are reproducible and can be achieved even for a significantly reduced number of RSS measurements. As a consequence, the proposed calibration-free single-anchor indoor localization method using an ESPAR antenna can be effectively implemented in real, cost-efficient deployments, providing reliable position estimation.

Author Contributions: Conceptualization, M.G. and L.K.; methodology, M.G.; software, M.G.; validation, M.G., K.N. and L.K.; writing—original draft preparation, M.G.; writing—review and editing, M.G., K.N. and L.K.; visualization, M.G.; supervision, L.K.; project administration, L.K.; funding acquisition, L.K. All authors have read and agreed to the published version of the manuscript.

Funding: This research was funded by the InSecTT (www.insectt.eu, (accessed on 30 April 2021)) project, which has received funding from the ECSEL Joint Undertaking (JU) under grant agreement no. 876038. The JU receives support from the European Union's Horizon 2020 research and innovation programme and Austria, Sweden, Spain, Italy, France, Portugal, Ireland, Finland, Slovenia, Poland, Netherlands, Turkey. The document reflects only the author's view; the commission is not responsible for any use that may be made of the information it contains. The EU expects us to include this disclaimer.

Institutional Review Board Statement: Not applicable.

Informed Consent Statement: Not applicable.

Data Availability Statement: Not applicable.

Conflicts of Interest: The authors declare no conflict of interest. The funders had no role in the design of the study; in the collection, analyses, or interpretation of data; in the writing of the manuscript, or in the decision to publish the results.

References

1. Kolomijeca, A.; Lopez-Salcedo, J.A.; Lohan, E.-S.; Seco-Granados, G. GNSS Applications: Personal Safety Concerns. In Proceedings of the 2016 International Conference on Localization and GNSS (ICL-GNSS), Barcelona, Spain, 28–30 June 2016; pp. 1–5.
2. Zhu, N.; Marais, J.; Betaille, D.; Berbineau, M. GNSS Position Integrity in Urban Environments: A Review of Literature. *IEEE Trans. Intell. Transp. Syst.* **2018**, *19*, 2762–2778. [CrossRef]
3. Ochiai, M.; Fujii, M.; Ito, A.; Watanabe, Y.; Hatano, H. A Study on Indoor Position Estimation Based on Fingerprinting Using GPS Signals. In Proceedings of the 2014 International Conference on Indoor Positioning and Indoor Navigation (IPIN), Busan, Korea, 27–30 October 2014; pp. 727–728.
4. Matz, C.J.; Stieb, D.M.; Brion, O. Urban-Rural Differences in Daily Time-Activity Patterns, Occupational Activity and Housing Characteristics. *Environ. Health* **2015**, *14*, 88. [CrossRef] [PubMed]
5. Stavrou, V.; Bardaki, C.; Papakyriakopoulos, D.; Pramatari, K. An Ensemble Filter for Indoor Positioning in a Retail Store Using Bluetooth Low Energy Beacons. *Sensors* **2019**, *19*, 4550. [CrossRef]
6. Zhang, C.; Qin, N.; Xue, Y.; Yang, L. Received Signal Strength-Based Indoor Localization Using Hierarchical Classification. *Sensors* **2020**, *20*, 1067. [CrossRef]
7. Li, F.; Liu, M.; Zhang, Y.; Shen, W. A Two-Level WiFi Fingerprint-Based Indoor Localization Method for Dangerous Area Monitoring. *Sensors* **2019**, *19*, 4243. [CrossRef]
8. Kwak, M.; Hamm, C.; Park, S.; Kwon, T.T. Magnetic Field Based Indoor Localization System: A Crowdsourcing Approach. In Proceedings of the 2019 International Conference on Indoor Positioning and Indoor Navigation (IPIN), Pisa, Italy, 30 September–3 October 2019; pp. 1–8.
9. Zhang, C. Sparse Visual Localization in GPS-Denied Indoor Environments. In Proceedings of the 2019 International Conference on Information Technology and Computer Application (ITCA), Guangzhou, China, 20–22 December 2019; pp. 87–91.
10. Vy, T.D.; Nguyen, T.L.N.; Shin, Y. Inertial Sensor-Based Indoor Pedestrian Localization for IPhones. In Proceedings of the 2019 International Conference on Information and Communication Technology Convergence (ICTC), Jeju Island, Korea, 16–18 October 2019; pp. 200–203.
11. Ren, A.; Zhou, F.; Rahman, A.; Wang, X.; Zhao, N.; Yang, X. A Study of Indoor Positioning Based on UWB Base-Station Configurations. In Proceedings of the 2017 IEEE 2nd Advanced Information Technology, Electronic and Automation Control Conference (IAEAC), Chongqing, China, 25–26 March 2017; pp. 1939–1943.
12. Luo, R.C.; Hsiao, T.-J. Indoor Localization System Based on Hybrid Wi-Fi/BLE and Hierarchical Topological Fingerprinting Approach. *IEEE Trans. Veh. Technol.* **2019**, *68*, 10791–10806. [CrossRef]
13. Tlili, F.; Hamdi, N.; Belghith, A. Accurate 3D Localization Scheme Based on Active RFID Tags for Indoor Environment. In Proceedings of the 2012 IEEE International Conference on RFID-Technologies and Applications (RFID-TA), Nice, France, 5–7 November 2012; pp. 378–382.
14. Ozdenizci, B.; Ok, K.; Coskun, V.; Aydin, M.N. Development of an Indoor Navigation System Using NFC Technology. In Proceedings of the 2011 Fourth International Conference on Information and Computing, Phuket, Thailand, 25–27 April 2011; pp. 11–14.
15. Yang, Z.; Zhou, Z.; Liu, Y. From RSSI to CSI: Indoor Localization via Channel Response. *ACM Comput. Surv.* **2013**, *46*, 1–32. [CrossRef]
16. Poulose, A.; Eyobu, O.S.; Han, D.S. An Indoor Position-Estimation Algorithm Using Smartphone IMU Sensor Data. *IEEE Access* **2019**, *7*, 11165–11177. [CrossRef]
17. Le, A.T.; Tran, L.C.; Huang, X.; Ritz, C.; Dutkiewicz, E.; Phung, S.L.; Bouzerdoum, A.; Franklin, D. Unbalanced Hybrid AOA/RSSI Localization for Simplified Wireless Sensor Networks. *Sensors* **2020**, *20*, 3838. [CrossRef] [PubMed]
18. Kaemarungsi, K.; Krishnamurthy, P. Properties of Indoor Received Signal Strength for WLAN Location Fingerprinting. In Proceedings of the First Annual International Conference on Mobile and Ubiquitous Systems: Networking and Services, 2004. MOBIQUITOUS 2004, Boston, MA, USA, 26 August 2004; pp. 14–23.
19. Li, X.; Zhu, J. Improved Indoor Positioning Method Based on CSI. In Proceedings of the 2019 International Conference on Intelligent Transportation, Big Data & Smart City (ICITBS), Changsha, China, 12–13 January 2019; pp. 274–277.
20. Qiang, C.; Van de Velde, S.; Steendam, H. How to Get the Best out of Your Fingerprint Database: Hierarchical Fingerprint Indoor Positioning for Databases with Variable Density. *IEEE Access* **2019**, 1. [CrossRef]

21. Chu, C.; Yang, S. A Particle Filter Based Reference Fingerprinting Map Recalibration Method. *IEEE Access* **2019**, *7*, 111813–111827. [CrossRef]
22. Fet, N.; Handte, M.; Marron, P.J. An Approach for Autonomous Recalibration of Fingerprinting-Based Indoor Localization Systems. In Proceedings of the 2016 12th International Conference on Intelligent Environments (IE), London, UK, 14–16 September 2016; pp. 24–31.
23. Lee, S.; Husen, M.N. Recalibration-Free Indoor Localization with Wi-Fi Fingerprinting of Invariant Received Signal Strength. In Proceedings of the 2016 IEEE/RSJ International Conference on Intelligent Robots and Systems (IROS), Daejeon, Korea, 9–14 October 2016; pp. 4649–4655.
24. Machaj, J.; Brida, P. Impact of Wi-Fi Access Points on Performance of RBF Localization Algorithm. In Proceedings of the 2012 ELEKTRO, Rajeck Teplice, Slovakia, 21–22 May 2012; pp. 70–74.
25. Maddio, S.; Cidronali, A.; Manes, G. Direction of Arrival Estimation of an Arbitrary Oriented Targets by a Highly Directive Antenna and Coarse RSSI Data. In Proceedings of the 2012 42nd European Microwave Conference, Amsterdam, The Netherlands, 29 October–1 November 2012; pp. 140–143.
26. Rzymowski, M.; Woznica, P.; Kulas, L. Single-Anchor Indoor Localization Using ESPAR Antenna. *Antennas Wirel. Propag. Lett.* **2016**, *15*, 1183–1186. [CrossRef]
27. Giorgetti, G.; Cidronali, A.; Gupta, S.; Manes, G. Single-Anchor Indoor Localization Using a Switched-Beam Antenna. *IEEE Commun. Lett.* **2009**, *13*, 58–60. [CrossRef]
28. Soewito, B.; Hassyr, F.A.; Geri Arisandi, T.G. A Systematic Literature Review of Indoor Position System Accuracy and Implementation. In Proceedings of the 2018 International Conference on Applied Science and Technology (iCAST), Manado, Indonesia, 26–27 October 2018; pp. 358–362.
29. Pakanon, N.; Chamchoy, M.; Supanakoon, P. Study on Accuracy of Trilateration Method for Indoor Positioning with BLE Beacons. In Proceedings of the 2020 6th International Conference on Engineering, Applied Sciences and Technology (ICEAST), Chiang Mai, Thailand, 1–4 July 2020; pp. 1–4.
30. Deng, B.; Huang, G.; Zhang, L.; Liu, H. Improved Centroid Localization Algorithm in WSNs. In Proceedings of the 2008 3rd International Conference on Intelligent System and Knowledge Engineering, Xiamen, China, 17–19 November 2008; pp. 1260–1264.
31. Kaemarungsi, K. Efficient Design of Indoor Positioning Systems Based on Location Fingerprinting. In Proceedings of the 2005 International Conference on Wireless Networks, Communications and Mobile Computing, Maui, HI, USA, 13–16 June 2005; Volume 1, pp. 181–186.
32. Gorski, K.; Groth, M.; Kulas, L. A Multi-Building WiFi-Based Indoor Positioning System. In Proceedings of the 2014 20th International Conference on Microwaves, Radar and Wireless Communications (MIKON), Gdansk, Poland, 16–18 June 2014; pp. 1–4.
33. Win, M.Z.; Scholtz, R.A. On the Robustness of Ultra-Wide Bandwidth Signals in Dense Multipath Environments. *IEEE Commun. Lett.* **1998**, *2*, 51–53. [CrossRef]
34. Zhang, Y.; Zhu, Y.; Xia, W.; Shen, L. A Single-Anchor Calibration Indoor Positioning System Using Heterogeneous Sensors. In Proceedings of the 2015 International Conference on Wireless Communications & Signal Processing (WCSP), Nanjing, China, 15–17 October 2015; pp. 1–5.
35. Ye, F.; Chen, R.; Guo, G.; Peng, X.; Liu, Z.; Huang, L. A Low-Cost Single-Anchor Solution for Indoor Positioning Using BLE and Inertial Sensor Data. *IEEE Access* **2019**, *7*, 162439–162453. [CrossRef]
36. Jespersen, M.H.; Serup, D.E.; Nielsen, M.H.; Hannesbo, M.H.; Williams, R.J.; Nielsen, K.S.; Mikkelsen, J.H.; Shen, M. An Indoor Multipath-Assisted Single-Anchor UWB Localization Method. In Proceedings of the 2018 IEEE MTT-S International Wireless Symposium (IWS), Chengdu, China, 6–10 May 2018; pp. 1–3.
37. Rath, M.; Kulmer, J.; Leitinger, E.; Witrisal, K. Single-Anchor Positioning: Multipath Processing With Non-Coherent Directional Measurements. *IEEE Access* **2020**, *8*, 88115–88132. [CrossRef]
38. Groth, M.; Rzymowski, M.; Nyka, K.; Kulas, L. ESPAR Antenna-Based WSN Node With DoA Estimation Capability. *IEEE Access* **2020**, *8*, 91435–91447. [CrossRef]
39. Ali, M.; Hur, S.; Park, Y. Wi-Fi-Based Effortless Indoor Positioning System Using IoT Sensors. *Sensors* **2019**, *19*, 1496. [CrossRef] [PubMed]
40. Wan, Y.; Tong, X.; Peng, Y.; Tian, X. Calibration-Free Localization Based on Fingerprinting of Channel State Information. In Proceedings of the 2019 IEEE 5th International Conference on Computer and Communications (ICCC), Chengdu, China, 6–9 December 2019; pp. 394–398.
41. Van Kleunen, W.A.P.; Viet-Duc, L.; Havinga, P.J.M. Calibration-Free Signal-Strength Localization Using Product-Moment Correlation. In Proceedings of the 2016 IEEE 13th International Conference on Mobile Ad Hoc and Sensor Systems (MASS), Brasilia, Brazil, 10–13 October 2016; pp. 210–218.
42. Patwari, N.; O'Dea, R.J.; Yanwei, W. Relative Location in Wireless Networks. In Proceedings of the IEEE VTS 53rd Vehicular Technology Conference, Spring 2001, Proceedings (Cat. No.01CH37202). Rhodes, Greece, 6–9 May 2001; Volume 2, pp. 1149–1153.
43. Elbakly, R.; Youssef, M. A Robust Zero-Calibration RF-Based Localization System for Realistic Environments. In Proceedings of the 2016 13th Annual IEEE International Conference on Sensing, Communication, and Networking (SECON), London, UK, 27–30 June 2016; pp. 1–9.
44. Harrington, R. Reactively Controlled Directive Arrays. *IEEE Trans. Antennas Propagat.* **1978**, *26*, 390–395. [CrossRef]

45. Kulas, L. Direction-of-Arrival Estimation Using an ESPAR Antenna with Simplified Beam Steering. In Proceedings of the 47th European Microwave Conference, Nuremberg, Germany, 10–12 October 2017; pp. 1–4.
46. Taillefer, E.; Hirata, A.; Ohira, T. Direction-of-Arrival Estimation Using Radiation Power Pattern with an ESPAR Antenna. *IEEE Trans. Antennas Propagat.* **2005**, *53*, 678–684. [CrossRef]
47. Kulas, L. RSS-Based DoA Estimation Using ESPAR Antennas and Interpolated Radiation Patterns. *Antennas Wirel. Propag. Lett.* **2018**, *17*, 25–28. [CrossRef]
48. Abdulkarim, H.D.; Sarhang, H. Normalizing RSS Values of Wi-Fi Access Points to Improve an Integrated Indoors Smartphone Positioning Solutions. In Proceedings of the 2019 International Engineering Conference (IEC), Erbil, Iraq, 23–25 June 2019; pp. 171–176.

sensors

MDPI

Article

Sensor Information Sharing Using a Producer-Consumer Algorithm on Small Vehicles

Rodrigo Vazquez-Lopez [1], Juan Carlos Herrera-Lozada [1], Jacobo Sandoval-Gutierrez [2,*], Philipp von Bülow [2] and Daniel Librado Martinez-Vazquez [2]

1 Instituto Politécnico Nacional, Centro de Innovación y Desarrollo Tecnológico en Cómputo (CIDETEC), Unidad Profesional Adolfo López Mateos, Ciudad de México 07700, Mexico; rvazquezl1800@alumno.ipn.mx (R.V.-L.); jlozada@ipn.mx (J.C.H.-L.)

2 Departamento de Procesos Productivos, Universidad Autónoma Metropolitana Unidad Lerma, Estado de México 52005, Mexico; p.von@correo.ler.uam.mx (P.v.B.); d.martinez@correo.ler.uam.mx (D.L.M.-V.)

* Correspondence: j.sandoval@correo.ler.uam.mx; Tel.: +52-728-282-7002 (ext. 3011)

Abstract: There are several tools, frameworks, and algorithms to solve information sharing from multiple tasks and robots. Some applications such as ROS, Kafka, and MAVLink cover most problems when using operating systems. However, they cannot be used for particular problems that demand optimization of resources. Therefore, the objective was to design a solution to fit the resources of small vehicles. The methodology consisted of defining the group of vehicles with low performance or are not compatible with high-level known applications; design a reduced, modular, and compatible architecture; design a producer-consumer algorithm that adjusts to the simultaneous localization and communication of multiple vehicles with UWB sensors; validate the operation with an interception task. The results showed the feasibility of performing architecture for embedded systems compatible with other applications managing information through the proposed algorithm allowed to complete the interception task between two vehicles. Another result was to determine the system's efficiency by scaling the memory size and comparing its performance. The work's contributions show the areas of opportunity to develop architectures focusing on the optimization of robot resources and complement existing ones.

Keywords: absolute position system; cooperative algorithm; intercepting vehicles; indoor positioning; robot framework; UWB sensors

check for updates

Citation: Vazquez-Lopez, R.; Herrera-Lozada, J.C.; Sandoval-Gutierrez, J.; von Bülow, P.; Martinez-Vazquez, D.L. Sensors Information Sharing Using a Producer-Consumer Algorithm on Small Vehicles. *Sensors* **2021**, *21*, 3022. https://doi.org/10.3390/s21093022

Academic Editor: Simon Tomažič

Received: 23 February 2021
Accepted: 21 April 2021
Published: 25 April 2021

Publisher's Note: MDPI stays neutral with regard to jurisdictional claims in published maps and institutional affiliations.

1. Introduction

The use of an operating system has made it possible to standardize the technologies for the problem of cooperativity between several robots or vehicles. To this end, the devices, the sensors, the communication modules, the protocols, and the programming languages are designed to be compatible with each other. This conception of unified platforms such as ROS, Kafka, AEROSTACK, MAVLink among others, resolves several hardware problems and allows a more user-friendly environment for software developers [1–4]. However, high compatibility leads to a significant increase in resource usage. Another consequence is to lose the diversity of solutions that are the result of particular case investigations.

In this research, a case study shows the importance of maintaining research with custom designs than generic designs. A basic platform scenario requires resources at three fundamental layers. The perception layer (indoor systems) uses a GoPro-type internal camera [5,6] or a VICON type external camera sophisticated system [7,8]. The communication layer involves the use of transmitters, routers, and standardized protocols such as WiFi, Zigbee, or Bluetooth [9]. The third layer of applications involves the use of a compatible information management system, i.e., master and slave workstations using a computer architecture with an operating system such as Ubuntu or similar. The above implies a high energy consumption in managing information compared to the energy required for actuators who achieve vehicles' displacement. However, it is essential to consider the advantages of these solutions. For example, end-users, developers, and interested people can program solutions in a more friendly way.

In another perspective, there are tailor-made solutions so specific, complex, and customized that they require specialized equipment and are not compatible with other systems. Therefore, the proposal is based on a methodology that combines customized solutions with generic ones to be implemented in low-performance vehicles or incompatible with operating systems. For this, a reduction or replacement of tools is proposed without losing compatibility with high-level technologies. Given the characteristics of this architecture, algorithms based on hardware or technology constraints should be implemented. Finally, the actual functionality should be validated through a known task and where the optimization of resources can be compared.

A significant difference between the design of a generic and a customized platform lies in taking advantage of the sensor's characteristics. In particular, UWB sensors allow sending messages and obtaining the position by triangulation. Consequently, known tools cannot be applied directly, and an intermediate algorithm has to be used. The libraries of programming languages based on operating systems are very advanced compared to embedded systems. It is important to remember that the migration of a tool or algorithm of custom technology to a general one is always possible. However, in the opposite direction, it is not always guaranteed. Therefore, it is essential not to miss the opportunity to investigate code optimization for embedded systems.

1.1. Contribution

The design and implementation of an alternative architecture for small vehicles is the main contribution of this work. Simultaneous localization and sharing information with a UWB technology reduced computing resources and consumption energy. A suitable producer–consumer algorithm gives a solution to manage the concurrent processes in reduced-scale hardware. Other contributions are listed as follow:

- Enable incompatible hardware with high-level architectures.
- Manage information packets between embedded systems.
- Achieve an interception task among vehicles avoiding the use of additional hardware or complex infrastructure.
- Make compatible an architecture that uses operating systems with one that does not.
- Adapt to the confinement conditions when performing experiments at home.
- The small vehicle platform allows test different autonomous navigation strategies, lowering the risk and cost of large-scale testing.

1.2. Organization

The article is organized as follows. Section 2 presents a brief review of the state-of-the-art in robotic systems architectures. The customized architecture for small vehicles and its underlying concepts are presented in Section 3. The experiments, results and discussions are developed during Section 4. Conclusions and future Work can finally be found in Section 5.

2. Background and Related Work

2.1. Robotic System Architectures

Software-oriented solutions in multi-vehicle tasks have consistently been standardized on architecture with at least three layers with particular purposes. These layers' functions are the quantification of physical variables, the communication between sensors and processing devices, and the applications that solve the tasks [10].

This architecture seeks to centralize the sensors' information in an integrated manner on the devices running the applications. Sophisticated communication systems are used to guarantee the exchange of information. The processing and storage devices use operating systems to program the solutions to the tasks. The above clearly shows that the software-oriented perspective allows for more excellent compatibility. However, poor performance or incompatibility can occur with specific problems.

2.1.1. Localization Sensors

The Inertial Navigation Systems (INS) are fed with data from their Inertial Measurement Unit (IMU), composed mainly of sensors such as gyroscopes and accelerometers. INS are cost-effective solutions that do not require additional infrastructure. The main disadvantage of these systems lies in the cumulative position error that grows during the operation of the vehicle [11]. However, the combination of the information from an INS with a global positioning system allows to improve the accuracy of the positioning estimates [12].

A variety of technologies applied to indoor global positioning systems have been sought. The results in [13] showed that accuracy depends on the implementation and methodology used. In particular, the technologies, which were applied in the development of autonomous vehicles, are global positioning systems by vision such as VICON [7,8], or Optitrack [14,15], Virtual Reality (VR) systems [16,17] and RF technology [18–20]. Meanwhile, the use of Ultra-Wide Band (UWB), part of the RF technologies, attracted interest during the last years [21], because with this technology it is also possible to obtain a good performance in terms of accuracy [22].

Typical tasks that autonomous vehicles must fulfill depend on the quality of the information about the vehicle's absolute and relative position within its workspaces, such as Simultaneous Localization and Mapping (SLAM), obstacle avoidance, trajectory tracking, and cooperation. The integration of absolute and relative positioning systems is, therefore, necessary. The most recent research efforts present various technologies to achieve a good precision in detecting the global and relative position and thus perform a satisfactory SLAM. On the other hand, UWB and vision systems have been combined to improve the estimation of the position in the workspace [23]. Other strategies bet on the fusion of UWB data with those coming from INS [24], and Optitrack [25]. Additionally, it has been tried to improve UWB systems' results on a platform of mobile robots using Gaussian processes [26] and neural networks [27].

2.1.2. Communication Protocols

Indoor Wireless Communications are technologies that allow devices to be uniquely identified and exchanged over a limited distance of 10 m to 100 m [9]. There are several criteria for selecting Bluetooth, UWB, ZigBee, or Wi-Fi technology. The most commonly used technologies are Bluetooth or WI-FI because most robots have it by default. ZigBee is identified as an alternative for users looking to work with a more significant number of devices, a low transfer rate, and requiring less energy consumption. Finally, UWB technologies are less present as a combination of high transfer rate and low normalized energy consumption.

2.1.3. Applications

Applications such as ROS, Kafka, and MAVLink are versatile tools to incorporate most devices, infrastructures, programming languages, and technologies. For example, ROS is defined as a robotic middleware to help manage the complexity and heterogeneity inherent in distributed systems [1]. Apache Kafka is a distributed messaging system widely used in big data applications [2] and MAVLink is a communication protocol used for the bidirectional communications between drones and ground stations over a wireless channel [4]. In this sense, by having approved research platforms, it is possible to leave aside hardware problems and focus on algorithms' performance.

There are different examples of applications that use ROS in its architecture as in [28–30]. In [31] Apache Kafka was used to parallelize a set learning architecture. In contrast, in [29], small drones are used to collect data to test obstacle avoidance algorithms.

Finally, MAVLink being a communication protocol, can be found implemented in several tools such as ROS. In [32] MAVLink was used with ROS to communicate with the pixhawk drone and perform autonomous flights. In [33], they used the protocol to communicate with real drones to validate the performance of the proposed architecture.

The element relationships of the Robotic System Architectures with the software developed are shown in Table 1.

Table 1. Comparison between similar applications and tools.

Software (Tool)	Computer Architecture	Supported System	Comm. Devices	Infrastructure
ROS [28]	x64 ARM	GNU/Linux	Wi-Fi	Network Camera FCU
Kafka, ROS [29]	ARM	GNU/Linux	4G	Network VPN Cloud Service
MAVLink [34]	ARM	Android 4.3	Zigbee	Cameras X-bee Modules Tablet
Proposed architecture	x64... STM, EV3	Linux... μC Based	Wi-Fi	Camera Network UWB

Table 1 shows that the proposed architecture incorporates non-compliant devices such as EV3 devices and UWB technology as localization, communication, and infrastructure devices.

Some architectures propose using cloud services to improve the computational capacity and intelligence in vehicles such as AGVs [35]. Some examples suggest using architectures based on the edge-fog model as seen in [30]. The authors used the publish-subscribe model of ROS to perform 3D scenario reconstruction through SLAM and proof of concept presented in [36]. In the case AVG [37] the authors give an application-oriented to AGV collaboration, while in [38] they use ROS. Other similar examples can be seen in [39–41].

2.1.4. Cooperative Tasks

Cooperative tasks have been investigated in various scenarios with software-oriented architectures. [42,43]. Information sharing using algorithms among aerial or ground vehicles has been studied. These algorithms are proposed to solve the path planning problem [44], for the surveillance of multiple moving ground targets [45] or intercepting intruders [46]. Tools of general-purpose have proposed the simulation of these algorithms. However, real implementation of the hardware of these algorithms requires a particular adaptation. The adaptation to solve this problem is shown in Section 3.

3. Customized Architecture for Small Vehicles

The previous section showed a typical software-oriented architecture with the most common technologies. In this section, an alternative hardware-oriented architecture will be developed. Figure 1 shown a general comparison of layers and technologies employed.

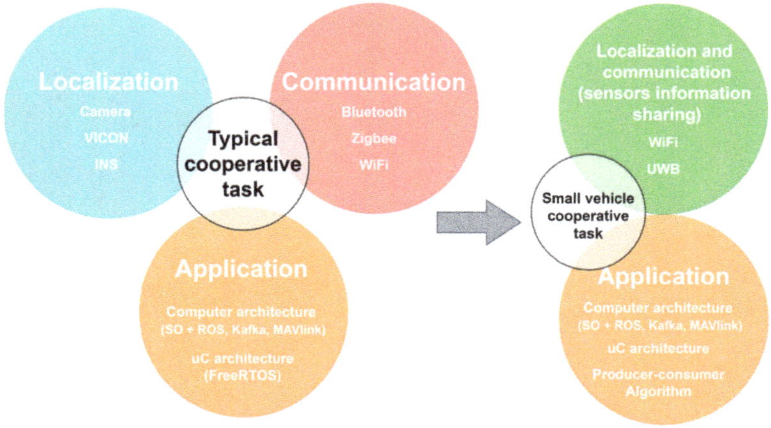

Figure 1. Comparison of software-oriented architecture (**left-side**) and proposed architecture (**right-side**).

The small vehicle platform allows test different autonomous navigation strategies, lowering the risk and cost of large-scale testing. The platform enables destructive testing at a meager cost, infrastructure and energy consumption reduced.

3.1. Small Vehicles Models

3.1.1. Ground Robot

We used a robot as a mobile reference point within the workspace. The robot with 3-DOF (x, y, θ) is of a differential configuration, so it has two independent, active wheels and a fixed wheel as a support point. The odometry of the vehicle uses encoders and a gyroscope as a relative reference system. The vehicle's operation is executed intrinsically in the controller using the dynamic model proposed in [47].

Figure 2 shows the robot with its differential system variables as it travels along a given path in an absolute coordinate system $X_A Y_A$. The mobile is composed of two wheels with diameter ϕ, each placed at a distance L from the intermediate point P. The velocities of the wheels are denoted by v_d (dextram = right) and v_s (sinestram = left) respectively, so that the system inputs are:

$$[L, \phi, \theta, v_s, v_d] \in \mathbb{R}^5. \quad (1)$$

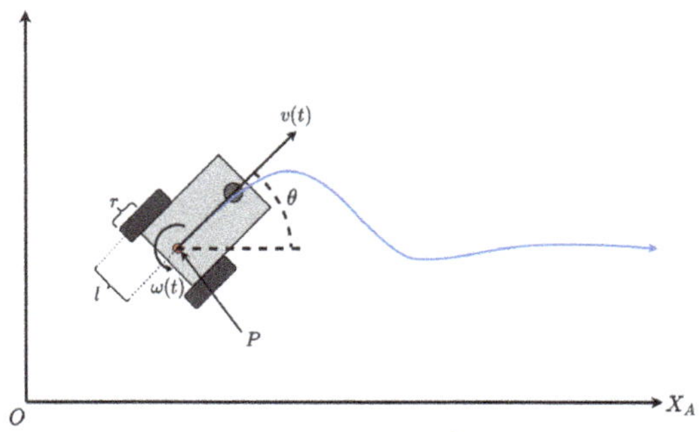

Figure 2. Ground vehicle model.

According to the model of direct kinematics presented in [47], the robot motion is given by the following matrix equation in (2).

$$\xi_A = \begin{bmatrix} (\omega_d + \omega_s)cos\theta \\ (\omega_d + \omega_s)sin\theta \\ \omega_d - \omega_s \end{bmatrix}, \tag{2}$$

3.1.2. Aerial Vehicle

A 6-DOF nano-quadcopter (x,y,z, pitch, roll, yaw) to follow the ground reference robot's trajectory was used. The internal odometry consisted of INS, whose IMU integrates an accelerometer, gyroscope (in three axes) and a barometric pressure sensor. Additionally, the INS data are merged with the height estimation obtained employing an optical flow sensor and the absolute positioning system's data. The operation of the vehicle is done in an autonomous, semi-autonomous way or by using teleoperation.

To operate the vehicle, we assumed the 12-state model, studied in [48,49] and validated in [50]. According to Figure 3, the model requires two reference systems: the non-inertial one $X'Y'Z'$ located at the center of gravity of the quadcopter, orientated as shown, and the inertial one x, y, and z relative to the center of the earth. Of the 12 parameters, the variables x, y and z represent the coordinates of the center of gravity, u, v and w are the linear velocities along the X', Y' and Z' axes, and ψ, θ and ϕ represent the rotation angles (pitch, roll and yaw) with the corresponding angular velocities. All variables are expressed within the inertial reference system. The model is non-linear and is described in [50,51].

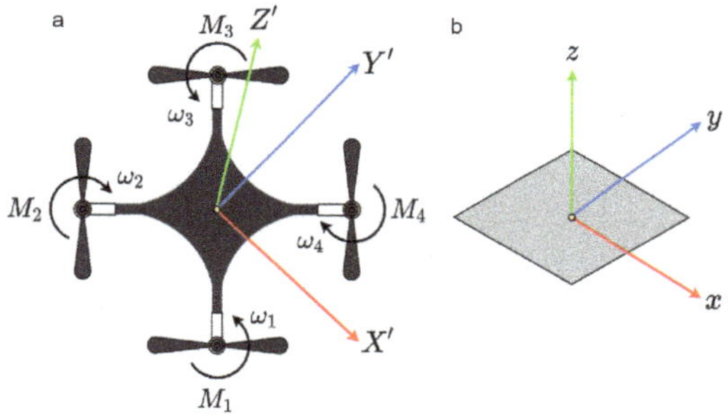

Figure 3. Nano-quadcopter. (**a**) Body frame. (**b**) Inertial frame.

3.2. *Sensor Information Sharing*

A particular research objective was proposed a 3D global positioning platform oriented to devices with limited hardware. The low-cost UWB sensors were used to perform simultaneous localization and information exchange between the small vehicles.

The system consists of anchors and tags that communicate with UWB pulses. The anchors within the workspace serve as absolute reference points for the tags. Likewise, the tags are used as a mobile reference point to triangulate the anchors' signals and, simultaneously, as a communication interface to other external devices. Two-Way Ranging (TWR) algorithms are used to detect a single tag, and a Time Difference of Arrival (TDoA) algorithm is used to detect the position of one or more tags inside the same workspace. The diagram of the operation is shown in Figure 4.

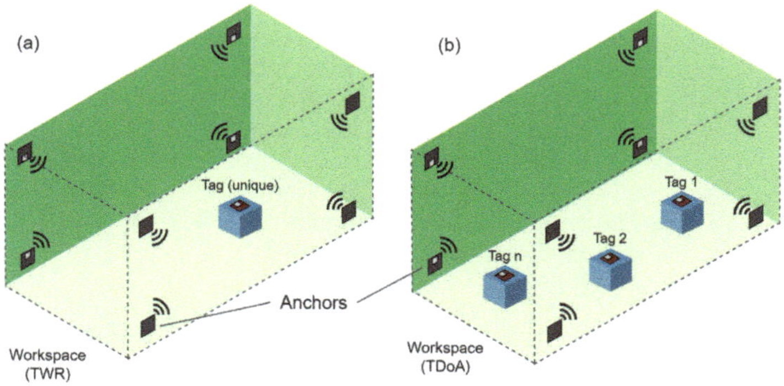

Figure 4. Different configurations of the absolute positioning system mounted within a workspace: (**a**) existence of only one tag: a location algorithm based on "two way ranging" TWR is used. (**b**)Two or more tags are present: a location algorithm based on "time difference of arrival" TDoA is used.

3.3. Producer-Consumer Algorithm

A reduced algorithm based on an analogy to the producer–consumer problem was another particular research objective. This consideration helps to perform tasks to share information with two or more complex sensors to localize small vehicles. Furthermore, this algorithm is reduced to achieve accessible communication among multiple robots and an extra path planning task was executed at the same time.

Collaboration in navigation among autonomous robots represents a problem of synchronization between the transmission and reception of positioning and movement data. In this problem, there appear two parallel processes (producer and consumer), which exchange information by a finite memory buffer [52]. Here the producer is in charge of generating and inserting data into the buffer, trying not to saturate it. Otherwise, the consumer extracts the data from the buffer one by one, preventing the buffer from emptying. Applications of this problem are commonly in concurrency, and message passing, although trivial, is extremely useful when needed [53]. Figure 5 shows the flowchart of the consumer–producer problem.

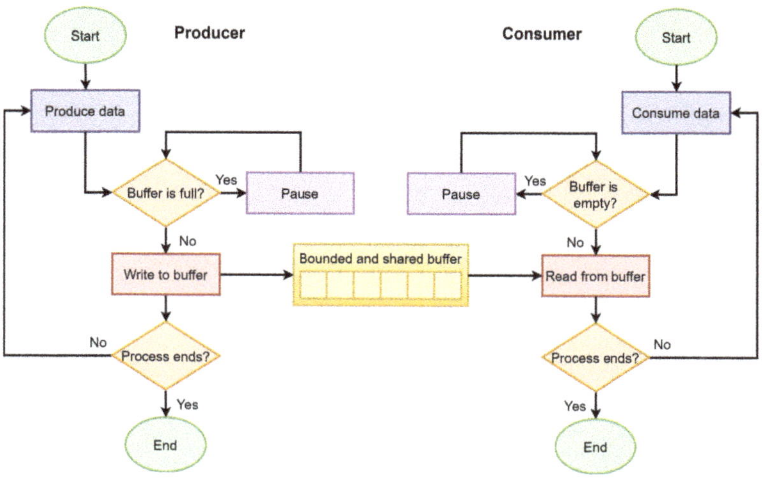

Figure 5. Flowchart of the producer–consumer problem.

The producer–consumer problem is somehow analogous to the problem of cooperative follow-up of a trajectory. In said problem, one or more dependent robots follow the trajectory generated by a master robot until finally coinciding in (almost) the same point. The master robot draws the trajectory in the workspace, so it operates as a producer of information on points. On the other hand, the dependent robots receive their information on the trajectory needed to get to the master, beginning from their starting points. It is, though, possible to view the dependent robots as consumers of information.

The FIFO (First In-First Out) buffer can be modelled as an ordered set D of q_i information elements. The state of this kind of memory is determined through the sequence $\{q_0, q_1, \cdots, q_m\}$, ordered by the moment insertion of its elements by (3) and (4). In this buffer, a new information element q_0 is always placed at the end of its queue by the insertion instruction of an $I(q_0, D)$:

$$I(q_0, D) = q_0 \cup D. \tag{3}$$

The operation raises all elements' position in the buffer by one place and increases, consequently, the cardinality by one unit: $I(q_0, D)| = |D| + 1$. On the other hand, the instruction of an extraction procedure $E(D)$, removes, when executed, the first element q_m from the queue of the m data elements in D:

$$E(D) = D \setminus q_m. \tag{4}$$

It is clear that this command lowers the cardinality of the set D by one unit: $|E(D)| = |D| - 1$.

For implementation of the algorithm in the robots, we used the following general considerations. Let $W \subset \mathbb{R}^3$ be the workspace and R the set of positions of the i robots, with $i \in \mathbb{Z}^+, i = 1, \ldots n$ and $n > 1$. The value of the instantaneous position at time t of each robot is represented by $R_i(t) = (x_i(t), y_i(t), z_i(t))$. The trajectory ${}_{t_0}^{t_f}R_i$ of the robot i between the starting point $R_i(t_0)$ and the endpoint $R_i(t_f)$ are characterized by a sequence of positions $R_i(t_j)$ which corresponds to the equidistant sequence of sampling moments $t_j = t_0 + j \cdot \Delta t$, within the interval between the initial t_0 and final time t_f, separated by a sampling interval Δt between each moment, such that: ${}_{t_0}^{t_f}R_i = \{R_i(t)\} \forall t \in [t_0, \ldots, t_f]$. To the master robot we assign, as identifier, the number 1. In this way the path of this robot is symbolized with ${}_{t_0}^{t_f}R_i$. The elements of the trajectory are then successively in D_1 loaded elements with the insertion command:

$$I(R_1(t_j), D_0). \tag{5}$$

The follower robots R_s, with $s \in \{2, 3, .., n\}$, reconstruct the path by successively extracting the leading elements in the D_1 buffer:

$$R_s \leftarrow (D_1). \tag{6}$$

The Algorithm 1 for the case of the master robot (producer) can be summarized as follows:

Algorithm 1 Producer algorithm

```
procedure PRODUCER(D_1, trajectory)
    Initialize |D_1| = 0 (empty)
    R_M starts trajectory at R_1(t_0)
    while trajectory is not finish do
        R_M inserts its position in memory D_1
        Delay Δt
```

The Algorithm 2 for the case of the follower robot (consumer) can be summarized as follows:

Algorithm 2 Consumer algorithm

procedure CONSUMER(R_M, D_1)
    if R_M starts *trajectory* at $R_1(t_0)$ **then**
        while position $R_1(t_j)$ is changing or D_1 is not empty **do**
            Extract an element from D_1 and send it to robot s, as its next position p_s
            Move to $p_s(x_s, y_S, z_s)$

4. Experiments, Results and Discussion

4.1. Experiments

The experiments performed were divided into two parts: (1) the characterization of the sensor information sharing with small vehicles and (2) intercepting and landing of the aerial vehicle on the ground vehicle with shared information using the producer–consumer algorithm.

4.1.1. Experiment 1: Characterization

This experiment aimed to characterize the absolute positioning system and, after that, estimate the tracking accuracy in 2D and 3D. The system was placed within a controlled environment, and functionality tests were carried out within the fixed position's polygon, defined by the anchors. The 2D tracking tests consisted of following prior defined trajectories using the ground vehicle. The corresponding robot was tagged to know in real-time its position in the xy plane, left side of Figure 6. In the xyz space, we used a nano quadcopter with a tag integrated, as shown on the right side of Figure 6. The data obtained was used to reconstruct the trajectory and to know the accuracy of the system.

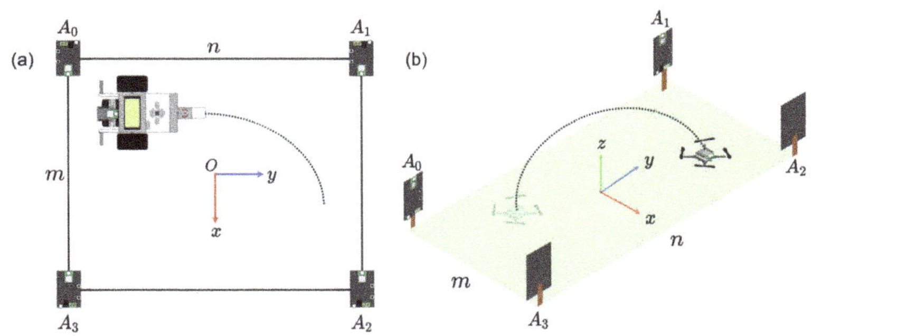

Figure 6. Diagram of the characterization experiment. (**a**) In 2D. (**b**) In 3D.

We used a Loco Positioning System (LPS), which employes a UWB DMW1000 module from Decaway and whose precision is estimated by the manufacturer to be within ± 0.1 m. The LPS is composed of a Loco Positioning Node (anchors) and a Loco Positioning Deck (tags). For triangulation of the signal, the system supports configurations with 4, 6 and 8 anchors and uses the TWR algorithm (a single tag) and TDoA in the existence of two or more tags. A minimum of four anchors must triangulate the signal within a 3D space, so we selected this configuration during our experiments. Four anchors (numbered from 0 to 3) were placed at a distance 0.16 m above the floor, forming a rectangle with dimensions m by n. We decided to place the absolute reference system's origin in the center of this $m \times n$ rectangle. Table 2 shows the values of m, n and the coordinates of each anchor.

Table 2. Positions of the anchors employed to characterize the positioning system.

Size		Anchor's Coordinates (x, y, z)			
m [m]	n [m]	Anchor 0 [m]	Anchor 1 [m]	Anchor 2 [m]	Anchor 3 [m]
1.86	5.0	(−0.93, −2.50, 0)	(0.93, 2.50, 0)	(−0.93, 2.50, 0)	(0.93, −2.50, 0)
1.90	5.0	(−0.95, −2.50, 0)	(0.95, 2.50, 0)	(−0.95, 2.50, 0)	(0.95, −2.50, 0)
5.20	1.9	(−2.50, −0.95, 0)	(2.50, 0.95, 0)	(−2.50, 0.95, 0)	(2.50, −0.95, 0)
2.00	2.0	(−1.00, −1.00, 0)	(1.00, 1.00, 0)	(−1.00, 1.00, 0)	(1.00, −1.00, 0)

A LEGO EV3 ground robot was used in its differential configuration. We employed a gyroscope with an accuracy of ±3 degrees and an optical encoder with an accuracy of ±1° to measure the wheels' rotation. The controller of the EV3 ground robot contains an ARM 9 processor with 64 MB of RAM and an SD card reader on which the GNU/Linux-based EV3dev Operating System (OS) was installed. The trajectories and odometry of the EV3 land mobile as described in [54] were implemented using Python as the programming language and the EV3dev OS libraries.

The airbone vehicle used is a Crazyflie 2.1 quadcopter, which has as main controller an STM32F405 and uses the nRF51822 for communication with the Crazyradio-PA telemetry system. The architecture of the Crazyflie quadcopter allows for the stacking of accessories, known as decks. We used the Flowdeck V2 as an optical flow sensor and the LPS Deck as a tag, compatible with LPS. The Crazyflie quadcopter's architecture is open to implement tools and algorithms made by the community of developers. Crazyflie quadcopter uses the Kalman Extended Filter, as proposed by [55,56] for the fusion of data of the IMU, the Flowdeck V2 and the LPS. The implemented control algorithms are based on [57,58].

For the individual tests, we worked with the TWR mode. We assigned tag 1 to the Crazyflie quadcopter and tag 2 to the LEGO EV3 robot. As a test to determine the UWB triangulation system's accuracy, we implemented a desired trajectory in the workspace for both robots. A computer was interfaced with a Crazyradio-PA to acquire the position data of each tag. For this purpose, a circle of 0.61 m radius and different heights, 0.02 m in the EV3 robot and 0.5 m in the Crazyflie quadcopter.

4.1.2. Experiment 2: Intercepting and Landing Task

The experiment's objective was to validate the operation of the proposed consumer-producer algorithm when solving the combined task of intercepting and landing, where the Crazyflie 2.1 quadcopter must intercept the LEGO EV3 lander and land on top of it. The initial position of the EV3 robot was $x = 0$ m, $y = 0.66$ m and $z = 0.02$ m, and its given path was an arc with a radius of 0.5 m and an arc length of also 0.5 m. The initial position of the Crazyflie quadcopter was at $x = 0$ m, $y = 0$ m and $z = 0$ m, and during the trajectory tracking phase, the quadcopter was asked to maintain a height of 0.5 m.

Figure 7 shows the flowchart of both the producer and consumer algorithms implemented on a computer to manage the dataflow of the EV3 robot, the Crazyflie quadcopter and the TDoA mode of the LPS. In this case, process A (producer) is the EV3 robot, and process B (consumer) is the Crazyflie quadcopter. The process starts when the computer has verified the communication between the EV3 robot and the Crazyflie quadcopter. The computer sends then to the EV3 robot the information about the trajectory to be followed. When the EV3 robot starts to move, it sends its position data to the computer. In parallel to the EV3 robot following the demanded path, the computer starts the producer-consumer algorithm and instructs the Crazyflie quadcopter to start its takeoff. When the Crazyflie quadcopter reaches the reference height of 0.50 m, it starts receiving the buffer position data of the EV3 robot from the computer.

The EV3 robot sends its position data continuously and independently to the D_1 buffer in the computer until it reaches the endpoint of the given trajectory. The Crazyflie quadcopter consumes the position data added by the EV3 robot in the buffer D_1 and computes its movements by comparing its position with the buffer's positions. At the moment, the Crazyflie quadcopter does not encounter any more position data in the buffer D_1 it starts a landing maneuver and touches down on top of the EV3 robot.

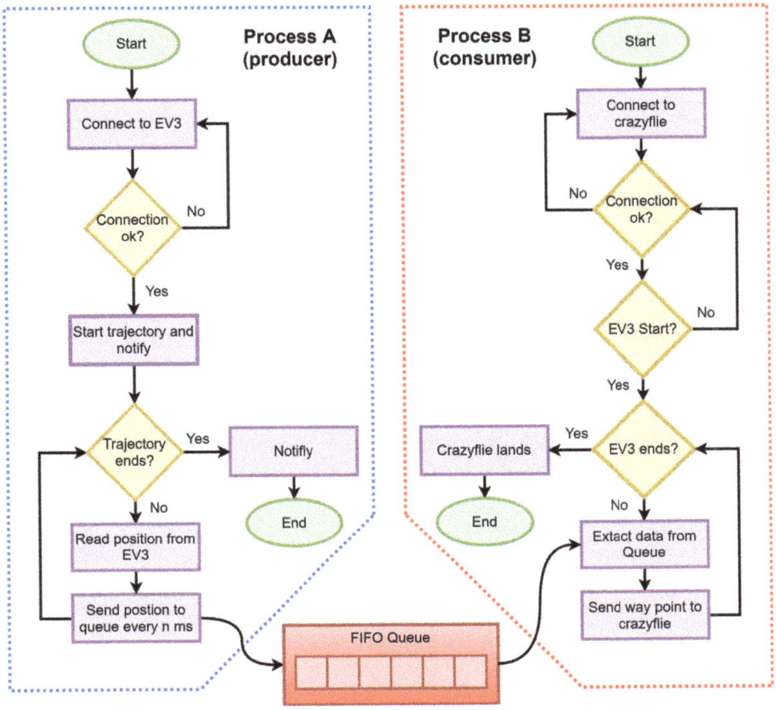

Figure 7. Flowchart of the process A (producer) implemented for the mobile robot EV3 and of the process B (consumer) implemented for the Crazyflie quadcopter.

The details about the algorithm's implementation can be viewed in the block diagram in Figure 8. It illustrates the relationship of the tools used between the computer, the EV3 robot and the Crazyflie quadcopter during the algorithm's execution. The white arrows indicate the internal communication flow, while the dotted arrows represent external communication flow.

The algorithm was implemented on a Python script using multi-threaded programming and executed on a computer with a Core i7 processor with 8 GB of RAM. The Python script sends orders over the local network to the EV3 robot using the MQTT protocol through an MQTT broker installed in the EV3dev OS in the EV3 robot. MQTT was used because it is ideal for machine-to-machine (M2M) communication due to this protocol's lightness. On the other hand, the EV3 robot was connected with a wireless network card.

The communication between the UWB tags and the computer was done using the Crazyradio-PA telemetry system, which communicates wirelessly, using the NRF24 protocol. One Crazyradio antenna can communicate with up to 8 tags.

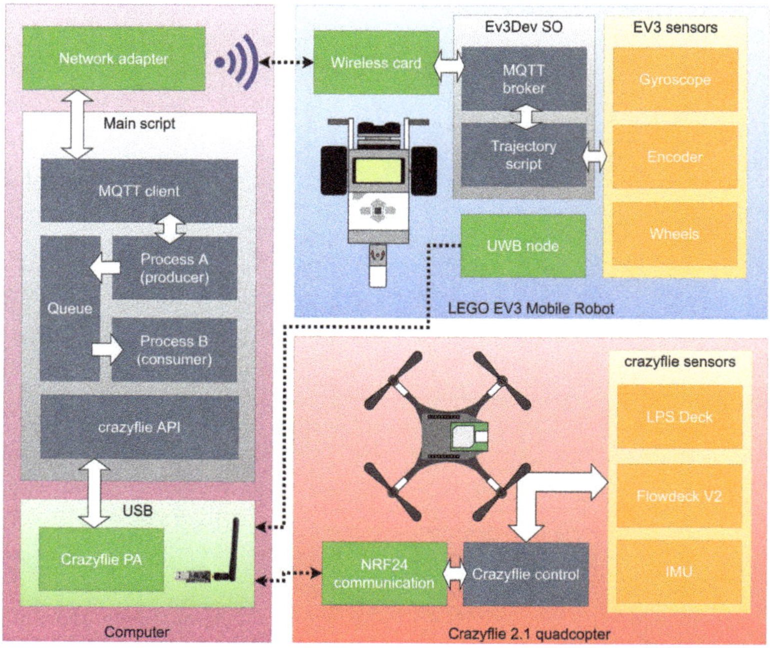

Figure 8. Block diagram illustrating the interaction between the components during the execution of the algorithm.

4.1.3. Experiment 3: Performance

This experiment aimed to evaluate the scaling of the system about the size of the consumer's buffer. The proposal consisted of generating a constant amount of information of 1.92 kbps by the producer. The tested trajectory was the same as in the previous experiments with a speed of 0.1 m/s. Buffer size started with 0, increments of 1 and up to 14 times (24 Bytes up to 360 Bytes). The units of measurement used were: the consumption rate in Kbps, the kinetic energy in $Kg \cdot m^2/s^2$, and the processing time of the information in the queue with s. The break-even point to determine the system's efficiency was to compare the increase of the energy and the processing time.

An additional test was to eliminate the information sent by the Ev3 robot to destabilize the operation of the quadcopter to 50% of the task.

4.2. Results and Discussion

During the characterization process for the anchor positions, as presented in Table 2 of Section 3, it was observed that a square-shaped distribution of the anchors in the workspace leads to a better performance than a rectangular one (Figure 9). Nevertheless, the manufacturer recommends a rectangular layout. The best configuration was a square workspace of 2 m by 2 m and with the anchors at its vertices after trials.

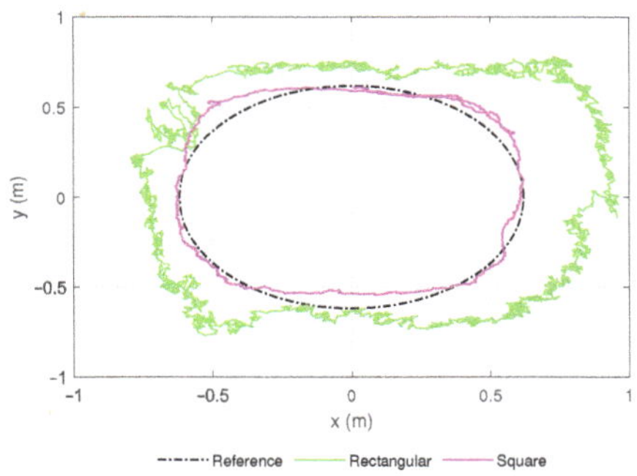

Figure 9. Performance of anchors in rectangular (green) and square (pink) configuration concerning a reference path.

Figure 10 shows the results obtained after characterization and tuning of the 2 m × 2 m workspace. Figure 10 shows the ideal trajectory (dashed line) and the one performed (green line) by the EV3 robot (green line). We found that the data obtained were consistent; however, when the robot moved near the anchors, the error increased significantly up to 0.1 m. Figure 10b,c show the EV3 robot's displacement on each of the x and y coordinate axes. The path carried out by the EV3 robot in the x-axis is much closer to the given one than in the y-axis, due to the tag's orientation. However, there are probabilities of perturbations that increase the error during some moments, for example, in t = 19.7 s.

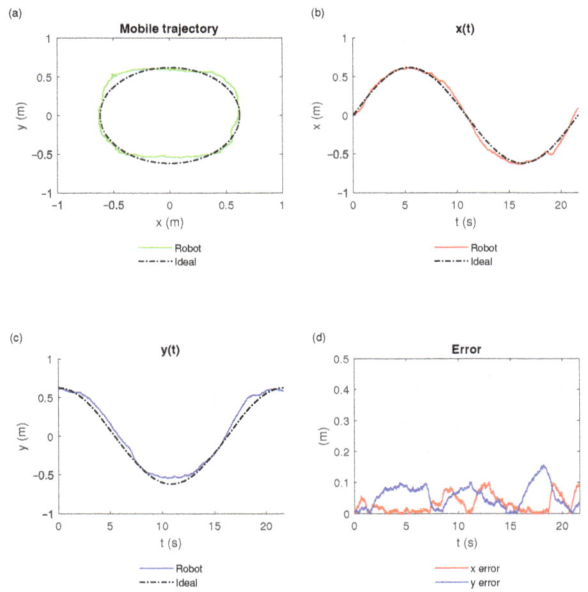

Figure 10. Results of the characterization run for a 2 × 2 m workspace. (**a**) Trajectory performed by the robot. (**b**) x-axis component of the movement. (**c**) y-axis component of the movement. (**d**) Quadratic error on x and y between the performed path and the given ideal trajectory.

Figure 10 shows the RMS-error calculation of the path in both coordinate axes. In the x-axis, the error is within a range below 0 to 0.1 m as claimed by the manufacturer of the UWB system; however, in the y-axis, the maximum error went up to 0.16 m. The average RMS-error in the x-axis was 0.042 m, again by a factor of 2/3 lower than the RMS-error of 0.062 m in the case of the y-axis.

The producer-consumer algorithm always kept the equilibrium of data. For this reason, the Crazyflie quadcopter intercepted the EV3 robot with acceptable performance. Figure 11 illustrates the best result we obtained when executing the algorithm; that is, it represents the sequence of the points $R_1(t_j)$ that were stored in D_1's queue, coming from the EV3 robot, which was later on converted into set points for the Crazyflie quadcopter. Figure 11a shows the trajectories performed by the EV3 robot (red line) and by the Crazyflie quadcopter (blue line) in 3D perspective. The ideal trajectory for the quadcopter is shown by the black dashed line. The projections of the trajectories on the x, y, and z axes are similarly represented in Figure 11b–d.

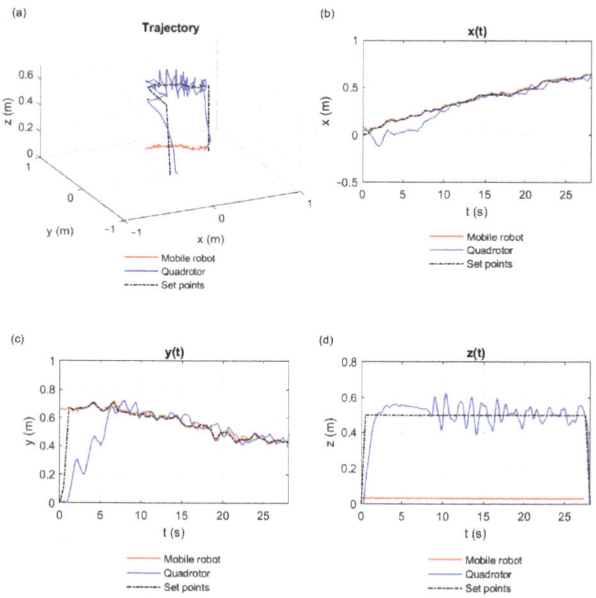

Figure 11. The paths taken by the EV3 (green line), and the Crazyflie quadcopter robot (red line) compared to the ideal path (black dashed line). (**a**) The trajectory performed in 3D perspective. (**b**) Displacement in x. (**c**) Displacement y. (**d**) Displacement in z.

Figure 12 shown the RMS-positioning-error of the EV3 robot and the Crazyflie quadcopter from the ideal path. We restricted the evaluation of the EV3 robot's performance to the xy-plane. In Figure 12a, it is seen that the RMS-error in x is always smaller than 0.22 m. When $(t_f - t_0) < 10$ s the error is greater than 0.1 m, for 10 s $< (t_f - t_0) <$ 19 s you get an average RMS-error of 0.04 m, and in the interval of 19 s $< (t_f - t_0) <$ 23 s the RMS-error is greater than 0.04 m but always less than 0.1 m, due to the proximity to the anchor three.

Figure 12b shown that the RMS-positioning-error in y is greater than the one in x. This error is because the distance between the two robots at the beginning is 0.61 m and is interpreted as an error. The Crazyflie quadcopter manages to reduce its RMS-positioning-error to less than 0.1 m after the first 6 seconds. Once a level of proximity under 0.1 m is reached, the average RMS-error in y drops down to a level of 0.032 m.

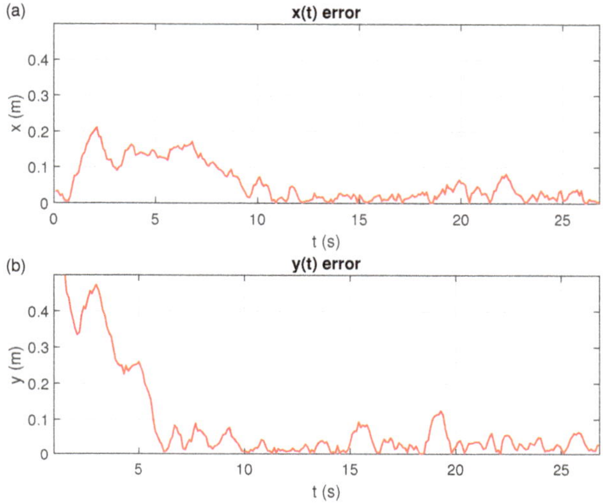

Figure 12. RMS-positioning-error between the EV3 robot and the Crazyflie quadcopter. (**a**) x-axis error. (**b**) y-axis error.

Figure 13a shows the performance results of the system about the scalability of the consumer buffer. The exponential behavior of the data rate with three segments can be observed. A moderate slope was observed in the first segment from 0 to 8 times the buffer size, from 8 to 12 a steeper slope and after 12 a sudden increase.

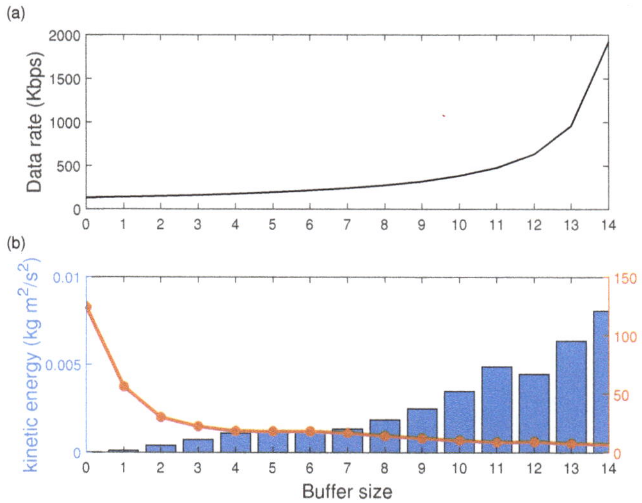

Figure 13. Buffer size vs. (**a**) consumption data rate and (**b**) kinetic energy and execution time comparison as a system's performance evaluation.

The results have allowed us to verify that the system can be scaled in terms of buffer size to increase the consumption data rate. The limitation of this increase is affected by the response of the quadcopter. Another scenario would be to increase the number of quadcopters sharing the information; in this sense, the rate is divided among the robots cooperating in the same task. The maximum operating limit of the platform is 6.8 Mbps and a packet length of 1023 bytes.

The results to find the balance point between the energy consumed by the quadcopter and the execution time are shown in Figure 13b. It was observed that with higher efficiency of the consumption data rate by the quadcopter, the performance and energy increases; however, for a rate lower than three and higher than 8, the results showed an excess or lack of time.

From the results, we can deduce that there is a relationship between the energy consumption of the quadcopter and the information efficiency. For this, by repeating the experiments, a buffer with the size of 6 is the optimal value for executing the task.

In buffer size 12, a perturbation was tested by eliminating information from the producer. Thus, it was verified that the lack of information decreased the energy consumption; therefore, the relationship between the quality of the information and the energy consumption was verified.

Figure 14 shows the results of the trajectories based on the scalability of the buffer size. A more stable behavior was observed when the information to be consumed is higher Figure 14(0–8). However, when the data is missing, there is a more unstable behavior as is shown in Figure 14(12).

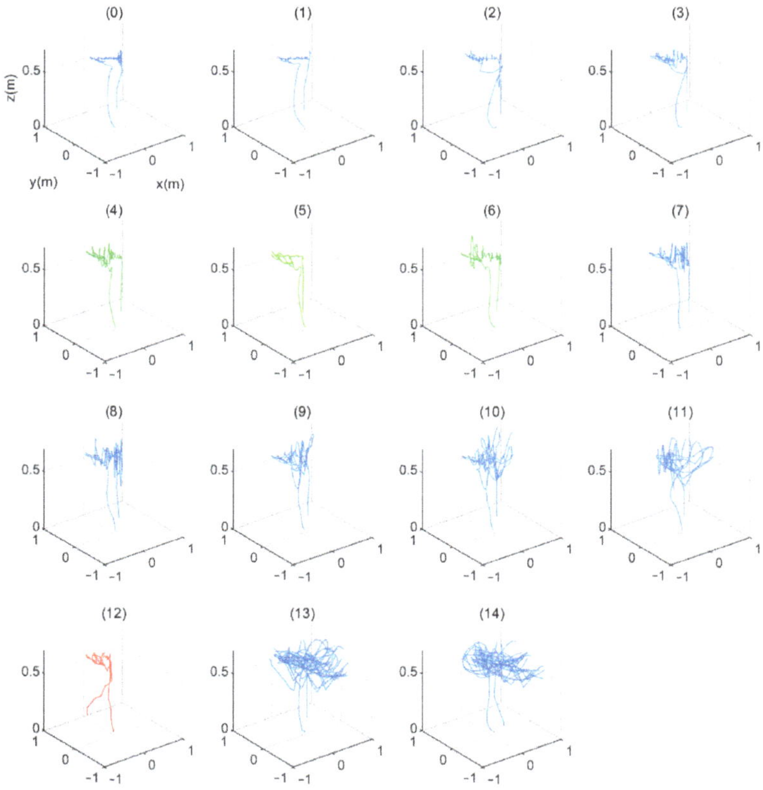

Figure 14. Different tests made using buffer sizes from 0 to 14.

Finally, a video of the experiments can be seen at https://rodrigovazquezlopez.github.io/Cooperation/ (accessed on 23 April 2021).

5. Conclusions and Future Work

The research highlighted the advantages of software-oriented robotic system architectures; however, the disadvantages with small vehicle resources were shown. The proposed hardware-oriented design succeeded in solving a small vehicle system's particular problem without losing the advantage of compatibility with widely used tools.

The producer-consumer algorithm successfully solved the problem of synchronization of messages shared by two or more vehicles. This algorithm was implemented in real hardware, while other similar algorithms only were simulated. This research also solved the problem of intercepting an airborne vehicle's trajectory to a level below 0.04 m and landing it safely on top of the ground vehicle. The EV3 robot and the Crazyflie quadcopter are affordable small-scale platforms and offer the experimenter reduced scale equipment that avoids unnecessary exposure of people to risks.

It was possible to reduce the number of anchors from 6 and 8, as mentioned in similar publications, to only 4. The square-shaped workspace gives better results than the recommended rectangular one. Furthermore, in parallel, perpendicular or diagonal paths, the RMS-positioning-error is reduced, compared to the one in curved paths, i.e., when the tags rotate in the plane, has to be interference that increases the error. The positioning error increased due to the proximity between a tag and anyone of the anchors. Even under these higher noise conditions, the results were satisfactory; that is, an average error of 0.04 m was achieved.

The LPS platform configuration has the following advantages: Experimental test in small environments of only 2 m $\times$ 2 m $\times$ 1 m (width, length and height), the platform is portable and does not require complex installations. Due to the ongoing epidemic, it is impossible to work in research laboratories in many places in the world. Therefore, the possibility of having a compact platform, like the one presented here, allows continuing developing research in our home-laboratories.

Performance tests showed the ability to scale the memory size and thus process more information. However, the efficiency of the system does not improve proportionally. A balance point was found to define a trade-off between data, power consumption, and processing time. Therefore, we can suggest considering the efficiency relationship between the size of the vehicles and their information management systems with the system's scalability.

For further development, a new intelligent and robust algorithm will be proposed to be executed intrinsically in the airborne and terrestrial robot and achieve the algorithm's implementation with a more significant number of cooperating robots.

Author Contributions: Conceptualization, R.V.-L., J.C.H.-L. and J.S.-G.; methodology, R.V.-L., J.C.H.-L. and J.S.-G.; software, R.V.-L.; validation, P.v.B., J.S.-G.; formal analysis, R.V.-L., J.C.H.-L. and J.S.-G.; investigation, R.V.-L., J.C.H.-L. and J.S.-G.; resources, D.L.M.-V.; data curation, D.L.M.-V.; writing—original draft preparation, R.V.-L., J.S.-G. and P.v.B.; writing—review and editing, R.V.-L., J.S.-G. and P.v.B.; visualization, P.v.B.; supervision, J.C.H.-L.; project administration, D.L.M.-V.; funding acquisition, D.L.M.-V. All authors have read and agreed to the published version of the manuscript.

Funding: This research received no external funding and the APC was funded by Sistemas Ciberfisicos research project by UAM-L.

Institutional Review Board Statement: Not applicable.

Informed Consent Statement: Not applicable.

Data Availability Statement: Not applicable.

Acknowledgments: This study was supported by the Secretaría de Investigación y Posgrado of the Instituto Politécnico Nacional under the research grant 20211103. R Vazquez-Lopez acknowledges financial support from Becas Nacionales-CONACYT. J. C. Herrera-Lozada, D. L. Martinez-Vazquez and J. Sandoval-Gutierrez thank the support from the Sistema Nacional de Investigadores-CONACYT.

Conflicts of Interest: The authors declare no conflict of interest.

References

1. Koseglu, M.; Celik, O.M.; Pektas, O. Design of an autonomous mobile robot based on ROS. In Proceedings of the 2017 International Artificial Intelligence and Data Processing Symposium (IDAP), Malatya, Turkey, 16–17 September 2017; pp. 1–5, doi:10.1109/IDAP.2017.8090199.
2. Aung, T.; Min, H.Y.; Maw, A.H. CIMLA: Checkpoint Interval Message Logging Algorithm in Kafka Pipeline Architecture. In Proceedings of the 2020 International Conference on Advanced Information Technologies (ICAIT), Yangon, Myanmar, 4–5 November 2020; pp. 30–35, doi:10.1109/ICAIT51105.2020.9261812.
3. Sanchez-Lopez, J.; Molina, M.; Bavle, H.; Sampedro, C.; Suárez Fernández, R.; Campoy, P. A Multi-Layered Component-Based Approach for the Development of Aerial Robotic Systems: The Aerostack Framework. *J. Intell. Robot. Syst. Theory Appl.* **2017**, *88*, 683–709, doi:10.1007/s10846-017-0551-4.
4. Atoev, S.; Kwon, K.; Lee, S.; Moon, K. Data analysis of the MAVLink communication protocol. In Proceedings of the 2017 International Conference on Information Science and Communications Technologies (ICISCT), Tashkent, Uzbekistan, 2–4 November 2017; pp. 1–3, doi:10.1109/ICISCT.2017.8188563.
5. Evers, R.; Masters, P. The application of low-altitude near-infrared aerial photography for detecting clandestine burials using a UAV and low-cost unmodified digital camera. *Forensic Sci. Int.* **2018**, *289*, 408–418, doi:10.1016/j.forsciint.2018.06.020.
6. Zahran, S.; Masiero, A.; Mostafa, M.M.; Moussa, A.M.; Vettore, A.; El-Sheimy, N. UAVS ENHANCED NAVIGATION IN OUTDOOR GNSS DENIED ENVIRONMENT USING UWB AND MONOCULAR CAMERA SYSTEMS. *Int. Arch. Photogramm. Remote Sens. Spat. Inf. Sci.* **2019**, *XLII-2/W13*, 665–672, doi:10.5194/isprs-archives-XLII-2-W13-665-2019.
7. Akçakoca, M.; Atıcı, B.; Gever, B.; Oğuz, S.; Demir, M.; Saldiran, E.; Yuksek, B.; Koyuncu, E.; Yeniceri, R.; Inalhan, G. A simulation-based development and verification architecture for micro uav teams and swarms. In Proceedings of the AIAA Scitech 2019 Forum, San Diego, CA, USA, 7–11 January 2019; doi:10.2514/6.2019-1979.
8. Dirir, A.; Elsayed, H. Object Tracking Framework for Unmanned Aerial Vehicles. In Proceedings of the 2019 IEEE Global Conference on Internet of Things (GCIoT), Dubai, United Arab Emirates, 4–7 December 2019; pp. 1–6, doi:10.1109/GCIoT47977.2019.9058406.
9. Lee, J.; Su, Y.; Shen, C. A Comparative Study of Wireless Protocols: Bluetooth, UWB, ZigBee, and Wi-Fi. In Proceedings of the IECON 2007—33rd Annual Conference of the IEEE Industrial Electronics Society, Taipei, Taiwan, 5–8 November 2007; pp. 46–51, doi:10.1109/IECON.2007.4460126.
10. Pinto, M.F.; Honório, L.M.; Marcato, A.L.M.; Dantas, M.A.R.; Melo, A.G.; Capretz, M.; Urdiales, C. ARCog: An Aerial Robotics Cognitive Architecture. *Robotica* **2021**, *39*, 483–502, doi:10.1017/S0263574720000521.
11. Xu, C.; Wang, X.; Duan, S.; Wan, J. Spatial-temporal constrained particle filter for cooperative target tracking. *J. Netw. Comput. Appl.* **2021**, *176*, doi:10.1016/j.jnca.2020.102913.
12. Onyekpe, U.; Kanarachos, S.; Palade, V.; Christopoulos, S.R., Vehicular localisation at high and low estimation rates during gnss outages: A deep learning approach. In *Deep Learning Applications, Volume 2*; Wani, M.A., Khoshgoftaar, T.M., Palade, V., Eds.; Springer: Berlin/Heidelberg, Germany, 2021; Volume 1232, pp. 229–248, doi:10.1007/978-981-15-6759-9_10.
13. Brena, R.; García-Vázquez, J.; Galván-Tejada, C.; Muñoz-Rodriguez, D.; Vargas-Rosales, C.; Fangmeyer, J. Evolution of Indoor Positioning Technologies: A Survey. *J. Sens.* **2017**, *2017*, doi:10.1155/2017/2630413.
14. Mohr, H. UAV Implementation of Distributed Robust Target Location in Unknown Environments. In Proceedings of the 2020 IEEE Aerospace Conference, Big Sky, MT, USA, 7–14 March 2020; pp. 1–10, doi:10.1109/AERO47225.2020.9172459.
15. Gargioni, G.; Peterson, M.; Persons, J.; Schroeder, K.; Black, J. A full distributed multipurpose autonomous flight system using 3D position tracking and ROS. In Proceedings of the 2019 International Conference on Unmanned Aircraft Systems (ICUAS), Atlanta, GA, USA, 11–14 June 2019; pp. 1458–1466, doi:10.1109/ICUAS.2019.8798163.
16. Yashin, G.; Trinitatova, D.; Agishev, R.; Ibrahimov, R.; Tsetserukou, D. AeroVr: Virtual reality-based teleoperation with tactile feedback for aerial manipulation. In Proceedings of the 2019 19th International Conference on Advanced Robotics (ICAR), Belo Horizonte, Brazil, 2–6 December 2019; pp. 767–772, doi:10.1109/ICAR46387.2019.8981574.
17. Wu, R.; Pandurangaiah, J.; Blankenship, G.; Castro, C.; Guan, S.; Ju, A.; Zhu, Z. Evaluation of Virtual Reality Tracking Performance for Indoor Navigation. In Proceedings of the 2020 IEEE/ION Position, Location and Navigation Symposium (PLANS), Portland, OR, USA, 20–23 April 2020; pp. 1311–1316, doi:10.1109/PLANS46316.2020.9110225.
18. Shin, J.M.; Kim, Y.S.; Ban, T.W.; Choi, S.; Kang, K.M.; Ryu, J.Y. Position tracking techniques using multiple receivers for anti-drone systems. *Sensors* **2021**, *21*, 35, doi:10.3390/s21010035.
19. Adjei, H.; Oduro-Gyimah, F.; Shunhua, T.; Agordzo, G.; Musariri, M. Developing a bluetooth based tracking system for tracking devices using arduino. In Proceedings of the 2020 5th International Conference on Computing, Communication and Security (ICCCS), Patna, India, 14–16 October 2020; pp. 1–5, doi:10.1109/ICCCS49678.2020.9276884.

20. Lie, K.; Yeo, K.; Ting, A.N.; Chieng, D.T. Indoor Tracking with Bluetooth Low Energy Devices Using K-Nearest Neighbour Algorithm. In Proceedings of the 2020 IEEE 10th Symposium on Computer Applications Industrial Electronics (ISCAIE), Shah Alam, Malaysia, 18–19 April 2020; pp. 155–159, doi:10.1109/ISCAIE47305.2020.9108790.
21. Shi, G.; Ming, Y. Survey of indoor positioning systems based on ultra-wideband (UWB) technology. *Wirel. Commun. Netw. Appl.* **2016**, *348*, 1269–1278, doi:10.1007/978-81-322-2580-5_115.
22. Silvia, Z.; Martina, C.; Fabio, S.; Alessandro, P. Ultra Wide Band Indoor Positioning System: analysis and testing of an IPS technology. *IFAC-PapersOnLine* **2018**, *51*, 1488–1492, doi:10.1016/j.ifacol.2018.08.292.
23. Lutz, P.; Schuster, M.; Steidle, F. Visual-inertial SLAM aided estimation of anchor poses and sensor error model parameters of UWB radio modules. In Proceedings of the 2019 19th International Conference on Advanced Robotics (ICAR), Belo Horizonte, Brazil, 2–6 December 2019; pp. 739–746, doi:10.1109/ICAR46387.2019.8981544.
24. Cao, Y.; Yang, C.; Li, R.; Knoll, A.; Beltrame, G. Accurate position tracking with a single UWB anchor. In Proceedings of the 2020 IEEE International Conference on Robotics and Automation (ICRA), Paris, France, 31 May–31 August 2020; pp. 2344–2350, doi:10.1109/ICRA40945.2020.9197345.
25. Gomaa, M.; Silva, O.D.; Mann, G.; Gosine, R.; Hengeveld, R. ROS Based Real-Time Motion Control for Robotic Visual Arts Exhibit Using Decawave Local Positioning System. In Proceedings of the 2020 American Control Conference (ACC), Denver, CO, USA, 1–3 July 2020; pp. 653–658, doi:10.23919/ACC45564.2020.9148022.
26. Ledergerber, A.; D'Andrea, R. Ultra-wideband range measurement model with Gaussian processes. In Proceedings of the 2017 IEEE Conference on Control Technology and Applications (CCTA), Maui, HI, USA, 27–30 August 2017; Volume 2017; pp. 1929–1934, doi:10.1109/CCTA.2017.8062738.
27. Zhao, W.; Goudar, A.; Panerati, J.; Schoellig, A.P. Learning-based Bias Correction for Ultra-wideband Localization of Resource-constrained Mobile Robots. *arXiv* **2020**, arXiv:2003.09371.
28. Pinto, M.F.; Honorio, L.M.; Melo, A.; Marcato, A.L.M. A Robotic Cognitive Architecture for Slope and Dam Inspections. *Sensors* **2020**, *20*, 4579, doi:10.3390/s20164579.
29. Munro, A.; Clayton, G. Drone Swarms, Communications Performance and Big Data. In Proceedings of the 2019 IEEE 90th Vehicular Technology Conference (VTC2019-Fall), Honolulu, HI, USA, 22–25 September 2019; pp. 1–5, doi:10.1109/VTCFall.2019.8891336.
30. Silva, L.A.Z.d.; Vidal, V.F.; Honório, L.M.; Dantas, M.A.R.; Pinto, M.F.; Capretz, M. A Heterogeneous Edge-Fog Environment Supporting Digital Twins for Remote Inspections. *Sensors* **2020**, *20*, 5296, doi:10.3390/s20185296.
31. Wallace, C.; Ackels, S.; Benavidez, P.; Jamshidi, M. Real-Time Distributed Ensemble Learning for Fault Detection of an Unmanned Ground Vehicle. In Proceedings of the 2020 IEEE 15th International Conference of System of Systems Engineering (SoSE), Budapest, Hungary, 2–4 June 2020; pp. 463–468, doi:10.1109/SoSE50414.2020.9130511.
32. Angel, M.; Abdulla, A.K.; David, M.; de la Escalera, A. 3D Trajectory Planning Method for UAVs Swarm in Building Emergencies. *Sensors* **2020**, *20*, 642, doi:10.3390/s20030642.
33. Koubâa, A.; Qureshi, B.; Sriti, M.F.; Allouch, A.; Javed, Y.; Alajlan, M.; Cheikhrouhou, O.; Khalgui, M.; Tovar, E. Dronemap Planner: A service-oriented cloud-based management system for the Internet-of-Drones. *Ad Hoc Netw.* **2019**, *86*, 46–62, doi:10.1016/j.adhoc.2018.09.013.
34. Basso, M.; Zacarias, I.; Tussi Leite, C.E.; Wang, H.; Pignaton de Freitas, E. A Practical Deployment of a Communication Infrastructure to Support the Employment of Multiple Surveillance Drones Systems. *Drones* **2018**, *2*, 26, doi:10.3390/drones2030026.
35. Damacharla, P.; Mehta, D.; Javaid, A.Y.; Devabhaktuni, V.K. Study on State-of-the-art Cloud Services Integration Capabilities with Autonomous Ground Vehicles. In Proceedings of the 2018 IEEE 88th Vehicular Technology Conference (VTC-Fall), Chicago, IL, USA, 27–30 August 2018; pp. 1–5, doi:10.1109/VTCFall.2018.8690650.
36. Salmeron-Garcia, J.J.; van den Dries, S.; Diaz-del Rio, F.; Morgado-Estevez, A.; Luis Sevillano-Ramos, J.; van de Molengraft, M.J.G. Towards a cloud-based automated surveillance system using wireless technologies. *Multimed. Syst.* **2019**, *25*, 535–549, doi:10.1007/s00530-017-0558-5.
37. Okumuş, F.; Dönmez, E.; Kocamaz, A.F. A Cloudware Architecture for Collaboration of Multiple AGVs in Indoor Logistics: Case Study in Fabric Manufacturing Enterprises. *Electronics* **2020**, *9*, 2023, doi:10.3390/electronics9122023.
38. Okumuş, F.; Kocamaz, A.F. Cloud Based Indoor Navigation for ROS-enabled Automated Guided Vehicles. In Proceedings of the 2019 International Artificial Intelligence and Data Processing Symposium (IDAP), Malatya, Turkey, 21–22 September 2019; pp. 1–4, doi:10.1109/IDAP.2019.8875993.
39. Cardarelli, E.; Digani, V.; Sabattini, L.; Secchi, C.; Fantuzzi, C. Cooperative cloud robotics architecture for the coordination of multi-AGV systems in industrial warehouses. *Mechatronics* **2017**, *45*, 1–13, doi:10.1016/j.mechatronics.2017.04.005.
40. Indri, M.; Lazzero, I. A new HW/SW architecture to move from AGVs towards Autonomous Mobile Robots. In Proceedings of the 2019 24th IEEE International Conference on Emerging Technologies and Factory Automation (ETFA), Zaragoza, Spain, 10–13 September 2019; pp. 1551–1554, doi:10.1109/ETFA.2019.8869193.
41. Yilmaz, A.; Temeltas, H. ROS Architecture for Indoor Localization of Smart-AGVs Based on SA-MCL Algorithm. In Proceedings of the 2019 11th International Conference on Electrical and Electronics Engineering (ELECO), Bursa, Turkey, 28–30 November 2019; pp. 885–889, doi:10.23919/ELECO47770.2019.8990622.

42. Li, J.; Dong, W.; Sheng, X.; Xu, S. Visual Servoing of Micro Aerial Vehicles with the Cooperation of Ground Vehicle. In Proceedings of the 2020 IEEE/ION Position, Location and Navigation Symposium (PLANS), Portland, OR, USA, 20–23 April 2020; pp. 150–155, doi:10.1109/PLANS46316.2020.9109994.
43. Destiarti, A.; Kristalina, P.; Sudarsono, A. Mobile cooperative tracking with RSSI ranging in EKF algorithm for indoor wireless sensor network. In Proceedings of the 2016 International Conference on Knowledge Creation and Intelligent Computing (KCIC), Manado, Indonesia, 15–17 November 2017; pp. 60–66, doi:10.1109/KCIC.2016.7883626.
44. Yulong, D.; Bin, X.; Jie, C.; Hao, F.; Yangguang, Z.; Guanqiang, G.; Lihua, D. Path Planning of Messenger UAV in Air-ground Coordination. *IFAC-PapersOnLine* **2017**, *50*, 8045–8051, doi:10.1016/j.ifacol.2017.08.1230.
45. Marzoughi, A.; Savkin, A. Autonomous navigation of a team of unmanned surface vehicles for intercepting intruders on a region boundary. *Sensors* **2021**, *21*, 297, doi:10.3390/s21010297.
46. Huang, H.; Savkin, A.; Li, X. Reactive autonomous navigation of UAVs for dynamic sensing coverage of mobile ground targets. *Sensors* **2020**, *20*, 3720, doi:10.3390/s20133720.
47. Siegwart, R.; Nourbakhsh, I.R.; Scaramuzza, D. *Introduction to Autonomous Mobile Robots*; MIT Press: Cambridge, MA, USA, 2011.
48. Greiff, M. Modelling and Control of the Crazyflie Quadrotor for Aggressive and Autonomous Flight by Optical Flow Driven State Estimation. Ph.D. Thesis, Department of Automatic Control, Lund University, Lund, Sweden, 2017. Student Paper.
49. Luis, C.; Ny, J.L. Design of a Trajectory Tracking Controller for a Nanoquadcopter. *arXiv* **2016**, arXiv:1608.05786.
50. Hustiu, S.; Lupascu, M.; Popescu, S.; Burlacu, A.; Kloetzer, M. Stable hovering architecture for nanoquadcopter applications in indoor environments. In Proceedings of the 2018 22nd International Conference on System Theory, Control and Computing (ICSTCC), Sinaia, Romania, 10–12 October 2018; pp. 659–663, doi:10.1109/ICSTCC.2018.8540759.
51. Budaciu, C.; Botezatu, N.; Kloetzer, M.; Burlacu, A. On the Evaluation of the Crazyflie Modular Quadcopter System. In Proceedings of the 2019 24th IEEE International Conference on Emerging Technologies and Factory Automation (ETFA), Zaragoza, Spain, 10–13 September 2019; Volume 2019; pp. 1189–1195, doi:10.1109/ETFA.2019.8869202.
52. Stallings, W. *Operating Systems: Internals and Design Principles*, 8th ed.; Prentice Hall: Hoboken, NJ, USA, 2012.
53. Dijkstra, E.W. Information streams sharing a finite buffer. *Inf. Process. Lett.* **1972**, *1*, 179–180, doi:10.1016/0020-0190(72)90034-8.
54. Feng, L.; Borenstein, J.; Everett, H.R. *Where Am I: Sensors and Methods for Mobile Robot Positioning*; University of Michigan: Ann Arbor, MI, USA, 1994.
55. Mueller, M.; Hamer, M.; D'Andrea, R. Fusing ultra-wideband range measurements with accelerometers and rate gyroscopes for quadrocopter state estimation. In Proceedings of the 2015 IEEE International Conference on Robotics and Automation (ICRA), Seattle, WA, USA, 26–30 May 2015; Volume 2015; pp. 1730–1736, doi:10.1109/ICRA.2015.7139421.
56. Mueller, M.; Hehn, M.; D'Andrea, R. Covariance correction step for Kalman filtering with an attitude. *J. Guid. Control Dyn.* **2017**, *40*, 2301–2306, doi:10.2514/1.G000848.
57. Smeur, E.; Chu, Q.; Croon, G.D. Adaptive incremental nonlinear dynamic inversion for attitude control of micro air vehicles. *J. Guid. Control Dyn.* **2016**, *39*, 450–461, doi:10.2514/1.G001490.
58. Mellinger, D.; Kumar, V. Minimum snap trajectory generation and control for quadrotors. In Proceedings of the 2011 IEEE International Conference on Robotics and Automation, Shanghai, China, 9–13 May 2011; pp. 2520–2525, doi:10.1109/ICRA.2011.5980409.

sensors

MDPI

Article

Appearance-Based Sequential Robot Localization Using a Patchwise Approximation of a Descriptor Manifold

Alberto Jaenal, Francisco-Angel Moreno * and Javier Gonzalez-Jimenez

Machine Perception and Intelligent Robotics Group (MAPIR), Department of System Engineering and Automation Biomedical Research Institute of Malaga (IBIMA), University of Malaga, 29071 Málaga, Spain; ajaenal@uma.es (A.J.); javiergonzalez@uma.es (J.G.-J.)
* Correspondence: famoreno@uma.es

Abstract: This paper addresses appearance-based robot localization in 2D with a sparse, lightweight map of the environment composed of descriptor–pose image pairs. Based on previous research in the field, we assume that image descriptors are samples of a low-dimensional Descriptor Manifold that is locally articulated by the camera pose. We propose a piecewise approximation of the geometry of such Descriptor Manifold through a tessellation of so-called *Patches of Smooth Appearance Change* (PSACs), which defines our *appearance map*. Upon this map, the presented robot localization method applies both a Gaussian Process Particle Filter (GPPF) to perform camera tracking and a Place Recognition (PR) technique for relocalization within the most likely PSACs according to the observed descriptor. A specific Gaussian Process (GP) is trained for each PSAC to regress a Gaussian distribution over the descriptor for any particle pose lying within that PSAC. The evaluation of the observed descriptor in this distribution gives us a likelihood, which is used as the weight for the particle. Besides, we model the impact of appearance variations on image descriptors as a white noise distribution within the GP formulation, ensuring adequate operation under lighting and scene appearance changes with respect to the conditions in which the map was constructed. A series of experiments with both real and synthetic images show that our method outperforms state-of-the-art appearance-based localization methods in terms of robustness and accuracy, with median errors below 0.3 m and 6°.

Keywords: appearance-based localization; computer vision; Gaussian processes; manifold learning; robot vision systems; indoor positioning; image manifold; descriptor manifold

Citation: Jaenal, A.; Moreno, F.-A.; Gonzalez-Jimenez, J. Appearance-Based Sequential Robot Localization Using a Patchwise Approximation of a Descriptor Manifold. *Sensors* **2021**, *21*, 2483. https://doi.org/10.3390/s21072483

Academic Editor: Simon Tomažič

Received: 4 March 2021
Accepted: 29 March 2021
Published: 2 April 2021

1. Introduction

Visual-based localization involves estimating the pose of a robot from a query image, taken with an on-board camera, within a previously mapped environment. The widely adopted approach relies on detecting local image features (e.g., points, segments) [1,2] that are projections of 3D physical landmarks. Though this feature-based localization has achieved great accuracy in the last years [3–5], it presents two major drawbacks that hinder long-term localization and mapping: (i) lack of robustness against image radiometric alterations; (ii) inefficiency of 2D-to-3D matching against large-scale 3D models [6]. A much less explored alternative to feature-based localization consists in localizing the robot through the scene appearance, represented by a descriptor of the whole image. According to this framework, localization is accomplished by comparing the appearance descriptor against a map composed of descriptor–pose pairs, without any explicit model of the scene's geometric entities [7,8]. This approach turns out to be particularly robust against perceptual changes and also appropriate for large-scale localization, as demonstrated by the fact that it is included in the front-end of state-of-the-art Simultaneous Localization and Mapping (SLAM) pipelines to perform relocalization and loop closure, typically in the form of Place Recognition (PR) [3,5].

The accuracy of appearance-based localization is, however, quite limited. Good results are reported only when the camera follows a previously mapped trajectory (i.e., in one dimension) [9–11] or when it is very close to any of the poses of the map [8,12]. In this work, we investigate whether localization based only on appearance can deliver *continuous* solutions that are accurate and robust enough to become a practical alternative to methods based on 3D geometric features. We restrict this work to planar motion, that is, we aim to estimate the camera pose given by its 2D position and orientation (3 d.o.f.).

To this end, we first assume that images acquired in a certain environment are samples of a low-dimensional Image Manifold (IM) that can be locally parameterized (or articulated) by the camera pose. This assumption has been justified by previous works [13,14], but only exploited under unrealistic conditions, where the IM was sampled from a fine grid of poses in the environment under fixed lighting conditions. This IM is embedded in an extremely high-dimensional space: $\mathbb{R}^{H\times W\times 3}$, where W and H stand for the width and height of the images, respectively. Patently, working in such extremely high-dimensional space is not only unfeasible, but also impractical since it lacks radiometric invariance. That is, the IM of a given environment might change drastically with the scene illumination and the automatic camera accommodation to light (e.g., gain and exposure time). Thus, it is primarily to project the images (i.e., samples of the IM) to a lower-dimensional space with a transformation that also provides such radiometric invariances [15]. This projection can be carried out by encoding the image into a descriptor vector, hence obtaining a new appearance space, the Descriptor Manifold (DM), which is still articulated by the camera pose. In this paper, we leverage Deep Learning (DL)-based holistic descriptors [8,16] to project the IM into a locally smooth DM. We are aware that such smoothness is not guaranteed for the descriptors employed here, since this feature has not been explicitly taken into account in their design. This issue will be addressed in future work, but in the context of our proposal, the selected DL-based descriptors perform reasonably well under this assumption.

Another capital aspect in appearance-based localization is that it requires an appropriate map, which, in our case, is built from samples of the DM that are annotated with their poses. In this paper, we assume that such samples, in the form of descriptor–pose image pairs, are given in advance and are representative of the visual appearance of the environment. Upon this set of pairs, we propose creating *Patches of Smooth Appearance Change* (PSACs), that is, regions that locally approximate the geometry of the DM using neighbor samples (see Figure 1). A tessellation of such PSACs results in a piecewise approximation of the DM that constitutes our *appearance map*, where pose data is only available at the vertices of the PSACs. The appearance smoothness within each PSAC allows us to accurately regress a descriptor for any pose within the pose space covered by the PSAC. This is accomplished through a Gaussian Process (GP), which delivers the Gaussian distribution of the regressed descriptor (refer to Figure 1).

Our proposal solves continuous sequential localization indoors by tracking the robot pose using a Gaussian Process Particle Filter (GPPF) [17,18] within the described *appearance map*. The particles are propagated with the robot odometry and weighted through the abovementioned Gaussian Process, which is implemented as the GPPF observation model for the image descriptor.

Pursuing to improve the robustness of our method against appearance changes, we model the descriptor variations in such situations as a white noise distribution that is introduced into the estimation of the observation likelihood. Finally, it is worth mentioning that our proposal can easily recover from the habitual PF particle degeneracy problem by launching a fast and multihypothesis camera relocalization procedure through global Place Recognition.

Our localization system has been validated with different indoor datasets affected by significant appearance changes, yielding notable results that outperform current state-of-the-art techniques, hence demonstrating its capability to reduce the gap between feature-based and appearance-based localization in terms of accuracy, while still leveraging the invariant nature of holistic descriptors.

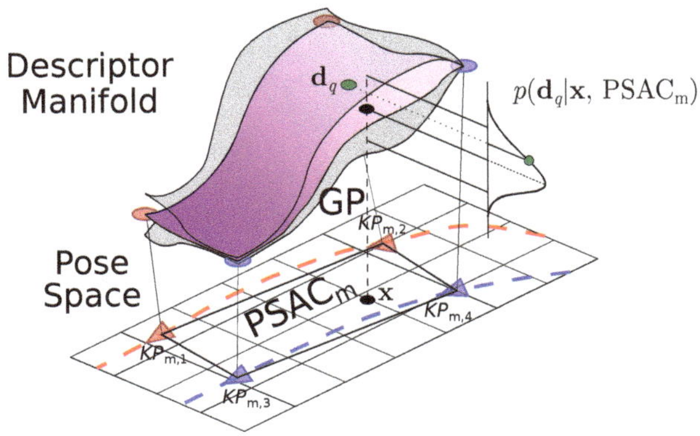

Figure 1. The Gaussian Process (GP) associated to a *Patch of Smooth Appearance Change* (PSACs) approximates the geometry of a neighborhood of the Descriptor Manifold (assumed to be locally smooth) with respect to the pose space, predicting the local likelihood $p(\mathbf{d}_q|\mathbf{x}, \mathrm{PSAC}_m)$ of the observation $\mathbf{d}_q$ in a given pose $\mathbf{x}$. In this example, the descriptor–pose pairs are extracted from two previous trajectories of the robot (in red and blue).

2. Related Work

This section reviews two concepts that are essential for the scope of this work: Global image descriptors and Appearance-based localization.

2.1. Global Image Descriptors

A well-founded way of getting a consistent dimensionality reduction from the Image Manifold to the Descriptor Manifold is through *Manifold Learning* tools, like LLE [19] or Isomap [20]. Their performance, however, is limited to relatively simple IMs that result from sequences of quasi-planar motions or deformations, like face poses, person gait, or hand-written characters [21]. Unfortunately, images taken in a real 3D scene give rise to complex, highly twisted IMs, which also present discontinuities due to occlusion borders [22]. Moreover, typical *Manifold Learning* tools are hardly able to generalize their learned representations to images captured under different appearance settings [15]. This prevents their application to generating low-dimensional embeddings adequate for camera localization.

Nevertheless, Deep Learning (DL)-based holistic descriptors have recently proven their suitability to enclose information from complete images, effectively reducing their dimensionality, while adding invariance to extreme radiometric changes [8,12,23,24]. This feature has made DL-based descriptors highly suitable for diverse long-term robot applications [9,25], e.g., Place Recognition (PR), where the goal is to determine if a certain place has been already seen by comparing a query image against a lightweight set of images [26]. In addition, these descriptors have proven to more sensitively reflect pose fluctuations than local features [26], which is described, for example, by the *equivariance* property [27,28]. Since we are targeting robust robot operation under different appearance conditions, global descriptors arise as the natural choice to address appearance-based localization.

2.2. Appearance-Based Localization

Appearance-based localization is typically formulated as a two-step estimation problem: first, PR is performed to find the most similar images within the map and, subsequently, the pose of the query image is approximated from the location of the retrieved ones [29,30]. In this scenario, DL-based works have proposed to improve the second stage through Convolutional Neural Network architectures that estimate relative pose transformations between covisible images [31–33].

The addition of temporal and spatial sequential information to appearance-based localization methods based on single instances provides more consistency to the estimation of the pose, as it reduces the possibility of losing camera tracking due to, for instance, perceptual aliasing [26]. Following this idea, SeqSLAM [10,34] proposes a sequence-to-sequence matching framework that reformulates PR with the aim to incorporate sequentiality, leading to substantial improvements under extreme appearance changes. Building upon SeqSLAM, SMART [11] integrates odometry readings to provide more consistent results. More recently, *Network-Flow*-based formulations have also been proposed to solve appearance-based sequence matching under challenging conditions, addressing camera localization [35,36] or position-based navigation and mapping [37]. Despite their relevant results, the nature of all these works is discrete, unlike our proposal, restricting all possible estimation to the locations present on the map.

Conversely, CAT-SLAM [38] employs image sequentiality as a source of topometric information to improve the discrete maps used by FAB-MAP [9], allowing interpolation within the sequence map through the association of continuous increments on appearance and pose. Although the estimates produced by this approach are continuous, they are restricted to the mapped trajectory. Our work overcomes this constraint by requiring multiple sequences or pose grids as a source for constructing the map PSACs. This way, we can perform localization even at unvisited map locations near the PSACs, achieving, consequently, more accuracy and reliability.

An interesting alternative to pose interpolation is the use of Gaussian Processes (GPs) regression [39], nonparametric, general-purpose tools that allow generalizing discrete representations to a continuous model, and, hence, can be adapted to perform continuous localization within discrete maps. For instance, [40,41] employed GPs to generate position estimates for omnidirectional images in indoor maps, achieving good performance although lacking applicability to robot rotations. Our approach is instead designed to work with both 2D positions and rotations for conventional cameras.

In turn, the authors of [18] proposed Gaussian Process Particle Filters (GPPFs) to solve appearance-based localization in maps of descriptor–pose pairs. The GP works as the observation model of the PF, estimating the likelihood of the observed holistic descriptor at each of the particle positions. This localization pipeline was later improved in [42] by using only the nearest map neighbors in the GP regression, allowing efficient localization within large environments. Despite being promising, both works have three major drawbacks: (i) they define a unique Gaussian Process between poses and descriptors for the whole environment, assuming that the manifold geometry has a similar shape for the entire environment, thus leading to inaccurate estimations, (ii) they do not propose a relocalization process in case of losing tracking, and (iii) they only consider localization under the same appearance than the map, lacking robustness to radiometric alterations.

Inspired by these works, we employ a GPPF to solve appearance-based localization in a continuous and sequential fashion within challenging indoor environments. We solve their first problem by locally modeling the mapping between poses and descriptors via specific GPs for each PSAC, providing refined estimates for each neighborhood. Our proposal solves the second issue through a fast and multihypothesis relocalization process based on global PR within the map. Finally, the last issue is addressed by incorporating a model of the appearance variation between the mapped and query images to the map.

3. System Description

This section describes our proposal for the process of appearance-based camera localization. First, we define the elements that form the *appearance map*, which are key contributions in this work, and then we address PR-based localization and camera tracking using a probabilistic formulation based on a GPPF.

3.1. Patches of Smooth Appearance Change

The *Patches of Smooth Appearance Change* (PSACs) are regions that locally model the interrelation between camera poses and image descriptors, and represent areas where the change in appearance is small.

3.1.1. Definition

The basic building unit of a PSAC is the pair $p_i = (\mathbf{d}_i, \mathbf{x}_i)$, composed by the global descriptor $\mathbf{d}_i \in \mathbb{R}^d$ of an image and the pose $\mathbf{x}_i \in SE(2)$ where it was captured.

We assume that these pairs are extracted from any of these two environmental representations (Figure 2): either from several **robot navigation sequences** (at least two) or from **pose grids** where the cameras have densely sampled the environment given fixed position and rotation increments (i.e., a regular grid). Optimally, a subset of these pairs should be selected so that they constitute the smallest number of samples from which the Descriptor Manifold (DM) can be approximated with sufficiently good accuracy. These *key* samples can be viewed as the equivalent to the *key-frames* in traditional, feature-based visual localization, and hence, we denote them *key-pairs* (*KPs*). Since determining such an optimal subset is a challenging issue itself, out of the scope of this work, the *KPs* are constantly sampled from the total collection of pairs.

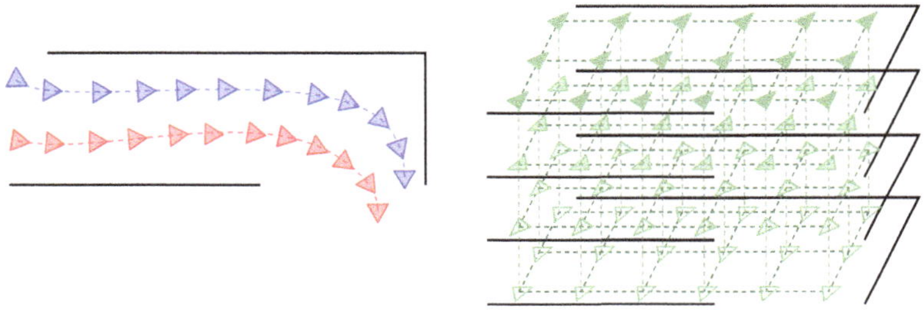

Figure 2. Each PSAC is constructed from descriptor–pose pairs that can be obtained from two different robot trajectories (left, blue and red) or a grid of poses (right, green). This is better seen in color.

Each PSAC is built from a group of adjacent *KPs* and approximates the DM in the region that they delimit. As explained later, the robot localization takes place within these PSACs by defining a suitable observation model for the GPPF (refer to Figure 1). Formally, let the m-th PSAC be

$$\mathrm{PSAC}_m = \left(\left\{ KP_{m,i} | \, i = 1, \ldots, Q \right\}, GP_m \right), \tag{1}$$

where $Q \geq 3$ is the number of *KPs* forming the PSAC. In turn, GP_m is a Gaussian Process specifically optimized for that particular PSAC that delivers a Gaussian distribution over the image descriptor for any pose nearby the PSAC (further explained in Section 3.1.2).

Thus, in order to determine the closeness between a query pair $p_q = (\mathbf{d}_q, \mathbf{x}_q)$ and a particular PSAC, we define two distance metrics as follows:

- The *appearance distance* $D^a_{m,q}$ from the query descriptor $\mathbf{d}_q$ to the m-th PSAC is defined as the average of the descriptor distances to each of its constituent *key-pairs*:

$$D^a_{m,q} = D^a(\mathrm{PSAC}_m, \mathbf{d}_q) = \frac{1}{Q} \sum_i^Q ||\mathbf{d}_q - \mathbf{d}_{m,i}||_2. \tag{2}$$

- Similarly, but in the pose space, we define the *translational distance* $D^t_{m,q}$ from $\mathbf{x}_q$ to the m-th PSAC as

$$D^t_{m,q} = D^t(\mathrm{PSAC}_m, \mathbf{x}_q) = \frac{1}{Q} \sum_i^Q ||\mathbf{t}_q - \mathbf{t}_{m,i}||_2, \tag{3}$$

with $\mathbf{t}_q$ being the translational component of the pose $\mathbf{x}_q = (\mathbf{t}_q, \theta_q)$.

Finally, the set of all PSACs covering the environment that has been sampled forms the **appearance map** $\mathcal{M}$:

$$\mathcal{M} = \{\text{PSAC}_m |\, m = 1, \ldots, M\}, \tag{4}$$

with M being the number of PSACs. This way, we achieve a much more accurate approximation of the relation between the pose space and the DM within each PSAC, ultimately modeling in $\mathcal{M}$, patchwise, the complete shape of the DM.

3.1.2. GP Regression

GPs are powerful regression tools [39] that have previously demonstrated their validity as observation models in Particle Filters [17,18,42].

In this work, we learn a specific GP for each PSAC from its vertices *KPs* and also using all the nearest pairs, in terms of translational distance. Then, for a certain query pose $\mathbf{x}_q$, the GP_m delivers an isotropic Gaussian distribution $\mathcal{N}(\boldsymbol{\mu}_{m,q}, \sigma^2_{m,q}\mathbf{I}_d)$, where $\boldsymbol{\mu}_{m,q} \in \mathbb{R}^d$ and $\sigma^2{}_{m,q} \in \mathbb{R}$ stand for its mean and uncertainty, respectively. This distribution is finally employed to estimate the likelihood $p(\mathbf{d}_q|\mathbf{x}_q, \text{PSAC}_m)$ of an observed image descriptor $\mathbf{d}_q$, given the query pose $\mathbf{x}_q$ within the PSAC_m.

For this, the GP regression employs a *kernel k*, which measures the similarity between two input 2-D poses $(\mathbf{x}_i, \mathbf{x}_j)$, with this structure:

$$k(\mathbf{x}_i, \mathbf{x}_j) = k_{RBF}(\mathbf{t}_i, \mathbf{t}_j) \cdot k_{RBF}(\theta_i, \theta_j) + k_W(\mathbf{x}_i, \mathbf{x}_j). \tag{5}$$

This *kernel k* first multiplies two Radial Basis Function (RBF) *kernels* $k_{RBF}(a_i, a_j) = \beta_a \exp(-\alpha_a ||a_i - a_j||^2_2)$ (α_a and β_a are optimizable parameters) for the separated translational and rotational components of the evaluated poses $\mathbf{x} = (\mathbf{t}, \theta)$. Then, a White Noise *kernel* $k_W(a_i, a_j) = \sigma^2_W \delta(a_i - a_j)$ is added, which models the variation suffered by the image descriptors taken at the same pose but under different appearances (refer to Figure 3). This is justified because, although global PR descriptors have demonstrated outstanding results in terms of invariance, such invariance is not ideal and small differences might appear. Thereby, since the construction of the map $\mathcal{M}$ is typically carried out considering just one particular appearance, and we aim for the robot localization to be operational under diverse radiometric settings, we propose the inclusion of a white noise distribution accounting for this circumstance in the regression. We model such descriptor variation with the variance σ^2_W, computed as the average discrepancy between the descriptor variances of pose adjacent pairs p_i, under the same $\sigma^2_{i,\text{same}}$ and different $\sigma^2_{i,\text{diff}}$ illumination settings:

$$\sigma^2_W = \frac{1}{N} \sum_i^N \left(\sigma^2_{i,\text{diff}} - \sigma^2_{i,\text{same}} \right). \tag{6}$$

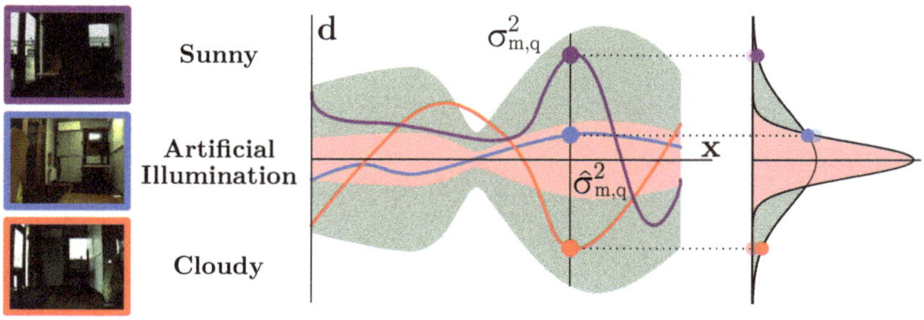

Figure 3. These three images have been captured at the same pose but with different appearances. Ideally, their descriptors (blue, red, and purple dots) should be identical but, in practice, certain inaccuracies appear. As the GP learns the descriptor distribution uniquely from the appearance of the map, this variation is not considered, leading to an underestimated GP uncertainty (red area, $\hat{\sigma}^2_{m,q}$). The inclusion of white noise expands such uncertainty (green area, $\sigma^2_{m,q}$), solving this issue.

3.2. Robot Localization

Once we have defined all the elements involved in the representation of the environment, we address here the process of localization within the appearance map $\mathcal{M}$. We aim to estimate the robot pose through appearance-based continuous tracking using a Gaussian Process Particle Filter (GPPF), namely, a PF that employs the GPs described above as observation models. Being well-known robot localization tools, we do not provide a deep description of Particle Filters here but instead refer to the reader to the seminar work [43] for further information.

At time-step t, each of the P particles in the filter represent a robot pose estimation $\mathbf{x}_p^{(t)}$ with an associated weight $w_p^{(t)}$, proportional to its likelihood. Besides, each particle is assigned to a certain region $\mathrm{PSAC}_{m,p}^{(t)}$, as explained next.

3.2.1. System Initialization

When the PF starts, we perform global localization based on Place Recognition to select the most similar PSAC in $\mathcal{M}$ to the query descriptor according to its appearance:

$$\mathrm{PSAC}_{\hat{m}} = \min_{\mathrm{PSAC}_m \in \mathcal{M}} D_{m,q}^{a}. \tag{7}$$

To account for multihypothesis initialization, we also consider as candidates those PSACs whose *appearance distance* is under a certain threshold proportional to $D_{\hat{m},q}^{a}$. Subsequently, the particles are uniformly assigned and distributed among all candidate PSACs, setting their initial weights to $w_p^{(t_0)} = \frac{1}{P}$.

Note that, if the robot tracking is lost during navigation, this procedure is launched again to reinitialize the system and perform relocalization.

3.2.2. Robot Tracking

Once each particle is assigned to a candidate PSAC, the robot pose estimation is carried out following the traditional *propagation-weighting* sequence:

Propagation. First, the particles are propagated according to the robot odometry:

$$\mathbf{x}_p^{(t)} = \mathbf{x}_p^{(t-1)} \oplus v^{(t)}, \tag{8}$$

with $v^{(t)} \sim \mathcal{N}(\overline{v}^{(t)}, \Sigma_v)$ representing noisy odometry readings, and $\oplus$ being the pose composition operator in $SE(2)$ [44].

Weighting. After the propagation, the *translational distance* between each particle's pose and all the PSACs is computed, so that the particle is assigned to the nearest PSAC ($\mathrm{PSAC}_{m,p}^{(t)}$). Then, we use the GP regressed in the PSAC to locally evaluate the likelihood of the observed descriptor $\mathbf{d}_q$ at the particle pose $\mathbf{x}_p^{(t)}$ as follows:

$$w_p^{(t)} = p(\mathbf{d}_q | \mathbf{x}_p^{(t)}, \mathrm{PSAC}_{m,p}^{(t)}) \sim \exp\Big(-\frac{d}{2}\ln({\sigma^2}_{m,p}^{(t)}) - \frac{||\mathbf{d}_q - \boldsymbol{\mu}_{m,p}^{(t)}||_2^2}{2({\sigma^2}_{m,p}^{(t)})} \Big), \tag{9}$$

with d being the dimension of the descriptor.

Finally, apart from *propagation* and *weighting*, two more operations can be occasionally applied to the particles.

Resampling: In order to prevent particle degeneracy, the GPPF resamples when the number of effective particles is too low, promoting particles with higher weights.

Reinitializing: During normal operation, the GPPF may lose the tracking of the camera, mainly due to extremely challenging conditions in the images (e.g., very strong change appearances, presence of several dynamic objects). We identify this situation by inspecting the *translational distance* between each particle and the *centroid* of its assigned

PSAC, defined as the average pose of all the key-pairs forming the PSAC. If all particles are at least twice farther from the centroid of their assigned PSAC than its constituent *key-pairs*, the tracking is considered lost. Consequently, the PF relocalizes by following the aforementioned initialization procedure.

4. Experimental Results

In this section, we present three experiments to evaluate the performance of our appearance-based localization system.

First, we carry out a verification of the regression outcome in Section 4.1, with the aim to experimentally validate the hypothesis of a smooth Descriptor Manifold within the regions covered by each PSAC. In Sections 4.2 and 4.3, we test our proposal with four different state-of-the-art global descriptors in two different datasets, with a combination of setups for the map sampling. This has provided us with an insight about the error incurred by our proposal and has allowed us to determine the best configuration for localization. The second experiment, in Section 4.4, compares the resulting setup with three appearance-based localization alternatives in terms of accuracy and robustness, revealing that our system equals or improves their performance in every scenario.

It is important to highlight that this evaluation does not include comparisons with feature-based localization methods, since they are not compatible with appearance changes in the images used for both mapping and localization, which is one of the main benefits of our proposal.

We employed two different indoor datasets for the experiments:

- The **COLD-Freiburg database** [45], which includes real images from an office gathered with a mobile robot under different appearance conditions. We have used only a representative subset of the sequences in *part A* of the dataset (Figure 4a).
- The synthetic **SUNCG Dataset** [46] rendered through the virtual **HOUSE3D** environment [47] allows us to test the localization on a grid map (see Figure 4b).

In turn, regarding the global image representation, we have tested the following state-of-the-art appearance descriptors:

- **NetVLAD**: VGG16 [48] based on off-the-shelf, 4096-sized NetVLAD features with Principal Component Analysis (PCA) whitening [8].
- **ResNet-101 GeM**: ResNet-101-based [49] fine-tuned generalized-mean features with learned whitening [12].
- **1M COLD Quadruplet** and **1M RobotCar Volume**: end-to-end learned condition invariant features with VGG16 NetVLAD [24] with quadruplet and volume loss functions in two different datasets.

Although none of these descriptors has been specifically designed to fulfill suitable properties for our pose regression approach, they have achieved promising results in terms of localization accuracy, as shown next. We used the GPy tool [50] to implement the proposed Gaussian Processes and empirically determined σ_W^2 from Equation (6), for each descriptor, by randomly sampling $N = 2000$ adjacent pairs with diverse illumination settings from the COLD-Freiburg database.

In this evaluation, we employ as metrics the median errors in translation and rotation (to inspect our method's accuracy), as well as the percentage of correctly localized frames (which illustrates the tracking and relocalization capabilities of our method). It is worth mentioning that traditional trajectory-based evaluation metrics as Absolute Trajectory Error (ATE) or Relative Pose Errors (RPE) are not applicable to this approach since our proposal, and appearance-based localization methods in general, yields global pose estimations that are not guaranteed to belong to a trajectory, due to possible tracking losses and relocalization situations.

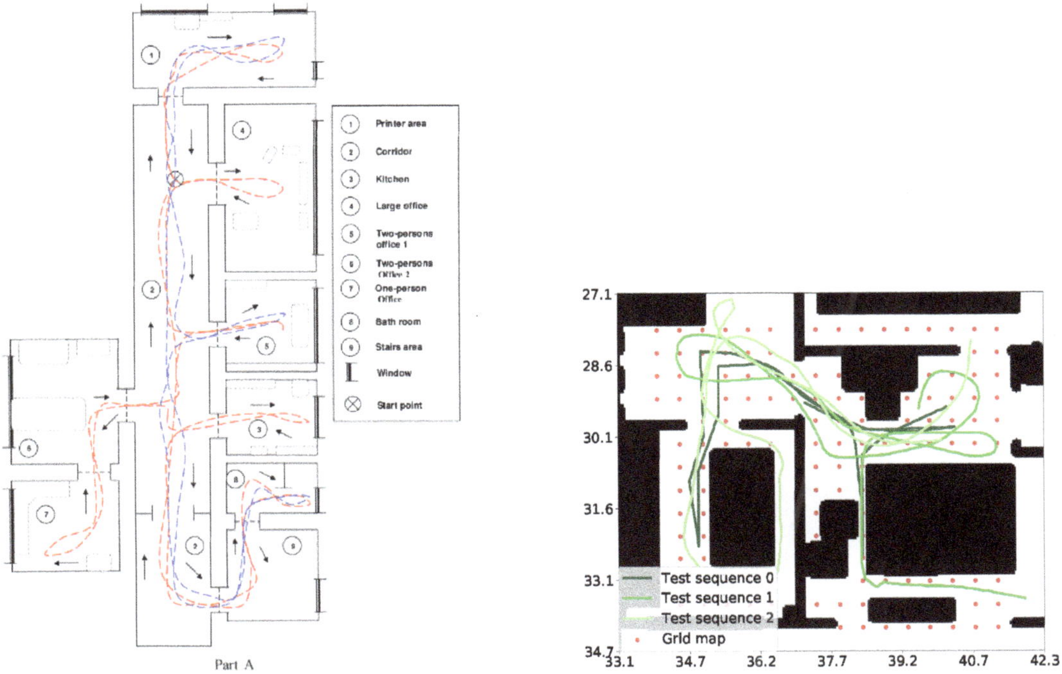

(a) (b)

Figure 4. Environments and sequences employed for evaluation. (**a**) Map of the COLD-Freiburg Part A environment. Samples of the standard and extended routes are depicted in blue and red, respectively (image from [45]). (**b**) Map of the house rendered by the SUNCG environment (the house employed was *034e4c22e506f89d668771831080b291*). The dense grid poses are shown in red and the test sequences in different shades of green. Black regions depict objects, where the robot is not able to navigate.

4.1. Corridor: Sanity Check

The main assumption of our proposal is the hypothesis of a locally smooth Descriptor Manifold with respect to the pose, on which PSACs are based. Since this assumption is not justified by previous work, we have conducted a basic test to evaluate the regression outcome of the PSACs in a simple scenario.

The proposed experiment studies the evolution of the image descriptor along a simple, linear trajectory, by comparing the observed descriptor and the mean of the descriptor distribution resulting from the regression within the PSAC. In this manner, the behavior of the descriptor can be examined along the corridor axis in order to prove its continuity and the validity of the PSAC approximation. For this, we have selected a portion of an artificially illuminated (*night*) sequence where the robot traveled along a $\sim$8 m-long corridor, as well as the NetVLAD image descriptor. For the PSACs, we used a map constructed with images with the same appearance selected every 20 frames.

In order to represent the evolution of the descriptors, we have applied Principal Component Analysis (PCA) to them and represented the first PCA element (that with larger variation). Thus, Figure 5 depicts the trajectory of the robot through the corridor along with the value of said first PCA element for both the observed descriptor and the mean of the descriptor distribution regressed by the GPs at each PSAC. The displayed results demonstrate that the descriptor has a continuous evolution along the corridor, almost lineal in the central part. Besides, the PSACs are proved to also have a continuous outcome and to approximate very accurately the values of the observed descriptor along the sampled trajectory.

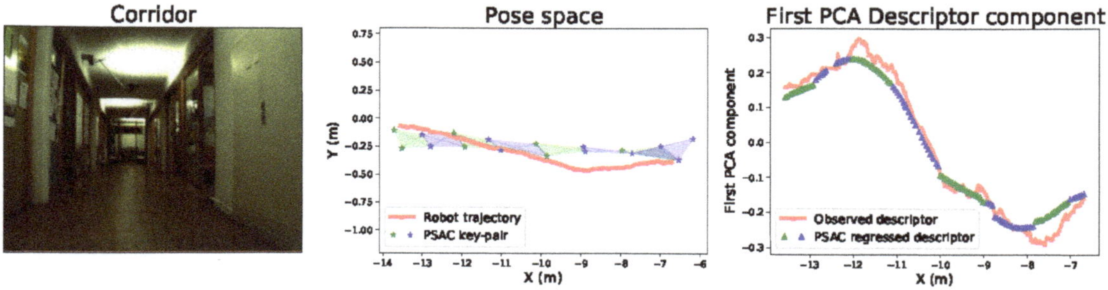

Figure 5. Experimental study of the global image descriptor behavior along a corridor of the COLD-Freiburg database (left image). The central figure depicts the robot pose trajectory and the PSACs (shadowed areas) through which it navigates. The rightmost figure compares, along the corridor, the behavior of the first Principle Component Analysis (PCA) component of the observed descriptor (coral) and the PSAC regression output (blue and green). This is better seen in color.

4.2. COLD: Sequential Map Testing

The COLD-Freiburg database (*part A*) provides odometry readings and real images for two different itineraries, namely, (i) **extended** (∼100 m-long), which covers the whole environment; (ii) **standard** (∼70 m), covering a subset of the environment, both depicted in Figure 4a. The dataset also provides images gathered under three different lighting conditions: at *night* (with artificial illumination), and on *cloudy* and *sunny* days.

In order to create the *appearance maps* for the experiments, we employed the first and second *night* sequences of the extended itinerary, since images captured under artificial illumination do not suffer from severe exposure changes or saturation like under the remaining conditions. From here on, we will refer to these as *map sequences*. Specifically, *key-pairs* from both *map sequences* have been obtained through Constant Sampling (CS) every 10, 20, and 30 pairs, resulting in three different maps with diverse density (described in more detail in Table 1), using $Q = 4$ *KPs* to construct every PSAC.

Table 1. Compared statistics of the created appearance maps. The dimension is calculated for a 2048-sized global descriptor.

Dataset	Area	Map	*Key-Pairs* (*KPs*)	PSACs	Size (Mb)	Construction Time (s)
COLD Database	∼900 m^2	Samp. 10	559	321	2.18	56.68
		Samp. 20	280	159	1.09	32.27
		Samp. 30	187	103	0.73	25.92
SUNCG	∼45 m^2	Dense	1203	679	4.69	140.94
		Sparse pos	451	227	1.76	129.36
		Sparse rot	723	432	2.82	136.95
		Sparse pos-rot	271	149	1.05	107.93

Finally, we have setup an extensive evaluation with six other sequences including different routes and illumination conditions: the first *night*; *cloudy* and *sunny* sequences of the **standard** part; the first *cloudy* and *sunny* sequences; and the third *night* sequence of the **extended** part.

Figure 6 shows a comprehensive test study depicting the localization performance of our proposal after twenty runs for all test sequences at each map, using the median translational (top) and rotational (down) errors as metrics. Note that the number of particles for the PF has been set to $P = 10^3$, as we have empirically found that increasing that number does not improve the accuracy results. The overall performance shows a median error predominantly below 0.3 m and 6°, which denotes promising results given the pure appearance nature of our approach, i.e., with no geometrical feature employed for localization.

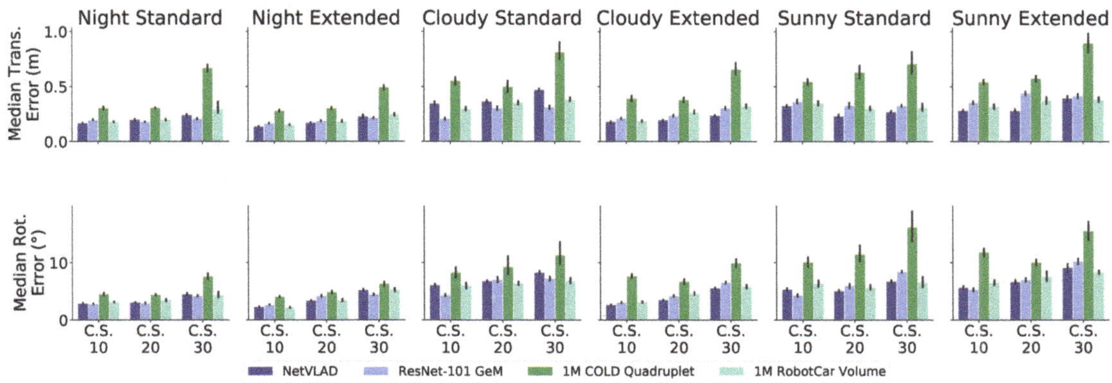

Figure 6. Comparison of the median translational and rotational errors of our proposal in the COLD-Freiburg dataset, tested at every sequence and with different Constant Sampling (CS) rates, using each holistic descriptors with 10^3 particles. Note that the maps were constructed employing sequences under similar conditions to the *night* sequences. This is better seen in color.

The results in Figure 6 show that scene appearance seems to be a key issue regarding the system's accuracy, as our proposal achieves better results in less-demanding lighting conditions like artificial illumination (night) or cloudy. Nevertheless, our system still demonstrates notable performance under challenging radiometric conditions, such as in sunny sequences (e.g., presence of lens flares and image saturation), hence proving its suitability for robust appearance-based localization.

On the other hand, the number of *KPs* that form the map is another factor influencing the performance, since the PSACs approximate the pose–descriptor interrelation the closer their *KPs* are. Although not particularly significant under advantageous conditions, this factor severely affects performance under challenging situations, as in *sunny* sequences, where localization is hindered as the sampling frequency decreases. Note that a more elaborate mapping technique than CS would improve these results, since an optimal selection of *KPs* would conform PSACs that achieve a more precise description of the DM geometry. Nevertheless, this will be explored in future work while, in this paper, we rely on CS to get still nonoptimal but notable results.

Finally, regarding the tested PR descriptors, the results show that in most cases, all perform similarly, with NetVLAD mostly achieving slightly better results. In turn, 1M COLD Quadruplet seems to struggle under complex illumination conditions, which might indicate that its empirically estimated white noise variance is unlikely to account adequately for these cases. The similar performance shown by all descriptors agrees with the fact that none of them has been specifically trained for appearance-based localization.

4.3. SUNCG: Grid Map Testing

The SUNCG Dataset provides a set of synthetic houses where a virtual camera can be placed at any pose. This feature allowed us to create a regular grid map in the space of planar poses with the camera and then to evaluate the impact of the map density in our proposal. Note that this dataset does not present appearance changes, and hence, the effect of such a characteristic cannot be evaluated in this experiment.

First, our *dense* grid map was created by selecting *KPs* with constant increments of 0.5 m in translation and 36° in rotation (refer to the red dots in Figure 4b). Then, we used subsampling to generate more grid maps for the evaluation, as described in Table 1, namely, the *sparse-position map* (subsampling half of the positions); the *sparse-rotation map* (subsampling half of the orientations); and the *sparse-position-rotation map* (subsampling both at the same time). In this case, we created PSACs with $Q = 8$ *KPs* for all maps. Additionally, we have recorded three ~30 m-long test sequences following the trajectories shown in shades of green in Figure 4b, generating a *synthetic odometry* corrupted by zero-mean Gaussian noise with $\sigma_u = (0.06\,\text{m},\ 1°)$.

The results of this experiment are shown in Figure 7, comparing the median errors in translation (left) and rotation (right) for all the descriptors employed in the previous experiment and for the described versions of the grid map. Again, we have set $P = 10^3$ particles for the PF. As can be seen, our proposal yields median errors under 0.2 m and 6° in the *dense map*, while using subsampled maps hinders the process of localization. It can be noted that subsampling exclusively on rotations does not worsen the accuracy, while subsampling positions has a noticeable impact on the overall performance. Consequently, PSACs prove to handle information sparseness more efficiently in orientation than in position. Not surprisingly, subsampling in both position and orientation clearly achieves the worst localization performance due to the combined loss of information.

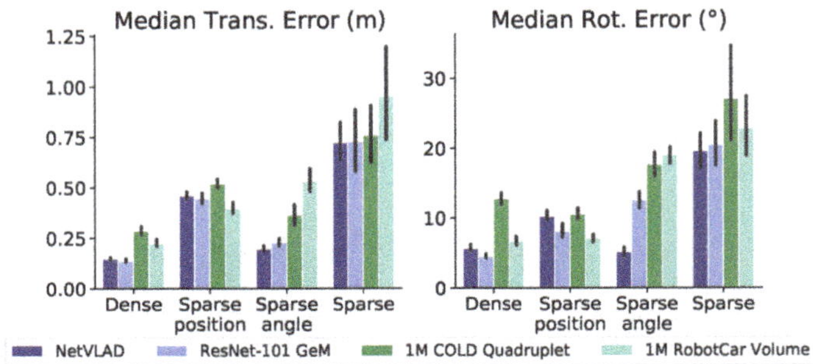

Figure 7. Median tracking errors of our proposal for the test sequences on the SUNCG generated maps, with 10^3 particles.

Regarding the holistic descriptors, all of them again demonstrate a similar behavior for each subsampling case, with 1M RobotCar Volume performing worst. ResNet-101 GeM and NetVLAD, in turn, achieve the best performance.

These results demonstrate that uniform grid sampling is a rough strategy for mapping environments, achieving results highly dependent on the sampling density. Besides, the construction of such maps with real robots becomes largely time-consuming, mostly being realizable when using virtual environments. Future work should investigate more elaborated strategies, designed to fulfill more adequate criteria concerning the map creation, ultimately pursuing an optimal approximation of the DM geometry.

4.4. Comparative Study

Finally, we compare the localization performance between our proposal and state-of-the-art appearance-based methods of diverse nature in both datasets. For the setup of our method, we selected the NetVLAD descriptor due to its performance against appearance changes, and added the *KPs* every 20 pairs for the COLD dataset, as it represents a fair trade-off between accuracy and the number of *KPs* employed.

These are the appearance-based localization methods involved in the comparison:

- **Gaussian Process Particle Filter (GPPF)** [42], configured with $P = 10^3$ particles.
- The **Pairwise Relative Pose estimator (PRP)** presented in [31]: a CNN-based regressor that estimates the pose transform between the query and the 5 most similar map images obtained through PR.
- The **Network flow** solution proposed in [37]: a sequential sparse localization method that includes *uniform* and *flow-based* mapping, both considered in this study. In order to make the results comparable, we modified its outcome, which is sparse, to pro-

duce continuous estimations. For that, we used the following weighting after the bipartite matching:

$$\mathbf{x}_i = \frac{\sum_j^{N_k} \frac{\mathbf{y}_i}{\mathbf{y}_{ij}} \mathbf{x}_i}{\sum_j^{N_k} \frac{\mathbf{y}_i}{\mathbf{y}_{ij}}} \tag{10}$$

where $\mathbf{x}_j$ is the pose of the each of the $N_k = 5$ most contributing *KPs*, $\mathbf{y}_{ij}$ is the flow connecting the *i*-th query and the *j*-th *KP*, and $\mathbf{y}_i = \sum_j^{N_k} \mathbf{y}_{ij}$ represents the query flow from the nearer *KPs* (refer to [37] for further info).

- **Our approach**, configured with $P = 10^3$ particles.

Table 2 shows the compared performance between all the described algorithms after twenty runs for each scenario. Note that, apart from the median errors, we included the percentage of correctly localized frames along the trajectory, showing the tracking robustness and relocalization potential. Concretely, a frame is considered to be correctly localized when the distance between the estimate and its true pose is below (0.5 m, 10°).

Table 2. Comparative median position and rotation errors, and % of correctly localized frames (m, $°$, (%)) of different state-of-the-art appearance-based localization methods. A frame is correctly localized when the distance between the estimate and its true pose is below (0.5 m, 10°) (L: sequences where the tracking got lost. N/A: not applicable). PRP—Pairwise Relative Pose estimator; PR—Place Recognition. In bold, best result for each sequence.

Dataset	Map	Sequence	GPPF [18,42] + Unif. Sampl	PRP CNN [31] + NetVLAD PR	Network Flow [37] + Unif. Sampl.	Network Flow [37] + Flow Sampl	Our Method C.S.
COLD Database	Samp. 20	Night std	L	1.17, 10.94 (8%)	**0.19**, 4.33 (66%)	0.26, 4.66 (60%)	0.2, **3.08 (85%)**
		Cloudy std	L	1.93, 14.59 (4%)	0.31, **4.76 (57%)**	0.36, 5.29 (46%)	**0.3**, 5.82 (56%)
		Sunny std	L	2.2, 14.75 (3%)	0.36, 6.91 (40%)	0.42, 7.7 (33%)	**0.23, 5.08 (66%)**
		Night ext	L	1.27, 11.07 (8%)	0.22, 3.88 (69%)	0.26, 4.36 (61%)	**0.17, 3.38 (82%)**
		Cloudy ext	L	1.46, 12.17 (6%)	0.22, 4.13 (60%)	0.31, 5.12 (52%)	**0.2, 3.48 (82%)**
		Sunny ext	L	2.11, 16.59 (2%)	0.3, 6.95 (50%)	0.35, 8.14 (39%)	**0.28, 6.67 (54%)**
SUNCG	Dense	Test sequence	1.07, 6.07 (17%)	0.75, 12.14 (13%)	N/A	N/A	**0.15, 5.69 (60%)**
	Sparse pos		1.21, 9.16 (7%)	1.14, 19.76 (2%)	N/A	N/A	**0.46, 4.30 (51%)**
	Sparse rot		1.24, 12.36 (2%)	0.89, 19.76 (6%)	N/A	N/A	**0.20, 5.15 (57%)**
	Sparse pos-rot		1.62, 20.16 (0%)	1.51, 24.51 (1%)	N/A	N/A	**0.72, 18.58 (19%)**

As can be seen, the challenging radiometric conditions in the COLD-Freiburg database caused the GPPF method to lose tracking, while the PRP estimator achieved low accuracy as a result of not exploiting the trajectory sequentiality, performing PR at every time-step instead. In turn, the solutions based on Network flow provide very accurate estimations in general, with the best results achieved by the uniformly sampled map under favorable conditions (i.e., *night* and *cloudy* sequences) and slightly worse in the case of severe appearance changes (i.e., *sunny* sequences). Our proposal, in contrast, demonstrates providing consistent results regardless of the appearance settings, achieving similar results to the Network flow solution in favorable conditions and outperforming all other methods under challenging radiometric settings.

In the case of the SUNCG dataset, the formulation proposed by the Network flow is incompatible with grid maps covering multiple rotations at the same location, as they are conceived to work only with positions. In turn, PRP and GPPF obtain low performance, even worsened in subsampled maps, while our proposal achieves the best results.

Despite its similarity with our approach, GPPF has shown to be unable to achieve robust localization due to the abovementioned issues: (i) deficiencies from considering a single pose–descriptor mapping for the whole environment, (ii) the absence of a relocalization process, and (iii) its limitation to environments without radiometric changes.

In conclusion, the presented comparison proves that these state-of-the-art localization methods based on appearance cannot provide both consistent and accurate localization estimations while operating within maps of diverse nature and captured under different appearance conditions. Our method, in turn, achieves higher performance in these conditions in terms of precision and robustness, showing notable results given its pure appearance nature.

5. Conclusions

We have presented a system for appearance-based robot localization that provides accurate, continuous pose estimations for camera navigation within a 2D environment under diverse radiometric conditions. Our proposal relies on the assumption that image global descriptors form a manifold articulated by the camera pose that adequately approximates the Image Manifold. This way, we gather pose–descriptor pairs from a lightweight map in order to create locally smooth regions called *Patches of Smooth Appearance Change* (PSACs) that shape, piecewise, the Descriptor Manifold geometry. Additionally, we robustly deal with appearance changes by modeling the descriptor variations under a white noise distribution.

We implemented a sequential camera tracking system built upon a Gaussian Process Particle Filter, which allows for multihypothesis pose estimation. Thus, our system optimizes a specific GP for each PSAC, subsequently being employed to define a local observation model of the descriptor for the Particle Filter. Furthermore, our method includes a relocalization process based on PR in case of tracking loss.

A first set of experiments has shown our proposal's error baseline in different environments and for a selection of holistic descriptors, revealing the most suitable configuration for our system. Finally, we have presented a comprehensive evaluation of the localization performance, showing that our approach outperforms state-of-the-art appearance-based localization methods in both tracking accuracy and robustness, even using images with challenging illuminations, yielding median errors below 0.3 m and 6°. Consequently, we have proven that pure appearance-based systems can produce continuous estimations with promising results in terms of accuracy, while working with lightweight maps and achieving robustness under strong appearance changes.

Future work includes research about (i) building the appearance map in an optimal way and wisely selecting where to sample the Descriptor Manifold; (ii) the design of a novel holistic descriptor that is more adequate to perform pose regression while keeping high invariance to radiometric changes.

Author Contributions: Conceptualization, A.J., F.-A.M., and J.G.-J.; methodology, A.J.; software, A.J.; validation, A.J.; investigation, A.J.; writing—original draft preparation, A.J., F.-A.M., and J.G.-J.; writing—review and editing, A.J., F.-A.M., and J.G.-J.; supervision, F.-A.M. and J.G.-J.; funding acquisition, J.G.-J. All authors have read and agreed to the published version of the manuscript.

Funding: This research was funded by: Government of Spain grant number FPU17/04512; by the "I Plan Propio de Investigación, Transferencia y Divulgación Científica" of the University of Málaga; and under projects ARPEGGIO (PID2020-117057) and WISER (DPI2017-84827-R) financed by the Government of Spain and European Regional Development's funds (FEDER).

Institutional Review Board Statement: Not applicable.

Informed Consent Statement: Not applicable.

Data Availability Statement: Not applicable.

Acknowledgments: We gratefully acknowledge the support of NVIDIA Corporation with the donation of the Titan X Pascal used for this research.

Conflicts of Interest: The authors declare no conflict of interest. The funders had no role in the design of the study; in the collection, analyses, or interpretation of data; in the writing of the manuscript, or in the decision to publish the results.

Abbreviations

The following abbreviations are used in this manuscript:

PSAC	Patch of Smooth Appearance Change
PR	Place Recognition
GP	Gaussian Process
GPPF	Gaussian Process Particle Filter
PF	Particle Filter
SLAM	Simultaneous Localization and Mapping
IM	Image Manifold
DM	Descriptor Manifold
DL	Deep Learning
KP	Key-pair
CS	Constant Sampling

References

1. Lowe, D.G. Object recognition from local scale-invariant features. In Proceedings of the Seventh IEEE International Conference on Computer Vision, Kerkyra, Greece, 20–27 September 1999; Volume 2, pp. 1150–1157.
2. Rublee, E.; Rabaud, V.; Konolige, K.; Bradski, G. ORB: An efficient alternative to SIFT or SURF. In Proceedings of the 2011 International Conference on Computer Vision, Barcelona, Spain, 6–13 November 2011; pp. 2564–2571.
3. Mur-Artal, R.; Montiel, J.M.M.; Tardos, J.D. ORB-SLAM: A versatile and accurate monocular SLAM system. *IEEE Trans. Robot.* **2015**, *31*, 1147–1163. [CrossRef]
4. Sattler, T.; Leibe, B.; Kobbelt, L. Efficient & effective prioritized matching for large-scale image-based localization. *IEEE Trans. Pattern Anal. Mach. Intell.* **2016**, *39*, 1744–1756.
5. Gomez-Ojeda, R.; Moreno, F.A.; Zuñiga-Noël, D.; Scaramuzza, D.; Gonzalez-Jimenez, J. PL-SLAM: A stereo SLAM system through the combination of points and line segments. *IEEE Trans. Robot.* **2019**, *35*, 734–746. [CrossRef]
6. Sattler, T.; Torii, A.; Sivic, J.; Pollefeys, M.; Taira, H.; Okutomi, M.; Pajdla, T. Are large-scale 3d models really necessary for accurate visual localization? In Proceedings of the IEEE Conference on Computer Vision and Pattern Recognition, Honolulu, HI, USA, 21–26 July 2017; pp. 1637–1646.
7. Oliva, A.; Torralba, A. Modeling the shape of the scene: A holistic representation of the spatial envelope. *Int. J. Comput. Vis.* **2001**, *42*, 145–175. [CrossRef]
8. Arandjelovic, R.; Gronat, P.; Torii, A.; Pajdla, T.; Sivic, J. NetVLAD: CNN architecture for weakly supervised place recognition. In Proceedings of the IEEE Conference on Computer Vision and Pattern Recognition, Las Vegas, NV, USA, 27–30 June 2016; pp. 5297–5307.
9. Cummins, M.; Newman, P. FAB-MAP: Probabilistic localization and mapping in the space of appearance. *Int. J. Robot. Res.* **2008**, *27*, 647–665. [CrossRef]
10. Milford, M.J.; Wyeth, G.F. SeqSLAM: Visual route-based navigation for sunny summer days and stormy winter nights. In Proceedings of the 2012 IEEE International Conference on Robotics and Automation, Saint Paul, MN, USA, 14–18 May 2012; pp. 1643–1649.
11. Pepperell, E.; Corke, P.I.; Milford, M.J. All-environment visual place recognition with SMART. In Proceedings of the 2014 IEEE International Conference on Robotics and Automation (ICRA), Hong Kong, China, 31 May–7 June 2014; pp. 1612–1618.
12. Radenović, F.; Tolias, G.; Chum, O. Fine-tuning CNN image retrieval with no human annotation. *IEEE Trans. Pattern Anal. Mach. Intell.* **2018**, *41*, 1655–1668. [CrossRef] [PubMed]
13. Crowley, J.L.; Pourraz, F. Continuity properties of the appearance manifold for mobile robot position estimation. *Image Vis. Comput.* **2001**, *19*, 741–752. [CrossRef]
14. Ham, J.; Lin, Y.; Lee, D.D. Learning nonlinear appearance manifolds for robot localization. In Proceedings of the 2005 IEEE/RSJ International Conference on Intelligent Robots and Systems, Edmonton, AB, Canada, 2–6 August 2005; pp. 2971–2976.
15. He, X.; Zemel, R.S.; Mnih, V. Topological map learning from outdoor image sequences. *J. Field Robot.* **2006**, *23*, 1091–1104. [CrossRef]
16. Gomez-Ojeda, R.; Lopez-Antequera, M.; Petkov, N.; Gonzalez-Jimenez, J. Training a convolutional neural network for appearance-invariant place recognition. *arXiv* **2015**, arXiv:1505.07428.
17. Ko, J.; Fox, D. GP-BayesFilters: Bayesian filtering using Gaussian process prediction and observation models. *Auton. Robot.* **2009**, *27*, 75–90. [CrossRef]
18. Lopez-Antequera, M.; Petkov, N.; Gonzalez-Jimenez, J. Image-based localization using Gaussian processes. In Proceedings of the 2016 International Conference on Indoor Positioning and Indoor Navigation (IPIN), Alcala de Henares, Spain, 4–7 October 2016; pp. 1–7.
19. Roweis, S.T.; Saul, L.K. Nonlinear dimensionality reduction by locally linear embedding. *Science* **2000**, *290*, 2323–2326. [CrossRef]
20. Tenenbaum, J.B.; De Silva, V.; Langford, J.C. A global geometric framework for nonlinear dimensionality reduction. *Science* **2000**, *290*, 2319–2323. [CrossRef] [PubMed]

21. Pless, R.; Souvenir, R. A survey of manifold learning for images. *IPSJ Trans. Comput. Vis. Appl.* **2009**, *1*, 83–94. [CrossRef]
22. Wakin, M.B.; Donoho, D.L.; Choi, H.; Baraniuk, R.G. The multiscale structure of non-differentiable image manifolds. In Proceedings of the Wavelets XI, Proceedings of the SPIE, San Diego, CA, USA, 17 September 2005; Volume 5914, p. 59141B.
23. Lopez-Antequera, M.; Gomez-Ojeda, R.; Petkov, N.; Gonzalez-Jimenez, J. Appearance-invariant place recognition by discriminatively training a convolutional neural network. *Pattern Recognit. Lett.* **2017**, *92*, 89–95. [CrossRef]
24. Thoma, J.; Paudel, D.P.; Chhatkuli, A.; Van Gool, L. Learning Condition Invariant Features for Retrieval-Based Localization from 1M Images. *arXiv* **2020**, arXiv:2008.12165.
25. Jaenal, A.; Zuñiga-Nöel, D.; Gomez-Ojeda, R.; Gonzalez-Jimenez, J. Improving Visual SLAM in Car-Navigated Urban Environments with Appearance Maps. In Proceedings of the 2020 IEEE/RSJ International Conference on Intelligent Robots and Systems (IROS), Las Vegas, NV, USA, 24 October–24 January 2021; pp. 4679–4685.
26. Lowry, S.; Sünderhauf, N.; Newman, P.; Leonard, J.J.; Cox, D.; Corke, P.; Milford, M.J. Visual place recognition: A survey. *IEEE Trans. Robot.* **2015**, *32*, 1–19. [CrossRef]
27. Lenc, K.; Vedaldi, A. Understanding image representations by measuring their equivariance and equivalence. In Proceedings of the IEEE Conference on Computer Vision and Pattern Recognition, Boston, MA, USA, 7–12 June 2015; pp. 991–999.
28. Jaenal, A.; Moreno, F.A.; Gonzalez-Jimenez, J. Experimental study of the suitability of CNN-based holistic descriptors for accurate visual localization. In Proceedings of the 2nd International Conference on Applications of Intelligent Systems, Las Palmas de Gran Canaria, Spain, 7–9 January 2019; pp. 1–6.
29. Sattler, T.; Weyand, T.; Leibe, B.; Kobbelt, L. Image Retrieval for Image-Based Localization Revisited. In Proceedings of the British Machine Vision Conference, Surrey, UK, 3–7 September 2012; Volume 1, p. 4.
30. Torii, A.; Arandjelovic, R.; Sivic, J.; Okutomi, M.; Pajdla, T. 24/7 place recognition by view synthesis. In Proceedings of the IEEE Conference on Computer Vision and Pattern Recognition, Boston, MA, USA, 7–12 June 2015; pp. 1808–1817.
31. Laskar, Z.; Melekhov, I.; Kalia, S.; Kannala, J. Camera relocalization by computing pairwise relative poses using convolutional neural network. In Proceedings of the IEEE International Conference on Computer Vision Workshops, Venice, Italy, 22–29 October 2017; pp. 929–938.
32. Balntas, V.; Li, S.; Prisacariu, V. RelocNet: Continuous Metric Learning Relocalisation using Neural Nets. In Proceedings of the European Conference on Computer Vision (ECCV), Munich, Germany, 8–14 September 2018; pp. 751–767.
33. Ding, M.; Wang, Z.; Sun, J.; Shi, J.; Luo, P. CamNet: Coarse-to-fine retrieval for camera re-localization. In Proceedings of the IEEE International Conference on Computer Vision, Seoul, Korea, 27 October–2 November 2019; pp. 2871–2880.
34. Sünderhauf, N.; Neubert, P.; Protzel, P. Are we there yet? Challenging SeqSLAM on a 3000 km journey across all four seasons. In Proceedings of the Workshop on Long-Term Autonomy, IEEE International Conference on Robotics and Automation (ICRA), Karlsruhe, Germany, 6–10 May 2013.
35. Naseer, T.; Spinello, L.; Burgard, W.; Stachniss, C. Robust visual robot localization across seasons using network flows. In Proceedings of the Twenty-Eighth AAAI Conference on Artificial Intelligence, Quebec City, QC, Canada, 27–31 July 2014; pp. 2564–2570.
36. Naseer, T.; Burgard, W.; Stachniss, C. Robust visual localization across seasons. *IEEE Trans. Robot.* **2018**, *34*, 289–302. [CrossRef]
37. Thoma, J.; Paudel, D.P.; Chhatkuli, A.; Probst, T.; Gool, L.V. Mapping, localization and path planning for image-based navigation using visual features and map. In Proceedings of the IEEE Conference on Computer Vision and Pattern Recognition, Long Beach, CA, USA, 15–20 June 2019; pp. 7383–7391.
38. Maddern, W.; Milford, M.; Wyeth, G. CAT-SLAM: Probabilistic localisation and mapping using a continuous appearance-based trajectory. *Int. J. Robot. Res.* **2012**, *31*, 429–451. [CrossRef]
39. Rasmussen, C.E. Gaussian processes in machine learning. In *Advanced Lectures on Machine Learning*; Springer: Berlin/Heidelberg, Germany, 2004; pp. 63–71.
40. Huhle, B.; Schairer, T.; Schilling, A.; Straßer, W. Learning to localize with Gaussian process regression on omnidirectional image data. In Proceedings of the 2010 IEEE/RSJ International Conference on Intelligent Robots and Systems, Taipei, Taiwan, 18–22 October 2010; pp. 5208–5213.
41. Schairer, T.; Huhle, B.; Vorst, P.; Schilling, A.; Straßer, W. Visual mapping with uncertainty for correspondence-free localization using Gaussian process regression. In Proceedings of the 2011 IEEE/RSJ International Conference on Intelligent Robots and Systems, San Francisco, CA, USA, 25–30 September 2011; pp. 4229–4235.
42. Lopez-Antequera, M.; Petkov, N.; Gonzalez-Jimenez, J. City-scale continuous visual localization. In Proceedings of the 2017 European Conference on Mobile Robots (ECMR), Paris, France, 6–8 September 2017; pp. 1–6.
43. Doucet, A.; de Freitas, N.; Gordon, N. *Sequential Monte Carlo Methods in Practice*; Statistics for Engineering and Information Science; Springer Science & Business Media: New York, NY, USA, 2001.
44. Blanco, J.L. *A Tutorial on SE(3) Transformation Parameterizations and On-Manifold Optimization*; Technical Report; Universidad de Málaga: Málaga, Spain, 2010; Volume 3, p. 6.
45. Pronobis, A.; Caputo, B. COLD: The CoSy localization database. *Int. J. Robot. Res.* **2009**, *28*, 588–594. [CrossRef]
46. Song, S.; Yu, F.; Zeng, A.; Chang, A.X.; Savva, M.; Funkhouser, T. Semantic scene completion from a single depth image. In Proceedings of the IEEE Conference on Computer Vision and Pattern Recognition, Honolulu, HI, USA, 21–26 July 2017; pp. 1746–1754.

47. Wu, Y.; Wu, Y.; Gkioxari, G.; Tian, Y. Building generalizable agents with a realistic and rich 3D environment. *arXiv* **2018**, arXiv:1801.02209.
48. Simonyan, K.; Zisserman, A. Very deep convolutional networks for large-scale image recognition. *arXiv* **2014**, arXiv:1409.1556.
49. He, K.; Zhang, X.; Ren, S.; Sun, J. Deep residual learning for image recognition. In Proceedings of the IEEE Conference on Computer Vision and Pattern Recognition, Las Vegas, NV, USA, 27–30 June 2016; pp. 770–778.
50. GPy. GPy: A Gaussian Process Framework in Python. Since 2012. Available online: http://github.com/SheffieldML/GPy (accessed on 30 March 2021).

sensors

MDPI

Article

Calibration of Visible Light Positioning Systems with a Mobile Robot

Robin Amsters [1,*], Eric Demeester [1], Nobby Stevens [2] and Peter Slaets [1]

[1] Department of Mechanical Engineering, KU Leuven, 3000 Leuven, Belgium; Eric.demeester@kuleuven.be (E.D.); peter.slaets@kuleuven.be (P.S.)
[2] Department of Electrical Engineering, KU Leuven, 3000 Leuven, Belgium; nobby.stevens@kuleuven.be
* Correspondence: robin.amsters@kuleuven.be

Abstract: Most indoor positioning systems require calibration before use. Fingerprinting requires the construction of a signal strength map, while ranging systems need the coordinates of the beacons. Calibration approaches exist for positioning systems that use Wi-Fi, radio frequency identification or ultrawideband. However, few examples are available for the calibration of visible light positioning systems. Most works focused on obtaining the channel model parameters or performed a calibration based on known receiver locations. In this paper, we describe an improved procedure that uses a mobile robot for data collection and is able to obtain a map of the environment with the beacon locations and their identities. Compared to previous work, the error is almost halved. Additionally, this approach does not require prior knowledge of the number of light sources or the receiver location. We demonstrate that the system performs well under a wide range of lighting conditions and investigate the influence of parameters such as the robot trajectory, camera resolution and field of view. Finally, we also close the loop between calibration and positioning and show that our approach has similar or better accuracy than manual calibration.

Keywords: indoor positioning; visible light positioning; sensor fusion; mobile robot; calibration

Citation: Amsters, R.; Demeester, E.; Stevens, N.; Slaets, P. Calibration of Visible Light Positioning Systems with a Mobile Robot. *Sensors* **2021**, *21*, 2394. https://doi.org/10.3390/s21072394

Academic Editor: Simon Tomažič

Received: 12 March 2021
Accepted: 27 March 2021
Published: 30 March 2021

1. Introduction

Since their introduction, Global Navigation Satellite Systems (GNSS) are the enabling technology for applications such as navigation, autonomous vehicles and emergency services. While GNSS can provide worldwide coverage and require only a receiver to use, they are typically not useful for indoor spaces. On one hand, building walls significantly reduce the signal strength, often making positioning impossible or reducing the accuracy [1]. However, even the nominal accuracy of GNSS (around 5 m [2]) is insufficient for indoor positioning, where an error of a couple of meters can mean that the user is located in one of several rooms. In order to provide indoor location, researchers have developed many Indoor Positioning Systems (IPS), yet a single standard like GNSS was not achieved. Indoor environments come in many different varieties and can favor different positioning technologies. Current systems are often based on Wi-Fi in order to reduce infrastructure cost; however, their accuracy is limited to a couple of meters [3]. Other technologies such as Ultra-WideBand (UWB) [4] and ultrasound [5] can provide much higher accuracy (centimeters), at the cost of additional specialized infrastructure.

With the introduction of solid-state lighting, a new type of indoor positioning has emerged. In Visible Light Positioning (VLP), light intensities are modulated at speeds imperceptible to the human eye, which allows for a one way transmission from transmitter to receiver. Similar to other positioning systems, the Received Signal Strength (RSS) or signal travel time can be used to determine the receiver location. Due to the local character of light, the influence of multipath is significantly reduced, resulting in an accuracy that can be as low as a couple of centimeters [6]. Existing lighting infrastructure can also be reused for positioning, thereby reducing the overall system cost significantly. These advantages led to an increasing research interest in recent years. However, the installation of new VLP systems remains an important issue. The majority of indoor positioning systems require some form of calibration. For example, systems that use range measurements to determine the receiver position via triangulation assume the locations of all transmitters to be known. Manually measuring transmitter locations can be a cumbersome process, as the transmitters are often mounted on ceilings and walls [7]. Fingerprinting systems also require a calibration procedure, in order to build an RSS map that can later be used for positioning. These site surveys can be lengthy and labor-intensive processes. Additionally, this RSS map may have to be updated when changes occur in the environment.

Several calibration procedures have been proposed for IPS using technologies such as UWB [8], Wi-Fi [9,10] and Radio Frequency Identification (RFID) [11,12]. In the field of visible light positioning, little literature is available on this subject. Our previous work [13] proposed a proof-of-concept calibration procedure with a mobile robot. In that procedure, the total number of lights needed to be determined manually [13]. Counting the number of transmitters is significantly less time-consuming compared to manually measuring the positions. However, it is still a tedious process that is prone to errors. In this work, we therefore introduce an improved calibration algorithm. Detected light sources are filtered based on their measured coordinates, as well as their place in the frequency spectrum. Using this new procedure, the number of light sources is no longer required. Additionally, accuracy is significantly improved. As [13] is a proof-of-concept, much remains unknown about the robustness of the approach. For example, which parameters have an effect on the accuracy of the procedure? To find out, we investigate the impact of a variety of factors on the calibration procedure. Finally, the goal of a calibration procedure is to prepare the system for positioning. The relation between calibration errors and positioning errors may be complex. In order to determine whether our system has satisfactory performance, we use the calibrated parameters for positioning. Following this approach, we close the loop between calibration and positioning and enable high-performance systems that are easy to deploy.

Our main contributions can therefore be summarized as follows:

- Improved calibration procedure with nearly double the accuracy compared to previous work [13].
- Extensive parameter study that investigates the influence of the transmitter waveform, lighting conditions, robot trajectory, camera resolution, Field Of View (FOV) and transmitter–receiver distance.
- The calibrated parameters are used for positioning. The accuracy of the resulting location data are evaluated independently from the calibration.

The rest of this paper is structured as follows: Section 2 describes related work, and Section 3 introduces the materials and methods used in this paper. Experimental results are presented in Section 4 and discussed in Section 5. Finally, a conclusion is drawn in Section 6.

2. Related Work

Table 1 provides an overview of calibration procedures proposed for different types of indoor positioning systems. In Table 1, "positioning technology" refers to the technology that is actually used for positioning (after the calibration has completed). During the calibration itself, other signals such as RGB-D cameras [10] or PDR [14] may be used. The following section will describe the broad categories of calibration methods in more detail.

Table 1. Calibration approaches for indoor positioning systems.

Reference	Positioning Technology	Calibration Method	Calibrated Parameters
		Calibration of fingerprinting systems	
[9]	Wi-Fi	Robot site survey	RSS map
[10]	Wi-Fi	Robot site survey	RSS map
[15]	Wi-Fi	Robot site survey	RSS map
[16]	Wi-Fi/LTE	signalSLAM	RSS map
[17]	Magnetic field	signalSLAM	RSS map
[14]	Wi-Fi	signalSLAM	RSS map
[18]	Wi-Fi + PDR	crowdsourcing	RSS map
[19]	Wi-Fi	crowdsourcing	RSS map
[20]	Wi-Fi/Magnetic/PDR	crowdsourcing	RSS map
[21]	Wi-Fi	crowdsourcing	RSS map
		Calibration of range based systems	
[7]	Ultrasound	Network optimization	Beacon coordinates
[8]	UWB	Autocalibration (initial solution), Followed by Network optimization	Range bias Beacon coordinates
[22]	Ultrasound	known locations	Beacon coordinates
[23]	Not specified	known locations	Beacon coordinates
[24]	UWB	known locations	Beacon coordinates
[25]	Wi-Fi	known locations	Beacon coordinates
[26]	UWB	Network optimization	Beacon coordinates
[27]	Ultrasound	Autocalibration Network optimization	Beacon coordinates
[28]	Ultrasound	known locations + Autocalibration	Beacon coordinates
[29]	Ultrasound	Network optimization	Beacon coordinates
		Calibration of VLP systems	
[30]	VLP	known locations	Receiver parameters
[31]	VLP	known locations	Receiver parameters + Beacon coordinates
[32]	VLP	known locations	Receiver parameters
[33]	VLP	known locations	Channel model
[34]	VLP	known locations	Channel model
[35]	Wi-Fi/ambient light/magnetic field	signalSLAM	RSS map
[36]	VLP	signalSLAM	RSS map
[37]	VLP	known locations	Channel model
This work	VLP	Robot site survey	Beacon coordinates

Fingerprinting-based IPS operate in two stages. In the first (offline) stage, a signal strength map is constructed. RSS values are measured at known locations throughout the entire space. It is possible to record just one type of signal (e.g., Wi-Fi). However, accuracy is generally improved by including multiple sources of information (e.g., magnetic field, Bluetooth, etc.) [38]. Signals already present in the environment are often used, in order to avoid the need for additional infrastructure. In the second (online) stage, the receiver location is unknown and one or more RSS values are measured. By matching the current signal fingerprint to the database, the receiver position is recovered. Contrary to triangulation-based IPS, fingerprinting approaches do not require transmitter coordinates. To ensure positioning accuracy, it is however important that the signal strength map is accurate. The map may also have to be updated periodically, if changes to the environment are made.

We distinguish four methods to construct the signal strength map:

- Manual site survey
- Robot site survey
- SignalSLAM
- Crowdsourcing

In manual site surveys, a trained expert records signal fingerprints at known locations. The entire space needs to be visited by the surveyor and as mentioned before, this process may have to be repeated. Manual site survey is time-consuming and labor-intensive and is thus not always practical in large indoor spaces [35]. The use of mobile robots has therefore been proposed to simplify this task. Mobile robots have been used to collect fingerprints for RFID [11,12] and Wi-Fi systems [10]. Some authors have even proposed algorithms that enable a robot to collect data without human intervention [9,15]. These navigation algorithms were relatively simple and did not follow the optimal trajectory (in terms of accuracy or time required), but they did succeed in covering the space eventually.

The goal of Simultaneous Localization And Mapping (SLAM) is to reconstruct a map of the environment, while simultaneously estimating the trajectory of the observer relative to that map. Solutions to the SLAM problem are most commonly based on Bayesian filtering [39]. SLAM algorithms have mainly been used for robotics applications, as earlier implementations required expensive sensors such as laser scanners (LIDAR) or depth cameras [40]. Recently, researchers started using sensors embedded in conventional smartphones to construct signal strength maps. This approach is sometimes also referred to as signalSLAM. Pedestrian Dead Reckoning (PDR) is often used to obtain a rough estimate of the user's trajectory, and drift is corrected by using absolute location fixes (for example, from GNSS signals or near-field communication tags) [16]. Alternatively, other signals of opportunity such as Wi-Fi, magnetic field or even ambient light [35] can be used to compensate PDR drift. When using signalSLAM to calibrate fingerprinting IPS, the main goal is to reconstruct the trajectory of the user and to add the measured signal strengths to the map based on that trajectory. Recent approaches tend to use a modified version of graphSLAM [36]. The main challenge with graph-based signalSLAM is the reduction of false positives when performing loop closures [14,16,17].

SignalSLAM calibration still requires surveyor to visit the entire indoor space. It is more efficient compared to manual site survey, as the surveyor can walk around continuously. In manual calibration, the surveyor has to stop and record his or her location periodically. Crowdsourcing approaches attempt to improve efficiency even further by removing the dedicated surveyor entirely. Initially, users can go about their regular tasks, while the systems collect both inertial and signal strength data from their smartphones in the background. As more data are collected, these systems obtain a more complete picture of the indoor environment, and position accuracy increases. In contrast to single site surveys, the map can continuously be updated. Crowdsourcing presents a number of interesting advantages, yet some challenges still remain. Kim et al. [18] assumed the initial location of the user was known and suggested it can be obtained from GNSS when the user enters the building. In contrast, the system described in [19] did not require the initial position, stride length or phone placement. Instead, a map of the environment was used to impose constraints that can filter improbable locations. The obtained trajectories were optimized through backpropagation, and Wi-Fi signal strength was added to the map based on the optimized path. In the work of Wang et al. [20], seed landmarks were extracted from the floor plan (e.g., doors), which can be used to obtain global observations. Additional landmarks were learned as more data entered the system. Yang et al. [21] first transformed the map into a stress-free floor plan, which is a high dimensional space in which the distance between two points reflects their walking distance (taking constraints such as walls into account). The similarity between the stress-free floor plan and the fingerprint space was used to label RSS signatures with their real locations. Crowdsourcing-based calibration does require users to give up their personal data, which may be an important barrier to some. Moreover, the approaches discussed above often required a floor plan, which may not always be available. Finally, the accuracy of both signalSLAM and crowdsourcing is typically low (in the range of several meters). Due to the relatively low quality input data (PDR and radio frequency signals), it is challenging to obtain robust and accurate systems with signalSLAM or crowdsourcing.

The calibration methods discussed so far are only applicable to fingerprinting-based IPS. Another category of positioning systems obtains the receiver position based on ranging. The travel time of a signal or the signal strength are used to determine the distance between transmitter and receiver. From the measured distances, the receiver position can then be obtained via triangulation. These types of IPS require accurate knowledge of the transmitter locations. Depending on the positioning technology used, additional parameters may also be required. For example, UWB systems often correct the bias on the distance measurements [8]. VLP systems based on RSS sometimes calibrate the gain [30] or Lambertian emission order [33]. For the calibration of range-based systems, we can again distinguish a few possible methods:

- Manual measurements
- Known locations of receiver(s) and/or transmitters
- Interbeacon ranging (autocalibration)
- Network optimization

Similar to fingerprinting systems, range-based IPS can be calibrated manually. In this case, the transmitter locations would be measured relative to some reference with rulers or laser-based measurement devices. While measuring the transmitter locations manually requires less work than performing a manual site survey for fingerprinting, it is still a tedious process. Transmitters are often mounted on the ceiling, which can make the process somewhat inconvenient. Ranging systems can also be calibrated based on known receiver locations, which may be easier to obtain than the transmitter coordinates [22–24]. However, ground truth measurements of the receiver locations are still required, which often requires an additional positioning system. Moreover, errors on the receiver location while calibrating will subsequently lead to errors on the transmitter locations. If sufficient transmitter positions are known, the others can be extrapolated without extra measurements [25].

Some IPS can use the same ranging techniques that enable receiver positioning to obtain the distance between transmitters, from which the transmitter locations can also be obtained [27,28]. These interbeacon ranging techniques (sometimes also referred to as autocalibration) do assume that beacons can communicate with each other. Additionally, the transmitters must be placed sufficiently close together such that they are within measurement range of each other, which may disqualify them from positioning technologies such as Bluetooth.

Finally, range-based IPS can also be calibrated based on a set of transmitter–receiver distances. If the quantity of data is large enough, no receiver or transmitter locations are required; a set of range measurements is sufficient. Calibration can then be formulated as an optimization problem that minimizes the residual of the trilateration equations [7,8,26,27,29]. Results from these approaches are not always unique, for example, in the case of rotational symmetry. Additionally, accuracy of the solution can be heavily dependent on the initial conditions [29].

Both ranging and fingerprinting can be used for VLP. Ranging is generally more accurate and robust. However, as the transmitters are lights that also illuminate the space, they are generally mounted on the ceiling and are pointing downwards. Therefore, transmitters likely do not have a line of sight (LOS) to each other. Even when VLP transmitters are within range of each other, they lack the necessary hardware for receiving signals. Therefore, autocalibration methods cannot be used by most conventional VLP systems. In fact, VLP calibration in general is not yet explored in depth. Rodríguez-Navarro et al. [30] proposed a method for calibrating the electrical parameters of a VLP amplification circuit. They performed an extensive parameter study and found that manufacturing tolerances on the resistors and capacitors contributed most to positioning errors due to incorrect calibration. By performing multiple intensity measurements at known locations, a system of equations can be constructed. The solution that minimizes the error provides the optimal calibration of the receiver parameters. In [31,32], calibration of transmitter coordinates based on known receiver locations was proposed. Similarly, Ref [33,37] were able to calibrate the channel model based on known receiver locations. However, these works either did not indicate how the receiver position should be obtained [31–33] or used an additional positioning system to obtain it [37]. Note that not all VLP systems require a calibrated channel model. Camera-based implementations such as [41] only detect the relative position of the light to the camera center, while photodiode-based systems use the signal strength to obtain the transmitter–receiver distance. Camera-based VLP systems therefore only need the location of each transmitter. However, the channel model of VLP is relatively well known; therefore, model-based fingerprinting is sometimes also possible given the transmitter locations [34].

In this work, we will focus on the calibration of light source locations and identifiers without prior knowledge or additional positioning systems, of which there are few examples. Liang and Liu [35] crowdsourced the construction of a signal strength map of opportunistic signals. Similar to [14,16], user trajectories were obtained with the help of a modified graphSLAM algorithm. Contrary to similar works, they also mapped the location of light sources and used them as landmarks in the positioning stage. However, as the lights were not modulated, their identity cannot be uniquely determined, resulting in a relatively low positioning accuracy (several meters) [35]. Additionally, unmodulated light sources are not easily distinguished from sunlight, as both increase the ambient lighting. In contrast, Yue et al. [36] did use modulated Light Emitting Diodes (LED). A modified version of graphSLAM was again used to construct the signal strength database. Absolute location fixes were obtained by detecting doors with changes in light intensity and magnetic field strength. Following calibration, positioning was performed by fusing PDR with fingerprint observations via a Kalman filter. Positioning accuracy after calibration was about 0.8 m on average, which is an improvement of approximately 70% over Wi-Fi-based fingerprinting under the same conditions. However, in rare occasions the positioning error can exceed 2 m.

3. Materials and Methods

In our proposed system, specific hardware was placed between the power lines and the lights, which modulated the intensity of each LED at a unique frequency (see Figure 1). Contrary to VLC, no data was transmitted. Instead, we used the modulated lights as a landmark. Detection and identification of the light itself is not the focus of this paper but was explained in our previous work [13]. The main variables of interest were the identity (i.e., frequency) and the coordinates in the camera frame of the light source. If the position of each light is known beforehand, this information can be used to obtain the receiver location. However, in this work, we will focus on the calibration itself. We chose frequency division for its easy implementation (see Section 4.2.1), but another modulation technique could also be used. So long as the light sources can be detected (within the field of view) and identified, the calibration procedure remains applicable.

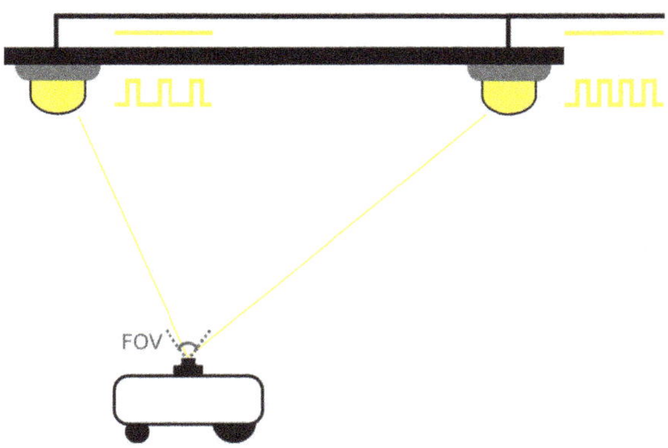

Figure 1. Model of the proposed Visible Light Positioning (VLP) system.

3.1. Experimental Setup

In order to evaluate the proposed calibration procedure, we used the same experimental setup as our previous works [13,41] (see Figure 2). Four VLP transmitters were mounted at a height of approximately 1.5 m. Light intensity of every transmitter was modulated by a Metal-Oxide Semiconductor Field-Effect Transistor (MOSFET) that was connected to a signal generator. Every LED had a unique frequency between approximately 1.5 kHz and 5 kHz, so that both low and high frequency modulation can be evaluated. The complete methodology used to obtain suitable transmitter frequencies was detailed in [13]. Table 2 lists the selected modulation frequencies, along with the other main hardware specifications.

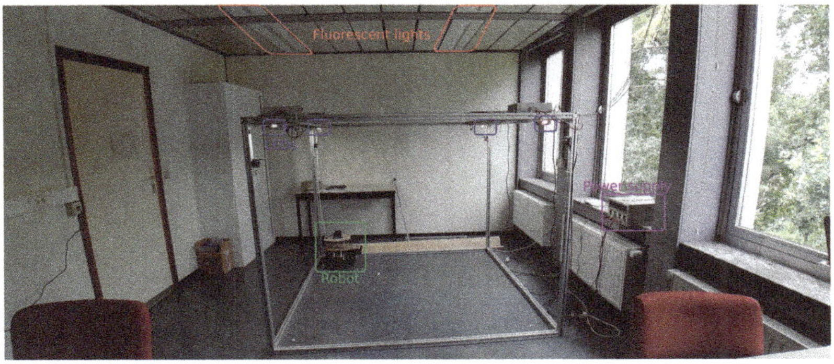

Figure 2. Experimental setup used in this paper, which was also used in our previous work [13,41].

Table 2. Hardware specifications.

Specifications	Value	Unit
	LED specifications	
Frequency LED 1	1.57	kHz
Frequency LED 2	2.03	kHz
Frequency LED 3	2.87	kHz
Frequency LED 4	4.92	kHz
	OpenMV camera specification	
Model	M7	
Resolution	640×480	pixels
Frame rate	30	fps
	RPI camera specification	
Model	v2	
Resolution	3280×2484	pixels
Frame rate	$\approx$0.75	fps
	LIDAR specifications	
Model	RPLIDAR A1	
Measurement range	12	m
Measurement frequency	5.5	Hz
Angular resolution	1	°
Distance resolution	0.2	cm
	Robot specifications	
Model	Kobuki	
Gyroscope measurement range	110	°/s
Odometry resolution	2578.33	tics/revolution
	Computer specifications	
Model	Dell OptiPlex 5050	
CPU	3.6×4	GHz
RAM	16	GB

A mobile robot was equipped with a custom sensor platform that contained a laser scanner and a camera (see Figure 3). Two different cameras will be investigated in this paper. Initially, we used the OpenMV M7 camera during experiments, as the low resolution allowed us to more quickly process the images and therefore speed up development. Later experiments used the Raspberry Pi (RPi) camera module. Similar to the OpenMV M7, the RPi camera allows a high flexibility over settings such as the exposure time. However, the RPi camera has a much greater resolution compared to the OpenMV camera. Section 4 will investigate whether this resolution improves accuracy. The sensor platform also contained a laptop, which recorded all data so that calibration could be performed offline. When the RPi camera module was used, a Raspberry Pi single-board computer recorded the images separately from the laptop, as the RPi camera does not have a Universal Serial Bus (USB) interface. The cost of the main components of the experimental setup is detailed in Appendix A. The robot platform was driven by a human operator via a remote control.

Figure 3. Mobile robot with custom sensor platform used in this paper, which was also used in our previous work [13,41].

3.2. Calibration Procedure

Figure 4 provides a graphical overview of the calibration procedure. The parameters used to obtain the results in Section 4 are listed in Table 3. The initial steps of the improved procedure were still the same as in [13]. Due to the rolling shutter of our Complementary Metal-Oxide-Semiconductor (CMOS) camera, modulated light sources are visible as stripe patterns in the images. The width of the stripes is proportional to the transmitter frequency [42]. The complete image processing pipeline is detailed in our previous work [13] and returns the frequency and pixel coordinates of the lights as output. The effects of lens distortion are largest at the edges in a picture. Therefore, only the images where a light was detected close to the image center were processed further. Measurements of the laser scanner were used to reconstruct the followed trajectory and a map of the environment by using the Google Cartographer SLAM algorithm [43]. By combining the trajectory of the robot with the detected light sources, we obtained a map of the environment with the light source locations relative to the map frame.

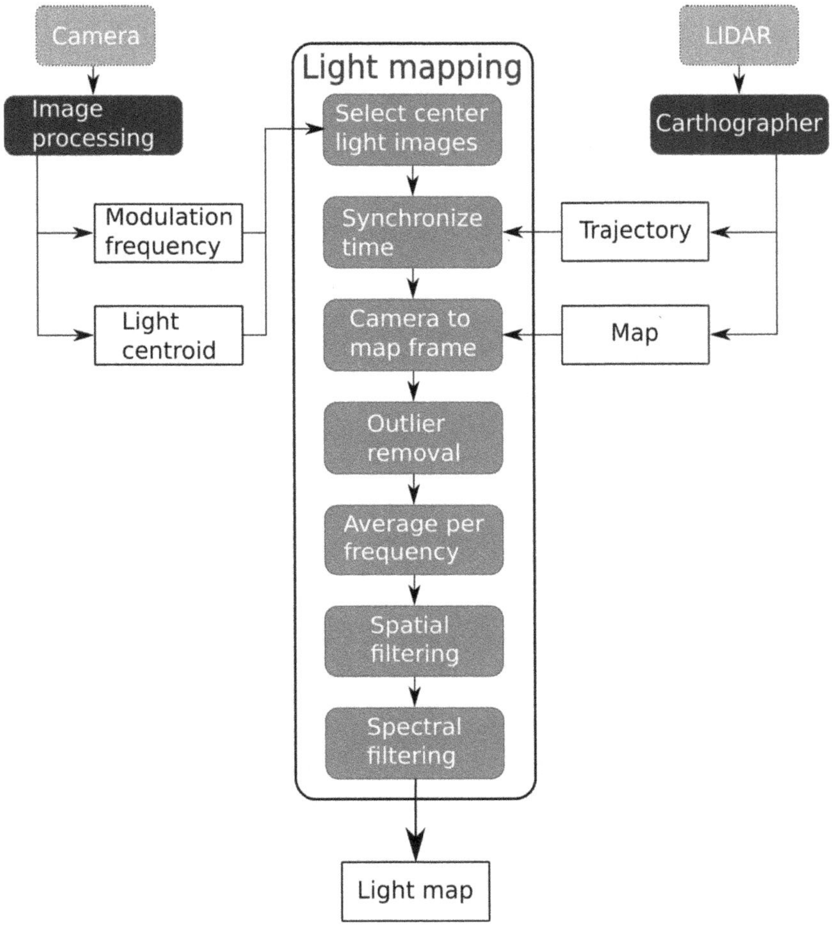

Figure 4. Calibration procedure overview.

Table 3. Calibration parameters.

Parameter	Value	Unit
Minimum physical distance between lights	1	m
Minimum distance in frequency spectrum	0.1	kHz
Minimum number of occurrences	2	/
Maximum horizontal distance from image center (OpenMV)	100	pixels
Maximum horizontal distance from image center (RPi)	300	pixels

At this point, we have obtained the world coordinates of detected light sources, in addition to their frequencies. Unfortunately, the same LED is occasionally labeled with different frequencies, depending on the image. The Canny edge detection step may calculate the stripe pattern to be one pixel larger or smaller than the actual value, resulting in a small spread in the frequency spectrum (see Figure 5). Additionally, the detected world coordinates of the light sources may also not be a single point, due to noise on the centroid detection and robot localization. The light may therefore appear as a cluster of coordinates in the light map (see Figure 5). Previously, we solved this problem by averaging the world coordinates per detected frequency. Then, the frequencies which were detected most often were kept, depending on the number of light sources. For example, in our experimental setup four lights are present; therefore, the calibration procedure selected the four frequencies that were detected most often, and the rest were discarded. This approach has the disadvantage that one first needs to know how many light sources are installed in the environment, and therefore some manual measurements may still be required.

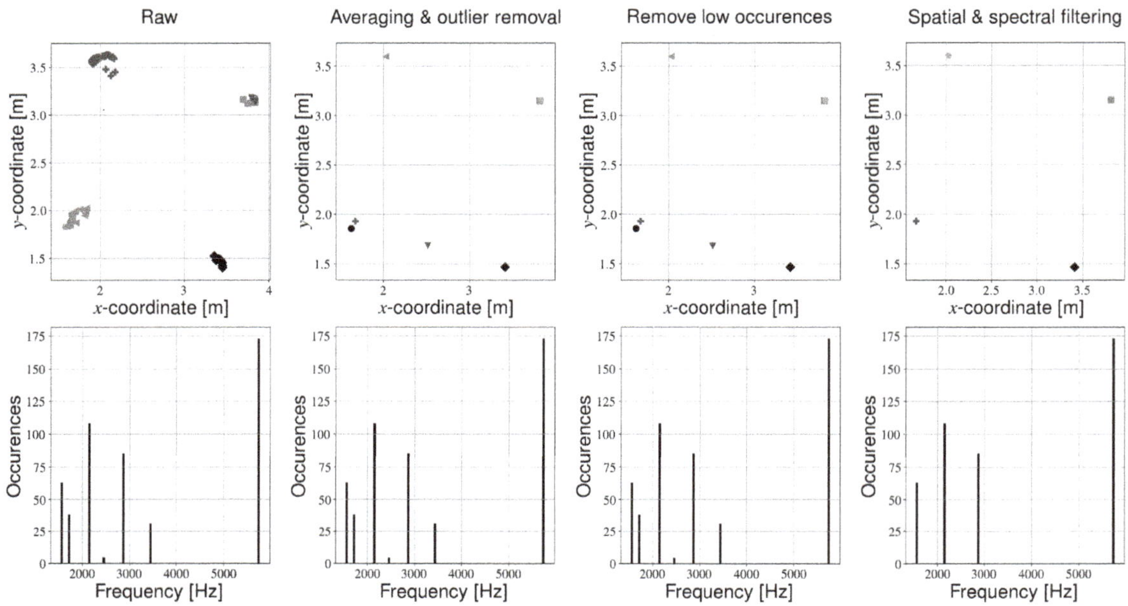

Figure 5. Filtering pipeline. The top row always indicates a light map, and a unique marker is used to indicate different frequencies. The bottom row shows the frequency spectrum corresponding to a certain processing step. From left to right: (1) raw output of the light mapping step, (2) light map after averaging coordinates of unique frequencies and removing outliers (3) light map after removing spurious observations (4) final results of spatial and spectral filtering.

We now propose a different method, whereby we filter the light sources based on their physical coordinates and the frequency spectrum. The intermediate results of the processing steps can be seen in Figure 5. First, outliers are removed from the detected coordinates per frequency. Outliers are defined as detected positions that have a distance of more than 2 standard deviations from the average position. Then, the coordinates of the remaining light sources are averaged per frequency. Light sources that were only observed a few times are removed. Next, light sources that are close to each other are combined. Filtering is first performed based on the coordinates of the light sources. Whenever the distance between two transmitters is below a certain threshold (Table 3), the LED with the lowest number of observations is removed. In case both lights have an equal amount of detections, their detected frequencies and positions are averaged. We call this step "spatial filtering". Finally, we combine light sources of approximately equal frequency. Whenever the distance in the frequency spectrum of two light sources is below a certain threshold (Table 3), the frequency with the largest number of observations is kept and the other is removed. Similar to the spatial filtering step, we average the frequencies and positions of light sources with an equal amount of detections. We call this final step "spectral filtering". The result of these additional processing steps is a light map with the correct number of transmitters. The number of light sources therefore no longer needs to be known beforehand. Instead, one only needs to know (approximately) how far lights are minimally spaced apart, which is much easier to obtain.

3.3. Data Processing

The result of the calibration procedure is a map of the environment that includes transmitter locations and their frequencies. Evaluating the accuracy of frequency detection is relatively straightforward and was performed by comparing the frequency applied by the signal generator to the frequency determined by the calibrating procedure. Evaluating the location of the light sources is more complex. Ideally, we could simply compare the coordinates determined by the calibration procedure to the coordinates in the physical setup. However, the calibration procedure produces coordinates relative to the map, which is not necessarily the same coordinate frame as the experimental setup, and obtaining the transformation between these frames is challenging. However, we can still compare the relative placement of the light sources. The distance between different light sources is irrespective of the coordinate frame. Therefore, in order to obtain the transmitter position accuracy, we subtracted the distance in the physical setup from the distance obtained in the light map. The distance error between two lights was therefore calculated by:

$$\varepsilon_{r,ij} = \left|d_{meas,ij} - d_{est,ij}\right| = \left|d_{meas,ij} - \sqrt{(x_{est,i} - x_{est,j})^2 + (y_{est,i} - y_{est,j})^2}\right| \quad (1)$$

where:

- $\varepsilon_{r,ij}$ is the error on the distance between lights i and j
- $d_{meas,ij}$ is the manually measured distance between lights i and j
- $d_{est,ij}$ is the distance between lights i and j as estimated by the calibration procedure.
- $x_{est,i}$ and $y_{est,i}$ are the estimated Cartesian coordinates of light source i
- $x_{est,j}$ and $y_{est,j}$ are the estimated Cartesian coordinates of light source j

In Section 4, we will investigate the performance of the system under a range of conditions. Unless explicitly mentioned otherwise, three experiments are conducted for every condition and the results from all three experiments are combined before further processing (for example, in order to obtain the cumulative error distribution).

4. Results

4.1. Baseline Results

Figure 6 compares the calibration results obtained in [13] with the method proposed in this paper. It is clear that the additional filtering steps significantly improve the calibration accuracy. On average, light sources can now be detected with an accuracy of approximately 6 cm, compared to 11 cm in [13]. Larger improvements are also visible in the higher percentiles of the distribution. More than 80% of light sources can be positioned with an accuracy of 10 cm, compared to 20 cm in [13]. Note that the same experimental data were used to obtain both error distributions; the difference is therefore purely due to improvements in data processing.

While this new method can localize the LEDs more accurately, the results of the frequency detection remain unchanged. As was the case in [13], it is challenging to calibrate the high frequency transmitter. As the frequency increases, the width of the stripes decreases, and a detection error of a few pixels results in a large frequency error (error of several hundred Hz). The lower and medium frequency sources can, however, be identified with relatively high accuracy (error of maximum 130 Hz).

4.2. Parameter Study

Using a mobile robot is a completely new approach to the calibration of VLP systems. As with any new technique, much is currently unknown about the effects of certain parameters on the calibration results. In the following sections, we will therefore investigate the influence of a number of factors on the calibration procedure. One parameter will be changed at a time, and unless otherwise specified, we will use the results from the previous section as a baseline to compare against. In doing so, we aim to create a better understanding of the strengths and limitations of the proposed approach.

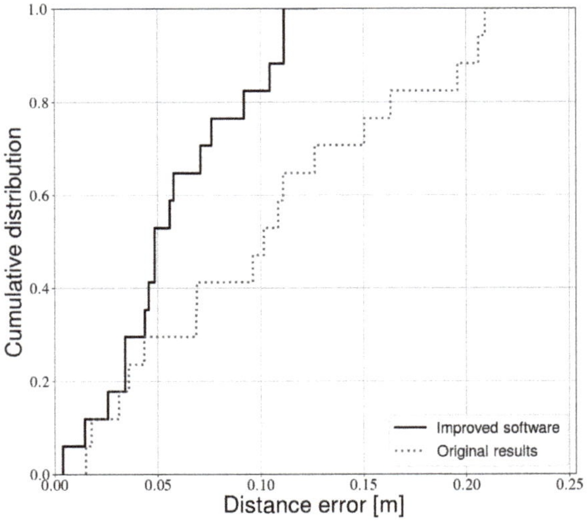

Figure 6. Cumulative error distribution of this paper compared with our previous work [13].

4.2.1. Transmitter Waveform

In visible light positioning, the light intensity of every LED is modulated in such a way that they are uniquely identifiable, even when multiple lights are in view at the same time. Generally, the lights continuously transmit a unique code or frequency [44]. The former is referred to as Code Division Multiple Access (CDMA), the latter as Frequency Division Multiple Access (FDMA). In this paper, we use FDMA as a multiple access technology. However, the calibration procedure can be adapted to support CDMA as well with relatively minor changes.

When using FDMA, both sine waves and square waves can be used as transmitter waveforms. A square wave is easier to generate and therefore the cost of the transmitters can be lower. Therefore, most VLP systems in literature use square waves. However, square waves have harmonics in the frequency spectrum. When selecting square wave frequencies, more care is needed to avoid interference. Photodiode-based VLP systems use the Fourier spectrum to separate the received signal into the components of each transmitter. Therefore, photodiode-based VLP systems are impacted most by harmonics. Camera-based VLP systems can use spatial multiplexing and are therefore less affected by this interference.

On the other hand, ideal sine waves have no harmonics, and therefore the available bandwidth can be used much more efficiently when photodiodes are used as receivers. The downside is that a sine wave is not as straightforward to generate with low-cost components. Additionally, the light intensity changes much more gradually with a sine wave, which makes frequency detection significantly more challenging (see Figure 7). In the following sections, we will therefore only use square waves, as our calibration procedure is not able to detect sine waves with sufficient accuracy.

4.2.2. Robot Trajectory

The motion of the robot platform may influence calibration results. If the robot can stay in motion, calibration time will be reduced. On the other hand, we expect motion blur may negatively impact results. During experiments, the robot was driven manually via a remote control. Two different types of trajectories were tested. In the first trajectory, the robot was continuously in motion and passed by every light source while covering the experimental space in a zigzag pattern. Figure 8a shows an example light map constructed from data recorded during a zigzag experiment. Performance is quite poor in this example—one light source was not even mapped at all. Other zigzag experiments occasionally resulted in even fewer light sources. Moreover, Figure 8c shows that the positioning accuracy of the fixtures that were detected is quite low. We therefore also tested a second type of trajectory, whereby the robot drives towards each light source sequentially. Once the camera is directly below the LED, data were recorded for a few seconds, before continuing to the next light. We call this as a "stop and go" trajectory, Figure 8b shows a light map constructed based on data from such an experiment. The light sources are now placed closer to the ITEM profiles (see Figure 2). Additionally, Figure 8c shows that the relative placement is significantly more accurate.

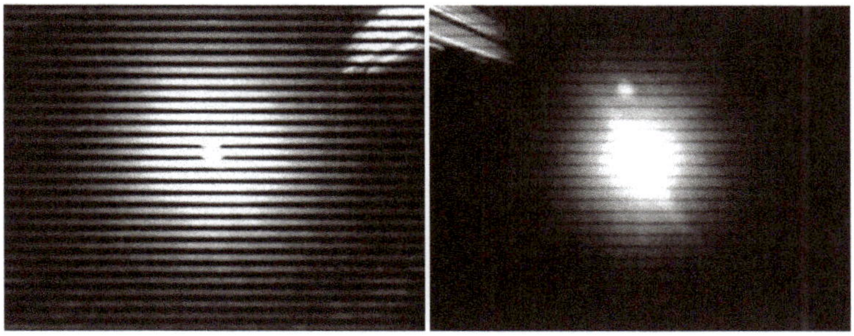

Figure 7. Comparison of 2 kHz square wave (**left**) and sine wave (**right**).

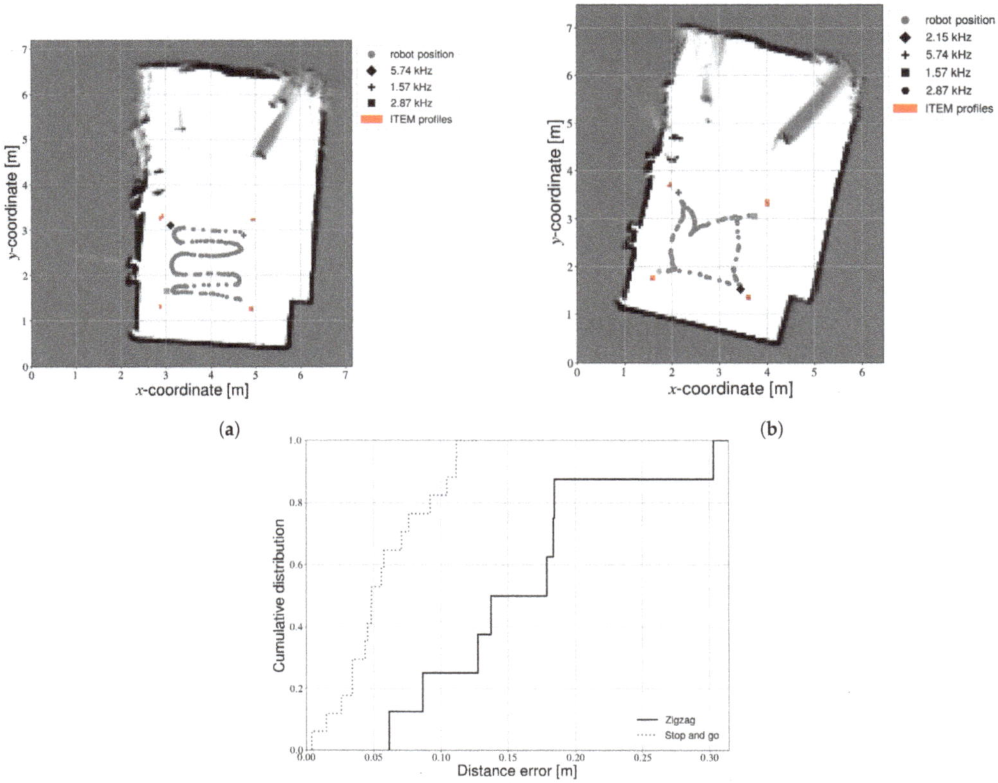

(c)

Figure 8. Calibration experiments with different trajectories. (**a**) Zigzag trajectory. (**b**) Stop and go trajectory. (**c**) Cumulative error distribution.

Table 4 compares the duration of the two types of trajectories, based on the average of three experiments for each type. Contrary to our expectations, we can observe that performing the calibration with the stop and go trajectory does not take significantly more time. The zigzag is an exhaustive search, and therefore takes a long time. In contrast, the stop and go trajectory is not a continuous motion but only gathers the data that is really required. Processing time for the stop and go trajectory is slightly increased compared to the zigzag. Fewer images were rejected, as more light sources were detected close to the center for this type of experiment. Consequently, more images needed to be processed and the processing time increased. However, the majority of this time was actually spent on LIDAR mapping (approximately 80%). Therefore, an increase in image processing only had a small impact on the overall time required. Additionally, the difference is less than 1 s. As the calibration can be performed offline, this time delay does not present an obstacle.

Table 4. Duration and processing time of calibration trajectories.

Trajectory	Average Duration [s]	Average Processing Time [s]
Zigzag	110.75	24.46
Stop and go	115.06	25.26

Many other types of trajectories could be considered. Determining the optimal calibration trajectory is outside the scope of this paper. With these results, we can however conclude that the robot should briefly stop at each LED in order to obtain accurate results. Unless otherwise specified, results in the following sections are obtained with a stop and go trajectory.

4.2.3. Lighting Conditions

Results from the previous section were all obtained under the same lighting conditions. It is well known that changing illumination levels can influence computer vision algorithms, and could thus negatively impact our proposed calibration procedure. In this section, we will therefore calibrate the experimental setup under a range of lighting conditions. More specifically, we distinguish 4 scenarios:

- Baseline: These lighting conditions applied to the results of the previous sections. More specifically, the shutters of the windows were closed, and no other light sources besides the LEDs were present.
- Other day: These experiments are conducted under the same circumstances as the baseline (closed shutters and no other light sources present) but on a different date approximately three months later. Baseline experiments took place in early spring when the sun sets earlier. In contrast, the "other day" experiments took place in summer, when the sky is clearer and the sun sets much later.
- Shutters open: During these experiments, the shutters of the windows were opened to allow sunlight to enter the room. These experiments also took place at the later date compared to the baseline.
- Fluorescent lights on: In addition to opening the shutters, the fluorescent lights are now also switched on. Fluorescent light is modulated and can in theory produce stripe patterns in our images. These experiments took place on the same day as the "Other day" and the "Shutters open" experiments.

For every condition, three experiments were again conducted with the stop and go trajectory. For every experiment, we determined the error on the light source location as explained in Section 3. Figure 9 shows that all conditions have similar performance. The "other day" experiments are very similar to the baseline, indicating that the parameters (e.g., exposure time) are not overfit to a specific point in time. The difference between both error distributions generally is not larger than 2 cm. Whether or not the shutters are opened also does not seem to negatively impact calibration accuracy, as unmodulated light sources are easily ignored by our calibration procedure. Similarly, we can observe that switching on the fluorescent light has little impact. While fluorescent light is modulated, the frequency is too low compared to the LEDs. Therefore, the additional light sources are simply ignored.

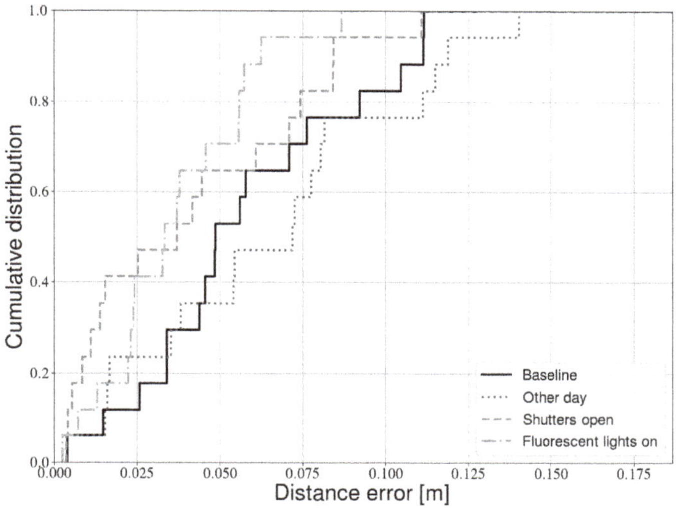

Figure 9. Influence of environmental conditions on calibration accuracy.

The results of the frequency detection were very similar under different lighting conditions. In fact, they were identical with only one exception. During one "shutters open" experiment, one medium frequency light source had a slightly larger error compared to the other experiments. This phenomenon did not occur during the experiments with the fluorescent lights, even though the shutters were also opened in this case.

4.2.4. Transmitter–Receiver Distance

Experiments so far were performed with LEDs mounted at a height of approximately 1.5 m. This makes them more easily accessible and thereby makes prototyping and experimenting easier. As the distance between transmitter and receiver increases, the LEDs will take up a relatively smaller portion of the image. Consequently, fewer stripes will be visible, and it will be more challenging to determine the transmitter frequency. This section will characterize the influence of increasing this distance on the accuracy of frequency estimation. To that end, we place the camera and an LED on a table and ensure that their normal planes were parallel. This horizontal setup (Figure 10) was different from how the LEDs are normally installed. However, whether the lights are mounted horizontally or vertically made no difference for this experiment; only the relative distance is important. In contrast, placing the light on a table rather than on the ceiling allowed us to change the distance much more easily and also enabled us to test performance at larger distances.

The distance between transmitter and receiver was increased from 1 m to 5 m. The light was modulated at a frequency of approximately 2 kHz. At every distance increment, images were recorded for approximately 30 s. In postprocessing, we determined the LED frequency using the process described in Section 3.2. Next, we calculated the difference between the true frequency (applied by the signal generator) and the frequency estimated by the calibration procedure. Table 5 shows the frequency estimation accuracy as a function of the transmitter–receiver distance. For short distances, the accuracy is approximately 95%, similar to Section 4.1. However, starting at a distance of 2 m, light sources can no longer be detected, hence the accuracy is 0%. The cause for this problem can be found by comparing images captured at different distances (Figure 11). At a distance of 1 m, several horizontal stripes are visible, from which the transmitter frequency can be calculated. At a distance of 3 m, only 1 stripe is visible, and the transmitter frequency can no longer be determined. Interestingly, the difference between 1 m and 3 m is much more pronounced than the difference between 3 m and 5 m.

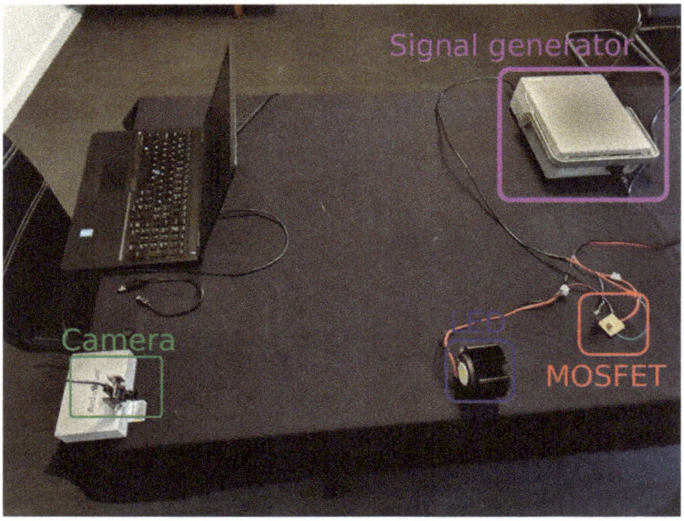

Figure 10. Experimental setup for transmitter–receiver distance experiments.

Table 5. Frequency estimation accuracy as a function of the transmitter–receiver distance. Field Of View (FOV) is kept fixed at 60 degrees.

Transmitter—Receiver Distance [m]	Detection Accuracy [%]
1.00	94.44
1.50	94.44
2.00	0.00
2.50	0.00
3.00	0.00
3.50	0.00
4.00	0.00
4.50	0.00
5.00	0.00

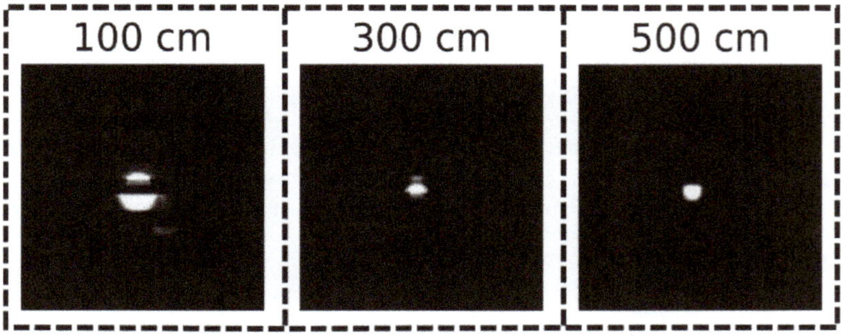

Figure 11. Example images taken at different transmitter–receiver distances. Images are cropped around the light source.

Results from the previous sections were obtained by using a camera lens with a field of view of 60 degrees. When a smaller FOV is used, the LED occupies a relatively larger portion of the image, and we may be able to detect it at greater distances. In order to test this hypothesis, we used a lens set (https://www.arducam.com/product/m12-mount-camera-lens-kit-arduino-raspberry-pi/, accessed on 12 March 2021) and varied the FOV from 10 to 60 degrees, at a fixed distance of 5 m. Again, the transmitter frequency was approximately 2 kHz, images were recorded during 30 s for every experiment, and the calibration procedure was used to estimate the transmitter frequency. Table 6 shows the accuracy of frequency estimation as a function of the FOV. It is clear that by sufficiently decreasing the FOV, the LED can still be detected. Even at a distance of 5 m, we can obtain the same accuracy of 95% as in Section 4.1 by using a lens with a FOV of 10 degrees. At shorter distances, a larger FOV can potentially also be used.

Table 6. Frequency estimation accuracy as a function of the FOV. Transmitter–receiver distance is kept fixed at 5 m.

FOV [°]	Detection Accuracy [%]
10	94.44
20	0.00
40	0.00
60	0.00

4.2.5. Camera Resolution

Similar to lighting conditions, camera resolution can have a significant impact on the performance of computer vision approaches. In the previous sections, the OpenMV M7 camera was used as an image sensor, which has a relatively low resolution of 640 × 480 pixels. We now hypothesize that a higher resolution will lead to a higher accuracy, for calibration of both the position and frequency of the LEDs. The increased resolution provides a higher granularity and thus potentially a greater accuracy in distinguishing the location of the light source. Additionally, a higher resolution provides additional stripes in the image, which may improve frequency detection. To verify this hypothesis, additional experiments were performed with a Raspberry Pi (RPi) camera sensor. Similar to Sections 4.2.2 and 4.2.3, the robot followed a stop and go trajectory, and data were recorded for offline postprocessing. Figure 12 compares images captured with both cameras. The RPi camera is designed for the low-cost single-board computer of the same name, which is particularly popular for embedded applications. The maximum resolution of 3280 × 2484 pixels was used, in order to amplify any effects related to the resolution. Images taken at such a high resolution take up a lot of space in memory. Therefore, pictures were not continually recorded. Rather, 10 images were taken when the robot was located underneath the light source. Both types of cameras were equipped with a 60 degree FOV lens. Due to an error in data recording, there are only two experiments with the RPi camera instead of the usual three.

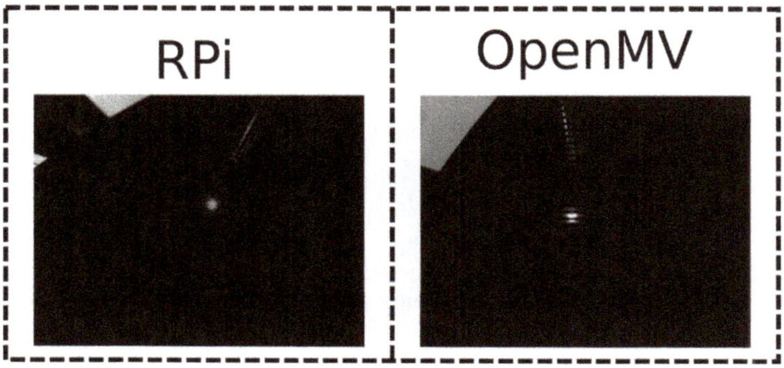

Figure 12. Example images of the same light taken with different cameras.

Table 7 contains the main quality metrics for calibration with both camera's. The results appear to support our previous hypothesis. The accuracy of transmitter positioning is improved, although the improvement is rather small. A larger improvement can be seen in the frequency identification. The high-frequency source can now be identified much more accurately, due to the increased bandwidth of the RPi camera.

Table 7. Calibration results for different camera sensors.

Camera	OpenMV	RPI
Average position error [m]	0.06	0.04
Maximum position error [m]	0.13	0.11
Average frequency error [Hz]	236.66	124.51
Maximum frequency error [Hz]	820.00	370.09
Processing time [s]	25.26	63.18
Trajectory time [s]	115.06	265.94

On the other hand, the time required to collect data with the RPi camera is more than double that of the OpenMV camera. The RPi camera is designed to be used with the Raspberry Pi single-board computer, which has significantly lower computational power compared to the laptop that collected the images from the OpenMV camera. Therefore, capturing each image takes a significantly longer amount of time. The main bottleneck here is the RPi itself, if one were to use a camera with a USB interface, a laptop could again be used to capture the images and calibration time would decrease. However, the duration of an experiment would likely still be higher with higher resolutions, yet less drastically so with the correct hardware. Processing time is also significantly increased when using the RPi camera. The higher resolution of the images means that the image processing takes a few seconds longer. The majority of the time increase can be attributed to the larger number of LIDAR measurements, which in turn is caused by the longer stationary time needed to capture the images with an RPi camera.

4.2.6. Field of View

Section 4.2.4 mentioned the effects of changing the field of view to improve the detection rate. Due to the nature of those experiments, we could not determine the error on the LED position. We therefore perform additional experiments with a changing FOV and multiple LEDs. These experiments were performed in the normal experimental setup (Figure 2). Section 4.2.5 showed that a larger resolution improves calibration performance. Therefore, we will again use this larger resolution camera in this section. Similar to Section 4.2.5, the RPi camera was used and images were only captured when the camera was approximately underneath a light source. By changing the lens, the FOV is again varied from 10 to 80 degrees. Figure 13 shows example pictures of the same light source for every FOV.

Figure 14 shows the mean accuracy of transmitter positioning and frequency detection as a function of the FOV. From 80 degrees onward, the calibration procedure starts failing, resulting in large errors. For the sake of clarity, these results are not included in these figures. In general, a smaller FOV leads to a better positioning accuracy, though the improvement is relatively small. The exception to this is the 40 degree lens, which actually has a larger error compared to 60 degrees. In contrast, a smaller FOV actually lowers frequency detection accuracy. The 40 degree lens fits the overall trend better in this case. Therefore, one can trade off the positioning and frequency accuracy. However, the 60 degrees from previous sections seems to have already been a good compromise for our setup. As discussed in Section 4.2.4, a larger height may still necessitate a smaller FOV, as the light sources will become difficult to detect otherwise.

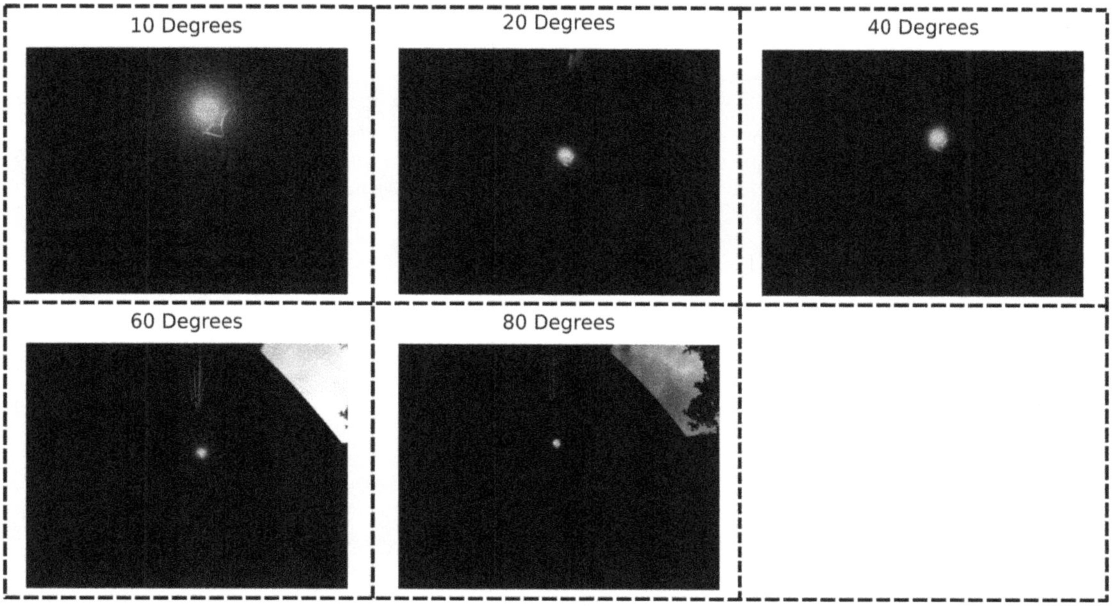

Figure 13. Example images for different FOV lenses.

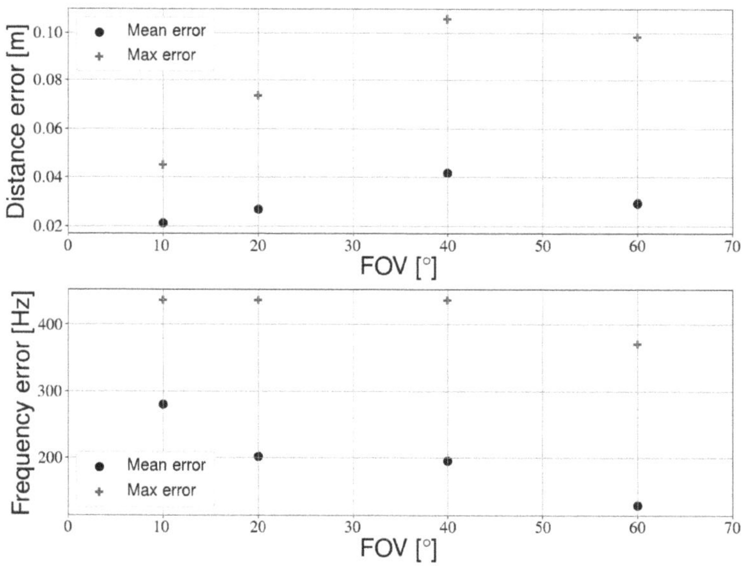

Figure 14. Top: positioning accuracy as a function of the camera FOV. Bottom: frequency accuracy as a function of the camera FOV.

4.3. *Influence on Positioning*

The goal of a calibration procedure is to accurately measure the environmental parameters needed for determining the receiver location. Calibration errors will likely result in positioning errors, though this is often not a simple linear relation. Therefore, it is challenging to determine how accurately the calibration needs to be performed in order to guarantee adequate positioning performance in a later stage. In this section, we evaluate the impact of calibration errors on visible light positioning, in order to determine if our calibration provides satisfactory results.

First, we calibrated the setup with the configuration that was determined to be the best trade-off between position and frequency accuracy. More specifically, we used an RPi camera with a field of view of 60 degrees and drove the robot in a stop and go trajectory. This setup was also calibrated manually as a point of reference. In our previous work, we described sensor fusion based robot positioning with three filters, namely an Extended Kalman Filter (EKF), a Particle Filter (PF) and a hybrid Particle/Kalman filter (PaKa) [41]. We now use the parameters from both the manual and robot calibration in these positioning approaches. We used all data from the conditions described in Section 4.2.3. Positioning accuracy results were obtained in the same way as described in [41]. Contrary to previous sections, "positioning accuracy" does not refer to the accuracy on the position of the transmitters. In this section specifically, "positioning accuracy" refers to the accuracy on the robot position, the calculation of which is described in [41].

The results from all experiments were combined into one cumulative distribution per filter, which are shown in Figure 15. It is clear that the proposed calibration procedure has little impact on positioning accuracy. Occasionally, the new method even improves accuracy. However, as explained in [41], the accuracy results of the PF and the PaKa can have a small variation due to the sampling of probability distributions. In general, the robot calibration is more accurate, albeit only slightly. Of the three positioning approaches, the hybrid filter seems to be least impacted by the new calibration method. The difference between the error distributions of the PaKa filter in Figure 15 is often less than 1 mm. All filters also have no trouble identifying the lights correctly. Even though the new calibration method introduces an error on the modulation frequency, it is not large enough to cause ambiguity among the transmitters.

Note that this calibration was performed with an RPi camera but that positioning was performed with the OpenMV camera. The above results therefore show that device heterogeneity is not an issue that needs to be specifically taken into account, contrary to some RSS-based positioning systems [45–48].

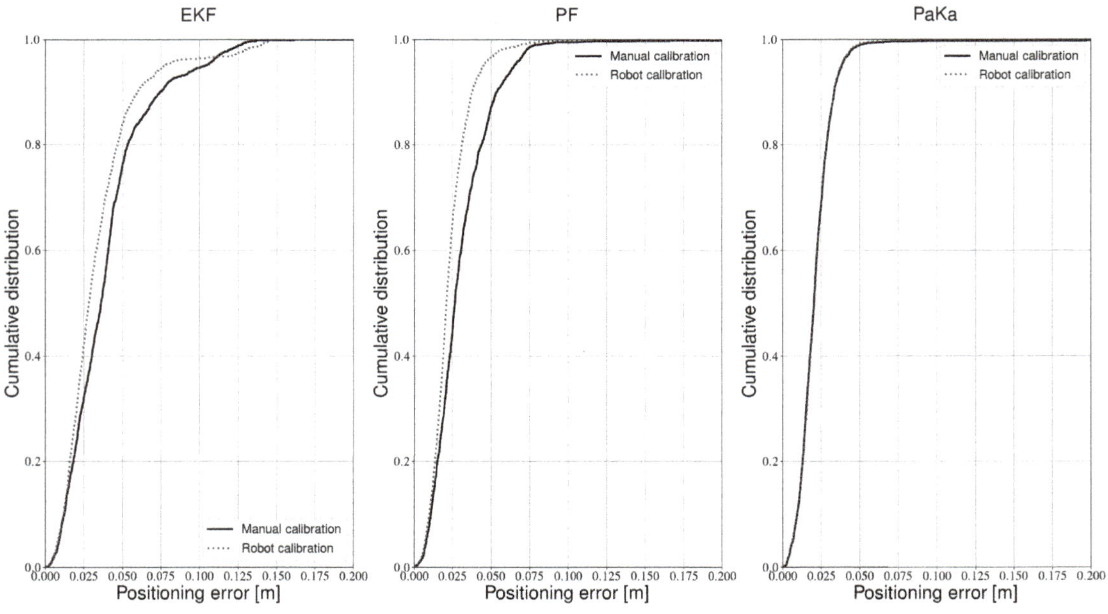

Figure 15. Cumulative positioning error distributions of calibration methods for different filters.

5. Discussion

Amsters et al. [13] proposed a proof of concept for a calibration procedure of VLP systems. Contrary to [13], the improved procedure described in this paper does not require prior knowledge of the number of light sources. In the vast majority of cases, the algorithm was able to determine the correct number of light sources. When using the zigzag trajectory, or a FOV of 80 degrees, the number of transmitters could be underestimated. In all the other tests that were performed (29 experiments in total), the calibration algorithm correctly determined the number of LEDs.

This type of calibration procedure for VLP systems has not been used before. Therefore, it was unclear how robust the approach is and which factors can influence the results. During our parameter study (Section 4.2), we obtained several key insights. For example, a limitation of the procedure is that it cannot be used for calibrating transmitters modulated with sine waves, which is a consequence of using a camera as a receiver. However, the majority of VLP systems described in literature use On-Off Keying (OOK) as a modulation scheme, even if only frequencies are transmitted [49]. In case code division is used, researchers also often opt for OOK. While we performed experiments with FDMA as a multiple access technology, it would be relatively straightforward to include code division multiplexing by expanding the image processing pipeline. Another limitation is that we can only determine the two-dimensional position of the LEDs. Some positioning approaches require knowledge of the ceiling height, which would have to be measured separately.

The proposed calibration procedure was also not influenced significantly by the ambient lighting, similar to the positioning approach used as an evaluation [41]. In contrast, the robot trajectory, height, FOV and resolution all had an impact on calibration accuracy. A large resolution should be used to increase accuracy of both frequency detection and transmitter positioning. However, we recommend the use of a USB camera in order to capture pictures faster. In our experiments, a FOV of 60 degrees provided a good trade-off between positioning and frequency detection accuracy. In case the distance between transmitter and receiver is large (as is the case with high ceilings), a smaller FOV may be required to detect the light sources. Finally, care should be taken to stop the robot at each light source, rather than using a continuous motion. The latter could lead to poor accuracy and an underestimation of the number of light sources.

The main objective of the technique is to calibrate the parameters of the system, so that these can be used for positioning in a later stage. The experimental results in Section 4.3 showed that the parameters of the experimental setup can be determined with sufficient accuracy. The error on the light source locations did not result in increased positioning errors. In the case of our experimental setup, the transmitter frequencies could also be determined with sufficient accuracy so as to not cause ambiguity. It is important to note that one should take care that the modulation frequencies are sufficiently far apart, as some error is introduced when calibrating the modulation frequency. We should also note that certain positioning approaches are more susceptible to calibration errors than others. The positioning approach used as an evaluation tool made use of sensor fusion. In case of large measurement errors, the filters can fall back on odometry data. However, this is only the case when the error on the observation is sufficiently large. More subtle disturbances such as errors on the transmitter coordinates cannot be filtered. Additionally, with this work we showed that it is possible to close the loop between calibration and positioning. That is, we can efficiently calibrate the setup with a mobile robot and then use the determined parameters for high-accuracy positioning. Manual calibration also leads to errors on the transmitter locations. As evidenced by our positioning case, these errors are likely of the same order of magnitude as the robot calibration.

Our work shares similarities with robot-based RFID calibration. Hähnel et al. [11] also used a mobile robot equipped with a LIDAR and used it to reconstruct a map of the environment. The position of RFID tags was later estimated based on the path of the robot. Similarly, Milella et al. [12] also mapped indoor spaces with a mobile robot in order to localize RFID tags. They used fuzzy logic to determine the likelihood of a tag location. Mirowski et al. [16] proposed the use of a mobile robot for calibration of Wi-Fi localization systems. Contrary to [11,12], they used Quick Response (QR) codes to aid with loop closures, which raises the question as to how these QR codes should be localized.

Literature on the subject of VLP calibration is limited. Most examples focused on obtaining the parameters of the channel model [33,37], which we cannot calibrate. However, the approach which we used as an evaluation tool does not require these parameters as the channel model is not used [41]. This does limit the calibration procedure to mostly camera-based positioning systems. It is possible to further extend the proposed calibration system by including a photodiode on the robot platform and using the intensity measurements to obtain the parameters of the channel model.

In order to obtain the transmitter locations and identities, we obtained the receiver position through SLAM, rather than the manual measurements used in [31,50]. Contrary to [35], we were also able to obtain light source identities. Yue et al. [36] did use modulated LEDs, yet they have significantly lower accuracy compared to our work. However, our approach required a dedicated procedure rather than crowdsourcing the required data. Additionally, our robot needed to be manually driven by a human operator. Nevertheless, it may be possible to let the robot perform the calibration autonomously, whereas crowdsourcing will always require the cooperation of humans.

6. Conclusions

In this work, we outlined an improved calibration procedure for VLP systems, based on data collection with a mobile robot. The new approach had significantly improved performance compared to previous work. Accuracy of LED localization was almost doubled. Additionally, whereas previous work remained a proof-of-concept, we performed an extensive parameter study to characterize the strengths and limitations of the approach. Based on these results, we suggested the use of high resolution camera, with a FOV of 60 degrees to further improve the accuracy of LED placement and frequency detection. We showed that ambient lighting has little influence on the proposed procedure. Through positioning experiments, we determined that the approach is also accurate enough to calibrate high-performance VLP systems. In doing so, an important barrier to entry is removed for visible light positioning systems.

Our approach required a dedicated site survey, rather than crowdsourcing. While less convenient, it did result in much greater accuracy. The procedure was also unable to calibrate the channel model. In future work, we could add a photodiode to the sensor platform in order to obtain the Lambertian emission parameters. Additionally, we could investigate the possibility of letting the robot perform the procedure autonomously, in order to reduce the human labor required.

Author Contributions: Conceptualization, R.A.; Funding acquisition, R.A., E.D., N.S. and P.S.; Investigation, R.A.; Methodology, R.A.; Project administration, R.A. and P.S.; Supervision, E.D., N.S. and P.S.; Writing—original draft, R.A.; Writing—review and editing, R.A., E.D., N.S. and P.S. All authors have read and agreed to the published version of the manuscript.

Funding: This research was funded by Research Foundation Flanders (FWO) under grant number 1S57720N.

Institutional Review Board Statement: Not applicable.

Informed Consent Statement: Not applicable.

Data Availability Statement: Not applicable.

Conflicts of Interest: The authors declare no conflict of interest. The funders had no role in the design of the study; in the collection, analyses, or interpretation of data; in the writing of the manuscript, or in the decision to publish the results.

Appendix A

Table A1. Cost of the experimental setup.

Component	Amount	Cost Per Unit [€]
Mobile robot	1	603.79
LIDAR	1	119.79
LED	4	9.08
ITEM profiles	1	298.74
Lab bench power supply	1	296.45
Signal generator	4	109.09
OpenMV camera	1	76.17
RPi camera	1	26.98
RPi	1	42.2895
laptop	1	459
lens kit	1	82.89
Computer	1	1363.46
Total		3842.2395

References

1. Liu, H.; Darabi, H.; Banerjee, P.; Liu, J. Survey of wireless indoor positioning techniques and systems. *IEEE Trans. Syst. Man Cybern. Part Appl. Rev.* **2007**, *37*, 1067–1080. [CrossRef]
2. Van Diggelen, F.; Enge, P. The worlds first gps mooc and worldwide laboratory using smartphones. In Proceedings of the 28th International Technical Meeting of the Satellite Division of the Institute of Navigation (ION GNSS+ 2015), Tampa, FL, USA, 14–18 September 2015; pp. 361–369.
3. Liu, F.; Liu, J.; Yin, Y.; Wang, W.; Hu, D.; Chen, P.; Niu, Q. Survey on WiFi-based indoor positioning techniques. *IET Commun.* **2020**, *14*, 1372–1383. [CrossRef]
4. Alarifi, A.; Al-Salman, A.; Alsaleh, M.; Alnafessah, A.; Al-Hadhrami, S.; Al-Ammar, M.A.; Al-Khalifa, H.S. Ultra Wideband Indoor Positioning Technologies: Analysis and Recent Advances. *Sensors* **2016**, *16*, 707. [CrossRef] [PubMed]
5. Priyantha, N.B. The Cricket Indoor Location System. Ph.D. Thesis, Massachusetts Institute of Technology, Cambridge, MA, USA, 2005.
6. Do, T.H.; Yoo, M. An in-depth survey of visible light communication based positioning systems. *Sensors* **2016**, *16*, 678. [CrossRef] [PubMed]
7. Duff, P.; Muller, H. Autocalibration algorithm for ultrasonic location systems. In Proceedings of the International Symposium on Wearable Computers, ISWC, White Plains, NY, USA, 21–23 October 2003; pp. 62–68. [CrossRef]
8. Preter, A.D.; Goysens, G.; Anthonis, J.; Swevers, J.; Pipeleers, G. Range bias modeling and autocalibration of an UWB positioning system. In Proceedings of the 2019 International Conference on Indoor Positioning and Indoor Navigation, IPIN 2019, Pisa, Italy, 30 September–3 October 2019. [CrossRef]
9. Lebreton, J.M.; Murad, N.; Lorion, R. Real-time radio signal mapping using an autonomous robot. In Proceedings of the 2015 IEEE Radio and Antenna Days of the Indian Ocean, RADIO 2015, Belle Mare, Mauritius, 21–24 September 2015; pp. 1–2. [CrossRef]
10. Mirowski, P.; Palaniappan, R.; Ho, T.K. Depth camera SLAM on a low-cost WiFi mapping robot. In Proceedings of the 2012 IEEE Conference on Technologies for Practical Robot Applications, TePRA 2012, Woburn, MA, USA, 23–24 Apri 2012, *1*, 1–6. [CrossRef]
11. Hähnel, D.; Burgard, W.; Fox, D.; Fishkin, K.; Philipose, M. Mapping and localization with RFID technology. In Proceedings of the IEEE International Conference on Robotics and Automation, New Orleans, LA, USA, 26 April–1 May 2004; Volume 2004, pp. 1015–1020. [CrossRef]
12. Milella, A.; Vanadia, P.; Cicirelli, G.; Distante, A. RFID-based environment mapping for autonomous mobile robot applications. In Proceedings of the IEEE/ASME International Conference on Advanced Intelligent Mechatronics, AIM, Zurich, Switzerland, 4–7 September 2007. [CrossRef]
13. Amsters, R.; Demeester, E.; Slaets, P.; Holm, D.; Joly, J.; Stevens, N. Towards automated calibration of visible light positioning systems. In Proceedings of the 2019 International Conference on Indoor Positioning and Indoor Navigation, IPIN 2019, Pisa, Italy, 30 September–3 October 2019. [CrossRef]
14. Gao, C.; Harle, R. Sequence-based magnetic loop closures for automated signal surveying. In Proceedings of the 2015 International Conference on Indoor Positioning and Indoor Navigation, IPIN 2015, Banff, AB, Canada, 13–16 October 2015; pp. 1–12. [CrossRef]
15. Palaniappan, R.; Mirowski, P.; Ho, T.K.; Steck, H.; MacDonald, M.; Whiting, P. Autonomous RF Surveying Robot for Indoor Localization and Tracking. In Proceedings of the Indoor Positioning and Indoor Navigation (IPIN), Guimarães, Portugal, 21–23 September 2011; pp. 21–23.

16. Mirowski, P.; Ho, T.K.; Yi, S.; MacDonald, M. SignalSLAM: Simultaneous localization and mapping with mixed WiFi, bluetooth, LTE and magnetic signals. In Proceedings of the 2013 International Conference on Indoor Positioning and Indoor Navigation, IPIN 2013, Montbeliard, France, 28–31 October 2013; pp. 1–10. [CrossRef]
17. Wang, S.; Wen, H.; Clark, R.; Trigoni, N. Keyframe based large-scale indoor localisation using geomagnetic field and motion pattern. In Proceedings of the IEEE International Conference on Intelligent Robots and Systems, Daejeon, Korea, 9–14 October 2016; pp. 1910–1917. [CrossRef]
18. Kim, Y.; Chon, Y.; Cha, H. Smartphone-based collaborative and autonomous radio fingerprinting. *IEEE Trans. Syst. Man Cybern. Part Appl. Rev.* **2012**, *42*, 112–122. [CrossRef]
19. Rai, A.; Chintalapudi, K.K.; Padmanabhan, V.N.; Sen, R. Zee: Zero-effort crowdsourcing for indoor localization. In Proceedings of the Annual International Conference on Mobile Computing and Networking, MOBICOM, Mobicom '12, Istanbul, Turkey, 22–26 August 2012; pp. 293–304. [CrossRef]
20. Wang, H.; Sen, S.; Elgohary, A.; Farid, M.; Youssef, M.; Choudhury, R.R. No need to war-drive: Unsupervised indoor localization. In Proceedings of the 10th International Conference on Mobile Systems, Applications, and Services, MobiSys'12, Ambleside, UK, 25–29 June 2012; pp. 197–210. [CrossRef]
21. Yang, Z.; Wu, C.; Liu, Y. Locating in fingerprint space: Wireless indoor localization with little human intervention. In Proceedings of the Annual International Conference on Mobile Computing and Networking, MOBICOM, Mobicom'12, Istanbul, Turkey, 22–26 August 2012; pp. 269–280. [CrossRef]
22. Mahajan, A.; Figueroa, F. Automatic self-installation and calibration method for a 3D position sensing system using ultrasonics. *Robot. Auton. Syst.* **1999**, *28*, 281–294. [CrossRef]
23. Cevher, V.; McClellan, J.H. Sensor array calibration via tracking with the extended Kalman filter. In Proceedings of the ICASSP, IEEE International Conference on Acoustics, Speech and Signal Processing, Salt Lake City, UT, USA, 7–11 May 2001; Volume 5, pp. 2817–2820. [CrossRef]
24. Djaja-Josko, V. A new anchor nodes position determination method supporting UWB localization system deployment. In Proceedings of the 2017 Signal Processing Symposium, SPSympo 2017, Jachranka, Poland, 12–14 September 2017; pp. 1–5. [CrossRef]
25. Choi, J.; Choi, Y.S.; Talwar, S. Unsupervised learning technique to obtain the coordinates of Wi-Fi access points. In Proceedings of the 2019 International Conference on Indoor Positioning and Indoor Navigation, IPIN 2019, Pisa, Italy, 30 Septeber–3 October 2019; pp. 1–6. [CrossRef]
26. Batstone, K.; Oskarsson, M.; Åström, K. Towards real-time time-of-arrival self-calibration using ultra-wideband anchors. In Proceedings of the 2017 International Conference on Indoor Positioning and Indoor Navigation, IPIN 2017, Sapporo, Japan, 18–21 September 2017; Volume 2017, pp. 1–8. [CrossRef]
27. Mautz, R.; Ochieng, W. A Robust Indoor Positioning and Auto-Localisation Algorithm. *J. Glob. Position. Syst.* **2007**, *6*, 38–46. [CrossRef]
28. Syafrudin, M.; Schweinzer, H.; Walter, C. An automated calibration method for the Local Positioning System LOSNUS. In Proceedings of the 2012 IEEE I2MTC—International Instrumentation and Measurement Technology Conference, Graz, Austria, 13–16 May 2012; pp. 1320–1325. [CrossRef]
29. Guevara, J.; Jiménez, A.R.; Prieto, J.C.; Seco, F. Auto-localization algorithm for local positioning systems. *Ad Hoc Netw.* **2012**, *10*, 1090–1100. [CrossRef]
30. Rodríguez-Navarro, D.; Lázaro-Galilea, J.L.; Bravo-Muñoz, I.; Gardel-Vicente, A.; Tsirigotis, G. Analysis and calibration of sources of electronic error in PSD sensor response. *Sensors* **2016**, *16*, 619. [CrossRef] [PubMed]
31. Rodríguez-Navarro, D.; Lázaro-Galilea, J.L.; Espinosa, F.; De-La-Llana-Calvo, A.; Santos, C. Simplified Calibration Process for a PSD-Based Optical IPS. In Proceedings of the IPIN 2018—9th International Conference on Indoor Positioning and Indoor Navigation, Nantes, France, 24–27 September 2018, Volume 4, pp. 1–6. [CrossRef]
32. Rodríguez-Navarro, D.; Lázaro-Galilea, J.L.; Gardel-Vicente, A.; Bravo-Muñoz, I.; De-La-Llana-Calvo, Á. Indoor positioning system based on PSD sensor. In *Geographical and Fingerprinting Data for Positioning and Navigation Systems: Challenges, Experiences and Technology Roadmap*; Academic Press: Cambridge, MA, USA, 2018; pp. 353–370. [CrossRef]
33. Alam, F.; Parr, B.; Mander, S. Visible Light Positioning Based on Calibrated Propagation Model. *IEEE Sens. Lett.* **2019**, *3*, 1–4. [CrossRef]
34. Bastiaens, S.; Plets, D.; Martens, L.; Joseph, W. Response Adaptive Modelling for Reducing the Storage and Computation of RSS-Based VLP. In Proceedings of the IPIN 2018—9th International Conference on Indoor Positioning and Indoor Navigation, Nantes, France, 24–27 September 2018; pp. 1–8. [CrossRef]
35. Liang, Q.; Liu, M. An Automatic Site Survey Approach for Indoor Localization Using a Smartphone. *IEEE Trans. Autom. Sci. Eng.* **2020**, *17*, 191–206. [CrossRef]
36. Yue, Y.; Zhao, X.; Li, Z. Enhanced and Facilitated Indoor Positioning by Visible-Light GraphSLAM Technique. *IEEE Internet Things J.* **2020**, *8*, 1183–1196. [CrossRef]
37. Amsters, R.; Demeester, E.; Stevens, N.; Slaets, P. In-depth analysis of unmodulated visible light positioning using the iterated extended kalman filter. *Sensors* **2019**, *19*, 5198. [CrossRef] [PubMed]

38. Azizyan, M.; Constandache, I.; Choudhury, R.R. SurroundSense: Mobile phone localization via ambience fingerprinting. In Proceedings of the Annual International Conference on Mobile Computing and Networking, MOBICOM, Beijing, China, 20–25 September 2009, pp. 261–272. [CrossRef]
39. Thrun, S. *Probabilistic Robotics*; MIT Press: Cambridge, MA, USA, 2002; Volume 45, pp. 52–57. [CrossRef]
40. Santos, J.M.; Portugal, D.; Rocha, R.P. An evaluation of 2D SLAM techniques available in Robot Operating System. In Proceedings of the 2013 IEEE International Symposium on Safety, Security, and Rescue Robotics, SSRR 2013, Linköping, Sweden, 21–26 October 2013. [CrossRef]
41. Amsters, R.; Holm, D.; Joly, J.; Demeester, E.; Stevens, N.; Slaets, P. Visible Light Positioning using Bayesian Filters. *J. Light. Technol.* **2020**, *38*, 5925–5936. [CrossRef]
42. Kuo, Y.S.; Pannuto, P.; Hsiao, K.J.; Dutta, P. Luxapose: Indoor positioning with mobile phones and visible light. In Proceedings of the Annual International Conference on Mobile Computing and Networking, MOBICOM, Maui, HI, USA, 7–11 September 2014, pp. 447–458. [CrossRef]
43. Hess, W.; Kohler, D.; Rapp, H.; Andor, D. Real-time loop closure in 2D LIDAR SLAM. In Proceedings of the IEEE International Conference on Robotics and Automation, Stockholm, Sweden, 16–21 May 2016; pp. 1271–1278. [CrossRef]
44. De Lausnay, S.; De Strycker, L.; Goemaere, J.P.; Nauwelaers, B.; Stevens, N. A survey on multiple access Visible Light Positioning. In Proceedings of the 2016 IEEE International Conference on Emerging Technologies and Innovative Business Practices for the Transformation of Societies, EmergiTech, Balaclava, Mauritius, 3-6 August 2016; pp. 38–42. [CrossRef]
45. Dong, F.; Chen, Y.; Liu, J.; Ning, Q.; Piao, S. A calibration-free localization solution for handling signal strength variance. In *Lecture Notes in Computer Science (Including Subseries Lecture Notes in Artificial Intelligence and Lecture Notes in Bioinformatics)*; Fuller, R., Koutsoukos, X.D., Eds.; Springer: Berlin/Heidelberg, Germany, 2009; Volume 5801 LNCS; pp. 79–90. [CrossRef]
46. Haeberlen, A.; Rudys, A.; Flannery, E.; Wallach, D.S.; Ladd, A.M.; Kavraki, L.E. Practical robust localization over large-scale 802.11 wireless networks. In Proceedings of the Annual International Conference on Mobile Computing and Networking, MOBICOM, MobiCom '04, Philadelphia, PA, USA, 26 September–1 October 2004; pp. 70–84. [CrossRef]
47. Kjærgaard, M.B. Indoor location fingerprinting with heterogeneous clients. *Pervasive Mob. Comput.* **2011**, *7*, 31–43. [CrossRef]
48. Mahtab Hossain, A.K.; Jin, Y.; Soh, W.S.; Van, H.N. SSD: A robust RF location fingerprint addressing mobile devices' heterogeneity. *IEEE Trans. Mob. Comput.* **2013**, *12*, 65–77. [CrossRef]
49. Zhuang, Y.; Hua, L.; Qi, L.; Yang, J.; Cao, P.; Cao, Y.; Wu, Y.; Thompson, J.; Haas, H. A Survey of Positioning Systems Using Visible LED Lights. *IEEE Commun. Surv. Tutor.* **2018**, *20*, 1963–1988. [CrossRef]
50. De-La-Llana-Calvo, Á.; Lázaro-Galilea, J.L.; Gardel-Vicente, A.; Rodríguez-Navarro, D.; Bravo-Muñoz, I. Indoor Positioning System Based on LED lighting and PSD Sensor. In Proceedings of the 2019 International Conference on Indoor Positioning and Indoor Navigation (IPIN), Pisa, Italy, 30 September–3 October 2019; pp. 1–8. [CrossRef]

sensors

MDPI

Article

Towards Precise Positioning and Movement of UAVs for Near-Wall Tasks in GNSS-Denied Environments

Félix Orjales [1,*], Javier Losada-Pita [2], Alejandro Paz-Lopez [2] and Álvaro Deibe [1]

1 Integrated Group for Engineering Research, Universidade da Coruña, 15403 A Coruña, Spain; adeibe@udc.es
2 CITIC Research Center, Universidade da Coruña, 15071 A Coruña, Spain; javier.losada@udc.es (J.L.-P.); alpaz@udc.es (A.P.-L.)
* Correspondence: felix.orjales@udc.es; Tel.: +34-981337400

Abstract: UAVs often perform tasks that require flying close to walls or structures and in environments where a satellite-based location is not possible. Flying close to solid bodies implies a higher risk of collisions, thus requiring an increase in the precision of the measurement and control of the UAV's position. The aerodynamic distortions generated by nearby walls or other objects are also relevant, making the control more complex and further placing demands on the positioning system. Performing wall-related tasks implies flying very close to the wall and, in some cases, even touching it. This work presents a Near-Wall Positioning System (NWPS) based on the combination of an Ultra-wideband (UWB) solution and LIDAR-based range finders. This NWPS has been developed and tested to allow precise positioning and orientation of a multirotor UAV relative to a wall when performing tasks near it. Specific position and orientation control hardware based on horizontal thrusters has also been designed, allowing the UAV to move smoothly and safely near walls.

Keywords: UAV; unmanned aerial vehicles; NWPS; indoor positioning systems; GPS denied; GNSS denied; autonomous vehicles

Citation: Orjales, F.; Losada-Pita, J.; Paz-Lopez, A.; Deibe, Á. Towards Precise Positioning and Movement of UAVs for Near-Wall Tasks in GNSS-Denied Environments. *Sensors* **2021**, *21*, 2194. https://doi.org/10.3390/s21062194

Academic Editor: Simon Tomažič

Received: 15 February 2021
Accepted: 19 March 2021
Published: 21 March 2021

1. Introduction

Most UAVs use a Global Navigation Satellite System (GNSS) to determine their position. GNSS receivers are small, readily available, and easy to use. Working in areas where a GNSS constellation is not accessible represents a drawback for UAV use. For example, using UAV in civil and structural inspection applications is often limited to visual inspections, and GNSS is required [1,2]. The literature on this subject shows efforts made to develop aerial platforms capable of performing sound-based [2], contact-based [3], or hammering inspection [4], even though positioning and navigation for these platforms have not yet been fully developed in these conditions.

Operations carried out touching the wall or at a short distance away, in the same order of magnitude as the vehicle size, are a challenge for a UAV's positioning control system. Proximity to walls or other objects distorts the aerodynamic currents generated by the UAV and causes changes in the UAV behavior, which, together with the proximity of the wall, increases the risk of collisions. Therefore, it is necessary to increase the control effort of the UAV under these conditions, and this also requires a more precise positioning system relative to the wall.

UAV positioning systems in applications without GNSS must be precise, with a high refresh rate and low latency. These positioning systems tend to be expensive. Moreover, its operation usually requires the deployment of precisely placed equipment and could require further calibration. These tasks tend to be cumbersome and time-consuming.

UAVs could change how some near-walls tasks are performed. Applications such as painting a high wall, installing anchor points without scaffolding, performing weld inspection in hard to reach areas, or carrying out wall thickness measurements. These tasks are usually expensive and dangerous, making the potential use of UAVs very valuable.

However, the previously mentioned limitations restrict the use of UAVs in this kind of applications.

1.1. Positioning without GNSS

Positioning is one of the first and more significant steps required to develop an automatic or autonomous UAV. A positioning system can typically make use of direct measurement sources (beacons, ad hoc signals, etc.), inertial/dead-reckoning measurements (such as Inertial Measurement Units (IMUs) or Magnetic, Angular Rate, and Gravity sensors (MARGs)) and environmental correlation measurements (LIDAR, Ultra-wideband (UWB), etc.). Outdoor applications mostly rely on satellite-based positioning systems [5]. When a GNSS constellation is not available, the UAV's navigation system needs to incorporate other ways to locate the vehicle.

Measurements from inertial sensors are relative readings and usually exhibit an accumulative positioning error. Environmental correlation measurement systems, on the other hand, deliver absolute measures. In this field, several wireless technologies and methods exist that can be employed to position a UAV flying indoors. Regarding the method used to evaluate distances, there are time-based, power-based, and angle-based approaches [6]. Time-based methodologies are among the most used, and most of them are based on Time of Arrival (ToA). Some variants differ on the synchronization method between transmitter and receiver and exhibit improved techniques for better accuracy [7]. Power-based approaches tend to be very sensitive to multipath or rebound generated by walls or obstacles. Received Signal Strength Indicator (RSSI) is among the best known, but the obtained measurements tend to be inaccurate [8]. Finally, angle-based approaches are uncommon as they require the use of specialized hardware.

Flying indoors usually has to be carried out with limited GNSS access or no access at all. In these cases, other Indoor Positioning Systems (IPS) must be used. Some of the characteristics of IPSs are determined by their physical layer. A survey of the most relevant IPSs currently available appears in [9] and more recently [10]. They include a comparison and a classification that can ease the IPS selection for the intended task. An IPS can be based on visible or invisible light (i.e., infrared), sound and ultrasound, magnetic fields, and RF. Widely used IPS technologies, such as Bluetooth [11], Wi-Fi, and UWB [12], fall into the last category.

1.2. Accurate Positioning and Movement Near Walls

To be viable in real-world scenarios, three key aspects need to be considered for the selection or development of a Near Wall Positioning System (NWPS) with UAV: low deployment complexity, high accuracy to avoid the obstacles in an enclosed and limited environment, and high refresh rate and low latency to control a fast and agile vehicle in a dynamically changing environment.

Navigation is another significant capability of an automatic or autonomous system, and it is deeply connected to the working environment. The requirements are different when navigating in a large, unobstructed environment or a confined space near walls. In the last case, moving from one point to another in a controlled and safe way implies dealing with obstacles, workspace limits, and aerodynamic distortions generated by the walls, the ceiling, or the ground. The orientation of the tool used to perform the intended task is another relevant aspect to consider, as it is linked with the UAV movement.

Multirotors are the preferred type of UAV for indoor applications because they can hold a static position and move indiscriminately in any horizontal direction. Horizontal movement is generated, in most cases, by tilting the vehicle. When a multirotor operates close to a wall, it is subject to changing aerodynamic forces. These forces vary with the distance and angle between the UAV and the wall. In general, these forces can change from repulsion to attraction with small changes in these parameters [13,14], making them unpredictable and, consequently, potentially leading to erratic behaviors. The ground and wall effect combination [15], as well as the tilt angle of the UAV [16], increase the effect of this phenomenon.

Both of the presented challenges, the positioning and control of the UAV, need to be addressed in order to obtain the full potential of the UAV in near-wall applications. Autonomous and automatic systems require reliable sensor inputs and adequate actuation outputs in order to be really effective. In addition, they need to be light-weight, simple, and low cost if they are intended to operate onboard little or medium-sized UAVs. These design limitations and requirements have been taken into account in the development of the proposed NWPS.

There are solutions, both commercial and in the research literature [17,18] that address these challenges. In most of these cases, they are concerned with detecting nearby walls or other objects and avoid colliding with them. The proposed solution, however, is different because it is focused on detecting the position and orientation of the wall relative to the UAV, navigating to reach closer to the wall, and making controlled contact with the tool to perform certain tasks.

Several tasks can be performed with a UAV that involve carrying some kind of probe to touch a wall in a controlled way. Tasks such as measuring the depth of carbonatation in concrete, sclerometry for determining the compressive strength of concrete, measuring dry film thickness, measuring metallic wall thickness, or detecting steel reinforcement bars in concrete structures, among others.

The intended task requires precise positioning of the tool or probe, making direct contact with the wall at a specific point and with a given angle. Both the position and orientation of the tool are essential factors for the success of the task. In the proposed architecture, the probe is rigidly attached to the body of the UAV. Therefore, the position and orientation of the UAV greatly influence the final result.

Regarding the positioning, what is presented in this document has focused on improving the accuracy in the distance and orientation estimations between vehicle and wall in the absence of GNSS signals. This requirement is critical to accomplish near-wall tasks. The new NWPS system merges two positioning solutions: LIDAR range finders and Ultra Wide Band (UWB) radio signals. At this point, this work has focused on indoor scenarios. However, this approximation could be applied outdoors after taking into account additional factors such as the weather conditions.

Regarding the control, multirotors need to tilt in pitch or roll axis to accomplish a horizontal displacement. This tilting action and the dynamic reactions related to it make difficult the precise positioning of the tool used to perform the task, it being a sensor or a manipulator. To overcome these problems, the proposed solution is based on the addition of small horizontal thrusters to the UAV and related hardware and software. With these thrusters, the UAV can make horizontal movements without tilting, thus easing the control of the tool.

The rest of the document has been structured as follows: Section 2 comprises the results of the literature review performed in relation to Indoor Positioning Systems and navigation near walls. In Section 3, the proposed positioning system based on UWB and LIDAR is presented. Section 4 explains in detail the proposed hardware system for horizontal movement close to walls. Section 5 summarizes the tests performed for both and highlights the more significant results. In the last section, a discussion of this work is presented.

2. Related Work

UAV research is a very active field with over 2500 scientific publications in the IEEE Xplore Digital Library during 2019 [19]. The two main challenges highlighted in the previous section are not an exception to this trend. This section will describe the most relevant efforts related to accurate interior positioning systems and navigation with multirotor UAVs near walls.

2.1. Precise Indoor Positioning

There are several alternatives available for indoor positioning. A survey and classification for all kinds of uses can be seen in [10]. In [20], there is a review of positioning systems for UAVs, as well as a discussion on their accuracy and characteristics. All in all, the prominent technologies used to perform accurate indoor position estimations (between 1 and 10 cm) are vision, ultrasonic, and UWB-based.

Vision-based systems as OptiTrack [21] or Vicon [22], based on motion capture, are very accurate, but their bulkiness and high cost make them inadequate. Simultaneous Location and Mapping (SLAM) is another option to consider. Thus, in [23] a UAV with stereo cameras is used, with a location error of less than 30 mm in the test set images, although in real-life scenarios higher errors are expected. In [24] an ultrasonic emitter is combined with Time of Flight (ToF) ranging cameras to improve the accuracy.

Regarding UltraWideBand (UWB) technology, in [25] a location system is used to perform indoor positioning with three drones at the same time, obtaining an error of around 250 mm. In subsequent work, this system is fused with SLAM techniques to improve the accuracy [26]. Similarly, works such as [27] follow the approach of merging UWB data with other sensors onboard the UAV platform, in this case inertial measuring units (IMU).

In the context of this work, UWB location systems present some advantages such as a high refresh rate and resistance to multipath propagation. Additionally, in comparison with methods such as SLAM, the UWB system requires less on-board data processing, saving weight and resulting in cheaper solutions.

2.2. UAV Movement Near Walls

Disturbances in aerodynamic forces generated by the UAV flight near walls need to be considered to perform the intended tasks successfully. The aerodynamic forces can be hard to predict. Small variations in UAV position or attitude can change interaction forces with the wall from repulsion to attraction. Both CFD techniques [28] and experimental data [29] have been analyzed to characterize these effects. They showed that flying very close to the wall can generate an attraction force to the wall and a rolling moment that tilts the UAV towards the wall. This problem can be mitigated using specific control algorithms on the UAV controller or making some hardware modifications. In [30], a controller that adapts in real-time against external disturbances is presented and tested in a near-wall flight.

Hardware modifications focus primarily on protecting the vehicle against damage caused by wall contacts, or in avoiding or controlling UAV tilting. Regarding hardware modifications, in [31] a shroud system used to absorb the impact energy is presented. In [32], a UAV with a spherical-shaped design comprised of an inner frame with a protective cage is described. For tilting control, there are systems capable of generating lateral forces without tilting the vehicle, such as in [33,34] that present two platforms capable of changing the orientation of the thrusters while in flight. This solution offers more accurate handling of the UAV attitude and a faster response time. However, on the other hand, it needs a complex and heavier control system. In [35], a payload with two horizontal propellers is used to perform inspections in vertical walls.

The tilting of the UAV affects also the tool orientation, and most applications require a specific tool orientation. In [4], the tool is reoriented using articulations and actuators. However, these types of mechanical contraptions add weight and complexity and decrease flight time.

3. Proposed NWPS

This section presents the most relevant characteristics of the proposed NWPS. The system uses the combination of two positioning systems based on different technologies that give complementary information. The correct blend of both allows obtaining a better estimate of the UAV position in the planned work scenarios.

Near-wall tasks are a challenge for most IPS as they tend to offer the best accuracy around the center of the room and worsen as the vehicle approaches the walls. The proposed system combines a UWB based IPS for global positioning with a combination of onboard LIDAR range finders to improve near-wall accuracy. It is an evolution of the one presented in [36], with improved reliability and a generalized geometry for different LIDAR sensor configurations.

3.1. UWB-Based IPS

UWB positioning systems are among the most used for accurate indoor positioning. They are based on measuring the ToF of RF signals in the UWB frequency range (3.1 to 10.6 GHz). They have good refresh rate as well as better accuracy and robustness than other RF based solutions.

The working principle consists of various anchors located at fixed and known positions and a moving tag. The tag position is computed using the distances to the anchors. These distances are measured using a Two Way Ranging (TWR) technique. Then, a triangulation algorithm uses these distances to compute the absolute position of the tag. Using TWR instead of Time of Arrival (ToA), both outward and return ToF are measured. This strategy does not need a synchronization procedure between anchors and tags; however, on the other hand, it requires more time to perform the measurements and limits the maximum number of anchors and tags within the desired refresh rate.

Anchor count and their position are to be considered to obtain the best positioning accuracy. Three are the minimum required to compute a three-dimensional position estimation, but at least four are recommended, as the UAV or the people around can shade out some of them. However, too many anchors can lead to suboptimal results and delays in the triangulation process. The anchors should be positioned carefully to maintain the line of sight with the tag, for optimal performance.

The proposed NWPS uses a commercially available UWB positioning system known as Pozyx [37]. It is smaller than other solutions and has some technical advantages. It can use up to eight anchors simultaneously. According to manufacturer specifications, positioning error should be below 150 mm near the room center in most scenarios, with a refresh rate of 25 Hz. However, when flying near walls some artifacts and positioning errors are expected.

3.2. LIDAR-Based Positioning System

To improve positioning accuracy when flying near walls, a LIDAR based relative positioning system was developed. A set of m onboard LIDAR sensors measures punctual distances to the wall. A specifically developed algorithm uses these distances to estimate the UAV position and orientation relative to the wall. A detailed description of the algorithm follows.

The m LIDAR sensors are placed in the local (UAV) reference frame at the following coordinates:

$$\mathbf{s}_i = (x_{is},\ y_{is}, z_{is}), i = 1,\ 2, \ldots, m \tag{1}$$

The direction in which the sensor i is aiming is given by the unit vector as follows:

$$\mathbf{a}_i = (x_{ia}, y_{ia}, z_{ia}),\ \|\mathbf{a}_i\| = 1, i = 1,\ 2, \ldots, m \tag{2}$$

Each sensor i. measures a distance $d_i > 0$ to the wall. The points in the wall sensed by each of the sensors are at distances d_i and directions $\mathbf{a}_i$, from the point $\mathbf{s}_i$. Thus, these points are at the following coordinates:

$$\mathbf{p}_i = (x_{ip}, y_{ip}, z_{ip}) = \mathbf{s}_i + d_i \cdot \mathbf{a}_i \tag{3}$$

Supposing the sensed surface (the wall) is a plane π, each of these m points should lie in it and, therefore, verify its equation in the form as follows:

$$\frac{x_{ip}}{a} + \frac{y_{ip}}{b} + \frac{z_{ip}}{c} = 1, i = 1,\ 2, \ldots, m. \tag{4}$$

where $a,\ b$, and c are the intersection points of the plane π with the axes x, y, and z of the reference system, respectively. Naming $r = 1/a,\ s = 1/b,\ t = 1/c$ Equation (4) becomes

$$x_{ip} \cdot \mathrm{r} + y_{ip} \cdot \mathrm{s} + z_{ip} \cdot \mathrm{t} = 1, i = 1, 2, \ldots, m \tag{5}$$

There are m equations such as (5), one for each sensor. They form a system of linear equations (SEL) that can be expressed in matrix-vector form as follows:

$$\begin{pmatrix} x_{1p} & y_{1p} & z_{1p} \\ x_{2p} & y_{2p} & z_{2p} \\ \vdots & \vdots & \vdots \\ x_{mp} & y_{mp} & z_{mp} \end{pmatrix} \cdot \begin{pmatrix} r \\ s \\ t \end{pmatrix} = \begin{pmatrix} 1 \\ 1 \\ \vdots \\ 1 \end{pmatrix} \tag{6}$$

Naming the three terms of this equation as **A**, **x**, and **b**, respectively, Equation (6) becomes

$$\mathbf{A} \cdot \mathbf{x} = \mathbf{b} \tag{7}$$

Each equation in the SEL depends on a sensor measurement, d_i. These measurements are contaminated with noise; thus, in general, $rank(\mathbf{A}) = 3$, and $rank(\mathbf{A}|\mathbf{b}) = 4$. Thus, the SEL will be overdetermined and inconsistent. Although there is no exact solution to this system, it is possible to compute a solution vector $\mathbf{x}^*$. that is optimum in the sense of least squares minimization of the error $e = ||\mathbf{A} \cdot \mathbf{x} - \mathbf{b}||$:

$$||\mathbf{A} \cdot \mathbf{x}^* - \mathbf{b}|| = \min_{\mathbf{x}} ||\mathbf{A} \cdot \mathbf{x} - \mathbf{b}|| \tag{8}$$

When $\mathbf{A} \in \mathbb{R}^{m \times 3}$. and $rank(\mathbf{A}) = 3$, the matrix $\mathbf{A}^T \cdot \mathbf{A}$. is regular, so pre-multiplying (7) by $\mathbf{A}^T$ leads to the following:

$$\left(\mathbf{A}^T \cdot \mathbf{A}\right) \cdot \mathbf{x}^* = \mathbf{A}^T \cdot \mathbf{b} \rightarrow \mathbf{x}^* = \left(\mathbf{A}^T \cdot \mathbf{A}\right)^{-1} \cdot \mathbf{A}^T \cdot \mathbf{b} \tag{9}$$

Normalization of $\mathbf{x}^*$ leads to the normalized equation of the optimum plane, as follows:

$$R = \frac{r}{||x^*||},\ S = \frac{s}{||x^*||},\ T = \frac{t}{||x^*||},\ D = \frac{1}{||x^*||} \ \rightarrow\ R \cdot x + \mathrm{S} \cdot \mathrm{y} + \mathrm{T} \cdot \mathrm{z} = \mathrm{D}. \tag{10}$$

where $\mathbf{n}^* = (R,\ S,\ T)$ is a unit vector normal to the estimated plane π^*.

The computed plane π^*. is optimum. However, due to the noise that contaminates the distances d_i, the points $\mathbf{p}_i$ do not lie in π^*. A global, normalized estimation of the error made by assuming the plan π^*. as the better solution can be made using the following:

$$\sigma = \frac{||\mathbf{A} \cdot \mathbf{x}^* - \mathbf{b}||}{||\mathbf{x}^*||} \tag{11}$$

This error estimation can be used as a quality indicator for the goodness of the estimation of π^* and to perform the fusion with the other information of the UWB.

In the current stage of development, the proposed LIDAR-based positioning subsystem has four sensors situated in the corners of a rectangle and facing the tool side of the UAV, the front in this case. They aim in the direction of the X-axis, as shown in Figure 1, and interact with the wall. The main requirements used for this LIDAR count and configuration were three: to allow position and orientation estimation of the UAV relative to the wall; to allow some failure tolerance; and to provide the quality estimator, σ. The selected sensors are "Benewake TFMini" and have an accuracy of ±6 cm from 0.3 to 6 m, ±1% from 6 to 12 m, and a 100 Hz refresh rate.

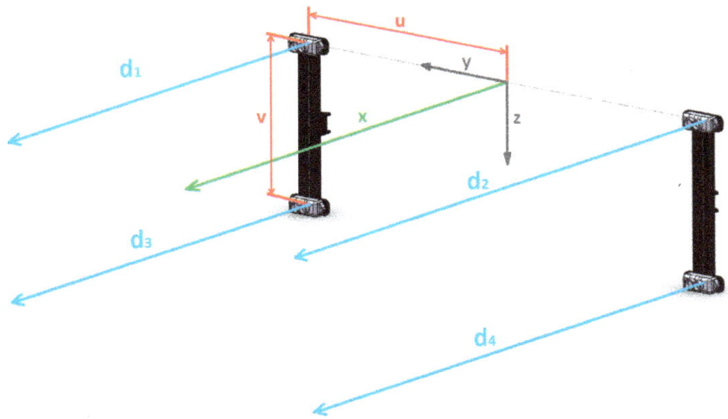

Figure 1. Frontal LIDAR-based range finders distribution in the UAV local reference frame.

From Figure 1, $\mathbf{s}_1 = (0, u, 0)^T$, $\mathbf{s}_2 = (0, -u, 0)^T$, $\mathbf{s}_3 = (0, u, -v)^T$, $\mathbf{s}_4 = (0, -u, -v)^T$, and $\mathbf{a}_i = (1, 0, 0)^T$, so

$$\mathbf{A} = \begin{pmatrix} d_1 & u & 0 \\ d_2 & -u & 0 \\ d_3 & u & -v \\ d_4 & -u & -v \end{pmatrix} \tag{12}$$

Solving (9) for $\mathbf{x}^* = (r^*, s^*, t^*)^T$,

$$\begin{aligned} r^* &= r_5 \\ s^* &= -\tfrac{r_3 \cdot r_5}{4 \cdot u} \\ t^* &= \tfrac{1}{v} \cdot \left(\tfrac{r_2 \cdot r_5}{2} - 1\right) \end{aligned} \tag{13}$$

where

$$\begin{aligned} r_1 &= d_1 + d_2 \\ r_2 &= d_3 + d_4 \\ r_3 &= d_1 - d_2 + d_3 - d_4 \\ r_4 &= \sum d_i{}^2, i = 1, .., 4 \\ r_5 &= \tfrac{4 \cdot r_1}{4 \cdot r_4 - r_3^2 - 2 \cdot r_2^2} \end{aligned} \tag{14}$$

Other simple mathematical relations compute the wall position and orientation relative to the UAV. This solution uses only simple algebraic operations, as shown in Equations (13) and (14). Therefore, it is easy to implement in a microcontroller. It allows real-time execution with a short delay time and minimal weight penalty for the UAV and generates, as has been shown, a quality measure of the estimation itself.

3.3. Data Fusion

Both the UWB and the LIDAR-based positioning subsystems deliver position estimations. However, the information that each one provides is different but complementary. The position from the UWB solution is absolute but without orientation information. The LIDAR subsystem delivers both distance and orientation information, but relative to the wall. The distance and orientation to the wall are critical parameters for near-wall tasks.

The fusion of this complementary information could offer an improvement over both subsystems but has to be adequately fused to unleash its full potential. The first step is to reference both measurements in the same coordinate system. Then, the data fusion itself can take place.

Both subsystems have been selected and designed to provide complementary information. However, there is no easy way to predetermine which one will offer the best estimation in real-life scenarios. For instance, when approaching the wall, UWB estimation should worsen, and LIDAR estimations should improve. However, unexpected situations can arise, and the estimations need to be evaluated in real-time to choose the optimal combination.

A specifically designed data fusion function based on fuzzy logic performs this task. This function uses the quality of the estimations from the LIDAR subsystem, σ, as the input parameter. Section 3.2 shows a detailed description of this function.

Figure 2 shows a schematic representation of the proposed NWPS estimation process that is executed on a PIC32 microcontroller. The output is a position estimation with the potential to improve the measurements of both subsystems. It gives more weight to the LIDAR information, especially when the UAV is near a wall, and the LIDAR delivers better accuracy. The performance of this system has been evaluated with several tests, as described in Section 5.

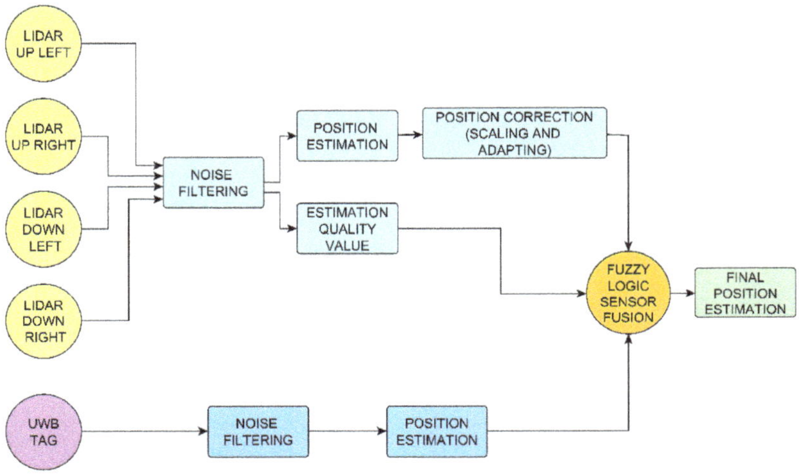

Figure 2. Conceptual diagram of the sensor fusion strategy based on fuzzy logic used for the position estimation.

4. Proposed Hardware for the Horizontal Movement of the UAV

This section presents the most relevant characteristics of the proposed modifications on the propulsion hardware of the multirotor.

One of the problems of near-wall movements is the aerodynamic distortions due to interactions between the wall and the prop wash. These distortions present a complex behavior, generating repulsive or attractive forces depending on the distance between wall and vehicle. Introducing the tilt angle of the UAV makes the interaction with the wall more unpredictable. This problem becomes worse when the distance to the wall becomes smaller. Additionally, flying near the floor or over other objects increases these forces due to the ground effect.

Another problem is related to the tool orientation. In several applications, the tool must interact with the wall at a certain angle. That is the case, for instance, of thickness measurement, which usually requires the tool to be positioned perpendicular to the measured surface. This task can be hard to accomplish with a conventional multirotor that needs to pitch or roll to change position.

To eliminate the tilt angle and allow a safer and more accurate horizontal movement, a horizontal hardware propulsion that consists of four small bidirectional motors and their propellers is added to the UAV. These thrusters are situated in a horizontal plane (XY), two of them facing the front, for forward/backward motion and the other two at a 90-degree angle for sideways motion, as shown in Figure 3.

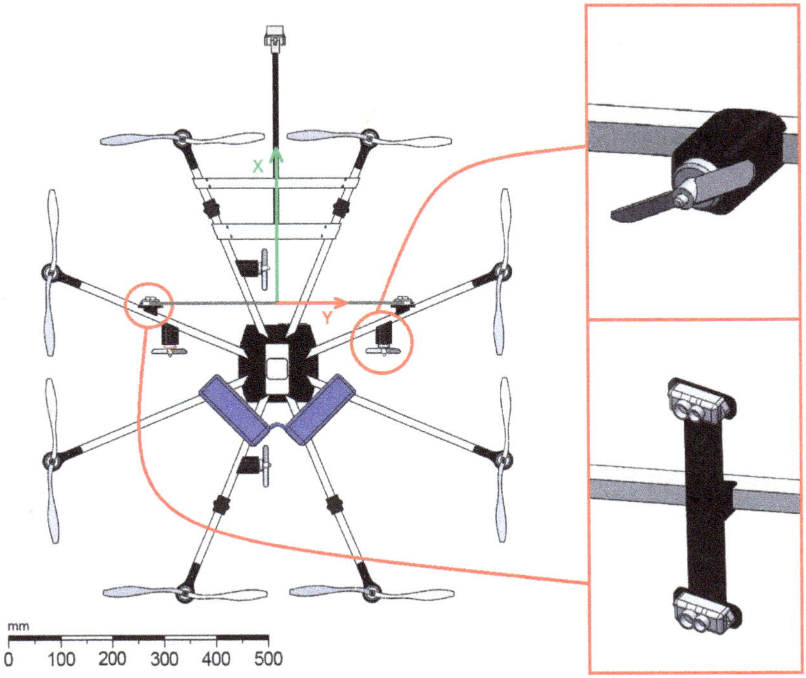

Figure 3. Placement of the LIDAR sensors and the horizontal movement hardware on the UAV platform.

In this case, a dummy tool was mounted at the front of the vehicle in order to make contact with the wall and simulate an inspection operation. It was mounted along the *X*-axis, between the front propellers, far enough to contact the wall while flying.

The four thrusters are controlled independently and share the power source of the UAV. A dedicated PIC32 microcontroller generates the control signals and is also responsible for all the NWPS calculations and fusion process explained in Section 3.3 The total weight added by the entire proposed system, including the propellers and associated electronics, is less than 200 g. With this configuration, the multirotor can move horizontally up to 3 m/s in X and Y directions without tilting. As vertical motion (in Z-axis) and yaw does not induce UAV tilting, with this strategy, the vehicle can yaw and displace in 3D without any induced tilting. The next section details the performance results of this propulsion system.

Lightweight and simplicity have a severe impact on the flying time of small UAVs. Keeping it light and simple helps in the final implementation in real-world scenarios. Both factors were considered from the initial design stages. The developed hardware is simple, so it does not require modifications to the original propulsion system. Only software configuration is needed to install it on a Commercial Off-The-Shelf (COTS) multirotor. Thus, the vehicle propulsion efficiency is not affected in the process. Furthermore, the little added weight enables its use even on small vehicles where other systems may not be feasible.

For all the tests in this paper, the proposed system was installed in an octocopter built using a HobbyKing X930 frame as a base. This frame has a glass fiber core structure, with aluminum arms for a total 895 mm of diameter. The main propulsion system is composed of eight Turnigy Aerodrive SK3 2836 brushless motors equipped with 305 × 105 mm propellers.

The power comes from two Lithium Polymer batteries with 3 cells in series and 5000 mAh each one, connected in parallel. All the embedded electronics are powered by the same battery using a dc-dc converter to reduce the voltage. A Raspberry Pi 3 microcomputer was also included to log and access easily to all the experimental data collected and synced by the PIC32 microcontroller using its integrated Wi-Fi connection.

Control Architecture

A newly designed control architecture controls both the main propulsion system of the UAV and the proposed horizontal thrusters. The new control logic comprises different UAV behavior modes and allows the addition of new flight modes or modification of the existing ones to suit the task requirements.

A switch in the remote control lets the human pilot select the flight mode in real-time. In all flight modes, the UAV main rotors control altitude and yaw. In horizontal displacement flight modes, the vehicle moves without tilting, and the horizontal thrusters control the horizontal displacement of the UAV. In its current form, there are three flight modes:

- Manual. It allows the usual operation of the UAV, with all movements controlled by the main rotors. This mode is intended for fast positioning of the UAV and for taking off and landing maneuvers.
- Manual horizontal. It uses the thrusters for horizontal displacements, manually controlled by the user. The functions of the sticks in the remote control remain the same as in typical UAV operations. With this flight mode, a human pilot can be easily trained for precise horizontal motions, while the UAV main rotors are in charge of the rest of the movements.
- Auto horizontal. It is adjustable and customizable for the application. In its current form, one of the sticks of the remote control is used to control the distance to the wall and the sideways displacement speed of the UAV, while the control system uses the NWPS fusion algorithm to keep the vehicle perpendicular to the wall surface and at a distance set by the pilot.

5. Tests and Results

This section shows and analyzes the results of the tests designed and carried out to evaluate the performance of the NWPS and the hardware for navigation near walls.

All tests were conducted in the same laboratory, a free space with a vertical wall 4 m long, 4 m wide, and 6 m tall with a second-floor platform that eases the flight observation. The setup for the UWB system was the same for all the experiments and consisted of five anchors deployed around the workspace at different heights, following the manufacturers' advice to obtain the best accuracy.

Three different test groups have been performed to evaluate all the proposed systems. The first one is comprised of the NWPS tests to evaluate the accuracy of the proposed positioning system. The second group assesses the horizontal movement hardware capabilities and advantages. Finally, a final test with both systems working together on a real UAV is also described and the results presented.

5.1. NWPS Tests Description

Two sets of tests were designed and performed to evaluate the accuracy of the proposed NWPS: one static and the other at a controlled speed.

On the static tests, the UAV was standing in five different positions over the intended working area, from 1 to 3 m from the wall in 0.5 m increments. Careful measurements resulted in precise values of distance to the wall in all static positions. The UWB and NWPS estimations were compared against that values, considered as ground truth.

The second set of tests consisted of wall approaching maneuvers. The dynamic tests have been designed considering the potential applications, and the approaching maneuver was selected as it is one of the most common for near-wall tasks. Wall approaching maneuvers were conducted at different, known and constant speeds. The UAV remained facing the wall in some of the tests. In the others, it maintained an angle of 20° to the wall.

The comparisons made between the outputs from UWB and NWPS served to assess the improvements in the NWPS achieved with the incorporation of the LIDAR.

A conveyor belt with regulated speed controlled the approach to the wall in the dynamic maneuvers. A Hohner 58 series incremental rotary encoder with 2000 pulses per turn was installed in the 65 mm diameter drive pulley, leading to a final resolution for the estimated position of 0.1 mm. Two photocell sensors at known and fixed distances to the wall were also used to provide a precise ground truth reference for both the speed and position of the UAV.

The microcontroller onboard the UAV captured all the data from the sensors in real-time to keep a precise synchronization between the different streams and logged it into the Raspberry Pi 3 equipped onboard.

5.2. NWPS Tests Results

The results for both the static and dynamic tests described previously are presented and analyzed in order to offer a broad vision of the proposed NWPS performance in different conditions.

5.2.1. NWPS Static Results

Figure 4 shows the result of the wall distance estimation on the static tests. The initial distance is 1 m, and it is increased by 0.5 m in each measurement. Both measuring systems offer very little dispersion over the considered 5000 samples for each test. The average estimation error rises to 122 mm. As can be seen, the proposed NWPS provides improved accuracy and stability of measurements over the entire range of distances.

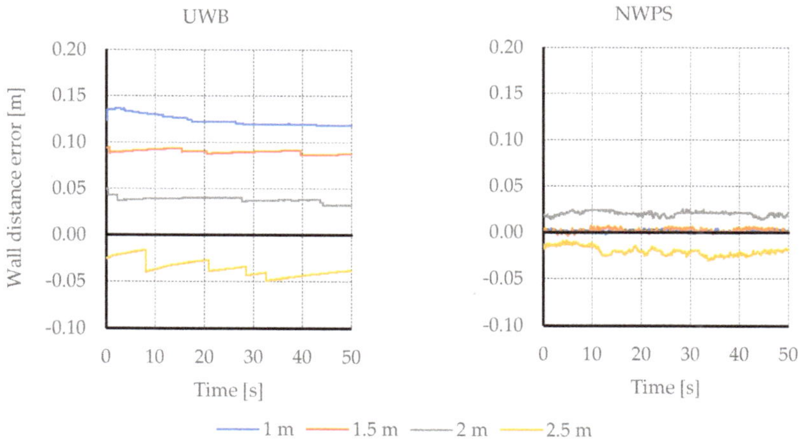

Figure 4. Results of the estimation of the distance to the wall in the static tests. Comparison between the Ultra-wideband (UWB) and the proposed Near-Wall Positioning System (NWPS).

5.2.2. NWPS Dynamic Results

A total of 20 tests were performed to evaluate the NWPS dynamic accuracy. A perpendicular and a 20-degree angle approach to the wall were tested. Each approach was repeated at five different speeds between 0.2 and 0.5 m/s using the conveyor belt and encoder system as a ground truth. All tests were repeated twice to reduce the chances of spurious errors. Figure 5 shows the distance estimation results for the fastest test performed perpendicular to the wall. As in the static tests, the UWB error increased with the proximity to the wall. As can be seen, the proposed NWPS behaves substantially better than the UWB alone.

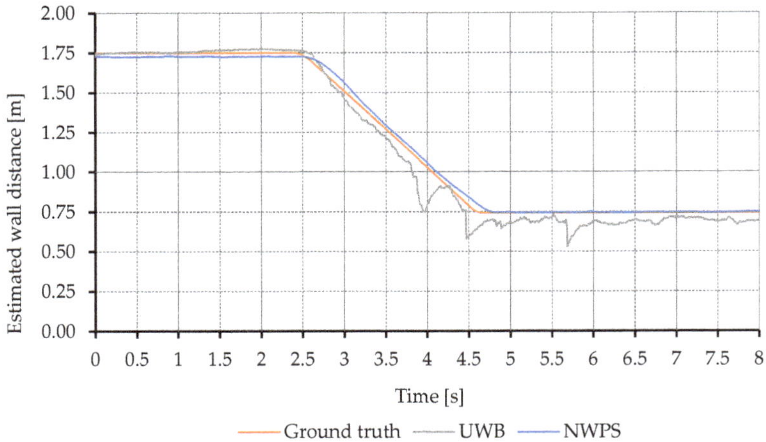

Figure 5. Estimation of the distance to the wall in perpendicular approaching maneuvers at 0.5 m/s. Comparison between conveyor belt measurements and the estimations from the UWB and the proposed NWPS.

Regarding the 20-degree angle tests, the results are very similar to the perpendicular approach with the UWB offering good estimations that worsen when the UAV is close to walls as can be seen in Figure 6. As in the previous case, the LIDAR information in the proposed NWPS improves the information from the UWB.

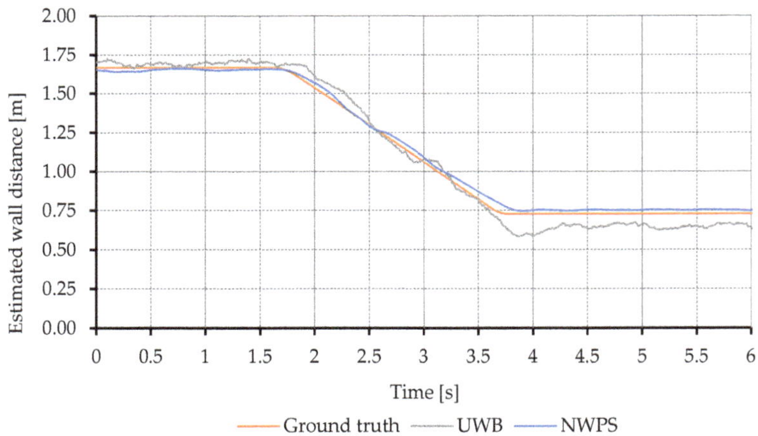

Figure 6. Estimation of the distance to the wall in 20-degree approaching maneuvers at 0.5 m/s. Comparison between conveyor belt measurements and the estimations from the UWB and the proposed NWPS.

There are some sudden error spikes in the UWB estimation, which can also be seen in Figure 5. That kind of error can be potentially dangerous for fast control loops.

Further analysis of the logged data can be seen in Figure 7, that shows a box plot of the estimation errors during dynamic tests compared to the conveyor belt ground truth, with data grouped by displacement speed in m/s. As expected, the error tends to increase with the speed, but the error magnitude and dispersion are higher for the UWB than for the proposed NWPS system. The high number of outliers in the UWB estimation evidence the spike-type errors seen in previous graphs.

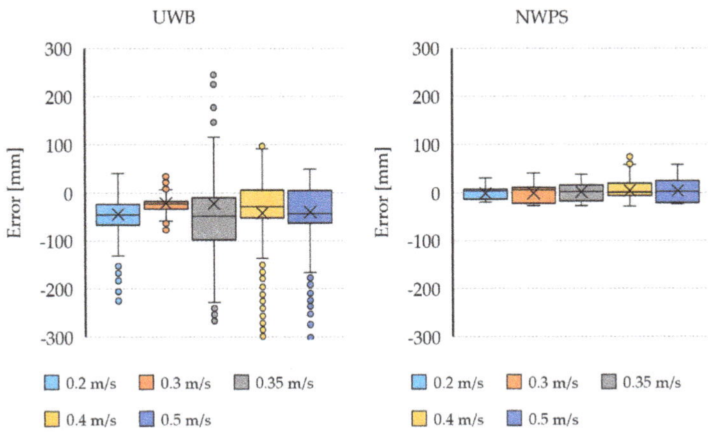

Figure 7. Estimation errors of the distance to the wall in the dynamic test, for the UWB and the proposed NWPS. Conveyor belt considered as ground truth.

5.3. Horizontal Movement Hardware Tests Description

A test was designed to demonstrate the effectiveness and evaluate the benefits of the horizontal position control hardware. This test, unlike all the others, was performed within an outdoor flight and was divided into two phases. In the first phase, the UAV performed a constant height linear movement in one direction and then returned to the origin. This phase was used to calculate the maximum horizontal speed that can be achieved using the horizontal thrusters. In the second phase, the UAV autonomously flew the same path as that of the previous phase and at the same speed. In this phase, however, only the main rotors were used.

This test was designed to be performed outdoors to be able to make high-speed movements without obstacles. Additionally, flying outdoors, GPS could be used as ground truth for the displacements and speeds. The test is used to reveal the expected enhancements of the proposed thruster-based navigation control schema. Moreover, the test shows the behavior of the proposed horizontal movement hardware in the face of disturbances such as wind or wind gusts.

5.4. Horizontal Movement Hardware Tests Results

The outdoor test flight was carried out in a UAV airfield, following all the legal and safety requirements. After the take-off, a flight was performed at a constant altitude of 20 m. The flight consisted of following a straight path back and forth. Using data from the onboard GPS, the estimated speed was 2.87 m/s one way and 2.8 m/s on the way back. The pitch angle during this displacement, measured using the UAV IMU was 1 degree.

The same speed was programmed for the second part of the test flight, but in this case, using the UAV main rotors. The UAV initially tilted 8 degrees in pitch to start the maneuver and then maintained the setpoint speed with a constant pitch angle of 6 degrees.

To perform wall-related tasks, the UAV could carry a tool, usually at the end of a pole. The tool movement created by this pitch angle is directly proportional to the distance of the tool to the multirotor center of rotation. This distance is 850 mm for the multirotor and dummy tool used in these tests. With that distance and the measured tilt angle, the resulting vertical tool displacement can be estimated at 118 mm for the start of the maneuver and 89 mm for the continuous movement. These unforeseen displacements make the precise positioning of the tool difficult.

This test demonstrates the capabilities and advantages of the proposed propulsion system to successfully move the multirotor at reasonably high speeds without undesired pitch or roll movements. It also tries to illustrate the tilt-induced tool movement, problematic when performing near-wall tasks, that the proposed navigation system can overcome.

5.5. Description of the Final Test

A final test was designed to verify the compliance with the expected capabilities of both the proposed NWPS and the navigation control system. The test consisted of real indoor flight in the absence of GNSS signals, with a UAV using the proposed NWPS and navigation control hardware. The test was run in four phases:

1. Takeoff and UAV stabilization at a distance of 2.5 m away from the wall.
2. Controlled approach maneuver to the wall using the horizontal thrusters, and the position estimation generated by the NWPS.
3. Stationary flight at a distance of 1 m to the wall (0.2 m from the tip of the tool).
4. Return to the origin and landing.

5.6. Final Test Results

The test was performed maintaining the same workspace and UWB setup and the maneuvers were recorded from an elevated platform to better estimate distances to the wall. A photogram of the recorded flight can be seen in Figure 8:

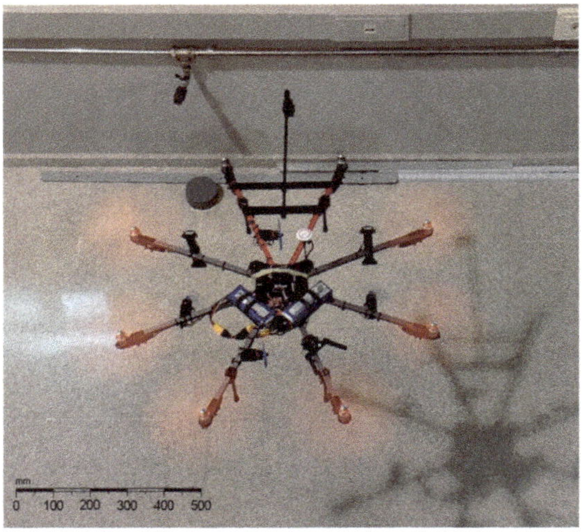

Figure 8. UAV flying close to a wall during the final test. The proposed NWPS and horizontal movement hardware are used to hover at a fixed distance to the wall.

Figure 9 shows the estimation of the distance to the wall, for both the proposed NWPS and the UWB. The same behavior is observed as in previous tests. The UWB offered good accuracy in general, with both systems estimations almost overlapping. However, the UWB system showed some spike-shaped errors and in the experiments performed underestimated the distance to the wall when flying close to it.

This underestimation problem could be caused by multipath propagation. When the vehicle is close to the wall part of the UWB signal travels directly, while some bounces off the wall, travelling a longer path. This may cause the UWB to estimate that the vehicle is further away from the anchors than what it really is and, consequently, closer to the wall.

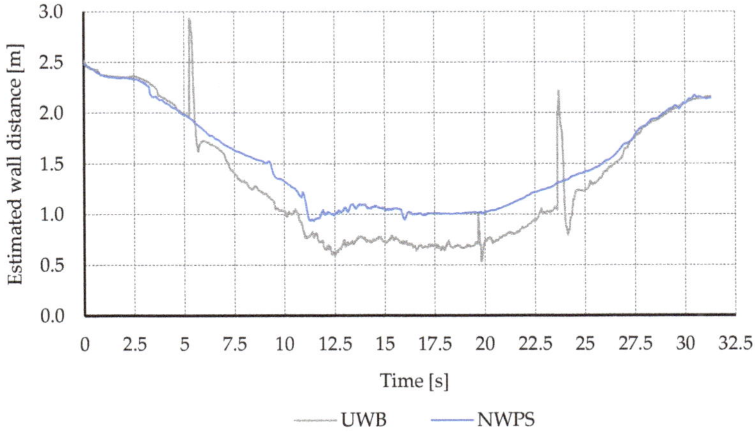

Figure 9. Estimation of the distance to the wall in an indoor flight test. Comparison between results from the UWB and the proposed NWPS.

6. Discussion

This paper presents a new near-wall positioning system (NWPS) that estimates the position of a multirotor UAV touching a wall or flying at a short distance away in the same order of magnitude as the vehicle size in GNSS-denied scenarios. A new strategy for controlling the position of the UAV and its associated tool near walls is also presented.

The proposed NWPS is based on the combination of UWB sensors and LIDAR range finders. It has been developed to improve the robustness and accuracy of commercially available IPS, particularly near walls. This NWPS adds very little weight to the UAV, and the selected estimation algorithm approach allows the computation of the position estimation using lightweight and low-cost microcontrollers. The range of the system and its accuracy is directly related to the LIDAR capabilities. The selected LIDAR sensors allow a range of up to 12 m. For the best accuracy, direct vision of the wall is required from all sensors.

The tilting of the UAV induces troublesome displacements of the working tool. This tilting can create turbulence and dangerous interaction forces with the wall. This paper presents a new strategy that allows the control of the UAV position without tilting. This strategy is based on the use of small thrusters and associated control algorithms and hardware.

Both systems have been tried separately in various tests to verify their correct operation and probe their advantages. Then, both systems were tried together, in a set of real flight tests on a GNSS denied indoor environment, flying near walls, with successful results. The outcome of the present work walks a few steps towards autonomous UAV usage in near-wall tasks. The use of UAV in these situations can reduce the operational costs of expensive processes, i.e., civil and structural inspections.

Author Contributions: Conceptualization, Á.D. and F.O.; methodology, F.O.; software, A.P.-L.; validation, J.L.-P. and F.O.; formal analysis, Á.D.; investigation, A.P.-L.; resources, J.L.-P.; data curation, J.L.-P.; writing—original draft preparation, F.O.; writing—review and editing, J.L.-P. and A.P.-L.; visualization, Á.D.; supervision, F.O.; project administration, Á.D.; funding acquisition, Á.D. All authors have read and agreed to the published version of the manuscript.

Funding: This work has been partially funded by the Ministerio de Ciencia, Innovación y Universidades of Spain/FEDER (grant RTI2018-101114-B-I00), Xunta de Galicia and FEDER (grant ED431C 2017/12). We wish to acknowledge the support received from the CITIC research center, funded by Xunta de Galicia and the European Union grant ED431G 2019/01.

Institutional Review Board Statement: Not applicable.

Informed Consent Statement: Not applicable.

Data Availability Statement: Not applicable.

Acknowledgments: We wish to acknowledge the support received from Navantia-UDC Joint Research Unit.

Conflicts of Interest: The authors declare no conflict of interest.

References

1. Hallermann, N.; Morgenthal, G. Visual inspection strategies for large bridges using Unmanned Aerial Vehicles (UAV). *Bridg. Maint. Saf. Manag. Life Ext. Proc. 7th Int. Conf. Bridg. Maint. Saf. Manag. IABMAS* **2014**, 661–667. [CrossRef]
2. Sugimoto, T.; Sugimoto, K.; Uechi, I.; Utagawa, N.; Kuroda, C. Efficiency Improvement of Outer Wall Inspection by Noncontact Acoustic Inspection Method using Sound Source Mounted Type UAV. *IEEE Int. Ultrason. Symp. IUS* **2019**, *2019*, 2091–2094. [CrossRef]
3. Bodie, K.; Brunner, M.; Pantic, M.; Walser, S.; Pfändler, P.; Angst, U.; Siegwart, R.; Nieto, J. An omnidirectional aerial manipulation platform for contact-based inspection. *arXiv* **2019**. [CrossRef]
4. Ichikawa, A.; Abe, Y.; Ikeda, T.; Ohara, K.; Kishikawa, J.; Ashizawa, S.; Oomichi, T.; Okino, A.; Fukuda, T. UAV with manipulator for bridge inspection—Hammering system for mounting to UAV. *SII IEEE/SICE Int. Symp. Syst. Integr.* **2018**, *2018*, 775–780. [CrossRef]
5. Wing, M.G.; Eklund, A.; Kellogg, L.D. Consumer-Grade Global Positioning System (GPS) Accuracy and Reliability. *J. For.* **2018**, *103*, 169–173. [CrossRef]
6. Dwiyasa, F.; Lim, M.H. A Survey of Problems and Approaches in Wireless-Based Indoor Positioning. In Proceedings of the 2016 International Conference on Indoor Positioning and Indoor Navigation, IPIN, Alcala de Henares, Spain, 4–7 October 2016; pp. 4–7.
7. Gueuning, F.E.; Varlan, M.; Eugène, C.E.; Dupuis, P. Accurate distance measurement by an autonomous ultrasonic system combining time-of-flight and phase-shift methods. *IEEE Trans. Instrum. Meas.* **1997**, *46*, 1236–1240. [CrossRef]
8. Wu, Z.H.; Han, Y.; Chen, Y.; Liu, K.J.R. A time-reversal paradigm for indoor positioning system. *IEEE Trans. Veh. Technol.* **2015**, *64*, 1331–1339. [CrossRef]
9. Liu, H.; Darabi, H.; Banerjee, P.; Liu, J. Survey of wireless indoor positioning techniques and systems. *IEEE Trans. Syst. Man Cybern. Part C Appl. Rev.* **2007**, *37*, 1067–1080. [CrossRef]
10. Brena, R.F.; García-Vázquez, J.P.; Galván-Tejada, C.E.; Muñoz-Rodriguez, D.; Vargas-Rosales, C.; Fangmeyer, J. Evolution of Indoor Positioning Technologies: A Survey. *J. Sens.* **2017**, *2017*. [CrossRef]
11. Faragher, R.; Harle, R. An Analysis of the Accuracy of Bluetooth Low Energy for Indoor Positioning Applications. In Proceedings of the 27th International Technical Meeting of the Satellite Division of the Institute of Navigation, ION GNSS 2014, Tampa, FL, USA, 8–14 September 2014; Volume 1, pp. 201–210.
12. Shi, G.; Ming, Y. Wireless Communications, Networking and Applications. *Lect. Notes Electr. Eng.* **2016**, *348*, 1269–1278. [CrossRef]
13. Davis, E.B. Aerodynamic Force Interactions and Measurements for Micro Quadrotors. Bachelor Thesis, Sardar Patel University, Anand, India, 2018. [CrossRef]
14. Robinson, D.C.; Chung, H.; Ryan, K. Computational investigation of micro helicopter near-wall effect. In Proceedings of the 19th Australasian Fluid Mechanics Conference, AFMC, Tampa, FL, USA, 8–12 September 2014; pp. 8–11.
15. Sanchez-Cuevas, P.J.; Heredia, G.; Ollero, A. Experimental approach to the aerodynamic effects produced in multirotors flying close to obstacles. *Adv. Intell. Syst. Comput.* **2018**, *693*, 742–752. [CrossRef]
16. Jardin, T.; Prothin, S.; Magana, C.G. Aerodynamic performance of a hovering microrotor in confined environment. *J. Am. Helicopter Soc.* **2017**, *62*, 1–7. [CrossRef]
17. DJI. DJI Guidance. Available online: https://developer.dji.com/guidance/ (accessed on 11 December 2020).
18. Sarim, M.; Nemati, A.; Kumar, M. Autonomous wall-Following Based Navigation of Unmanned Aerial Vehicles in Indoor Environments. In Proceedings of the AIAA Infotech at Aerospace, Kissimmee, FL, USA, 5–9 January 2015.
19. IEEE. IEEE Xplore. Available online: https://ieeexplore.ieee.org/Xplore/home.jsp (accessed on 6 February 2020).
20. Pérez, M.C.; Gualda, D.; De Vicente, J.; Villadangos, J.M.; Ureña, J. Review of UAV positioning in indoor environments and new proposal based on US measurements. *CEUR Workshop Proc.* **2019**, *2498*, 267–274.
21. NaturalPoint, I. OptiTrack. Available online: https://optitrack.com/motion-capture-robotics/ (accessed on 12 December 2020).
22. Ltd, V.M.S. Vicon. 2020. Available online: https://www.vicon.com (accessed on 12 December 2020).
23. Fu, C.; Carrio, A.; Campoy, P. Efficient Visual Odometry and Mapping for Unmanned Aerial Vehicle Using ARM-Based Stereo Vision Pre-Processing System. In Proceedings of the 2015 International Conference on Unmanned Aircraft Systems (ICUAS), Denver, CO, USA, 9–12 June 2015; pp. 957–962.
24. Paredes, J.A.; Álvarez, F.J.; Aguilera, T.; Villadangos, J.M. 3D indoor positioning of UAVs with spread spectrum ultrasound and time-of-flight cameras. *Sensors* **2018**, *18*, 89. [CrossRef]
25. Tiemann, J.; Wietfeld, C. Scalable and precise multi-UAV indoor navigation using TDOA-based UWB localization. *Int. Conf. Indoor Position. Indoor Navig. IPIN* **2017**, *2017*, 1–7. [CrossRef]

26. Tiemann, J.; Ramsey, A.; Wietfeld, C. Enhanced UAV Indoor Navigation through SLAM-Augmented UWB Localization. In Proceedings of the 2018 IEEE International Conference on Communications Workshops, ICC Workshops 2018—Proceedings, Kansas City, MO, USA, 20–24 May 2018; pp. 1–6.
27. Fresk, E.; Ödmark, K.; Nikolakopoulos, G. Ultra WideBand enabled Inertial Odometry for Generic Localization. *IFAC-PapersOnLine* **2017**, *50*, 11465–11472. [CrossRef]
28. Tanabe, Y.; Sugiura, M.; Aoyama, T.; Sugawara, H.; Sunada, S.; Yonezawa, K.; Tokutake, H. Multiple rotors hovering near an upper or a side wall. *J. Robot. Mechatron.* **2018**, *30*, 344–353. [CrossRef]
29. Conyers, S.A. Empirical Evaluation of Ground, Ceiling, and Wall Effect for Small-Scale Rotorcraft. *Electron. Theses Diss.* **2019**, *1570*.
30. Lee, D.; Awan, A.; Kim, S.; Kim, H.J. Adaptive Control for a VTOL UAV Operating near a Wall. In Proceedings of the AIAA Guidance, Navigation, and Control Conference, Minneapolis, MI, USA, 13–16 August 2012; American Institute of Aeronautics and Astronautics: Reston, VA, USA, 2012; pp. 1–12.
31. Patel, C.A. *Building a Testbed for Mini Quadrotor Unmanned Aerial Vehicle With Protective Shroud With Protective Shroud*; Dotoral Thesis, The University of Queensland: Queensland, Australia, 2002.
32. Briod, A.; Kornatowski, P.; Zufferey, J.-C.; Floreano, D. A Collision-resilient Flying Robot. *J. Field Robot.* **2014**, *31*, 496–509. [CrossRef]
33. Kamel, M.; Verling, S.; Elkhatib, O.; Sprecher, C.; Wulkop, P.; Taylor, Z.; Siegwart, R.; Gilitschenski, I. The Voliro Omniorientational Hexacopter: An Agile and Maneuverable Tiltable-Rotor Aerial Vehicle. *IEEE Robot. Autom. Mag.* **2018**, *25*, 34–44. [CrossRef]
34. Myeong, W.; Myung, H. Development of a Wall-Climbing Drone Capable of Vertical Soft Landing Using a Tilt-Rotor Mechanism. *IEEE Access* **2019**, *7*, 4868–4879. [CrossRef]
35. González-deSantos, L.M.; Martínez-Sánchez, J.; González-Jorge, H.; Arias, P. Active UAV payload based on horizontal propellers for contact inspections tasks. *Meas. J. Int. Meas. Confed.* **2020**, *165*. [CrossRef]
36. Orjales, F.; Losada-Pita, J.; Paz-Lopez, A.; Deibe, A. Positioning System for UAV Precision Tasks Near Walls in GPS Denied and Metallic Environments. In *Proceedings of the Robot 2019: Fourth Iberian Robotics Conference*; Silva, M.F., Luís Lima, J., Reis, L.P., Sanfeliu, A., Tardioli, D., Eds.; Springer International Publishing: Cham, Switzerland, 2020; pp. 315–326.
37. NV, P. Pozyx. Available online: https://www.pozyx.io/ (accessed on 12 December 2020).

sensors

MDPI

Communication

A Novel Location Source Optimization Algorithm for Low Anchor Node Density Wireless Sensor Networks

Zhongliang Deng, Shihao Tang *, Xiwen Deng, Lu Yin and Jingrong Liu

School of Electronic Engineering, Beijing University of Posts and Telecommunications, Beijing 100876, China; dengzhl@bupt.edu.cn (Z.D.); dengxiwen@bupt.edu.cn (X.D.); inlu_mail@bupt.edu.cn (L.Y.); jingrongliu@bupt.edu.cn (J.L.)

* Correspondence: swift71116@bupt.edu.cn; Tel.: +86-010-6119-8509

Abstract: Location information is one of the basic elements of the Internet of Things (IoT), which is also an important research direction in the application of wireless sensor networks (WSNs). Aiming at addressing the TOA positioning problem in the low anchor node density deployment environment, the traditional cooperative localization method will reduce the positioning accuracy due to excessive redundant information. In this regard, this paper proposes a location source optimization algorithm based on fuzzy comprehensive evaluation. First, each node calculates its own time-position distribute conditional posterior Cramer-Rao lower bound (DCPCRLB) and transfers it to neighbor nodes. Then collect the DCPCRLB, distance measurement, azimuth angle and other information from neighboring nodes to form a fuzzy evaluation factor set and determine the final preferred location source after fuzzy change. The simulation results show that the method proposed in this paper has better positioning accuracy about 33.9% with the compared method in low anchor node density scenarios when the computational complexity is comparable.

Keywords: cooperative localization; location source optimization; fuzzy comprehensive evaluation; DCPCRLB

Citation: Deng, Z.; Tang, S.; Deng, X.; Yin, L.; Liu, J. A Novel Location Source Optimization Algorithm for Low Anchor Node Density Wireless Sensor Networks. *Sensors* **2021**, *21*, 1890. https://doi.org/10.3390/s21051890

Academic Editor: Simon Tomažič

Received: 21 January 2021
Accepted: 5 March 2021
Published: 8 March 2021

1. Introduction

The Internet of Things (IoT) is a booming new industry and wireless sensor networks (WSN), as the perception layer of the IoT system [1], provide data collection [2], information transmission [3], scene recognition [4] and other functions to ensure the normal operation of the networked system. Wireless sensor networks can replace humans working in harsh natural environments and complete complex and tedious tasks. They are widely used in precision agriculture [5], elderly care [6], air monitoring [7], smart home [6], disaster supervision [8] and many other fields.

Taking into account the actual needs of IoT systems, WSN usually have the characteristics of real-time communication [9], random and irregular distribution of nodes [10], dynamic topology [11], large scale [12], complex deployment environment [13], etc., which brings a certain degree of difficulty to information collection, processing and analysis. Obtaining accurate location and time information of data sources is a prerequisite for sensor network analysis and application data [9–13]. Therefore, the positioning and time synchronization technology of WSN is an important part of the application of the IoT, which has attracted the enthusiasm of researchers at home and abroad.

Common positioning principles include Approximate Perfect Point-In-Triangulation (APIT) [14], distance vector hop (DV-Hop) [15], received signal strength indicator (RSSI) fingerprint [16] that are range-free method, and RSSI ranging [17], times of arrival (TOA) [18] and time difference of arrival (TDOA) [19] based on ranging information. The positioning accuracy of the range-base localization algorithms are usually better than the range-free positioning localization algorithms. The localization algorithms based on TOA ranging information is one of the main research directions of sensor network positioning due to its low cost and high positioning accuracy [20]. However, due to the relatively high cost of anchor nodes, it is difficult to deploy them in large numbers in actual applications. Cooperative localization method can rely on the location information and communication channels of other nodes to provide coordinated information to improve the positioning performance of the system [21]. However, the coordinated information in the scene of low anchor node density may also cause a decrease in positioning accuracy [22], as shown in Figure 1.

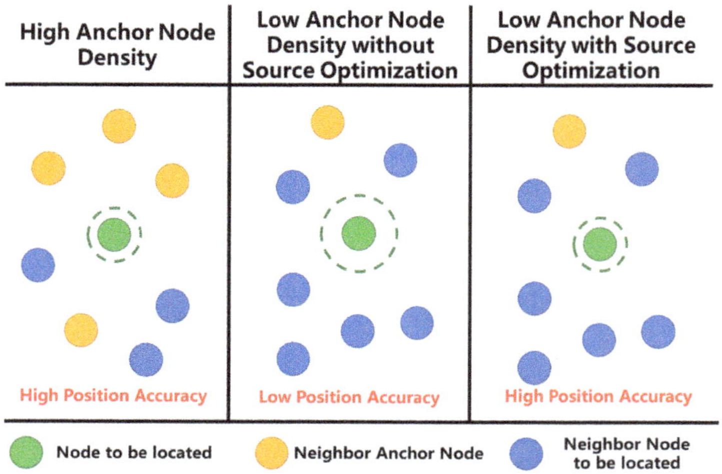

Figure 1. Different position accuracy with different anchor node density. Under low anchor node density, the number of neighbor nodes to be located in the position source increases, and the uncertainty of the localization information of the position source increases, resulting in a decrease in the position accuracy of the node to be located.

Therefore, it is necessary to optimize and screen out the location source that is helpful to improve the positioning performance. The main contributions of this paper are summarized as follows:

(1) Frist of all, the system model of low anchor node density is defined. Nodes calculate location and time skew with TOA method in this model.
(2) Then, a novel location source optimization algorithm is proposed for low anchor node density scenario. In the proposed method, a location source select structure is established with fuzzy comprehensive evaluation. Distribute conditional posterior Cramer-Rao lower bound (DCPCRLB), distance measurement and direction angle is considered as the most significant factors to select location source.
(3) The validity and rationality of the proposed method are verified by experiments.

The structure of this paper is as follows: Section 1 is the introduction, which summarizes the background knowledge and explains the significance of studying the location source select in cooperative localization. Section 2 lists the related work of this research direction in recent years. Section 3 gives out the system model of cooperative localization in low anchor node density, then proposes a novel location source select method based on fuzzy comprehensive evaluation. Section 4 shows the simulation scenario, analyses the performances of the proposed method in this paper and discusses the future work of this studying. Section 5 presents a summary of the research content of this paper.

2. Related Work

In recent years, anchor node selection has been a research topic in cooperative localization. Researchers pay lot of attention to anchor node selection, but the ambiguity of nodes position and distance measurement leads to a low accuracy of node position.

So far, domestic and foreign experts and scholars have done a lot of research on anchor node selection from different aspects and have achieved good results. The main work is as follow:

In [23], a localization method called mobile-beacon based iterative localization (MBIL) is proposed. In this method, the position confidence of the node is calculated with the number of iterations, the residual energy and the deviation degree error of the localized node's estimated location. The confidence is used to optimize the location source, and could achieve a high positioning performance is achieved in a short time.

Reference [24] analyzes the localization error caused by the selection of anchor nodes first. Based on above analysis conclusion, they proposes an improved least square localization algorithm based on the selection of anchor nodes with distance clustering (LSL-DC). With distance clustering, the anchor nodes are chosen. The simulation results indicate LSL-DC algorithm can improve the localization precision.

In [25], distance measure errors are also considered as the most significant factor which affect position accuracy, a optimizing method called node segmentation with improved particle swarm optimization (NS-IPSO) is proposed to filter the positioning source by the distance between nodes and the communication frequency, so as to avoid large distance measurement errors and improve positioning accuracy.

Reference [26] considers a localization problem in non-uniformly and holes in application environment, which affect the accuracy of distance estimation and causes large position errors in node positioning. They proposed a localization method called boundary-based anchor selection method for WSNs node localization (BASL). In this method, nodes first explore WSN connectivity to confirm whether they are boundary region nodes. Then, the node to be locate selects anchor nodes by checking the number of boundary region nodes in their shortest path between itself and anchor nodes. The results show that the BASL method can alleviate localization error which is caused by the hole in the scenarios.

Reference [27] analyzed the impact of horizontal dilution of precision (HDOP) on positioning accuracy in underwater scenes, and selected the node with the smallest HDOP as the location source to obtain the highest positioning accuracy, which called generalized second-order time-difference-of-arrival (GSTDOA) algorithm.

In [28], an enhanced three-dimensional DV-hop algorithm is proposed, which enhance its location accuracy. Coplanarity is used to select an optimal set of beacon nodes around an unidentified node for its location estimation.

Reference [29] proposed social network analysis based localization technique with closeness centrality (SNA-CC). Closeness centrality is obtained by calculating the average distance value between the node and all its neighbor nodes. This paper uses this as the importance evaluation criterion of nodes to screen the nodes. After screening, the positioning accuracy has been improved to a certain extent, and its essence is still in the screening of distance measurement errors.

In [30], a method called dynamic reference selection-based self-localization algorithm (DRSL) is proposed which combine location sources with the smallest least square error were selected to achieve the best positioning accuracy.

The following Table 1 summarizes the above-mentioned location source selection algorithm literature, the screening factors used, and the method of fusion between factors, and compares them with the method proposed in this article.

Table 1. Comparative of different location source selection algorithms.

Location Source Selection Algorithm in Localization	Factors of Location Source Selection	Method of Fusion between Factors
MBIL [23]	the number of iterations, the rediual energy and the deviation degree error	linear combination
LSL-DC [24]	the distance measure error	single factor
NS-IPSO [25]	the distance measure error and the communication frequency	linear combination
BASL [26]	the number of boundary region nodes	single factor
GSTDOA [27]	HDOP	single factor
enhanced three-dimensional DV-hop [28]	coplanarity	single factor
SNA-CC [29]	closeness centrality	single factor
DRSL [30]	the smallest least square error	single factor
Proposed method	the DCPCRLB, the distance measurement and direction angle	the fuzzy comprehensive evaluation

The above methods only consider part of the factors that affect the positioning performance, and do not include all factors in the positioning source optimization framework. The main contribution of this paper is a novel optimization algorithm of the fuzzy comprehensive evaluation [31] framework, which forms a flexible and fast positioning source selection framework. At the same time, it is based on many factors to select the location source such as the DCPCRLB considering the influence of clock, distance observation, direction angle, etc.

3. Location Source Selection Algorithm Based on Fuzzy Comprehensive Evaluation

In this section, first a two-dimensional positioning scene is shown, and the positioning principle and positioning method used in this article are basically explained. Then the DCPCRLB is introduced to pave the way for the proposed location source optimization method. Finally, the node selection algorithm proposed in this article is introduced, and the steps and operation process of the algorithm are introduced.

As shown in the Figure 2, the positioning node selection is an optional step after calculating the distance measurement in the overall positioning process and before the positioning solution. The addition of positioning node selection can reduce the computational complexity of subsequent steps, thereby reducing the time-consuming of the overall positioning process. At the same time, since the node selection screens low-quality positioning sources, the positioning accuracy can be improved.

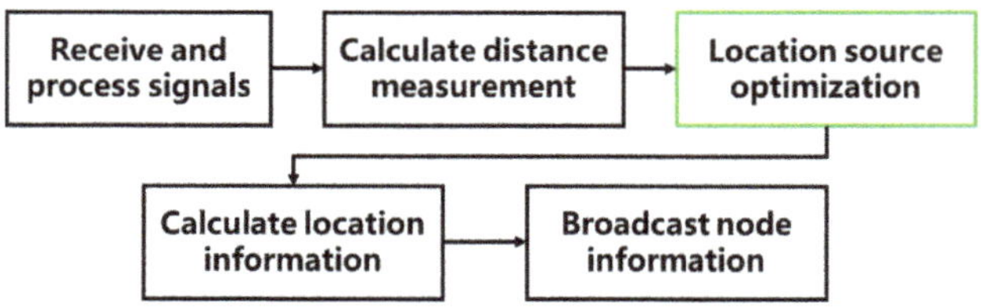

Figure 2. Flow chart of cooperative localization with location source optimization.

3.1. System Model

In the actual layout, the anchor nodes in the same scene are usually at the same height, so there is a big error in height measurement. Height measurement is usually achieved by other methods, so two-dimensional scenario is considered. The positioning scene includes anchor nodes and nodes to be positioned. The position vector of node i at time l is represented by $\boldsymbol{x}_{i,l} = \begin{bmatrix} x_{i,l} & y_{i,l} \end{bmatrix}^T$, and the clock slope is represented by $a_{i,l} \triangleq (\tilde{t}_{i,l} - \tilde{t}_{i,l-1})/(T_l - T_{l-1})$, $\tilde{t}_{i,l}$ represents the local time of node i, and T_l represents the real time. Assume that the anchor node time is synchronized with the real time, that is, for anchor node i, $a_{i,l}=1$. The estimated vector of the node i to be located at time l is $\boldsymbol{\theta}_{i,l}=\begin{bmatrix} x_{i,l} & y_{i,l} & a_{i,l} \end{bmatrix}^T$, its line-of-sight neighbor node set is defined as $N_{i,l}$. The set of vectors to be estimated for all nodes to be located at time l is $\boldsymbol{\Theta}_l=\{\cdots,\boldsymbol{\theta}_{i,l},\cdots\}$.

Node i obtains TOA observations by receiving information from neighboring node j at time l as follows:

$$\widetilde{r}_{ij,l}=||\boldsymbol{x}_{i,l}-\boldsymbol{x}_{j,l}||+cT\left(a_{i,l}-a_{j,l}\right)+n_{ij,l}, \tag{1}$$

where $||\boldsymbol{x}_{i,l}-\boldsymbol{x}_{j,l}||$ represents the Euclidean distance between node i and node j at time l, c is the speed of light, T is the time difference between adjacent moments, and $n_{ij,l}$ is TOA observation noise, assuming that it conforms to Gaussian distribution as $n_{ij,n}\sim\mathcal{N}(0,\sigma_r^2)$. Then the probability density function of $\widetilde{r}_{ij,l}$ with respect to $\boldsymbol{\theta}_{i,l},\boldsymbol{\theta}_{j,l}$ can be expressed as:

$$p\left(\widetilde{r}_{ij,l}\middle|\boldsymbol{\theta}_{i,l},\boldsymbol{\theta}_{j,l}\right)=\frac{1}{\sqrt{2\pi\sigma_r^2}}\exp\left(\frac{\left(\widetilde{r}_{ij,l}-||x_{i,l}-x_{j,l}||-cT\left(a_{i,l}-a_{j,l}\right)\right)^2}{2\sigma_r^2}\right) \tag{2}$$

Assuming that the change process of the vector $\boldsymbol{\theta}_{i,l}$ to be estimated at node i conforms to the Gaussian Markov process [32], according to the Bayesian formula:

$$p\left(\boldsymbol{\theta}_{i,l}\middle|\widetilde{r}_{ij,l}\right)\propto p(\boldsymbol{\theta}_{i,l}|\boldsymbol{\theta}_{i,l-1})\prod_{j\in N_{i,l}}p\left(\widetilde{r}_{ij,l}\middle|\boldsymbol{\theta}_{i,l},\boldsymbol{\theta}_{j,l}\right), \tag{3}$$

According to the MAP criterion, the estimated value $\hat{\boldsymbol{\theta}}_{i,l}$ of the vector $\boldsymbol{\theta}_{i,l}$ to be estimated at node i can be expressed as:

$$\begin{aligned}\hat{\boldsymbol{\theta}}_{i,l}&=\underset{\boldsymbol{\theta}_{i,l}}{argmax}\,p\left(\boldsymbol{\theta}_{i,l}\middle|\widetilde{r}_{ij,l},\boldsymbol{\theta}_{j,l}\right)\\&=\underset{\boldsymbol{\theta}_{i,l}}{argmax}\,p(\boldsymbol{\theta}_{i,l}|\boldsymbol{\theta}_{i,l-1})\prod_{j\in N_{i,l}}p\left(\widetilde{r}_{ij,l}\middle|\boldsymbol{\theta}_{i,l},\boldsymbol{\theta}_{j,l}\right),\end{aligned} \tag{4}$$

3.2. Distributed Cramer-Rao Lower Bound

Cramer-Rao lower bound (CRLB) is the inverse matrix of the Fisher Information Matrix (FIM) of the random variable $\boldsymbol{\theta}_{i,l}$, and is a theoretical lower bound of the mean square error of the target state estimation. The conditional posterior CRLB (CPCRLB) of the estimated state $\hat{\boldsymbol{\theta}}_{i,l}$ of node i can be expressed as:

$$\begin{aligned}MSE\left(\hat{\boldsymbol{\theta}}_{i,l}\middle|\widetilde{r}_{ij,l}\right)&=E\left\{[\hat{\boldsymbol{\theta}}_{i,l}-\boldsymbol{\theta}_{i,l}][\hat{\boldsymbol{\theta}}_{i,l}-\boldsymbol{\theta}_{i,l}]^T\middle|\widetilde{\boldsymbol{r}}_{i,l}\right\}\\&\geq F^{-1}(\hat{\boldsymbol{\theta}}_{i,l}|\widetilde{\boldsymbol{r}}_{i,l}),\end{aligned} \tag{5}$$

Among them, $F(\hat{\boldsymbol{\theta}}_{i,l}|\widetilde{\boldsymbol{r}}_{i,l})$ represents the conditional FIM of the target state estimated value $\hat{\boldsymbol{\theta}}_{i,l}$. [33] gives the global Fisher information matrix $F(l)$ iterative calculation formula of centralized CPCRLB, which can be applied to a centralized network structure with a central fusion center to realize a centralized cooperative localization. This article is aimed at a distributed network structure, and each node cannot obtain the global Fisher information matrix. Therefore, we adjust the above formula to obtain the local Fisher information matrix iterative formula corresponding to the DCPCRLB:

$$\boldsymbol{F}_i(l)\approx\boldsymbol{B}_i^{22}(l-1)-\boldsymbol{B}_i^{21}(l-1)\left(\boldsymbol{B}_i^{11}(l-1)+\boldsymbol{F}_i(l-1)\right)^{-1}\boldsymbol{B}_i^{12}(l-1), \tag{6}$$

where:

$$\boldsymbol{B}_i^{11}(l-1)=E\left(-\Delta_{\boldsymbol{\theta}_{i,l-1}}^{\boldsymbol{\theta}_{i,l-1}}\ln p(\boldsymbol{\theta}_{i,l}|\boldsymbol{\theta}_{i,l-1})\right), \tag{7}$$

$$\boldsymbol{B}_i^{12}(l-1)=E\left(-\Delta_{\boldsymbol{\theta}_{i,l-1}}^{\boldsymbol{\theta}_{i,l}}\ln p(\boldsymbol{\theta}_{i,l}|\boldsymbol{\theta}_{i,l-1})\right)=\left(\boldsymbol{B}_i^{21}(l-1)\right)^T, \tag{8}$$

$$B_i^{22}(l-1) = E\left(-\Delta_{\theta_{i,l}}^{\theta_{i,l}}\left(\ln p(\theta_{i,l}|\theta_{i,l-1}) + \ln p(\widetilde{r}_{i,l}|\theta_{i,l})\right)\right), \tag{9}$$

where $\Delta_{\mathbf{\Theta}_{l-1}}^{\mathbf{\Theta}_l} = \nabla_{\mathbf{\Theta}_{l-1}}\nabla_{\mathbf{\Theta}_l}^T, \nabla_{\mathbf{\Theta}_l} = \left[\cdots, \frac{\partial}{\partial\theta_{i,l}}, \cdots\right]^T$. $\mathbf{F}(l)$ is a global Fisher information matrix. The DCPCRLB of node i can be obtained by inverting the Fisher information matrix, that is, $CRLB\left(\theta_{j,l}\right) = \mathbf{F}_i^{-1}(l)$.

3.3. Location Source Optimization Algorithm

In a 2-dimensional wireless sensor network, the node to be located needs to establish communication with at least three anchor nodes to complete positioning. In the location-time joint estimation problem, it is necessary to establish communication with at least four anchor nodes to complete the calculation of the vector to be estimated. In the scenario of low anchor node density, the cooperative localization method provides neighbor nodes as pseudo-anchor nodes for the node to be located, but the pseudo-anchor node itself has low accuracy as the node to be located, and the inappropriate introduction of too many pseudo-anchor nodes will affect the final estimation result has a serious impact, so it is necessary to optimize the introduced pseudo-anchor nodes.

In this section, we will introduce the fuzzy comprehensive evaluation method to design the location source selection algorithm. For node $j \in N_{i,l}$ to participate in the optimal process of positioning and settlement, the DCPCRLB $posCRLB_{j,l}$, node distance measurement $\widetilde{r}_{ij,l}$, direction angle $\alpha_{ij,l}$ will participate in the judgment as a set of factors:

$$posCRLB_{j,l} = CRLB_{1,1}\left(\theta_{j,l-1}\right) + CRLB_{2,2}\left(\theta_{j,l-1}\right) + Z \cdot c^2 CRLB_{3,3}\left(\theta_{j,l-1}\right), \tag{10}$$

$$\alpha_{ij,l} = arctan\frac{x_{j,l-1} - x_{i,l-1}}{y_{j,l-1} - y_{i,l-1}}, \tag{11}$$

$CRLB_{1,1}\left(\theta_{j,l-1}\right)$ represents the element in the first row and first column of $CRLB\left(\theta_{j,l-1}\right)$, and Z is the scaling constant. The raw data $\mathbf{D}_{i,l}$ of the candidate node for node i is:

$$\mathbf{D}_{i,l} = \begin{bmatrix} posCRLB_{1,l} & posCRLB_{2,l} & \cdots & posCRLB_{N,l} \\ \widetilde{r}_{i1,l} & \widetilde{r}_{i2,l} & \cdots & \widetilde{r}_{iN,l} \\ \alpha_{i1,l} & \alpha_{i2,l} & \cdots & \alpha_{iN,l} \end{bmatrix}, \tag{12}$$

where M is the number of nodes to be selected for node i. Then standardize the data and determine the membership function of each factor according to the number of neighbor anchor nodes and the number of nodes participating in the positioning solution, and perform fuzzy evaluation on the selected nodes to obtain the evaluation matrix $\mathbf{A}_{i,l}$:

$$\mathbf{A}_{i,l} = \begin{bmatrix} a_{11} & a_{12} & \cdots & a_{1M} \\ a_{21} & a_{22} & \cdots & a_{2M} \\ a_{31} & a_{32} & \cdots & a_{3M} \end{bmatrix}, \tag{13}$$

where:

$$a_{1j} = \frac{\max_j posCRLB_{j,l} - posCRLB_{j,l}}{\max_j posCRLB_{j,l} - \min_j posCRLB_{j,l}}, \tag{14}$$

$$a_{2j} = \frac{\max_j \widetilde{r}_{ij,l} - \widetilde{r}_{ij,l}}{\max_j \widetilde{r}_{ij,l} - \min_j \widetilde{r}_{ij,l}}, \tag{15}$$

$$a_{3j} = \frac{\max_j \|\varphi_{i,l} - \alpha_{ij,l}\| - \|\varphi_{i,l} - \alpha_{ij,l}\|}{\max_j \|\varphi_{i,l} - \alpha_{ij,l}\| - \min_j \|\varphi_{i,l} - \alpha_{ij,l}\|}, \tag{16}$$

Among them, $\varphi_{i,l}$ is jointly determined by the direction angle of neighbor anchor nodes and the number of undetermined sources. The weight function matrix $\boldsymbol{P}_{i,l} = \begin{bmatrix} w_1 & w_2 & w_3 \end{bmatrix}$ can be determined by methods such as entropy weight method [34] and analytical hierarchy process (AHP) [35]. Finally, a fuzzy transformation $\boldsymbol{Q}_{i,l} = \boldsymbol{P}_{i,l} \cdot \boldsymbol{A}_{i,l}$ is performed to obtain the evaluation results of each candidate node, and the one with the largest value is selected as the preferred source for positioning.

As shown in Figure 3, after obtaining neighbor node information and distance measurement, the positioning process enters the positioning source optimization algorithm proposed in this article. First, a certain number of anchor nodes are included in the preferred location source $\boldsymbol{S}_{i,l}$, and then according to Equations (10) and (11) to calculate the raw data $\boldsymbol{D}_{i,l}$ of all other neighbor nodes. The evaluation matrix $\boldsymbol{A}_{i,l}$ is obtained after normalization processing according to Equations (14)–(16), and the weight function matrix $\boldsymbol{P}_{i,l}$ is obtained according to the method mentioned in [34]. Finally, the evaluation value matrix $\boldsymbol{Q}_{i,l}$ of each node is obtained through fuzzy transformation, and the node with the highest evaluation value is selected from it and included in $\boldsymbol{S}_{i,l}$. If the preferred positioning source is sufficient, the location source optimization ends and the positioning solution is entered, otherwise $\boldsymbol{D}_{i,l}$ is recalculated.

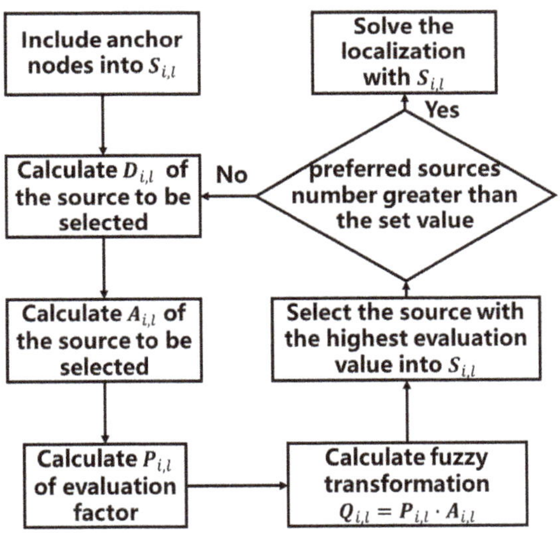

Figure 3. Flow chart of location source selection algorithm based on fuzzy comprehensive evaluation.

4. Simulation Scenario and Result Analysis

4.1. Simulation Scenario Set

In order to make our simulation environment close to the actual scene, the simulation scene will be set according to the zone 1 scene in Figure 4a, which is underground parking lot in Beijing University of Posts and Telecommunications, and Figure 4b is the real scene of zone 1.

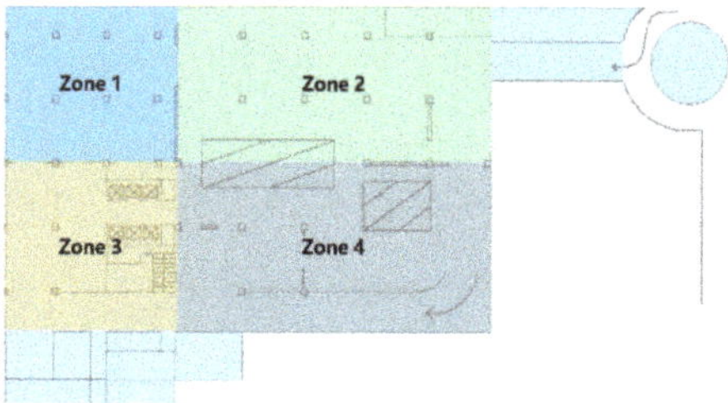

(**a**) (**b**)

Figure 4. (**a**) Simulation scenario. Zone 1 is the used simulation scene. (**b**) The real scene of zone 1 [22].

The simulation scene is set as a rectangular area of 20 m × 24 m according to the real size of zone 1. To set the average number of anchor node connections around 3, the number of anchor nodes is 4, and the maximum communication distance is set to 10 m. The number of nodes to be located is set to 20 and all nodes conform to the uniform distribution in the simulation scene. All TOA observations are the line of sight (LOS), and the distance measurement error conforms to Gaussian distribution with standard deviation of 0.3 m according to the maximum error of DW1000 (Decawave, Dublin, Ireland). All simulation results are the average of 1000 independent runs.

The initial position measurement error of nodes conforms to the Gaussian distribution, the standard deviation of the anchor node is 0.1 m, the node to be located is 10 m. In the simulation process, the clock slope of node to be located is set as 1 ppm according to the performance of the crystal oscillator used in the hardware (TG5032CFN, EPSON, Nagano-ken, Japan). Anchor nodes are static and the velocity of node to be located is set to 3 m/s as almost the fastest speed of human walking. Root mean square error (RMSE) and cumulative distribution functions (CDF) are used to evaluate the performance of proposed method.

4.2. Simulation Result Analysis

It can be seen from Figure 5 that when the number of neighbor anchor nodes is small, both the DRSL method and the method proposed in this article can effectively improve the positioning accuracy, further increase the number of adjacent anchor nodes, and the positioning source optimization algorithm can significantly improve the positioning accuracy. In addition, the different performance is noticed when all neighbor nodes are used in the graph to participate in positioning. When the neighbor anchor nodes are sufficient, the joining of neighbor cooperative nodes is difficult to improve the positioning accuracy. When the number of neighbor anchor nodes is small, the addition of neighbor cooperative nodes has a positive impact on the positioning accuracy. The proposed algorithm optimizes its role in the positioning process by screening collaborative nodes.

These results in Figure 6 show that the position accuracy affected by communication distance. Obviously, the position accuracy will increase as the communication distance increases. This is because the number of nodes that can be selected increases after the communication distance is increased, and the higher the possibility of selecting a higher-quality node combination, so various methods can obtain better positioning results. At the same time, the position accuracy of the proposed method is better than other methods. This is because the method in this article comprehensively considers many factors that affect the positioning accuracy. The improvement of the communication distance is very effective for improving the positioning accuracy, but as the communication distance increases, the improvement effect of the positioning accuracy becomes smaller. This is because it is difficult for nodes that are too far away to provide effective information for positioning. Therefore, in practical applications, a rigorous analysis of the communication distance, that is, the transmission and reception power, is performed according to the requirements for positioning accuracy, and a reasonable communication distance is obtained.

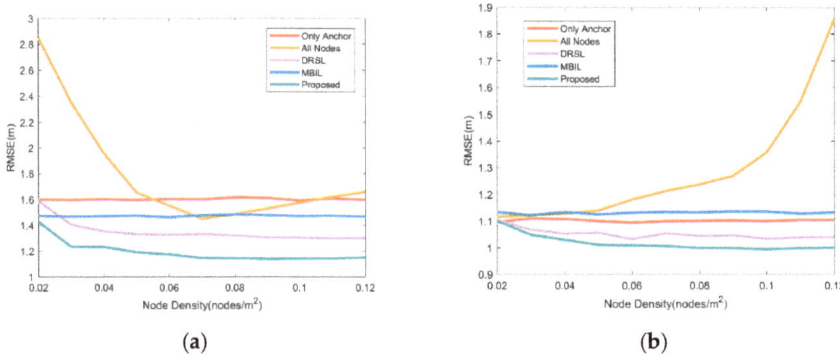

Figure 5. Comparison of positioning accuracy of various source selection methods under different node densities. (**a**) is the comparison when the average number of anchor node connections is 3. (**b**) is the comparison when the average number of anchor node connections is 10. The red line represents the non-cooperative localization using only anchor nodes, the orange represents the cooperative localization using all neighbor nodes as the source, the purple line represents the DRSL source selection location algorithm proposed in [30], the blue line represents the MBIL algorithm in [23] and the green line represents the source optimization algorithm proposed in this article for cooperative localization.

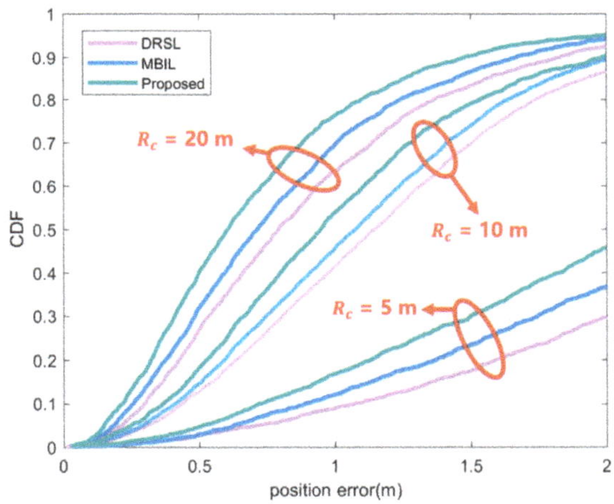

Figure 6. CDF of position error under different communication distance R_c. In the figure, the purple line represents the DRSL source selection location algorithm proposed in [30], the blue line represents the MBIL algorithm in [23] and the green line represents the source optimization algorithm proposed in this article.

Figure 7 shows the positioning accuracy of the node to be located under different initial positioning errors. It can be seen from the figure that the positioning accuracy improves with the decrease of the initial positioning error, because the quality of the position information of the neighboring nodes will affect the positioning accuracy of the nodes. The positioning accuracy of the method proposed in this paper is better than the other two comparison methods. This is because this paper uses DCPCRLB as the evaluation basis for the location information, which has a better estimate of the quality of the location information.

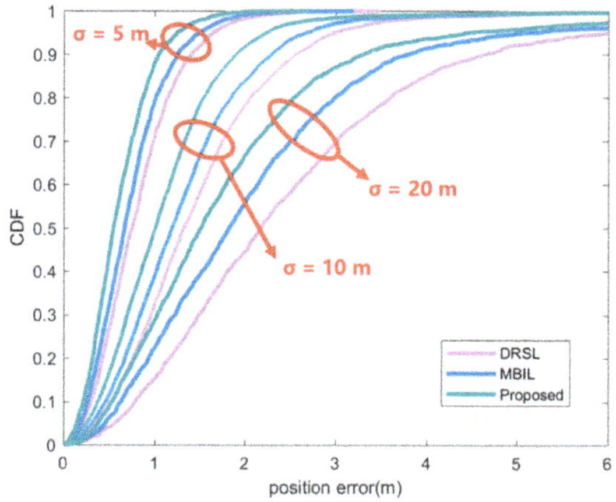

Figure 7. CDF of position error under different initial position error. In the figure, the purple line represents the DRSL source selection location algorithm proposed in [30], the blue line represents the MBIL algorithm in [23] and the green line represents the source optimization algorithm proposed in this article.

In Figure 8, compared with the other three cases, the position accuracy performs best when $v_{max} = 3$ m/s. Although the motion state will cause a certain error in the position estimation of the node to be located, it also has the advantage of changing the topology and optimizing the node distribution. Therefore, when $v_{max} = 3$ m/s, the position accuracy is slightly improved compared to the static situation. When the speed is too high ($v_{max} = 8$ m/s or $v_{max} = 15$ m/s), the error caused by the movement cannot be compensated by optimizing the node distribution. The error caused by motion can be compensated by combining positioning with sensor data such as accelerometer or gyroscope [36].

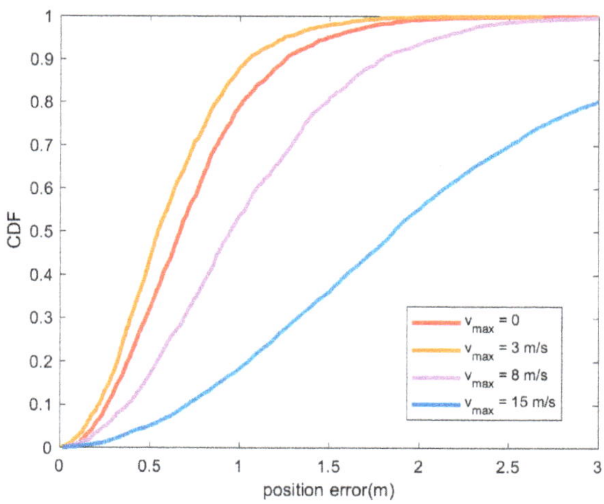

Figure 8. CDF of position error under different maximum node speed v_{max}. In the figure, the red line represents the position error CDF when all nodes are static, the orange line represents $v_{max} = 3$ m/s, and the purple line represents $v_{max} = 8$ m/s. The blue line indicates the situation where $v_{max} = 15$ m/s. In all cases, anchor node is static, the nodes to be located conform to the uniform distribution with the maximum speed v_{max}, and the direction of movement is random.

Figure 9 compares the positioning performance when using different numbers of preferred position sources in a low anchor node density scenario. It can be seen that as the number of preferred positioning sources increases, the position accuracy is improved. As the number of preferred positioning sources increases, the signal quality of the position sources and the geometric distribution of the position sources will be improved, and the position accuracy will be improved to a certain extent, but the improvement effect will decrease as the number of equipotential sources increases. By comparing the results with Figure 5, it can be found that there is a limit to improving the positioning effect by increasing the number of position sources. Too many position sources may also cause a decrease in position accuracy, which is the fact reflected by the yellow line in Figure 5.

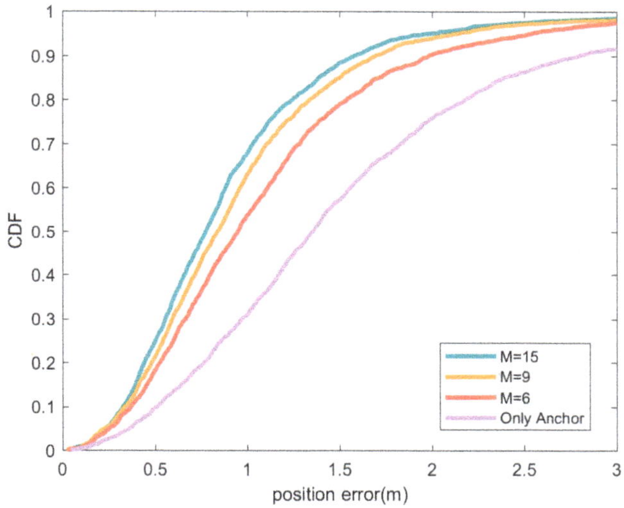

Figure 9. CDF of position error under different number of preferred positioning sources. In the figure, the green line represents the location error CDF when the preferred location source is 15, the orange line represents the preferred location source is 9, and the red line represents the preferred location source is 6. The purple line indicates the situation where only anchor nodes are used for positioning.

Figure 10 reflects the average positioning time of the proposed method and DRSL method under different numbers of neighbor nodes. The time complexity of the location source optimization algorithm is mainly related to both the average number of neighbor nodes N_1 and the number of preferred sources N_2. For the calculation of the original data $\boldsymbol{D}_{i,l}$, the evaluation matrix $\boldsymbol{A}_{i,l}$ and the fuzzy transformation $\boldsymbol{Q}_{i,l}$ in the algorithm proposed in this paper, the time complexity is $2O(N_1)$, $5O(N_1)$ and $3O(N_1)$. Finally, a preferred positioning source is obtained after comparison. Repeating the above process N_2 times is the time complexity of the node. By optimizing part of the algorithm process, the time complexity is finally $5O(N_1N_2)$. In the proposed method, nodes transmit a localization vector including $\boldsymbol{\theta}_{i,l} = \begin{bmatrix} x_{i,l} & y_{i,l} & a_{i,l} \end{bmatrix}^T$ and a the DCPCRLB $posCRLB_{i,l}$, so the communication cost is $2O(1)$. The computational complexity, run-time and communication overhead of the three algorithms is shown in Table 2.

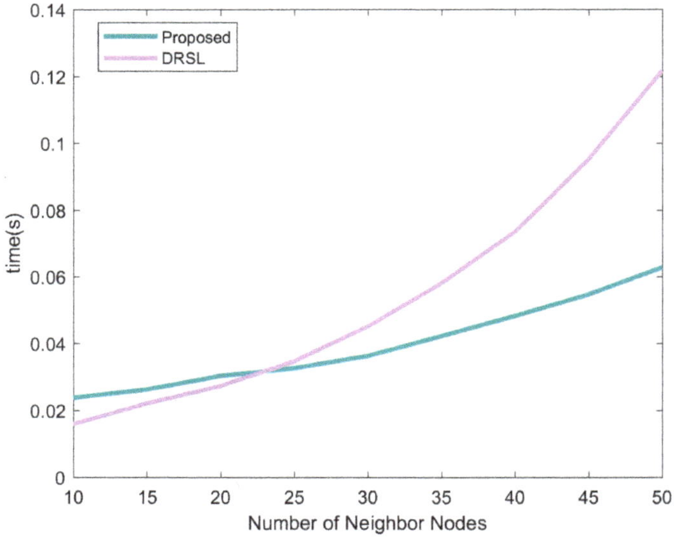

Figure 10. The average positioning time of each node under different numbers of neighbor nodes.

Table 2. Comparisons of different methods for each node at each iteration.

Method	Computational Complexity	Run-Time	Communication Overhead
Proposed method	$5O(N_1N_2)$	42.319 ms	$2 \times O(1)$
DRSL [30]	$O(N_1^2) + 2O(N_2)$	84.532 ms	$O(N_2)$+ $O(1)$
MBIL [23]	$O(N_1N_2^2)$	326.443 ms	$O(N_1) + 2 \times O(N_1)$

4.3. Future Research Directions

In the next stage of research, the following aspects will be mainly focused on: First, we plan to analyze other factors which affect position accuracy. Secondly, we shall implement the proposed method based on hardware platform and apply the proposed method in realistic scenarios. The measured results in realistic scenarios will be compared with the simulation results to improve the performances of the proposed method.

5. Conclusions

We propose a new positioning source optimization method for low-anchor node density wireless sensor networks, which comprehensively considers the positioning performance, distance, location and other factors that affect the positioning accuracy of cooperative nodes to select positioning sources. First, each node calculates its own CRLB and transmits it to neighboring nodes through collaborative information. Neighboring nodes calculate the CRLB, distance measurement and direction angle of neighboring cooperative nodes to obtain the optimal evaluation matrix, and obtain the fuzzy evaluation result through weight addition, and obtain the final optimal location source. In the fuzzy comprehensive evaluation framework, the weights of various evaluation factors can be flexibly configured. Therefore, the proposed method can screen out different neighbor nodes as positioning sources according to requirements, and then obtain a combination of positioning sources with good location performance and distribution, which can improve the positioning accuracy of the node. Compared with the DRSL and MBIL method, the positioning signal source obtained by this method can obtain higher positioning accuracy about 33.9% and 19.4% under low anchor node density, and the sacrificed calculation time is almost negligible.

Author Contributions: Conceptualization, S.T.; methodology, Z.D., S.T. and L.Y.; software, S.T. and X.D.; validation, S.T. and J.L.; formal analysis, S.T.; investigation, S.T., X.D. and J.L.; resources, Z.D. and S.T.; data curation, S.T.; writing—original draft preparation, S.T.; writing—review and editing, S.T., X.D., L.Y. and J.L.; visualization, S.T.; supervision, X.D. and J.L.; project administration, Z.D. and L.Y.; funding acquisition, Z.D. All authors have read and agreed to the published version of the manuscript.

Funding: This research was funded by the National Key Research & Development Program: 2016YFB0502003.

Institutional Review Board Statement: Not applicable.

Informed Consent Statement: Not applicable.

Data Availability Statement: Not applicable.

Conflicts of Interest: The authors declare no conflict of interest. The funders had no role in the design of the study; in the collection, analyses, or interpretation of data; in the writing of the manuscript, or in the decision to publish the results.

References

1. Capella, J.V.; Campelo, J.C.; Bonastre, A.; Ors, R. A reference model for monitoring IoT WSN-based applications. *Sensors* **2016**, *16*, 1816. [CrossRef] [PubMed]
2. Carvalho, C.; Gomes, D.G.; Agoulmine, N.; De Souza, J.N. Improving prediction accuracy for WSN data reduction by applying multivariate spatio-temporal correlation. *Sensors* **2011**, *11*, 10010–10037. [CrossRef] [PubMed]
3. Lu, W.; Gong, Y.; Liu, X.; Wu, J.; Peng, H. Collaborative energy and information transfer in green wireless sensor networks for smart cities. *IEEE Trans. Ind. Inf.* **2017**, *14*, 1585–1593. [CrossRef]
4. Bagula, A.B.; Osunmakinde, I.; Zennaro, M. On the relevance of using Bayesian belief networks in wireless sensor networks situation recognition. *Sensors* **2010**, *10*, 11001–11020. [CrossRef]
5. Kumar, S.A.; Ilango, P. The impact of wireless sensor network in the field of precision agriculture: A review. *Wirel. Wirel. Pers. Commun.* **2018**, *98*, 685–698. [CrossRef]
6. Adame, T.; Bel, A.; Carreras, A.; Melia-Segui, J.; Oliver, M.; Pous, R. CUIDATS: An RFID–WSN hybrid monitoring system for smart health care environments. *Future Gener. Comput. Syst.* **2018**, *78*, 602–615. [CrossRef]
7. Boubrima, A.; Bechkit, W.; Rivano, H. Optimal WSN deployment models for air pollution monitoring. *IEEE Trans. Wirel. Commun.* **2017**, *16*, 2723–2735. [CrossRef]
8. Han, G.; Yang, X.; Liu, L.; Zhang, W.; Guizani, M. A disaster management-oriented path planning for mobile anchor node-based localization in wireless sensor networks. *IEEE Trans. Emerg. Top. Comput.* **2017**, *8*, 115–125. [CrossRef]
9. Radmand, P.; Talevski, A.; Petersen, S.; Carlsen, S. Comparison of industrial WSN standards. In Proceedings of the 4th IEEE International Conference on Digital Ecosystems and Technologies, Dubai, UAE, 13–16 April 2010; pp. 632–637.
10. Kenniche, H.; Ravelomananana, V. Random geometric graphs as model of wireless sensor networks. In Proceedings of the 2010 2nd International Conference on Computer and Automation Engineering (ICCAE), Singapore, 26–28 February 2010; pp. 103–107.
11. Elhoseny, M.; Hassanien, A.E. Extending homogeneous WSN lifetime in dynamic environments using the clustering model. In *Dynamic Wireless Sensor Networks*; Springer: Berlin/Heidelberg, Germany, 2019; pp. 73–92.
12. Popescu, D.; Dragana, C.; Stoican, F.; Ichim, L.; Stamatescu, G. A collaborative UAV-WSN network for monitoring large areas. *Sensors* **2018**, *18*, 4202. [CrossRef]
13. Elhoseny, M.; Hassanien, A.E. Expand mobile WSN coverage in harsh environments. In *Dynamic Wireless Sensor Networks*; Springer: Berlin/Heidelberg, Germany, 2019; pp. 29–52.
14. Yuan, Y.; Huo, L.; Wang, Z.; Hogrefe, D. Secure APIT localization scheme against sybil attacks in distributed wireless sensor networks. *IEEE Access* **2018**, *6*, 27629–27636. [CrossRef]
15. Cheikhrouhou, O.; M Bhatti, G.; Alroobaea, R. A hybrid DV-hop algorithm using RSSI for localization in large-scale wireless sensor networks. *Sensors* **2018**, *18*, 1469. [CrossRef]
16. Yoo, J. Change Detection of RSSI Fingerprint Pattern for Indoor Positioning System. *IEEE Sens. J.* **2019**, *20*, 2608–2615. [CrossRef]
17. Li, G.; Geng, E.; Ye, Z.; Xu, Y.; Lin, J.; Pang, Y. Indoor positioning algorithm based on the improved RSSI distance model. *Sensors* **2018**, *18*, 2820. [CrossRef] [PubMed]
18. Wu, S.; Zhang, S.; Huang, D. A TOA-based localization algorithm with simultaneous NLOS mitigation and synchronization error elimination. *IEEE Sens. Lett.* **2019**, *3*, 1–4. [CrossRef]
19. Díez-González, J.; Álvarez, R.; Sánchez-González, L.; Fernández-Robles, L.; Pérez, H.; Castejón-Limas, M. 3D Tdoa problem solution with four receiving nodes. *Sensors* **2019**, *19*, 2892. [CrossRef]
20. Zhang, L.; Chen, M.; Wang, X.; Wang, Z. TOA estimation of chirp signal in dense multipath environment for low-cost acoustic ranging. *IEEE Trans. Instrum. Meas.* **2018**, *68*, 355–367. [CrossRef]
21. Wymeersch, H.; Lien, J.; Win, M.Z. Cooperative localization in wireless networks. *Proc. IEEE* **2009**, *97*, 427–450. [CrossRef]

22. Deng, Z.; Tang, S.; Jia, B.; Wang, H.; Deng, X.; Zheng, X. Cooperative Localization and Time Synchronization Based on M-VMP Method. *Sensors* **2020**, *20*, 6315. [CrossRef] [PubMed]
23. Su, Y.; Guo, L.; Jin, Z.; Fu, X. A Mobile-beacon Based Iterative Localization Mechanism in Large-scale Underwater Acoustic Sensor Networks. *IEEE Internet Things J.* **2020**, *8*, 3653–3664. [CrossRef]
24. Chen, X.; Chen, J.; Chen, C.; He, J.; Lei, B. A node localization algorithm for wireless sensor networks using distance clustering to select the anchor nodes. *Sens. Lett.* **2013**, *11*, 745–748. [CrossRef]
25. Phoemphon, S.; So-In, C.; Leelathakul, N. A hybrid localization model using node segmentation and improved particle swarm optimization with obstacle-awareness for wireless sensor networks. *Expert Syst. Appl.* **2020**, *143*, 113044. [CrossRef]
26. Zhang, K.; Zhang, G.; Yu, X.; Hu, S. Boundary-Based Anchor Selection Method for WSNs Node Localization. *Arab. J. Sci. Eng.* **2021**. [CrossRef]
27. Sun, S.; Qin, S.; Hao, Y.; Zhang, G.; Zhao, C. Underwater Acoustic Localization of the Black Box Based on Generalized Second-Order Time Difference of Arrival (GSTDOA). *IEEE Trans. Geosci. Remote Sens.* **2020**. [CrossRef]
28. Kaushik, A.; Lobiyal, D.K. Enhanced Three-Dimensional DV-Hop Algorithm. In *ICT Systems and Sustainability*; Springer: Singapore, 2019; pp. 269–278.
29. Ahmad, T.; Li, X.J.; Seet, B.-C.; Cano, J.-C. Social network analysis based localization technique with clustered closeness centrality for 3d wireless sensor networks. *Electronics* **2020**, *9*, 738. [CrossRef]
30. Gao, J.; Shen, X.; Mei, H.; Zhang, Z. Dynamic Reference Selection-Based Self-Localization Algorithm for Drifted Underwater Acoustic Networks. *Sensors* **2019**, *19*, 3920. [CrossRef] [PubMed]
31. Zhang, T.N.; Li, J.L. Application of multi-step fuzzy comprehensive evaluation. *J. Harbin Eng. Univ.* **2002**, *3*, 132–135.
32. Barker, A.L.; Brown, D.E.; Martin, W.N. Bayesian estimation and the Kalman filter. *Comput. Math. Appl.* **1995**, *30*, 55–77. [CrossRef]
33. Zheng, Y.; Ozdemir, O.; Niu, R.; Varshney, P.K. New Conditional Posterior Cramér-Rao Lower Bounds for Nonlinear Sequential Bayesian Estimation. *IEEE Trans. Signal Process.* **2012**, *60*, 5549–5556. [CrossRef]
34. Anees, J.; Zhang, H.-C.; Baig, S.; Guene Lougou, B.; Robert Bona, T.G. Hesitant Fuzzy Entropy-Based Opportunistic Clustering and Data Fusion Algorithm for Heterogeneous Wireless Sensor Networks. *Sensors* **2020**, *20*, 913. [CrossRef] [PubMed]
35. Wang, Y.-M.; Luo, Y.; Hua, Z. On the extent analysis method for fuzzy AHP and its applications. *Eur. J. Oper. Res.* **2008**, *186*, 735–747. [CrossRef]
36. Luo, J.; Zhang, C.; Wang, C. Indoor multi-floor 3D target tracking based on the multi-sensor fusion. *IEEE Access* **2020**, *8*, 36836–36846. [CrossRef]

sensors

Review

A Survey of Smartphone-Based Indoor Positioning System Using RF-Based Wireless Technologies

Santosh Subedi and Jae-Young Pyun *

Department of Information and Communication Engineering, Chosun University, Kwangju 501-759, Korea; santoshmsubedi@chosun.kr
* Correspondence: jypyun@chosun.ac.kr

Received: 8 October 2020; Accepted: 10 December 2020; Published: 17 December 2020

Abstract: In recent times, social and commercial interests in location-based services (LBS) are significantly increasing due to the rise in smart devices and technologies. The global navigation satellite systems (GNSS) have long been employed for LBS to navigate and determine accurate and reliable location information in outdoor environments. However, the GNSS signals are too weak to penetrate buildings and unable to provide reliable indoor LBS. Hence, GNSS's incompetence in the indoor environment invites extensive research and development of an indoor positioning system (IPS). Various technologies and techniques have been studied for IPS development. This paper provides an overview of the available smartphone-based indoor localization solutions that rely on radio frequency technologies. As fingerprinting localization is mostly accepted for IPS development owing to its good localization accuracy, we discuss fingerprinting localization in detail. In particular, our analysis is more focused on practical IPS that are realized using a smartphone and Wi-Fi/Bluetooth Low Energy (BLE) as a signal source. Furthermore, we elaborate on the challenges of practical IPS, the available solutions and comprehensive performance comparison, and present some future trends in IPS development.

Keywords: indoor positioning system; fingerprinting localization; Bluetooth low energy; Wi-Fi; performance metrics; positioning algorithms

---

1. Introduction

Localization is the discovery of a location of a user, which is a basic need for pervasive applications such as behavior recognition, smart medication, and smart building that require accurate position information of the users to yield accurate and timely services. The demand for location-based service (LBS) has gradually increased at present owing to the rapid development and popularization of smart devices and technologies. The extensively used technology for LBS is global navigation satellite systems (GNSS). GNSS based LBS are employed in consumer products such as vehicle navigation, navigation services on the smartphone and geotagging, and scientific observation systems such as variations in the earth's rotation and monitoring the tectonic plates. The global positioning system (GPS) [1], the Russian GLONASS [2], the European GALILEO [3], and the Chinese BeiDou Satellite Navigation System [4] are some emerging GNSS. Although GNSS-based LBS are widely used, their performance is limited to the outdoor environment only. Besides, indoor environments are often complex due to obstacles and environment changes, resulting in signal fluctuation or noise. Hence, it invites extensive research on indoor LBS or indoor positioning system (IPS) with alternative wireless technology.

The IPS is realized with the different signal sources or access points (APs). There are two choices of signal source: Either already deployed APs like Wi-Fi [5,6] and geomagnetic fields [7] or deploy a new signal source like Bluetooth low energy (BLE) and ultra-wideband (UWB) radio signal tags. The signal-free solutions in IPS is a dead-reckoning technique that uses off-the-shelf mobile sensors to detect position changes. Some of the wireless signal measuring principles in IPS are received signal strength (RSS), time of arrival (TOA), time difference of arrival (TDOA), and angle of arrival (AOA). RSS has been widely used for designing IPS owing to its nonrequirement of extra hardware and easy implementation. Indoor LBS is applicable at asset management, people tracking, trade fairs and events, etc.

As the indoor environment is complex, multipath propagation and shadowing effect on a radio signal is common [8]. Hence, the received signal can contain line-of-sight (LOS) and non-line-of-sight (NLOS) signal components. It results in less accurate time synchronization and propagation time measurement, which poses a problem on IPS that relies on signal measurement principles like TOA, TDOA [9], and AOA [10]. Moreover, the RSS is also unstable, owing to the superimposition of multipath signals of varying phases. Meanwhile, the magnetic signal has a very limited discernibility in addition to the requirement of proper calibration of the magnetometer in the smartphone.

Since the indoor environment is much complex and less characterized than the outdoor environment, it is not easy to model the indoor radio signal propagation. The indoor signal propagation model is usually based on a propagation path and the known obstacles where minor indoor changes can render the signal propagation model invalid. The localization methods like trilateration [11] and weighted centroid (WC) localization [12] rely on the signal propagation model to estimate the distance from the RSS. Moreover, these methods require precise calibration of path loss exponent for every indoor environment. Fingerprinting localization is the most widely employed IPS that rely on radio fingerprint to produce localization result. However, this method's training phase is labor-intensive and time-consuming, and the time complexity of the execution phase grows with the size of the localization area. Moreover, the instability of the RSS in the indoor environment enforces frequent update of the radio map database.

The signal-free localization method relies on mobile sensors such as accelerometers, gyroscope, magnetometer, and barometer and can track the users by continuously estimating their displacement from a known starting point. The necessity of a known starting point and integrated sensor readings to measure a position resulting in an unacceptable accumulated error are dead reckoning-based IPS problems.

As multiple approaches of techniques and technologies have put forward while realizing an IPS, there is no fixed set of rules that guide designing an IPS. In recent years, an initiative has been put forward to set a common benchmark for IPS. For example, the *EvAAL* framework [13,14] and the Microsoft competition [15] offer researchers a real and challenging test site with independent evaluation. Francesco P. et al. have elaborated on the benchmarks for IPS standardization in [16]. Although IPS are yet to achieve standardization, some research works, such as [5,17,18] that are based on radio frequency (RF), have shown surprising localization accuracy (decimeter level accuracy). In [5], an approach is presented that can compute subnanosecond time-of-flight employing commodity Wi-Fi cards. Here, packets are transmitted on multiple Wi-Fi bands and stitched their information together, mimicking a wideband radio. Reference [17] put forward an approach to jointly estimate the AOA and time-of-flight by combining channel state information (CSI) values across subcarriers and antennas. Similarly, [18] presents the fingerprinting localization based on 60 GHz impulse radio to reduce the effects of NLOS propagation. This work exploits 60 GHz technology's ability to provide accurate temporal and spatial information for TOA estimation.

A practical IPS should bear the properties like easy implementation, acceptable (meter level) localization accuracy, feasible system cost, scalable/robust system, and minimum computational complexity. However, the localization solutions at hand are more focused only on acquiring better positioning accuracy. In this paper, we discuss the strength and weaknesses of various state-of-art approaches compared with performance metrics of practical wireless technology-based IPS. Therefore, the objective of this survey is to provide a comprehensive outline of available systems and solutions of smartphone-based IPS that uses RF (particularly Wi-Fi and BLE) as its signal source , so that readers may be educated in this rapidly growing area.

This survey paper is organized as follows. Section 2 summarizes some published survey literature on RF-based IPS. In Section 3, a brief description of some RF-based localization technologies and some localization solutions are presented. Signal measurement principles and the performance matrices are elaborated in Sections 4 and 5, respectively. Section 6 presents the challenges of practical IPS, and Section 7 presents positioning algorithms and a survey of some available solutions. A conclusion is drawn in Section 8 with discussion of future research trends.

2. Typical Survey Papers on RF-Based IPS

The available IPS put forward in the past decade are designed for various applications utilizing diverse techniques and technologies. The variations in localization methods, signal sources, and target applications have resulted in a systematical investigation of IPS. However, the indoor localization problem is still waiting for a satisfactory and reliable solution. Meanwhile, the literature on different IPS approaches is summarized in many survey papers based on different research topics [19–30]. This section outlines the typical survey papers on IPS to differentiate our work from the existing literature.

Pavel et al. presented a comprehensive study on smartphone-based IPS [19]. The main focus of their discussion is RF and magnetic field-based fingerprinting along with map aided navigation and inertial sensors. They have elaborated on the fingerprinting algorithm in prospective to wireless local area networks (WLAN); however, the significant issues of fingerprinting, such as offline workload and computational complexity, are not discussed. Moreover, this work lacks the evaluation of the IPS based on comparison criteria. The next review on device-based IPS is presented in [22,25], where they deal with seamless outdoors-indoors localization solutions for smartphones and indoor localization for various devices, respectively. As a topic of discussion of [25] is seamless localization, it briefly mentions the approaches of indoor/outdoor localization and their integration and lacks a clear focus on IPS. Similarly, [22] reviews IPS that are realized employing a smartphone and tag devices compatible with various technologies. This work highlights the use of various technologies such as Wi-Fi, BLE, UWB, ultrasound, etc. in IPS and presents their performance comparison. Nevertheless, the localization methods for IPS and their issues are not addressed in the paper.

Another focus of survey papers on IPS is based on the classification of localization systems [21,26,28]. A review of IPS' multiple techniques and technologies was carried out in [26], which also focuses on the internet of things (IoT). The authors list out the localization system's applications and tabulate some of the existing IPS to show their advantages and disadvantages. They review the wide variety of techniques and technologies for IPS; however, the presented challenges do not wholly reflect the practical limits and challenges of a pragmatic IPS. Hui Liu et al. have presented a survey of wireless indoor positioning techniques and systems to summarize the positioning algorithms and the existing systems and solutions [28]. This work illustrates an IPS's performance metrics, which can be a basis for judging any IPS. Besides, a graphical outline of wireless-based positioning systems depicting localization accuracy (resolution), use areas (scale), and wireless technologies is presented. However, this work does not address the challenges and future trends of IPS. A short review of IPSs and the employed algorithms are presented in [21]. Here, a comparison of a few IPSs based on parameters like used technology, accuracy, and robustness are tabulated.

Wi-Fi fingerprinting-based IPS has been neatly reviewed in [29,30]. In [29], Sunning et al. elaborate fingerprinting localization based on spatial and temporal signal patterns wherein they further discuss the various offline workload reduction algorithms for fingerprinting localization. In addition, they summarize the collaborative localization model of IPS that employ distance and proximity-based approaches. Moreover, they review the motion assisted Wi-Fi localization, energy efficiency for smartphones, and list out some future directions. Meanwhile, [30] further elaborates on Wi-Fi-based fingerprinting by formulating the localization problem and detailing the conventional localization approaches. Unlike [29,30] explains the available data clustering modules in IPS for computational complexity reduction. Although both the works are impressive, none of them address other promising technologies (e.g., BLE, etc.) and techniques (e.g., neural network-based fingerprinting, etc.) for fingerprinting localization.

Some other survey papers on IPS are [23,24,27]. The review study in [24] focuses on IPS for emergency responders. Here, the IPS that were designed for emergency responders is categorized as a radio signal, IMU, and hybrid-based approach wherein various localization solutions are discussed. Hakan and Shuang present a survey on IPS that rely on RF, infrared, and ultrasonic technologies [27]. They highlight the benefits of RFID positioning systems. As RSS has been used one way or the other in most of the RF-based IPS, [23] presents a survey paper on IPS that employs CSI wherein the benefits of CSI over RSS are highlighted.

In our survey work, we focus on practical IPSs that rely on RF (Wi-Fi or BLE or both) and smartphones for realizing the real-time localization system. We elaborate on the localization algorithms targeted for such practical IPSs, highlighting the fingerprinting localization due to its promising accuracy. We review the challenges that practically limit an RF-based IPS to achieve all the performance metrics. Moreover, we discuss and tabulate the existing localization solutions that intend to solve a reliable IPS's single or multiple challenges. Furthermore, we highlight the future research direction in RF-based IPS.

3. Wireless Technologies for IPS

Many representative technologies have been employed to develop IPS. The technologies can be categorized into RF, lightwave, acoustic wave, and mechanical, as shown in Figure 1.

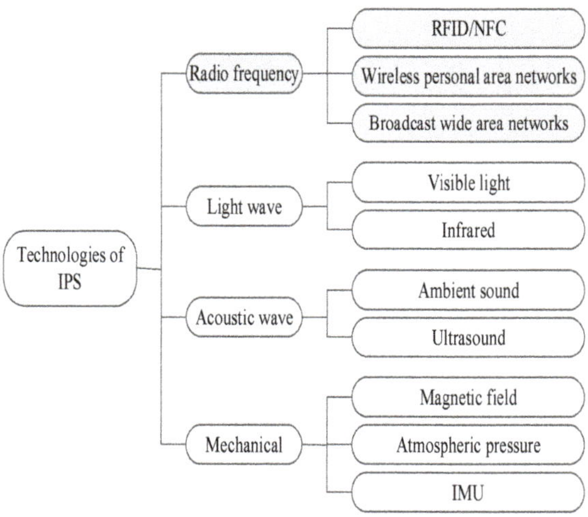

Figure 1. Categorization of major technologies used in indoor positioning system (IPS) development.

RF technologies such as Wi-Fi and BLE are the most used technologies in IPS development. Similarly, the magnetic field, inertial measurement unit (IMU), and atmospheric pressure can be employed for IPS development. Apart from them, lightwave such as visible light and infrared and acoustic waves such as ambient sound and ultrasound are also utilized in indoor localization. Since this survey aims to study IPS approaches based on wireless technologies, an overview of the localization solutions based on some selected RF technologies is emphasized. As Wi-Fi and BLE are the dominating technologies for IPS development, we present their relative comparison in terms of IPS in Table 1.

The RF technology can be further categorized into broadcast wide area network (WAN), wireless personal area network (WPAN), and RFID/NFC. The cellular network, TV/FM radio signals, and GPS repeaters belong to the WAN family, whereas Wi-Fi, BLE, WSN, and UWB belong to the WPAN family.

3.1. RFID/NFC

The radio frequency identification (RFID) systems rely on two main components, namely RFID tag and RFID reader to fulfill their objective. The RFID reader wirelessly acquires the electronically stored information of RFID tags. The reader contains a transceiver to transmit RF signals and read the data emitted from the tags. The tags can be categorized as passive and active. The passive tags get energy from incoming radio signals, whereas a battery powers active tags. RFID systems operate in four frequency bands: Low frequency (125 kHz), high frequency (13.56 MHz), ultra-high frequency (433, 868–915 MHz), and microwave frequency (2.45 GHz, 5.8 GHz). The property of RFID to detect and recognize the nearby tag enabled it to be used for IPS.

Some of the RFID localization systems using passive tags are [31–34] where the tags are deployed on the floor at a fixed distance forming a grid and estimate localization results by detecting multiple tags. The work in [35,36] are based on active tags where RSS is used to estimate the location of the user. RFID has also been combined with other technologies for IPS. For example, [37] combines it with the ultrasonic sensor, [36,38] combine with image sensors to detect the location of objects.

Near Field Communication (NFC) is the short-range (5 cm or less) wireless communication technology. It is mainly a specialized branch within the family of RFID technologies (high-frequency band of RFID). Localization can be realized with NFC by deploying several tags at places of interest, where a location is estimated simply by touching the tag with the NFC equipped device [39,40].

3.2. UWB

UWB uses very low energy for short-range and high-bandwidth communications over a large portion of the radio spectrum. In general, an emitted radio wave is considered UWB if its bandwidth exceeds 500 MHz or 20% of the carrier frequency. The properties of UWB, such as very less power consumption, effective penetration through dense materials, and less sensitive to the multipath effect owing to a very short duration of UWB pulses make the UWB suitable for IPS development. As of this writing, *Apple* (iPhone 11 and 12) and *Samsung* (Galaxy Note 20 Ultra) have launched their new smartphones that have UWB chip on it.

IPS based on UWB can estimate location accurately owing to the possibility of precise time measurements of the propagation time of UWB pulses. Yanjia et al. proposed a robust method to mitigate the path overlapping effects that induce TOA and AOA based positioning inaccuracy [41]. Their method is based on the spectral observation of beamforming and yields the least-squares estimation of joint TOA and AOA with a low computational cost. As the performance of UWB based IPS deteriorates in the NLOS channel, [42] proposed a method to identify NLOS by measuring signal strengths in the first path and multipath. RSS based UWB IPS system has also been put forward to have an accuracy between 0.1 to 0.2 m [43].

3.3. Wireless Sensor Networks

Wireless sensor networks (WSN) are the group of spatially dispersed and dedicated sensor nodes for monitoring and recording the environment's physical conditions and organizing the collected data at a central location [44,45]. The nodes of WSN are equipped with a processor, storage, a power supply, a transceiver, and one or many sensors, with an actuator. WSN operates at an unlicensed band of 2.4 GHz, and it can use several off-the-shelf wireless technologies like Bluetooth, UWB, and ZigBee where most applications use IEEE 802.15.4 and ZigBee [46]. Some sensor nodes in WSN, called anchor nodes, are aware of their position information. Therefore, the localization problem in WSN-based IPS is to determine the location of other nodes based on location information obtained from the anchor nodes. IPS using WSN normally consists of distance or angle estimation between nodes or their combination to produce localization results.

3.4. Wi-Fi

Wi-Fi is a technology for radio wireless local area networking (WLAN) of devices based on 802.11 IEEE network standard, operating in the 2.4 and 5 GHz ISM radio bands. The devices that can use Wi-Fi include PCs, smartphones/tablets, smart TVs, video game consoles, digital audio players, cars, and printers. Wi-Fi is the most popular means of communicating data wirelessly and is increasingly deployed everywhere, including home and public indoor environments. Wi-Fi-based IPS are being intensively studied owing to the widespread deployment of Wi-Fi hot spots.

The localization methods like fingerprinting [47,48] and trilateration [49] can be realized using Wi-Fi. Similarly, signal measurement principles like RSS [50], CSI [51], TOA [52], and their hybrid combination [53] can be used to provide Wi-Fi-based localization service. Although existing Wi-Fi APs can be employed to design a Wi-Fi-based IPS, those Wi-Fi networks are deployed for wireless communication, and localization is not their primary purpose. In other words, the Wi-Fi APs are not dedicated to localization; hence, an enhanced and efficient localization algorithm is required for practical localization results.

Note that Wi-Fi can be used only on the Android platform, and iOS does not provide a public API [54,55] at present. Furthermore, some restrictions regarding permissions and allowed frequency of Wi-Fi scans have been introduced from Android 8.0 (API level 26). Such restrictions have been further tightened in Android 9 (API level 28) and Android 10 (API level 29). For example, in Android 10, a successful call to *WifiManager.startScan()* requires *ACCESS_FINE_LOCATION*, *ACCESS_COARSE_LOCATION* or *ACCESS_FINE_LOCATION*, and *CHANGE_WIFI_STATE* permissions [56].

Moreover, a Wi-Fi module usually requires 3–4 s to process the *startscan* command from the positioning app to acquire a new scan result [57]. In practice, repeated RSS data are acquired until new scan results are produced to perform sampling per second in Wi-Fi AP. In such a case, subsequent filtering must be performed on the obtained Wi-Fi data to remove the erroneous statistical analysis.

3.5. BLE

BLE was released as Bluetooth version 4.0 in June 2010. The BLE is designed for devices that do not require large amounts of data transfer and is intended for short-range wireless transmission with low power consumption and cost [46,58]. It is reported that the power draw of the smartphone is lower for BLE than for Wi-Fi [59]. Similar to Wi-Fi, BLE operates at an ISM band of 2.4 GHz. The frequency band is divided into 40 channels spaced at 2 MHz apart, among which the three channels (37, 38, and 39) are used for broadcasting advertisement [60]. It is noteworthy that the three advertising channels are strategically placed to avoid interference with coexisting technologies such as IEEE 802.11 and ZigBee [61]. Moreover, the signal fluctuation in BLE is a consequence of the random use of the advertisement channels [62].

Similar to Wi-Fi, the localization methods like proximity [63], trilateration [64], and fingerprinting [65] can be realized using BLE. The tag device can estimate the RSS from a nearby BLE beacon by intercepting the advertisement packets transmitted by the beacons. The advertisement interval can range from 100 to 2000 ms. The typical advertisement interval of BLE beacons used in IPS is 300 ms by considering the normal walking speed (1.3 m/s). Moreover, the scan interval also can be set in the positioning application. A typical value of the scan interval is 1000 ms (1 s) to produce positioning results every second.

In contrast to Wi-Fi, the BLE beacon is generally compatible with Android and iOS platforms. However, the compatibility may differ depending on different beacon packages, e.g., iBeacon and Eddystone. Apple develops the iBeacon profile (natively supported in iOS), and its signal contains three main pieces of information, namely, *UUID*, *Major*, and *Minor* [66]. Whereas, Eddystone format is developed by Google (with Android users in mind) as an open-source protocol for BLE beacons. The Eddystone broadcasters advertise fields (referred to as frame types), namely, *Eddystone-UID* (unique static ID), *Eddystone-URL* (includes a compresses URL), and *Eddystone-TLM* (telemetry data) [67]. When a beacon region is detected by an iOS application (that monitors iBeacon's signal), some action (e.g., a push alert to the home screen) is triggered, which can be helpful in real-time scenarios (e.g., stores in a mall). However, iOS does not have such background operating system support for Eddystone triggering [68]. In recent times, cross-platform app SDK like FlutterBlue [69], and Universal Bluetooth Beacon Library [70] have come into the picture that support Android and iOS platforms.

Table 1. Comparison of Wi-Fi and Bluetooth Low Energy (BLE) technology in terms of IPS.

Parameters	Wi-Fi	BLE
Deployment cost	Low	High
AP reliability	Not-dedicated to IPS	Dedicated to IPS
Hardware efficiency	Requires ≥ 3 s to scan new RSS data [57]	RSS sample acquired every second
AP differentiating parameters	SSID, BSSID (MAC)	UUID, MAC [65]
Transmission range	High (∼50 m)	Low (∼30 m)
Power consumption [tag]	High [71]	Low [71]
Power source [AP]	Plugged into mains	Powered by coin shaped battery
Channel availability	Three independent channels at most (2.4 GHz band) [57]	Three advertisement channels [72]
Proximity detection	Normally final location estimation is available	Immediate, Near,and Far proximity available [73]
Implementation platform	Only on Android devices [54,55]	iOS and Android devices

3.6. Cellular Networks

The cellular network refers to a long-range wireless network distributed over the cells where each cell is served by at least one fixedly located transceiver known as a base station. The cellular network can be classified by the technical standards that have been evolved from 1G (analog) to the latest 5G (digital). The cellular networks-based IPS benefit from cellular signals such as wide coverage, existing infrastructure, multiple frequency bands, and supported by a large number of mobile communication devices.

As the technical standard of cellular networks has evolved with generations, the localization estimation accuracy has also been increased accordingly. For example, the cell-ID localization in 2G helped to improve the accuracy of hundreds to tens of meters [74]. Similarly, localization based on timing via synchronization signals in 3G and reference signals dedicated to localization in 4G has helped increase localization accuracy. Furthermore, it is expected to attain an accuracy of localization estimation in the range of centimeters using 5G-based devices where the 5G networks are expected to use precise localization estimation through all layers of the communication protocol stack [75]. In particular, 5G cellular technology is expected to have a large signal bandwidth (mm-wave) and beamforming capabilities, making the localization more robust and efficient [76].

4. Signal Measurement Principles

4.1. RSS

The RSS is a measurement of the power present in a received radio signal. The RSS value is measured in decibel-milliwatt (dBm) and has a typical negative value ranging from nearly 0 dBm (excellent signal) to less than -100 dBm (poor signal). As the distance between the transmitter and receiver increases, the RSS gets attenuated due to many factors including the antenna of transmitting and receiving devices, the number of walls and floors, the number of people and furniture, etc. Note that the RSS does not decrease linearly as the distance increases [77].

RSS modeling is usually done by the combined effects of large-scale fading and small-scale fading [78]. The large-scale fading component depicts the signal attenuation as the signal travels over a distance and is absorbed by objects such as walls and floors along the way to the smartphone. This fading component predicts the mean of the RSS and usually has a log-normal distribution [79]. Similarly, the small-scale fading describes the fluctuation of signal due to multipath fading. For the NLOS component, the small-scale fading is modeled with a Rayleigh distribution, whereas, for the LOS component, it is modeled by Rician distribution. In IPS, the fluctuating RSS are filtered using many approaches such as Gaussian filter [80], moving average filter [12,81], and exponential averaging [82].

Owing to walls and other objects between the transmitter and the receiver, NLOS signals in an indoor environment are common, which can significantly degrade the localization accuracy. The RSS, while used in parameter estimation (e.g., path loss exponent) methods, is converted to distance employing a path-loss model. In a practical indoor environment, the localization scenario is so complex (includes both LOS and NLOS signal propagation) that it is difficult to establish an accurate model to work. On the other hand, the fingerprinting localization that stores the RSS as a radio map is less sensitive to the NLOS conditions [83]. Reference [84] classified the methods for reducing the NLOS error as direct and indirect methods. Here, the direct method refers to directly processing the measurement results reducing the NLOS propagation error employing Kalman filtering, particle filtering, etc. The indirect method refers to the fingerprinting localization.

RSS value can be acquired without any extra hardware using off-the-shelf smartphones. Moreover, RSS does not require any time synchronization between the transmitter and the receiver. Most importantly, RSS values can be implied to realize any indoor positioning methods, where it can be converted to distance for lateral approaches and stored in a database for scene analysis. Hence, RSS has been a prime choice of signal measurement principle in IPS.

4.2. TOA

The TOA is the travel time or time of flight of a radio signal from a transmitter to a receiver. As the signal travels with a known velocity, the distance can be directly calculated from the TOA. Figure 2 illustrates a TOA measurement-based localization system.

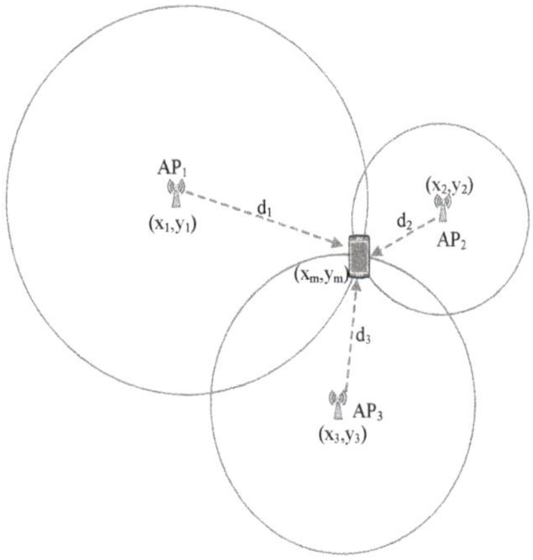

Figure 2. Localization based on time of arrival (TOA) measurement.

Let c be the speed of light, then the distance between ith AP and the tag device can be estimated by the following relation [85]:

$$d_i = (t_i - t_0) \times c, \tag{1}$$

where t_0 and t_i are the time instant of signal transmission and signal reception respectively, and $c = 3 \times 10^8$ m/s. The TOA technique requires precise time synchronization for transmitters and receivers. The estimated distance can be utilized for the trilateration algorithm to estimate user location. TOA has been used with various wireless technologies like UWB [86] and Wi-Fi [87].

4.3. TDOA

For TDOA measurement, the difference in arrival time from multiple APs is employed. In TDOA based localization, the distance difference between the tag device and APs is calculated based on time difference measurements as shown in Figure 3.

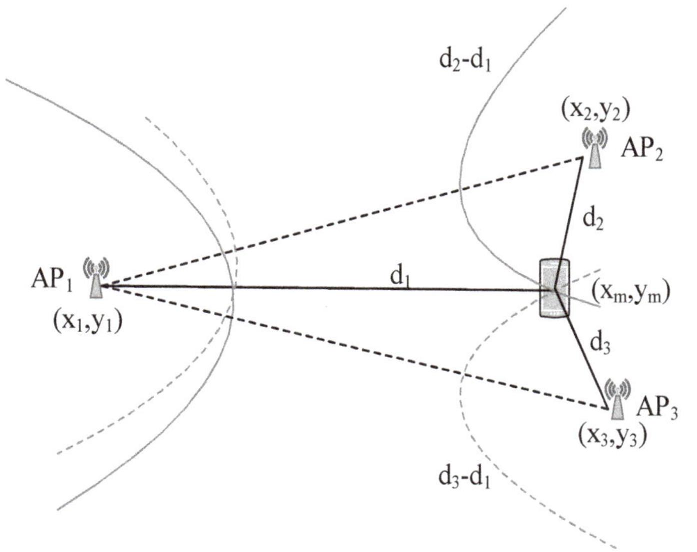

Figure 3. Localization based on time difference of arrival (TDOA) measurement.

Here, the difference of distance to APs and to the AP where the signal first arrives is [88]:

$$d_{ij} = (t_i - t_j)c = \sqrt{(x_i - x_m)^2 + (y_i - y_m)^2} - \sqrt{(x_j - x_m)^2 + (y_j - y_m)^2}, \quad (2)$$

where t_i and t_j are the time instant of signal reception from AP i and j, respectively. Geometrically, with a given TDOA measurement, the tag device must lie on a hyperboloid with a constant range difference between the two APs. Apart from TOA, TDOA needs time synchronization of APs only. Besides, the timing measurements at TOA and TDOA can be achieved down to a fraction of chip duration assuming a LOS condition; however, NLOS can cause information loss during the time measurement. To incorporate the NLOS condition, [89] suggests modeling the error as a mixture of LOS and NLOS models for a robust algorithm. Furthermore, [90,91] put forward a three-dimensional least-square positioning technique and NLOS error estimation approach for positioning in the NLOS environment, respectively.

4.4. AOA

The AOA information is extracted employing the directionally sensitive antennas [89]. The AOA measurement determines the direction of propagation of a radio wave incident on an antenna array. It can be done by measuring the TDOA at individual elements of the antenna array [92]. AOA-based localization system estimates the location of the tag device as the intersection point of pairs of hypothetical signal paths particular angles as shown in Figure 4.

At the 2D plane, the AOA approach requires only two APs to determine the location of a tag device [93]. In AOA-based IPS, time synchronization between the APs and the tag device is not required. However, it may require relatively complex hardware to obtain angle measurement [94]. For instance, Ubicarse [95] leverages antenna arrays and computes AOA of signals from the APs to estimate the device orientation.

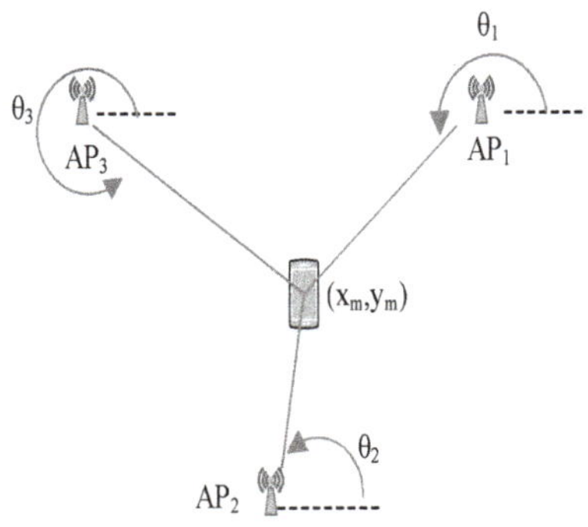

Figure 4. Localization based on angle of arrival (AOA) measurement.

4.5. CSI and RTT

The CSI is an emerging approach that tries to replace RSS information in IPS [23,96–98]. The CSI explains how a signal propagates from a transmitter to the receiver. In other words, it describes the information that represents a combined effect of scattering, fading, and power decay with the distance. It is reported that CSI achieves higher robustness compared to Wi-FI RSS information [99]. Moreover, the SpotFi combines CSI values across subcarriers and antennas to jointly estimate the AOA and time-of-flight for decimeter level localization [17]. However, current smartphones are not compatible with CSI data collection, making it impossible to implement CSI-based positioning solutions on the present-day smartphone.

The RTT stands for Round Trip Time that can measure the distance without requiring time synchronization between the communicating nodes. The time spent by a specific frame is measured while traveling from a transmitter to a receiver and back again to the transmitter. Wi-Fi RTT was introduced in Android 9 (API level 28), which is specified by IEEE 802.11 mc standard and built on a new packet type known as fine timing measurement (FTM) frame [100]. Here, the Wi-Fi RTT API provides Wi-Fi location functionality to measure the distance to nearby RTT-capable Wi-Fi APs and peer *Wi-Fi Aware* devices [101].

The FTM protocol is shown in Figure 5, where a tag device initiates the FTM process by sending an FTM request to an AP. The AP (that supports FTM protocol) responds to the FTM request either to agree or to disagree with the ranging process. If the AP agrees, it starts to send an FTM message, waits for its acknowledgment (ACK), and transmits the FTM result afterward. The transmitting timestamp of the FTM message and the reception timestamp of its ACK is utilized to infer the propagation delay between the tag device and the AP. Here, the AP can send multiple FTM messages for averaging the estimated distances [102]. Table 2 differentiates FTM- and UWB-based approaches.

Table 2. Comparison of fine timing measurement (FTM)- and ultra-wideband (UWB)-based approaches.

Parameters	FTM-Based Approach	UWB-Based Approach
Time transfer	Reference Broadcast Infrastructure Synchronization (RBIS)	Precision Time Protocol (PTP)
Ranging	Fine Timing Measurement (FTM)	Two-Way Ranging (TWR)
Cost	Low	High
Power consumption	High	Low
Distance estimation accuracy	>1 m [103]	5–10 cm
Smartphone compatibility	Wi-Fi RTT introduced in Android 9 (API level 28)	Samsung Galaxy Note 20 Ultra and iPhone 11/12 contain a chip for UWB

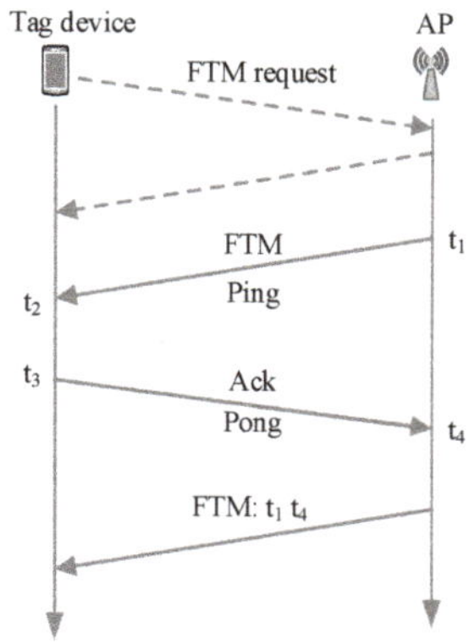

Figure 5. The fine timing measurement (FTM) protocol.

From Figure 5, RTT is calculated for an FTM message as [104,105]

$$RTT = (t_4 - t_1) - (t_3 - t_2) \tag{3}$$

The distance (d_{RTT}) between the transmitter and receiver can be estimated by multiplying the RTT with the speed of light (c) as follows

$$d_{RTT} = \frac{RTT}{2} \times c \tag{4}$$

Multiple trilaterations can be employed with the estimated distance to localize a tag device. The main challenge on RTT is the NLOS that increases uncertainty in the time measurement [106]. It can be minimized by using LOS/NLOS identification approaches [107,108].

5. The Performance Metrics

Since there are different technologies and methods to realize an IPS, the most important performance metric is the localization accuracy. In addition to localization accuracy, other performance indicators of an IPS are complexity, scalability/robustness, and cost.

5.1. Accuracy and Precision

The localization accuracy can be defined as a difference between an estimated location and the tag device's actual location. Similarly, the precision indicates the degree to which repeated location estimates produce identical results under unchanged conditions.

Usually, the mean squared error (MSE) is used as the accuracy indicator. Indoor environments are often complex due to different obstacles and environmental changes, resulting in signal fluctuation. In the meantime, high localization accuracy is often expected for adequate location-based service. The localization accuracy of the IPS depends on the used technology and techniques.

The precision yields information like how convergent the localization result can be over many trials or how consistently the system works. The cumulative distribution function (CDF) is used as a precision indicator. The CDF represents the distance error distribution between the estimated location and the tag device's actual location. The MSE, as well as CDF, should be exploited while comparing two or more localization algorithms.

5.2. Complexity

For an IPS, its complexity can be categorized in terms of hardware and software. The adopted technology and the signal measurement principle for the IPS account for hardware complexity.

For example, most present-day smartphones supported technologies like Wi-Fi, BLE, and the geomagnetic field. However, standard mobile devices do not support technologies like UWB and ultrasound, and the IPS using such technologies should use a dedicated system that requires proprietary equipment. Moreover, geomagnetic based IPS can produce localization results without deploying any hardware; however, Wi-Fi and BLE need to be deployed.

As for the chosen signal measurement principle, obtaining RSS from Wi-Fi and BLE are relatively easy with standard mobile phones since such devices typically need to scan RSS for their routine functioning. However, it is not easy to obtain accurate time and angle measurements that increase the IPS complexity.

The complexity depends on the computation load represented by the calculations required to perform localization regarding the software. In a server-based IPS, the localization algorithm's execution is carried out on a centralized server where the positioning could be calculated quickly due to its powerful processing capability and abundant power supply. Here, the computational complexity mainly depends on the indoor localization area [109]. However, if the positioning algorithm is executed in the tag device, it may increase the complexity. Moreover, the complexity concerned with the software also depends on the technique used for IPS development. For example, fingerprinting localization has larger complexity, and it grows as the localization environment increases. Here, the complexity can be minimized using clustering.

5.3. Scalability and Robustness

Scalability in IPS refers to the localization system's ability to perform well even when any change in the area of interest for localization and/or on signal source occurs. The changes can be an extension of the localization area and/or an extension of signal coverage. If an IPS need not be taken down in such a scenario, it is considered excellent scalability. For example, when any IPS is constructed, it provides services in a limited area of interest, and an increase in the localization area might be needed in some future time. In some cases, the transmitting power and signal-broadcasting rate can be increased for good signal coverage. In such situations, the positioning techniques like proximity, WCL, and trilateration are easy to expand by merely adding the identical signal sources and updating the system with the location coordinate information of the added hardware. However, the fingerprinting based system needs an offline site survey for every change in the localization area or signal source. When the localization area is expanded, the extended area's radio map needs to be freshly constructed. Moreover, when the transmitting power at the APs or signal broadcasting rate (e.g., advertisement packet broadcasting interval in BLE) is changed, a new site survey for the whole localization area is required. Hence, fingerprinting localization has relatively low scalability.

Robustness is also an essential factor in IPS that allows the system to function normally without human intervention when the localization environment changes. For example, some signal sources could be out of service occasionally, or testbed layout changes could cause some signal to no longer support LOS propagation. In this scenario, the IPS has to provide localization services with incomplete or noisy information (RSS fluctuation). The robustness can be gained by introducing redundant information into a localization estimation. For example, rather than using the only three APs for trilateration, an IPS can include many supplementary APs to make the system more robust. Moreover, fingerprinting based IPS can adopt a larger set of RSS samples to increase robustness.

5.4. Cost

The IPS cost factor represents the infrastructure cost and the time and effort for system installation and maintenance procedures. Particularly, the cost depends on factors like the size of the localization area, the required accuracy, the used technology, power consumption, etc. The IPS system, such as PDR, geomagnetic-based, and barometer-based, can be realized with only the smartphone and do not need any additional infrastructure. Some signal sources like Wi-Fi and BLE require APs deployment; however, Wi-Fi is already deployed for other purposes. Hence, if the IPS requires no additional infrastructure or is based on existing infrastructure, the cost can be substantially saved.

Moreover, if an IPS is robust and scalable, saving in time and labor is possible, which reduces the system's cost. In addition, an increase in localization space requires the deployment of more APs in localization techniques like proximity. Power consumption at both the AP and the tag device is also a critical cost issue. For example, BLE consumes less power than Wi-Fi, where BLE operates with a coin-shaped battery, but Wi-Fi needs to be plugged into mains. Furthermore, when the localization operation is carried out on the server-side, the tag device's power consumption can be reduced. Lastly, the location update rates, signal broadcasting rate, and the desired system accuracy can also affect power consumption.

6. The Problems of Practical IPS

6.1. Complex Indoor Environment and Unstable RSS

The indoor area consists of multiple floors, walls, furniture, and human beings, which results in a complex radio environment. Hence, RSS exhibits high variability in space and time, even in a fixed indoor environment owing to various noise factors, interferences, and attenuation. A probable radio environment at an indoor location is displayed in Figure 6. This instability of RSS results in increased localization estimation error. Therefore, irrespective of the localization technique, RSS filtration/ smoothing is required to minimize the localization estimation error. Figure 7 shows the fluctuation of Wi-Fi and BLE RSS in a fixed indoor environment (without any change in the radio environment).

From Figure 7, it is seen that RSS fluctuates a lot even at a fixed point in the indoor environment. To minimize this problem, a low-pass smoothing filter can be employed. Some of the representative smoothing filters are moving average filter [12,81], Kalman filter [110–112], Gaussian filter [72,113], and exponential averaging [114].

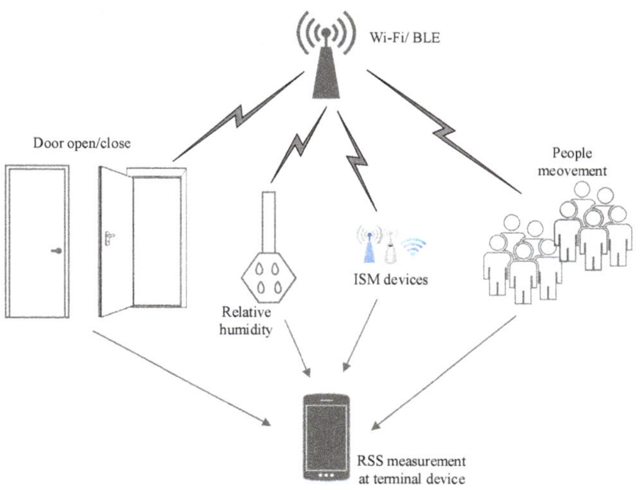

Figure 6. Graphical representation of interference at indoor environment.

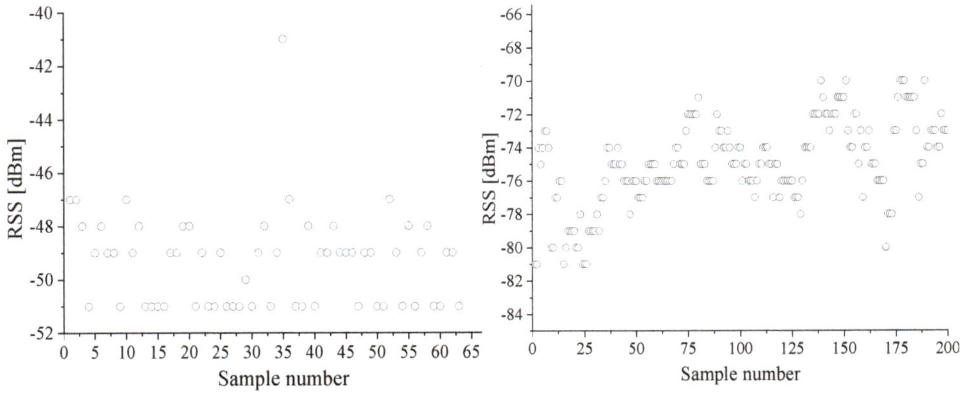

(a) Wi-Fi (b) BLE

Figure 7. Variation of Wi-Fi and BLE RSS at a fixed indoor environment.

6.2. Terminal Device Heterogeneity and Battery Efficiency

Most positioning techniques require either calibrated environmental parameters or the radio map construction to provide localization service. Such parameters/radio-map, when calibrated/constructed by a terminal device and localized by another kind of terminal device, may result in an adverse effect on position estimation owing to the different gains of receivers and antennas [115–117]. For example, the calibrated signal strength at a unit distance for the log-distance path loss model by two different terminal devices may not yield the same value. Similarly, the stored signal strength for any RP on a testbed acquired by a terminal device may differ by certain decibels when measured by a different terminal device at the same measurement place. Therefore, it is required to address the issue of terminal device heterogeneity while designing a practical IPS.

The available indoor localization solutions intend to provide high accuracy service with the terminal device's consumption of high battery energy. The existing IPSs require active monitoring of the wireless channels to listen to specific beacon message or advertisement signals periodically. On top of this, some IPS integrate different technologies for better positioning results. For example, Reference [118] employs the magnetic field, PDR, and QR code and the BLE signals to enhance the localization estimation. While it is practical performance-wise, it is not ideal in terms of battery efficiency. Note that localization service is not the primary task of any terminal device; hence, the battery drainage can lead to consumer dissatisfaction. Thus, it is worthwhile to focus on energy consumption simultaneously to the localization accuracy.

6.3. Learning Methodology of Radio Signals in Scene Analysis

The fingerprinting localization technique is predominantly realized in IPS applications owing to its high reliability. However, the learning methodology of the radio signals in fingerprinting localization is costly in terms of time-consumption and workload. In other words, a data collector has to visit every hypothetical grid or the RP that is typically one meter apart to acquire the offline training dataset. Generally, the training dataset at an RP consists of an average of RSS samples (typically 35 samples) from each AP along with the RP's coordinate. Moreover, the radio map has to be updated repeatedly owing to the RSS fluctuation and possible change in the radio environment. This issue on fingerprinting-based localization intensifies practical limits and challenges in realizing a reliable and scalable IPS to meet the required accuracy of practical IPS. Many research works have endeavored to reduce the offline workload of fingerprinting localization [119,120]. Some of the approaches are the use of a self-guided robot, simultaneous localization and mapping (SLAM) [121,122], machine learning [123,124], and crowdsourcing [125,126] . In addition, the signal propagation model has been employed for generating a fingerprinting database to reduce the offline workload [127–129]. However, an efficient solution for the data collection problem is yet to discover.

6.4. Computational Time and System Cost

Apart from good positioning accuracy, fast position estimation is also favored for a practical IPS. Typically, a new position estimation at each second is demanded. Hence, computational time should also be considered while improvising the positioning method for better positioning results [130]. For example, the positioning result of probability-based fingerprinting is better than deterministic-based fingerprinting (Wk-NN); however, the computational complexity of the probabilistic approach of fingerprinting is higher than the deterministic approach [30]. Furthermore, as two or more technologies are integrated for better localization solution, the computational complexity increases.

Not every novel and efficient localization algorithm may be commercially successful owing to the system cost. Considering the Wi-Fi and BLE, Wi-Fi APs are already deployed in every building for communication purposes, whereas BLE needs to be deployed that increases the system cost. Furthermore, positioning methods requiring lengthy data learning and frequent data updating procedures raise the system cost. The increased system cost can make the concerned parties reluctant to adopt IPS for better service.

7. Positioning Algorithms and Survey of Available Solutions

This section presents some positioning algorithms, focusing on fingerprinting localization and summarizing various state-of-art approaches intended to solve IPS problems. A comparison of available systems and solutions against the performance metrics is presented in Table 3.

7.1. Proximity-Based

The word proximity is defined as nearness in space, time, or relationship. As the definition suggests, proximity in IPS provides symbolic location information if an object is present within an AP's vicinity where the received signal strength determines the vicinity. A proximity-based IPS is illustrated in Figure 8.

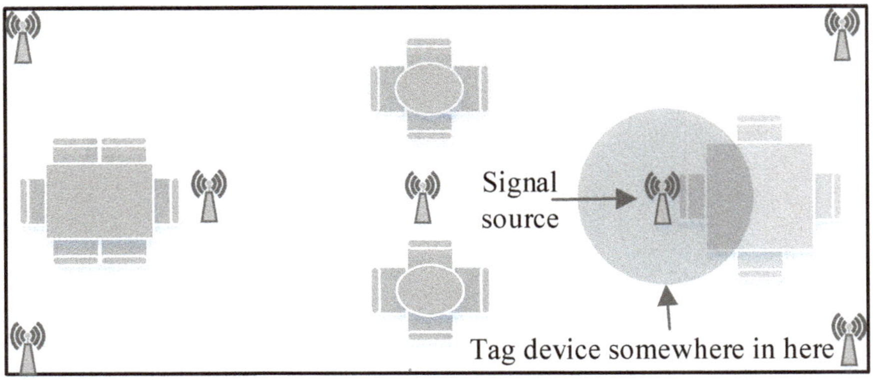

Figure 8. Graphical representation of proximity-based IPS.

In proximity-based IPS, when a tag device detects an AP, the tag's position is associated with the AP's location. In this scenario, when the tag device detects more than one APs in its vicinity, the tag device's location can be referred to the AP's real location having the strongest signal. The proximity-based IPS is the simplest among all the algorithms and is very easy to implement. However, for better localization accuracy or high resolution, a dense deployment of APs is mandatory. Generally, the IPS using wireless technologies like WSN, BLE [131,132], RFID [133], and NFC [39] is employed for proximity-based IPS development. Reference [134] uses pedestrian dead reckoning (PDR) and BLE beacons to estimate the user location. This work employs the proximity information of the beacons for correction of the estimated position of PDR.

7.2. Lateral/Angular

The lateral technique estimates the position of a tag device by measuring the distances from multiple APs. The distance can be obtained from signal measuring principles like RSS, TOA, and TDOA. Similarly, the angulation technique estimates the tag device's location by computing angles relative to multiple APs using AOA. WC localization and trilateration are the lateral methods, whereas triangulation is the angulation method.

7.2.1. WC Localization

In proximity-based IPS, when the tag device detects multiple APs in its vicinity, the location can be estimated as a centroid of the real location of the detected APs. Furthermore, a certain weight can be assigned to each detected AP based on their signal strength to estimate a weighted centroid. The simplest WC localization equation is defined by the following set of equations [12]:

$$\begin{gathered} x_w = \frac{\sum_{j=1}^{u} x_j \times w_j}{\sum_{j=1}^{u} w_j} \\ y_w = \frac{\sum_{j=1}^{u} y_j \times w_j}{\sum_{j=1}^{u} w_j} \\ w_j = \frac{1}{d_j^g}, \end{gathered} \tag{5}$$

where (x_w, y_w) is the estimated WC, (x_j, y_j) is the previously known AP coordinate, d_j is the estimated distance between the tag device and jth AP, g is the degree of weight, and u is the total number of APs considered for WC localization.

Figure 9 illustrates the WC localization procedure.

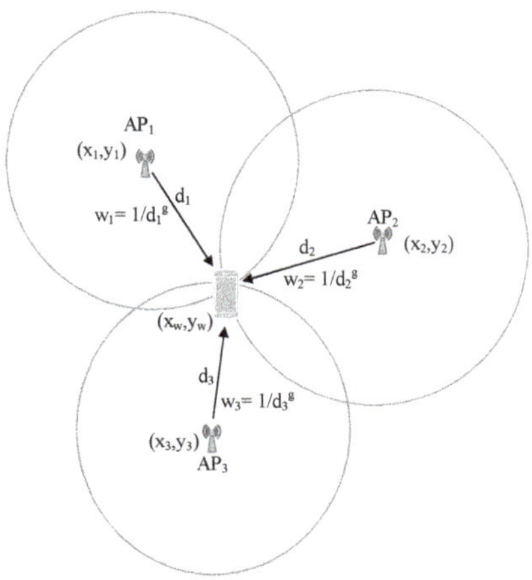

Figure 9. Procedure for estimating weighted centroid (WC) localization at a tag device.

The WC localization has the following characteristics [72]:

1. The estimated location is confined inside the APs' real location only.
2. The estimated location is dragged towards the nearest AP owing to its largest weight.

The degree of weight (g) can be adjusted as per the distance between the deployed APs. A large value of g drives the WC location very close to the real location of the AP with the strongest signal, whereas a very low value (close to zero) yields a geometrical centroid among the u APs.

7.2.2. Trilateration

Trilateration is based on measured distances between a tag device and several APs with their known real location coordinates. Given the distance to an AP, it is known that the tag device must be along the circumference of a circle centered at the AP and radius equal to the tag-AP distance. For a 2D localization, at least three noncollinear APs are needed, whereas, for 3D localization, at least four noncoplanar APs are required to perform trilateration operation.

Let us consider B APs with their real location coordinate $x_i = (x_i, y_i)$ $(i = 1, 2, \ldots . B)$ and unknown location of the tag device be $\mathrm{x} = (x, y)$. The distances between the tag device and the APs is d_i $(i = 1, 2, \ldots, B)$. The relationship between APs/tag positions and their distances in 2D can be written as [135]:

$$\begin{bmatrix} (x_1 - x)^2 + (y_1 - y)^2 \\ (x_2 - x)^2 + (y_2 - y)^2 \\ \vdots \\ (x_B - x)^2 + (y_B - y)^2 \end{bmatrix} = \begin{bmatrix} d_1^2 \\ d_2^2 \\ \vdots \\ d_B^2 \end{bmatrix} \tag{6}$$

Equation (6) can be represented as $A\mathrm{x} = b$ where A and b are defined as:

$$A = \begin{bmatrix} 2(x_B - x_1) & 2(y_B - y_1) \\ 2(x_B - x_2) & 2(y_B - y_2) \\ \vdots & \vdots \\ 2(x_B - x_{B-1}) & 2(y_B - y_{B-1}) \end{bmatrix} \tag{7}$$

$$b = \begin{bmatrix} d_1^2 - d_B^2 - x_1^2 - y_1^2 + x_B^2 + y_B^2 \\ d_2^2 - d_B^2 - x_2^2 - y_2^2 + x_B^2 + y_B^2 \\ \vdots \\ d_{B-1}^2 - d_B^2 - x_{B-1}^2 - y_{B-1}^2 + x_B^2 + y_B^2 \end{bmatrix} \tag{8}$$

The location of the tag device can be estimated based on the least squares system using $\mathrm{x} = (A^T A)^{-1} A^T b$ [136].

7.2.3. Triangulation

In contrast to trilateration, triangulation uses angle measurements in addition to distance measurements to estimate the position of the tag device. Two angles and one length are required for a 2D localization. Particularly, triangulation utilizes the geometric properties of triangles to estimate the tag location.

Given the known length between the APs (known location coordinates of the APs) and after estimating the AOA as shown in Figure 10, the location of the tag device can be estimated as follows [137]:

$$\begin{aligned} x_m &= \frac{y_2 - y_1 + x_1 \tan\theta_1 - x_2 \tan\theta_2}{\tan\theta_1 - \tan\theta_2} \\ y_m &= y_2 - \frac{(x_2 - x_1)\tan\theta_1 - (y_2 - y_1)}{\tan\theta_1 - \tan\theta_2} \tan\theta_2, \end{aligned} \tag{9}$$

where θ_1 and θ_2 are the estimated angle of incident at two APs with their location coordinates (x_1, y_1) and (x_2, y_2), respectively.

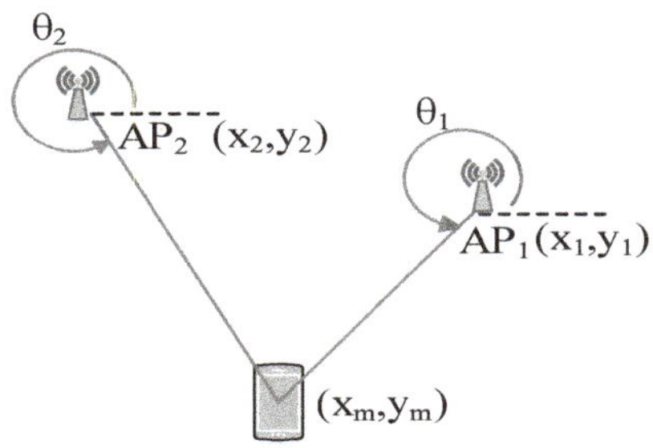

Figure 10. Triangulation localization using two known APs and AOA.

7.3. Fingerprinting

Fingerprinting is also called scene analysis, where signal strength at reference points (RPs) is measured and stored in the database along with the location of the coordinate of the RPs. For localization, new signal strength is measured and compared with the saved ones to estimate a location. Hence, a fingerprinting localization has two phases of operations as illustrated by Figure 11. In the offline phase, the area of interest is divided into nonoverlapping hypothetical grids. The typical grid size is 1 m. The data collector goes from one grid to another to collect RSS from hearable APs. In the online phase, the freshly acquired RSS is compared to the stored one to estimate the user's position.

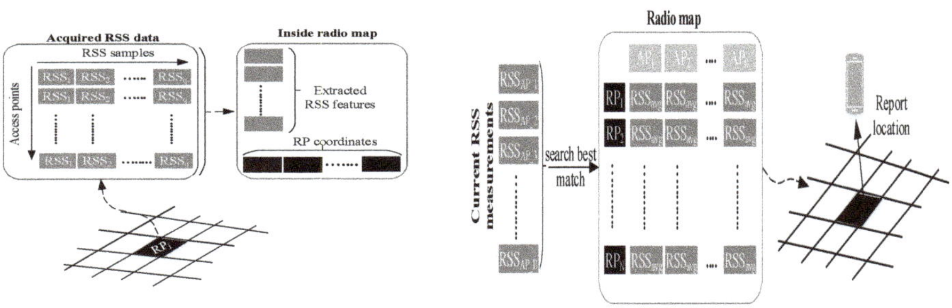

(**a**) offline phase (**b**) online phase

Figure 11. The working procedure of typical fingerprinting localization.

Fingerprinting is the most widely used indoor localization method due to its good localization accuracy and nonrequirement of LOS measurements of APs. The technologies like Wi-Fi, BLE, and geomagnetic field can be used to realize the fingerprinting localization. Although fingerprinting localization has good localization accuracy, it comes with a time-consuming and labor-intensive offline phase. In particular, conventional fingerprinting localization can be categorized as deterministic and probabilistic. The former approach implements fingerprinting data comparison algorithms to find the estimated position, whereas the latter approach yields localization information by estimating a probability distribution over the RPs. In addition, neural networks have also been utilized resembling the scene analysis to produce localization estimation.

7.3.1. Deterministic Fingerprinting Localization

In deterministic fingerprinting localization, the observed RSS is compared against the stored one in the database, and then the coordinate with the closest match is considered the estimated tag device's location. This approach is the basic fingerprinting that is commonly termed as the nearest neighbor (NN) fingerprinting localization [138]. To improvise the localization result, *k*-nearest neighbor

(KNN) can be employed where k nearest RPs are selected based on their closest match for an enhanced result [130,139]. Furthermore, individual weights can be assigned to the selected k RPs to have a weighted k-nearest neighbor (Wk-NN) fingerprinting [72]. Let the positioning distance (D_j) between the stored RSS at jth RP and the online observed RSS be given by [140]:

$$D_j = \sum_{i=1}^{B} \sqrt{(RSS_{i_{online}} - RSS_{i_{offline}})^2}\ j = 1, 2, \ldots.., N, \tag{10}$$

where i is the number of APs ranging from 1 to B. The RPs are arranged with ascending order of D_j and the first k RPs with their known positions $J_z[x_z, y_z]$ are selected to estimate the final location (T_{Wk-NN}) using the following relation [141]:

$$T_{Wk-NN} = \frac{\sum_{z=1}^{k} J_z \times W_z}{\sum_{z=1}^{k} W_z},\ \text{where } W_z = \frac{1}{D_z} \tag{11}$$

7.3.2. Probabilistic Fingerprinting Localization

The probabilistic approach of fingerprinting yields localization information by estimating a probability distribution over the RPs. Here, a matching probability is calculated between the online-observed RSS readings and the prestored fingerprinting data in the radio map database.

Since the statistical distribution of RSS at a particular RP can be thought of as Gaussian probability distribution, the RSS values should obey the normal distribution $N\ (\mu, \sigma^2)$ where μ and σ^2 are mean and variance of RSS data [142], respectively. Now, the likelihood function ($L(\mu, \sigma^2)$) is given by the following relation [143]:

$$L(\mu, \sigma^2) = \prod_{i=1}^{q} \frac{1}{\sqrt{2\pi}\sigma} \exp^{-\frac{(RSS_i - \mu)^2}{2\sigma^2}} = (2\pi\sigma^2)^{\frac{-q}{2}} \exp^{\left(-\frac{1}{2\sigma^2}\sum_{i=1}^{q}(RSS_i - \mu)^2\right)} \tag{12}$$

We can obtain the logarithmic equation of (12) as follows:

$$\log[L(\mu, \sigma^2)] = -\frac{q}{2}\log(2q) - \frac{q}{2}\log(\sigma^2) - \frac{q}{2\sigma^2}\sum_{i=1}^{q}(RSS_i - \mu)^2 \tag{13}$$

Hence, the likelihood equations can be written as:

$$\begin{aligned} \frac{\partial \log[L(\mu, \sigma^2)]}{\partial \mu} &= \frac{1}{\sigma^2}\sum_{i=1}^{q}(RSS_i - \mu)^2 = 0 \\ \frac{\partial \log[L(\mu, \sigma^2)]}{\partial \sigma^2} &= -\frac{q}{2\sigma^2} + \frac{1}{2\sigma^4}\sum_{i=1}^{q}(RSS_i - \mu)^2 = 0 \end{aligned} \tag{14}$$

From (14), we get,

$$\mu* = \overline{RSS} = \frac{1}{q}\sum_{i=1}^{q} RSS_i \tag{15}$$

$$\sigma^{*2} = \frac{1}{q}\sum_{i=1}^{q}(RSS_i - \mu)^2 \tag{16}$$

The unique solution (μ^*, σ^{*2}) of the likelihood equations should also be a local maximum point. In other words, when $|\mu| \to \infty$ or $\sigma^2 \to \infty$ or $\sigma^2 \to 0$, the non-negative function $L(\mu, \sigma^2) \to 0$. Hence, the maximum likelihood equation of μ and σ^2 will be (15) and (16), respectively.

Hence, with this idea, the average of RSS reading and its corresponding variance from each AP is calculated for the construction of a radio map in the offline phase of the probability-based fingerprinting localization. Later, in the online phase, after acquiring new RSS readings ($RSS_i, i = 1, 2, 3, \ldots, B$) at an unknown location from B APs, we can estimate the probability of the RP (x, y) with respect to ith AP $(P_i(x, y))$ as follows:

$$P_i(x,y) = \frac{1}{\sqrt{2\pi}\sigma_i} \exp^{\frac{-(RSS_i-\mu)^2}{2\sigma_i^2}}, \tag{17}$$

where μ_i and σ_i^2 are the stored average RSS and its corresponding variance from the ith AP, respectively. This way, the probability of each RP can be established where the tag's location could be located at the RP with maximum probability.

7.3.3. Neural Networks-Based Fingerprinting Localization

The indoor localization problem has been attempted to solve using neural networks too. The radio map is used as inputs and the targets for training the neural network during the offline state to obtain the appropriate weights. In the online stage, the input signal strength is multiplied with a trained weight matrix at different layers to yield either 2D/3D location or probabilities depending on the activation function at the output layer. In general, an artificial neural network (ANN) with one hidden layer (shallow) or multiple hidden layers (deep) can be employed in neural networks based localization. The basic architecture of multilayer perception (MLP) is given in Figure 12.

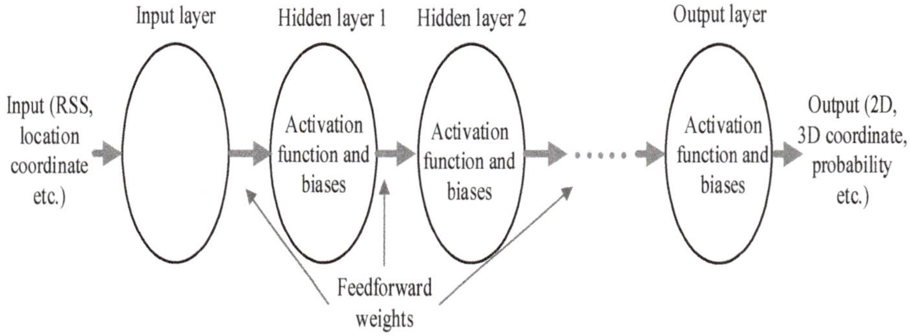

Figure 12. A typical architecture of multilayer perception (MLP).

In [144], an indoor localization solution based on deep neural networks (DNN) is presented. This work employs a four-layered DNN-based probabilistic estimator and RSS preprocessing, where the raw RSS is normalized to the range between 0 to 1. For offline DNN training, the preprocessed RSS readings of the radio map database are used. Online positioning consists of two layers: coarse and fine positioning. At coarse positioning, trained DNN is used to yield probabilities of all the RPs, and the Hidden Markov model is used to refine the coarse positioning estimate. ANN-based fingerprinting is used in [145] to estimate the location coordinate. This work uses clustering where a separate ANN is trained for each cluster or region.

7.3.4. Survey of Available Fingerprinting Localization Solutions

- **Fingerprinting localization using Wi-Fi and BLE signals**: Pavel et al. present an IPS research work based on Wk-NN positioning method using BLE beacons [146]. The k-nearest fingerprints are found in a radio map database by employing the Euclidean distance between the observed RSS and the database's referred one. This work further compares the localization methods based on Wi-Fi and a combination of BLE and Wi-Fi. They recommend that the combination of wireless technologies help to increase the localization accuracy. Next work based on BLE beacons using fingerprinting technique is reported in [80] where a Gaussian filter is used to preprocess the received RSS. This work proposes a distance-weighted filter based on the triangle theorem of trilateral relations to filter out the wrong distance value caused by an abnormal RSS.

The traditional Wk-NN fingerprinting has also been realized with Wi-Fi signals. Reference [29] elaborates recent advances on Wi-Fi fingerprinting localization. They overview advanced localization techniques and efficient system development utilizing Wi-Fi technology in their survey work. An improvisation over the conventional Wk-NN fingerprinting using Wi-Fi signals is put forward in [147,148]. The former approach uses average RSS and standard deviation of Wi-Fi signals at the RPs from the APs to construct a fingerprint radio map. Both the average RSS and the standard deviation are processed to estimate a Euclidean distance in the online phase. With the Euclidean distance, k RPs are selected to estimate a coarse location. Furthermore, a joint probability for each RP is calculated, based on which the k RPs are selected to estimate another coarse location. Later, both the coarse localization estimations are fused, employing the shortest Euclidean distance and the largest joint probability to yield a final localization estimation. Meanwhile, the later approach proposes to use Manhattan distance instead of Euclidean distance to compare the closeness of acquired Wi-Fi signal strength with the stored database.
Some of the examples of probability-based fingerprinting localization using Wi-Fi as a signal source are illustrated in [123,142,149,150]. Similarly, the literature that employ BLE for probability-based fingerprinting are presented in [59,151].

- **Machine learning-based methods**: The machine learning algorithm extracts valuable information from the raw data and represents it as a model or hypothesis, which can be used for other unseen data to infer things. Although Gaussian process regression (GPR) is widely used in geostatistics as a Bayesian kringing, it has drawn a lot of attention in the machine learning community in recent decades. The GPR can be defined as a supervised learning task, which can predict the RSSs at arbitrary coordinates based on acquired training data. The prediction of RSS across the testbed with little training data helps to reduce the human workload significantly. Reference [123] presents a GPR-based fingerprinting IPS using indoor Wi-Fi APs. This work uses a few data points to train the Gaussian process (GP), where the firefly algorithm is used to estimate the GP's hyperparameters. Moreover, it also shows that the probabilistic-based localization performs better than deterministic-based localization using the predicted radio map. Liu et al. proposed a GPR-plus method with Bluetooth transmitters using a naïve Bayes algorithm [152]. They compare their method with [123] and claim that their method is computationally cheaper. Another example of GPR-based fingerprinting is put forward in [149]. This work estimated the hyperparameters by using the subspace trust-region method and shows that location estimation with a radio map built using GPR is better than that of Horus fingerprinting method [153]. The GPR-based IPS in [151] utilizes BLE beacons for localization where the Hlhyperparameters are optimized employing limited memory BFGS-B [154]. Here, the predicted RSS data is further preprocessed for RSS clustering, where the final localization result is obtained with the minimized offline workload and reduced online computational complexity.
  In [155], the use of a support vector machine (SVM) is proposed to estimate the Wi-Fi signal strength at non-sight-surveyed locations on the testbed. This system creates an RSS reference surface for each AP using discrete train data with SVM. During the testing phase, the sampled online RSS from each AP is searched on the corresponding surfaces. Here, the coordinate that is found in the higher number of such surfaces is estimated as the tag device's location.
- **Crowdsourcing techniques**: Although machine learning approaches like GPR are intended to solve the offline workload problem, they still require a little training data that are manually acquired from the localization area. Hence, recent literature on solving the offline workload problem of fingerprinting localization is more focused on the crowdsourcing [156,157]. Here, the main concept is to crowdsource the RSS data from freely moving users across the testbed. It is straightforward to understand that unlabeled RSS data are easy to acquire from various users. However, the main concern is to find a plausible way to label the crowdsourced RSS data with the ground-truth location.
  In [158], a smartphone-based crowdsourcing approach is proposed that employs an accelerometer

as a pedometer. Here, multidimensional scaling (MDS) is used to create a map that displays the relative positions of several objects employing only a table of distance values among them. The walking distance between two RPs is estimated using the accelerometer to form a distance matrix. The MDS utilizes the distance matrix as its input to map all the RPs into a d-dimensional Euclidean space forming a "stress-free" floor plan. Meanwhile, a next distance matrix is also formed utilizing the walking distance between two fingerprint positions, where again the MDS maps all fingerprints to a d-dimensional Euclidean space to form a fingerprint space. Finally, the stress-free floor plan and the fingerprint space are mapped to form a radio map database. X. Tong et al. suggested a *FineLoc* system for indoor radio map construction employing BLE beacons and PDR as the source of reference information [159]. The *FineLoc* system generated the tag's trace and then determines the map for the trace. For online positioning, this system merges the tag's trace into the existing floorplan.

Similarly, [160] has put forward a trajectory learning method utilizing crowdsourcing measurements to support the absence of a map. Here, the k-nearest neighbor is used to perform a classification model with linear discriminant analysis (LDA) and principal component analysis (PCA) for floor detection. The combination of LDA and PCA employs the acquired training data to make a classification model. Moreover, [161] uses a commercial software called Trusted Positioning Navigator (T-PN) for crowdsourcing based IPS. This method forms a crowdsourced fingerprinting database employing the RSS values and position information from the T-PN software.

- **Clustering-based approaches**: The conventional fingerprinting is also termed as flat fingerprinting and can be converted to two-step fingerprinting using clustering or segmentation.The two-step fingerprinting is realized with a coarse localization step and fine localization step, as the name suggests. Clustering reduces the searching space of RPs in the online phase of fingerprinting, which eventually reduces the system's computational cost. Moreover, it also helps to reduce the localization estimation error by removing the outliers.

  Clustering on IPS can be realized using either hardware (Wi-Fi or BLE) or the RSS clustering. An example of a clustering module using hardware is Horus [153]. Here, the clustering module is employed where any cluster is a set of RPs sharing a common set of Wi-Fi APs. This approach estimates the tag's position based on the largest posterior probability by Bayesian interference [162]. Similarly, [163] uses BLE beacon proximity to reduce the searching space in the online phase. Here, BLE's proximity provides coarse localization, and for fine localization, a selected set of RPs is used with Wi-Fi fingerprint datasets.

  The performance of indoor fingerprinting positioning can be improved with RSS clustering [30]. An RSS clustering method chooses a set of cluster centers to reduce the sum of squared distances between the RSS value and their corresponding centers. For example, a K-means clustering [164] begins by choosing both the number of output clusters and the corresponding set of initial cluster heads, where the clustering algorithm iteratively refines the output clusters to decrease the sum of squared distances [165]. Hence, K-means clustering has a requirement of an arbitrary selection of initial cluster centers. On the other hand, affinity propagation clustering (APC) starts by assigning each point (RP in this study) the same chance to become a cluster center where all the points are joined in the large space [166]. Reference [167] uses APC for clustering the testbed using Wi-Fi RSS data. Here, the cluster-head is determined on the coarse localization, and Wk-NN is used for fine localization. In addition to APC and K-means, other clustering methods in IPS include fuzzy c-means and hierarchical clustering strategy (HCS) [168–171]. APC has been a widely used clustering technique in IPS owing to its initialization-independent and better cluster head selection characteristics. Many kinds of literature on IPS have employed APC for RSS clustering where their fine localization is either probabilistic-based or deterministic-based [150,167,172,173].

Table 3. Wireless technology (Wi-Fi and BLE)-based localization system and solution.

System	Tech.	Signal	Positioning Algorithm	Performance Metrics					
				Accuracy	Precision	Complexity	Scalability/ Space Dimension	Robustness	Cost
Horus [174]	Wi-Fi	RSS	Probabilistic method	2 m	90% (2.1 m)	Moderate	Good/2D	Good	Low
RADAR [175]	Wi-Fi	RSS	deterministic method	3–5 m	50% (2.5 m), 90% (5.9 m)	Moderate	Good/2D,3D	Good	Low
Robot assistive [176]	Wi-Fi	RSS	Bayesian approach	1.5 m	Over 50% (1.5 m)	Medium	Good/2D	Good	Medium
Ekahau [177]	Wi-Fi	RSS	Probabilistic method	1 m	50% (2 m)	Moderate	Good/2D	Good	Low
IPS in [178]	Wi-Fi, IMU	RSS	Fingerprinting+PDR	2.4 m	88% (3 m)	High	Good/2D	Weak	Low
IPS in [123]	Wi-Fi	RSS	Probabilistic method	3 m	90% (9 m)	Moderate	Excellent/2D	Good	Low
IPS in [149]	Wi-Fi	RSS	Probabilistic method	2.3 m	N/A	Moderate	Excellent/2D	Good	Low
IPS in [179]	Wi-Fi, IMU	RSS	Fingerprinting+PDR	2.2 m	90% (1 m)	High	Good/2D	Weak	Low
IPS in [124]	Wi-Fi	RSS	Support vector regression	0.68 m	90% (1.4 m)	Moderate	Excellent/2D	Good	Low
IPS in [150]	Wi-Fi	RSS	Probabilistic method	2.23 m	80% (3 m)	Moderate	Good/2D	Weak	Low
IPS in [172]	Wi-Fi	RSS	Probabilistic method	1.94 m	94% (3 m)	Moderate	Good/2D	Weak	Low
IPS in [180]	BLE, IMU	RSS	Proximity+PDR	0.28 m	N/A	High	Excellent/2D	Weak	Medium
IPS in [72]	BLE	RSS	WCL+Fingerprinting	1.6 m	90% (2 m)	Moderate	Good/2D	Good	Medium
IPS in [134]	BLE, IMU	RSS	Proximity+PDR	2.26 m	N/A	High	Good/2D	Weak	Medium
IPS in [59]	BLE	RSS	Probabilistic method	N/A	95% (2.6 m)	Moderate	Good/2D	Good	Medium
IPS in [151]	BLE	RSS	Probabilistic method	2.25 m	76% (3 m)	Moderate	Excellent/2D	Good	Medium
IPS in [181]	BLE, IMU	RSS	Fingerprinting+PDR	0.65 m	95% (1.5 m)	High	Good/2D	Weak	Medium
IPS in [114]	BLE	RSS	WCL+Fingerprinting	1.06 m	95% (1.5 m)	Moderate	Good/2D	Good	Medium
IPS in [182]	BLE	RSS	Rank-based fingerprinting	0.78 m	70% (1 m)	Moderate	Good/2D	Good	Medium
IPS in [118]	BLE, Magnetic field, IMU	RSS, Magnetic Flux	Wk-NN+PDR	1.31 m	85% (2 m)	Very High	Good/2D	Weak	Low
FineLoc [159]	BLE IMU	RSS	Crowdsourcing, trace merge, PDR	1.21 m	80% (1.57 m)	High	Excellent/2D	Good	Medium
LiFS [158]	Wi-Fi, IMU	RSS	Crowdsourcing, deterministic method	2 m	90% (4 m)	High	Excellent/2D	Good	Low
IPS in [125]	BLE	RSS	Crowdsourcing, probabilistic method	2.34 m	76% (4 m)	Moderate	Excellent/2D	Good	Medium

8. Conclusions and Future Research Trends

We conduct this survey to review the RF-based indoor localization solutions presented in recent literature. As modern smartphones and wireless technologies are evolving, the paper focuses more on practical IPS realized in a smartphone and uses Wi-Fi and BLE as their primary signal source. We described different available wireless technologies and techniques for IPS development. Furthermore, the paper explains the different performance metrics of IPS and their trade-off. As the available localization solutions focus on achieving good localization accuracy/precision, their computational complexity should not be forgotten.

Fingerprinting localization is a promising technique of IPS. However, it is doomed by the requirement of the offline training phase. Although machine-learning and crowdsourcing approaches are put forward to solve the issue of data collection, the IPS still awaits feasible solutions. The RF-based wireless technologies, particularly Wi-Fi and BLE, are widely used for indoor LBS owing to their characteristics like signal penetration, power consumption, localization accuracy, and convenient deployment.

Future research direction on RF-based IPS may lead towards a hybrid system integrating multiple techniques and alternative technologies, efficient learning methodology of radio signals, and deep learning approaches. As 5G technology is rising across the world and UWB chip is available on the latest smartphones, future IPS can exploit and integrate these technologies to develop a better IPS. Future research works can be focused on finding a better solution for easy data acquisition. For example, machine-learning/hardware (Wi-Fi, BLE, and UWB) assistive crowdsourcing approaches are worth considering. Since DNN is used successfully in designing IPS, DNN-based IPS is also a good research topic. Moreover, as there is a lack of standardization (a set of rules that serve as a guide to designing an IPS), there is also no fixed wireless technology widely accepted as a major technology for future IPS. Most of the available systems and solutions are disjoint, where no ubiquitous IPS exists. Hence, the IPS requires a standardization that can narrow down the techniques and technologies, which fulfills the performance metrics of practical IPS.

We believe that the timely and comprehensive overview of the recent works in this survey will further encourage new research efforts into the practical IPS.

Author Contributions: All authors contributed to the conceptualization and methodology of the manuscript; S.S. contributed in writing; J.-Y.P. reviewed and edited the manuscript. All authors have read and agreed to the published version of the manuscript.

Acknowledgments: This study was supported by research fund from Chosun University, 2020.

Conflicts of Interest: The authors declare no conflict of interest.

References

1. GPS. GPS: The Global Positioning System. 2020. Available online: https://www.gps.gov/(accessd on 28 August 2020).
2. Information Analytical Centre of GLONASS and GPS Controlling. 2020. Available online: https://www.glonass-iac.ru/en/ (accessd on 28 August 2020).
3. European GNSS Service Centre | European GNSS Service Centre. 2020. Available online: https://www.gsc-europa.eu/ (accessd on 28 August 2020).
4. BeiDou Navigation Satellite System. 2020. Available online: http://en.beidou.gov.cn/ (accessd on 28 August 2020).
5. Vasisht, D.; Kumar, S.; Katabi, D. Decimeter-level localization with a single WiFi access point. In Proceedings of the 13th USENIX Symposium on Networked Systems Design and Implementation (NSDI '16), Santa Clara, CA, USA 17–18 March 2016; pp. 165–178.
6. Liu, H.H.; Yang, Y.N. WiFi-based indoor positioning for multi-floor environment. In Proceedings of the TENCON 2011—2011 IEEE Region 10 Conference, Bali, Indonesia, 21–24 November 2011; pp. 597–601.
7. Bhattarai, B.; Yadav, R.K.; Gang, H.S.; Pyun, J.Y. Geomagnetic Field Based Indoor Landmark Classification Using Deep Learning. *IEEE Access* **2019**, *7*, 33943–33956. [CrossRef]

8. Xiong, J.; Jamieson, K. Arraytrack: A fine-grained indoor location system. Presented at the 10th USENIX Symposium on Networked Systems Design and Implementation (NSDI '13), Lombard, IL, USA, 2–5 April 2013; pp. 71–84.
9. Xiong, J.; Sundaresan, K.; Jamieson, K. Tonetrack: Leveraging frequency-agile radios for time-based indoor wireless localization. In Proceedings of the 21st Annual International Conference on Mobile Computing and Networking, Paris, France, 7–11 September 2015; pp. 537–549.
10. Xiong, J. Pushing the Limits of Indoor Localization in Today's Wi-Fi Networks. Ph.D. Thesis, UCL (University College London), London, UK, 2015.
11. Rusli, M.E.; Ali, M.; Jamil, N.; Din, M.M. An improved indoor positioning algorithm based on rssi-trilateration technique for internet of things (IoT). In Proceedings of the 2016 International Conference on Computer and Communication Engineering (ICCCE), Kuala Lumpur, Malaysia, 25–27 July 2016; pp. 72–77.
12. Subedi, S.; Kwon, G.R.; Shin, S.; Hwang, S.s.; Pyun, J.Y. Beacon based indoor positioning system using weighted centroid localization approach. In Proceedings of the 2016 Eighth International Conference on Ubiquitous and Future Networks (ICUFN), Vienna, Austria, 5–8 July 2016; pp. 1016–1019.
13. Potortì, F.; Barsocchi, P.; Girolami, M.; Torres-Sospedra, J.; Montoliu, R. Evaluating indoor localization solutions in large environments through competitive benchmarking: The EvAAL-ETRI competition. In Proceedings of the 2015 International Conference on Indoor Positioning and Indoor Navigation (IPIN), Banff, AB, Canada, 13–16 October 2015; pp. 1–10.
14. Potortì, F.; Park, S.; Jiménez Ruiz, A.; Barsocchi, P.; Girolami, M.; Crivello, A.; Lee, S.; Lim, J.; Torres-Sospedra, J.; Seco, F.; et al. Comparing the performance of indoor localization systems through the EvAAL framework. *Sensors* **2017**, *17*, 2327. [CrossRef] [PubMed]
15. Lymberopoulos, D.; Liu, J.; Yang, X.; Choudhury, R.R.; Sen, S.; Handziski, V. Microsoft indoor localization competition: Experiences and lessons learned. *GetMob. Mob. Comput. Commun.* **2015**, *18*, 24–31. [CrossRef]
16. Potortì, F.; Crivello, A.; Barsocchi, P.; Palumbo, F. Evaluation of indoor localisation systems: Comments on the ISO/IEC 18305 standard. In Proceedings of the 2018 International Conference on Indoor Positioning and Indoor Navigation (IPIN), Nantes, France, 24–27 September 2018; pp. 1–7.
17. Kotaru, M.; Joshi, K.; Bharadia, D.; Katti, S. Spotfi: Decimeter level localization using wifi. In Proceedings of the 2015 ACM Conference on Special Interest Group on Data Communication, London, UK, 17–21 August 2015; pp. 269–282.
18. Wang, J.; Liang, X.; Shi, W.; Niu, Q.; Gulliver, T.A.; Liu, X. Fingerprinting localization based on 60 GHz impulse radio. In Proceedings of the 2015 IEEE Pacific Rim Conference on Communications, Computers and Signal Processing (PACRIM), Victoria, BC, Canada, 24–26 August 2015; pp. 491–495, [CrossRef]
19. Davidson, P.; Piché, R. A survey of selected indoor positioning methods for smartphones. *IEEE Commun. Surv. Tutor.* **2016**, *19*, 1347–1370. [CrossRef]
20. Harle, R. A survey of indoor inertial positioning systems for pedestrians. *IEEE Commun. Surv. Tutor.* **2013**, *15*, 1281–1293. [CrossRef]
21. Al Nuaimi, K.; Kamel, H. A survey of indoor positioning systems and algorithms. In Proceedings of the 2011 International Conference on Innovations in Information Technology, Abu Dhabi, UAE, 25–27 April 2011; pp. 185–190.
22. Xiao, J.; Zhou, Z.; Yi, Y.; Ni, L.M. A survey on wireless indoor localization from the device perspective. *ACM Comput. Surv. CSUR* **2016**, *49*, 1–31. [CrossRef]
23. Yang, Z.; Zhou, Z.; Liu, Y. From RSSI to CSI: Indoor localization via channel response. *ACM Comput. Surv. CSUR* **2013**, *46*, 1–32. [CrossRef]
24. Ferreira, A.F.G.; Fernandes, D.M.A.; Catarino, A.P.; Monteiro, J.L. Localization and positioning systems for emergency responders: A survey. *IEEE Commun. Surv. Tutor.* **2017**, *19*, 2836–2870. [CrossRef]
25. Maghdid, H.S.; Lami, I.A.; Ghafoor, K.Z.; Lloret, J. Seamless outdoors-indoors localization solutions on smartphones: Implementation and challenges. *ACM Comput. Surv. CSUR* **2016**, *48*, 1–34. [CrossRef]
26. Zafari, F.; Gkelias, A.; Leung, K.K. A survey of indoor localization systems and technologies. *IEEE Commun. Surv. Tutor.* **2019**, *21*, 2568–2599. [CrossRef]
27. Koyuncu, H.; Yang, S.H. A survey of indoor positioning and object locating systems. *IJCSNS Int. J. Comput. Sci. Netw. Secur.* **2010**, *10*, 121–128.
28. Liu, H.; Darabi, H.; Banerjee, P.; Liu, J. Survey of wireless indoor positioning techniques and systems. *IEEE Trans. Syst. Man Cybern. Part Appl. Rev.* **2007**, *37*, 1067–1080. [CrossRef]

29. He, S.; Chan, S.H.G. Wi-Fi fingerprint-based indoor positioning: Recent advances and comparisons. *IEEE Commun. Surv. Tutor.* **2016**, *18*, 466–490. [CrossRef]
30. Khalajmehrabadi, A.; Gatsis, N.; Akopian, D. Modern WLAN fingerprinting indoor positioning methods and deployment challenges. *IEEE Commun. Surv. Tutor.* **2017**, *19*, 1974–2002. [CrossRef]
31. Subramanian, S.P.; Sommer, J.; Schmitt, S.; Rosenstiel, W. RIL—Reliable RFID based indoor localization for pedestrians. In Proceedings of the 2008 16th International Conference on Software, Telecommunications and Computer Networks, Milan, Italy, 14–16 November 2008; pp. 218–222.
32. Choi, B.S.; Lee, J.W.; Lee, J.J. Localization and map-building of mobile robot based on RFID sensor fusion system. In Proceedings of the 2008 6th IEEE International Conference on Industrial Informatics, Daejeon, Korea, 13–16 July 2008; pp. 412–417.
33. Choi, B.S.; Lee, J.W.; Lee, J.J. An improved localization system with RFID technology for a mobile robot. In Proceedings of the 2008 34th Annual Conference of IEEE Industrial Electronics, Orlando, FL, USA, 10–13 November 2008; pp. 3409–3413.
34. Tesoriero, R.; Gallud, J.A.; Lozano, M.D.; Penichet, V.M. Tracking autonomous entities using RFID technology. *IEEE Trans. Consum. Electron.* **2009**, *55*, 650–655. [CrossRef]
35. Koutsou, A.D.; Seco, F.; Jiménez, A.R.; Roa, J.O.; Ealo, J.L.; Prieto, C.; Guevara, J. Preliminary localization results with an RFID based indoor guiding system. In Proceedings of the 2007 IEEE International Symposium on Intelligent Signal Processing, Alcala De Henares, Spain, 3–5 October 2007; pp. 1–6.
36. Chae, H.; Han, K. Combination of RFID and vision for mobile robot localization. In Proceedings of the 2005 International Conference on Intelligent Sensors, Sensor Networks and Information Processing, Melbourne, Australia, 5–8 December 2005; pp. 75–80.
37. Hepeng, D.; Donglin, S. Indoor location system using RFID and ultrasonic sensors. In Proceedings of the 2008 8th International Symposium on Antennas, Propagation and EM Theory, Kunming, China, 2–5 November 2008; pp. 1179–1181.
38. Wang, C.S.; Cheng, L.C. RFID & vision based indoor positioning and identification system. In Proceedings of the 2011 IEEE 3rd International Conference on Communication Software and Networks, Xi'an, China, 27 May 2011; pp. 506–510.
39. Ozdenizci, B.; Ok, K.; Coskun, V.; Aydin, M.N. Development of an indoor navigation system using NFC technology. In Proceedings of the 2011 Fourth International Conference on Information and Computing, Phuket, Thailand, 25–27 April 2011; pp. 11–14.
40. Ozdenizci, B.; Coskun, V.; Ok, K. NFC internal: An indoor navigation system. *Sensors* **2015**, *15*, 7571–7595. [CrossRef]
41. Luo, Y.; Law, C.L. Indoor positioning using UWB-IR signals in the presence of dense multipath with path overlapping. *IEEE Trans. Wirel. Commun.* **2012**, *11*, 3734–3743. [CrossRef]
42. Kim, D.H.; Kwon, G.R.; Pyun, J.Y.; Kim, J.W. NLOS identification in UWB channel for indoor positioning. In Proceedings of the 2018 15th IEEE Annual Consumer Communications & Networking Conference (CCNC), Las Vegas, NV, USA, 12–15 January 2018; pp. 1–4.
43. Waadt, A.; Wang, S.; Kocks, C.; Burnic, A.; Xu, D.; Bruck, G.H.; Jung, P. Positioning in multiband OFDM UWB utilizing received signal strength. In Proceedings of the 2010 7th Workshop on Positioning, Navigation and Communication, Dresden, Germany, 11–12 March 2010; pp. 308–312.
44. Subedi, S.; Das, S.; Shet, N.S.V. Dynamic Spectrum Allocation in Wireless sensor Networks. *Int. J. Mod. Eng. Res.* **2014**, *4*, 18–25.
45. Sohraby, K.; Minoli, D.; Znati, T. *Wireless Sensor Networks: Technology, Protocols, and Applications*; John Wiley & Sons: Hoboken, NJ, USA, 2007.
46. Siekkinen, M.; Hiienkari, M.; Nurminen, J.K.; Nieminen, J. How low energy is bluetooth low energy? Comparative measurements with zigbee/802.15.4. In Proceedings of the 2012 IEEE wireless communications and networking conference workshops (WCNCW), Paris, France, 1 April 2012; pp. 232–237.
47. Luo, R.C.; Hsiao, T.J. Indoor Localization System Based on Hybrid Wi-Fi/BLE and Hierarchical Topological Fingerprinting Approach. *IEEE Trans. Veh. Technol.* **2019**, *68*, 10791–10806. [CrossRef]
48. Khandker, S.; Torres-Sospedra, J.; Ristaniemi, T. Analysis of Received Signal Strength Quantization in Fingerprinting Localization. *Sensors* **2020**, *20*, 3203. [CrossRef]

49. Mathivannan, S.; Srinath, S.; Shashank, R.; Aravindh, R.; Balasubramanian, V. A Dynamic Weighted Trilateration Algorithm for Indoor Localization Using Dual-Band WiFi. In *International Symposium on Web and Wireless Geographical Information Systems*; Springer: Berlin/Heidelberg, Germany, 2019; pp. 174–187.
50. Ismail, A.H.; Kitagawa, H.; Tasaki, R.; Terashima, K. WiFi RSS fingerprint database construction for mobile robot indoor positioning system. In Proceedings of the 2016 IEEE International Conference on Systems, Man, and Cybernetics (SMC), Budapest, Hungary, 9–12 October 2016; pp. 1561–1566.
51. Wang, X.; Gao, L.; Mao, S.; Pandey, S. CSI-based fingerprinting for indoor localization: A deep learning approach. *IEEE Trans. Veh. Technol.* **2016**, *66*, 763–776. [CrossRef]
52. Ciurana, M.; Cugno, S.; Barcelo-Arroyo, F. WLAN indoor positioning based on TOA with two reference points. In Proceedings of the 2007 4th Workshop on Positioning, Navigation and Communication, Dresden, Germany, 11–12 March 2007; pp. 23–28.
53. Kodippili, N.; Dias, D. Integration of fingerprinting and trilateration techniques for improved indoor localization. In Proceedings of the 2010 Seventh International Conference on Wireless and Optical Communications Networks-(WOCN), Colombo, Sri Lanka, 6–8 September 2010; pp. 1–6.
54. iOS Wi-Fi Management APIs. 2017. Available online: developer.apple.com/library/archive/qa/qa1942/_index.html (accessed on 27 January 2020).
55. Berkovich, G. Accurate and reliable real-time indoor positioning on commercial smartphones. In Proceedings of the 2014 International Conference on Indoor Positioning and Indoor Navigation (IPIN), Busan, Korea, 27–30 October 2014; pp. 670–677.
56. Wi-Fi Scanning Overview. 2020. Available online: https://developer.android.com/guide/topics/connectivity/wifi-scan#wifi-scan-restrictions (accessed on 28 November 2020).
57. Liu, H.H.; Liu, C. Implementation of Wi-Fi signal sampling on an android smartphone for indoor positioning systems. *Sensors* **2018**, *18*, 3. [CrossRef] [PubMed]
58. Dahlgren, E.; Mahmood, H. Evaluation of Indoor Positioning Based on Bluetooth Smart Technology. Master's Thesis, Chalmers University of Technology, Goteborg, Sweden, 2014.
59. Faragher, R.; Harle, R. Location fingerprinting with bluetooth low energy beacons. *IEEE J. Sel. Areas Commun.* **2015**, *33*, 2418–2428. [CrossRef]
60. Spachos, P.; Plataniotis, K. BLE Beacons in the Smart City: Applications, Challenges, and Research Opportunities. *IEEE Internet Things Mag.* **2020**, *3*, 14–18, [CrossRef]
61. Corbacho Salas, A. Indoor Positioning System Based on Bluetooth Low Energy. Bachelor's Thesis, Universitat Politècnica de Catalunya, Barcelona, Spain, 2014.
62. Powar, J.; Gao, C.; Harle, R. Assessing the impact of multi-channel BLE beacons on fingerprint-based positioning. In Proceedings of the 2017 International Conference on Indoor Positioning and Indoor Navigation (IPIN), Sapporo, Japan, 18–21 September 2017; pp. 1–8, [CrossRef]
63. Mackey, A.; Spachos, P.; Song, L.; Plataniotis, K.N. Improving BLE Beacon Proximity Estimation Accuracy Through Bayesian Filtering. *IEEE Internet Things J.* **2020**, *7*, 3160–3169, [CrossRef]
64. Kluge, T.; Groba, C.; Springer, T. Trilateration, Fingerprinting, and Centroid: Taking Indoor Positioning with Bluetooth LE to the Wild. In Proceedings of the 2020 IEEE 21st International Symposium on "A World of Wireless, Mobile and Multimedia Networks" (WoWMoM), Rome, Italy, 21 July 2020; pp. 264–272, [CrossRef]
65. Tomažič, S.; Dovžan, D.; Škrjanc, I. Confidence-Interval-Fuzzy-Model-Based Indoor Localization. *IEEE Trans. Ind. Electron.* **2019**, *66*, 2015–2024, [CrossRef]
66. Jeon, K.E.; She, J.; Soonsawad, P.; Ng, P.C. BLE Beacons for Internet of Things Applications: Survey, Challenges, and Opportunities. *IEEE Internet Things J.* **2018**, *5*, 811–828, [CrossRef]
67. Akinsiku, A.; Jadav, D. BeaSmart: A beacon enabled smarter workplace. In Proceedings of the NOMS 2016—2016 IEEE/IFIP Network Operations and Management Symposium, Istanbul, Turkey, 25–29 April 2016; pp. 1269–1272, [CrossRef]
68. Beacon Compatibility with iOS and Android. 2020. Available online: https://www.beaconzone.co.uk/blog/beacon-compatibility-with-ios-and-android/ (accessed on 28 November 2020).
69. FlutterBlue. 2020. Available online: https://pub.dev/packages/flutter_blue (accessed on 28 November 2020).
70. Universal Bluetooth Beacon Library. 2020. Available online: https://github.com/andijakl/universal-beacon (accessed on 28 November 2020).

71. Putra, G.D.; Pratama, A.R.; Lazovik, A.; Aiello, M. Comparison of energy consumption in Wi-Fi and bluetooth communication in a Smart Building. In Proceedings of the 2017 IEEE 7th Annual Computing and Communication Workshop and Conference (CCWC), Las Vegas, NV, USA, 9–11 January 2017; pp. 1–6.
72. Subedi, S.; Pyun, J.Y. Practical fingerprinting localization for indoor positioning system by using beacons. *J. Sens.* **2017**, *2017*. [CrossRef]
73. Newman, N. Apple iBeacon technology briefing. *J. Direct Data Digit. Mark. Pract.* **2014**, *15*, 222–225. [CrossRef]
74. Yassin, A.; Nasser, Y.; Awad, M.; Al-Dubai, A.; Liu, R.; Yuen, C.; Raulefs, R.; Aboutanios, E. Recent advances in indoor localization: A survey on theoretical approaches and applications. *IEEE Commun. Surv. Tutor.* **2016**, *19*, 1327–1346. [CrossRef]
75. Slock, D. Location aided wireless communications. In Proceedings of the 2012 5th International Symposium on Communications, Control and Signal Processing, Roma, Italy, 2–4 May 2012; pp. 1–6.
76. Witrisal, K.; Hinteregger, S.; Kulmer, J.; Leitinger, E.; Meissner, P. High-accuracy positioning for indoor applications: RFID, UWB, 5G, and beyond. In Proceedings of the 2016 IEEE International Conference on RFID (RFID), Orlando, FL, USA, 3–5 May 2016; pp. 1–7.
77. Anagnostopoulos, G.G.; Deriaz, M. Accuracy enhancements in indoor localization with the weighted average technique. *SensorComm* **2014**, *2014*, 112–116.
78. Sklar, B. Rayleigh fading channels in mobile digital communication systems. I. Characterization. *IEEE Commun. Mag.* **1997**, *35*, 90–100. [CrossRef]
79. Kaemarungsi, K. Design of Indoor Positioning Systems Based on Location Fingerprinting Technique. Ph.D. Thesis, University of Pittsburgh, Pittsburgh, PA, USA, 2005.
80. Jianyong, Z.; Haiyong, L.; Zili, C.; Zhaohui, L. RSSI based Bluetooth low energy indoor positioning. In Proceedings of the 2014 International Conference on Indoor Positioning and Indoor Navigation (IPIN), Busan, Korea, 27–30 October 2014; pp. 526–533.
81. Chai, S.; An, R.; Du, Z. An indoor positioning algorithm using bluetooth low energy RSSI. In Proceedings of the 2016 International Conference on Advanced Materials Science and Environmental Engineering, Chiang Mai, Thailand, 26–27 June 2016; Atlantis Press: Amsterdam, The Netherlands, 2016.
82. Brown, R.G.; Meyer, R.F. The fundamental theorem of exponential smoothing. *Oper. Res.* **1961**, *9*, 673–685. [CrossRef]
83. Bshara, M.; Orguner, U.; Gustafsson, F.; Van Biesen, L. Fingerprinting localization in wireless networks based on received-signal-strength measurements: A case study on WiMAX networks. *IEEE Trans. Veh. Technol.* **2009**, *59*, 283–294. [CrossRef]
84. Song, B.; Zhang, S.; Long, J.; Hu, Q. Fingerprinting localization method based on toa and particle filtering for mines. *Math. Probl. Eng.* **2017**, *2017*. [CrossRef]
85. Joung, J.; Jung, S.; Chung, S.; Jeong, E. CNN-based Tx–Rx distance estimation for UWB system localisation. *Electron. Lett.* **2019**, *55*, 938–940, [CrossRef]
86. Heidari, M.; Pahlavan, K. Identification of the absence of direct path in toa-based indoor localization systems. *Int. J. Wirel. Inf. Netw.* **2008**, *15*, 117–127. [CrossRef]
87. Yang, C.; Shao, H.R. WiFi-based indoor positioning. *IEEE Commun. Mag.* **2015**, *53*, 150–157. [CrossRef]
88. Khalaf-Allah, M. Particle Filtering for Three-Dimensional TDoA-Based Positioning Using Four Anchor Nodes. *Sensors* **2020**, *20*, 4516. [CrossRef]
89. Gustafsson, F.; Gunnarsson, F. Mobile positioning using wireless networks: Possibilities and fundamental limitations based on available wireless network measurements. *IEEE Signal Process. Mag.* **2005**, *22*, 41–53. [CrossRef]
90. Wei, X.; Palleit, N.; Weber, T. AOD/AOA/TOA-based 3D positioning in NLOS multipath environments. In Proceedings of the 2011 IEEE 22nd International Symposium on Personal, Indoor and Mobile Radio Communications, Toronto, ON, Canada, 11–14 September 2011; pp. 1289–1293.
91. Wang, G.; Zhu, W.; Ansari, N. Robust TDOA-based localization for IoT via joint source position and NLOS error estimation. *IEEE Internet Things J.* **2019**, *6*, 8529–8541. [CrossRef]
92. Wang, G.; Razul, S.G.; See, C.M. DOA estimation using multiple antenna arrays. In Proceedings of the 2017 Sensor Signal Processing for Defence Conference (SSPD), London, UK, 6–7 December 2017; pp. 1–5.
93. Sadhukhan, P.; Das, P.K. MGALE: A modified geometry-assisted location estimation algorithm reducing location estimation error in 2D case under NLOS environments. In *International Workshop on Mobile Entity Localization and Tracking in GPS-less Environments*; Springer: Berlin/Heidelberg, Germany, 2009; pp. 1–18.

94. Oumar, O.A.; Siyau, M.F.; Sattar, T.P. Comparison between MUSIC and ESPRIT direction of arrival estimation algorithms for wireless communication systems. In Proceedings of the The First International Conference on Future Generation Communication Technologies, London, UK, 12–14 December 2012; pp. 99–103.
95. Kumar, S.; Gil, S.; Katabi, D.; Rus, D. Accurate indoor localization with zero start-up cost. In Proceedings of the 20th Annual International Conference on Mobile Computing and Networking, Houston, TX, USA, 29–30 May 2014; pp. 483–494.
96. Wang, Y.; Liu, J.; Chen, Y.; Gruteser, M.; Yang, J.; Liu, H. E-eyes: Device-free location-oriented activity identification using fine-grained wifi signatures. In Proceedings of the 20th Annual International Conference on Mobile Computing and Networking, Houston, TX, USA, 29–30 May 2014; pp. 617–628.
97. Wu, K.; Xiao, J.; Yi, Y.; Gao, M.; Ni, L.M. Fila: Fine-grained indoor localization. In Proceedings of the 2012 Proceedings IEEE INFOCOM, Orlando, FL, USA, 25–30 March 2012; pp. 2210–2218.
98. Xiao, J.; Wu, K.; Yi, Y.; Wang, L.; Ni, L.M. Pilot: Passive device-free indoor localization using channel state information. In Proceedings of the 2013 IEEE 33rd International Conference on Distributed Computing Systems, Philadelphia, PA, USA, 8–11 July 2013; pp. 236–245.
99. Farid, Z.; Nordin, R.; Ismail, M. Recent advances in wireless indoor localization techniques and system. *J. Comput. Netw. Commun.* **2013**, *2013*. [CrossRef]
100. Wi-Fi Location: Ranging with RTT. 2019. Available online: https://developer.android.com/guide/topics/connectivity/wifi-rtt (accessed on 4 November 2019).
101. Wi-Fi Aware Overview. 2020. Available online: https://developer.android.com/guide/topics/connectivity/wifi-aware (accessed on 28 November 2020).
102. Gentner, C.; Ulmschneider, M.; Kuehner, I.; Dammann, A. WiFi-RTT Indoor Positioning. In Proceedings of the 2020 IEEE/ION Position, Location and Navigation Symposium (PLANS), Portland, OR, USA, 20–23 April 2020; pp. 1029–1035, [CrossRef]
103. Horn, B.K. Doubling the Accuracy of Indoor Positioning: Frequency Diversity. *Sensors* **2020**, *20*, 1489. [CrossRef] [PubMed]
104. Guo, G.; Chen, R.; Ye, F.; Peng, X.; Liu, Z.; Pan, Y. Indoor Smartphone Localization: A Hybrid WiFi RTT-RSS Ranging Approach. *IEEE Access* **2019**, *7*, 176767–176781. [CrossRef]
105. Bullmann, M.; Fetzer, T.; Ebner, F.; Ebner, M.; Deinzer, F.; Grzegorzek, M. Comparison of 2.4 GHz WiFi FTM-and RSSI-Based Indoor Positioning Methods in Realistic Scenarios. *Sensors* **2020**, *20*, 4515. [CrossRef] [PubMed]
106. Positioning in Wlan Systems. 2019. Available online: patents.google.com/patent/US20180295601A1/en (accessed on 4 November 2019).
107. Si, M.; Wang, Y.; Xu, S.; Sun, M.; Cao, H. A Wi-Fi FTM-Based Indoor Positioning Method with LOS/NLOS Identification. *Appl. Sci.* **2020**, *10*, 956. [CrossRef]
108. Cao, H.; Wang, Y.; Bi, J.; Xu, S.; Si, M.; Qi, H. Indoor Positioning Method Using WiFi RTT Based on LOS Identification and Range Calibration. *ISPRS Int. J. Geo-Inf.* **2020**, *9*, 627. [CrossRef]
109. Taki, S.U.; Chakrabarty, A.; Piran, M.J.; Pham, Q.V.; Suh, D.Y. An Indoor Positioning and Navigation System Using Named Data Networking. *IEEE Access* **2020**, *8*, 196408–196424, [CrossRef]
110. Welch, G.; Bishop, G. *An Introduction to the Kalman Filter*; University of North Carolina: Chapel Hill, NC, USA, 1995.
111. Apte, V.; Powar, Y.A. Improving the accuracy of wireless LAN based location determination systems using Kalman filter and multiple observers. In Proceedings of the IEEE Wireless Communications and Networking Conference, Las Vegas, NV, USA, 3–6 April 2006; Volume 1, pp. 463–468.
112. Terejanu, G.A. *Discrete Kalman Filter Tutorial*; University at Buffalo: New York, NY, USA, 2013; p. 14260.
113. Farhad, A.; Kim, D.H.; Subedi, S.; Pyun, J.Y. Enhanced LoRaWAN Adaptive Data Rate for Mobile Internet of Things Devices. *Sensors* **2020**, *20*, 6466. [CrossRef]
114. Subedi, S.; Gang, H.S.; Ko, N.Y.; Hwang, S.S.; Pyun, J.Y. Improving Indoor Fingerprinting Positioning With Affinity Propagation Clustering and Weighted Centroid Fingerprint. *IEEE Access* **2019**, *7*, 31738–31750. [CrossRef]
115. Zou, H.; Jin, M.; Jiang, H.; Xie, L.; Spanos, C.J. WinIPS: WiFi-based non-intrusive indoor positioning system with online radio map construction and adaptation. *IEEE Trans. Wirel. Commun.* **2017**, *16*, 8118–8130. [CrossRef]
116. Zou, H.; Huang, B.; Lu, X.; Jiang, H.; Xie, L. A robust indoor positioning system based on the procrustes analysis and weighted extreme learning machine. *IEEE Trans. Wirel. Commun.* **2015**, *15*, 1252–1266. [CrossRef]

117. Hossain, A.M.; Jin, Y.; Soh, W.S.; Van Nguyen, H. SSD: A robust RF location fingerprint addressing mobile devices' heterogeneity. *IEEE Trans. Mob. Comput.* **2011**, *12*, 65–77. [CrossRef]
118. Gang, H.S.; Pyun, J.Y. A Smartphone Indoor Positioning System Using Hybrid Localization Technology. *Energies* **2019**, *12*, 3702. [CrossRef]
119. Wang, X.; Qin, D.; Guo, R.; Zhao, M.; Ma, L.; Berhane, T.M. The Technology of Crowd-Sourcing Landmarks-Assisted Smartphone in Indoor Localization. *IEEE Access* **2020**, *8*, 57036–57048, [CrossRef]
120. Lee, G.; Moon, B.; Park, M. Crowdsourcing-Based Learning Data Collection for Real-Time Sensor Error Correction in Indoor Environments. *IEEE Access* **2020**, *8*, 127353–127367, [CrossRef]
121. Mirowski, P.; Ho, T.K.; Yi, S.; MacDonald, M. SignalSLAM: Simultaneous localization and mapping with mixed WiFi, Bluetooth, LTE and magnetic signals. In Proceedings of the International Conference on Indoor Positioning and Indoor Navigation, Montbeliard, France, 28–31 October 2013; pp. 1–10.
122. Montemerlo, M.; Thrun, S.; Koller, D.; Wegbreit, B. FastSLAM: A factored solution to the simultaneous localization and mapping problem. In Proceedings of the Eighteenth National Conference on Artificial Intelligence (AAAI-02), Edmonton, AB, Canada, 28 July–1 August 2002; pp. 593–598.
123. Yiu, S.; Yang, K. Gaussian process assisted fingerprinting localization. *IEEE Internet Things J.* **2016**, *3*, 683–690. [CrossRef]
124. Shi, K.; Ma, Z.; Zhang, R.; Hu, W.; Chen, H. Support vector regression based indoor location in IEEE 802.11 environments. *Mob. Inf. Syst.* **2015**, *2015*. [CrossRef]
125. Subedi, S.; Gang, H.S.; Pyun, J.Y. Regression Assisted Crowdsourcing Approach for Fingerprint Radio Map Construction. In Proceedings of the 2019 International Conference on Indoor Positioning and Indoor Navigation (IPIN), Pisa, Italy, 30 September–3 October 2019; pp. 1–7.
126. Rai, A.; Chintalapudi, K.K.; Padmanabhan, V.N.; Sen, R. Zee: Zero-effort crowdsourcing for indoor localization. In Proceedings of the 18th Annual International Conference on Mobile Computing and Networking, Istanbul, Turkey, 22–26 August 2012; pp. 293–304.
127. Peng, M.; Zhang, Z.; Kai, C.; Wang, L. Wi-Fi Fingerprint Database Construction Algorithm based on Regional Environment Parameters. In Proceedings of the 2019 International Symposium on Signal Processing Systems, Beijing, China, 20–22 September 2019; pp. 51–55.
128. Bi, J.; Wang, Y.; Li, Z.; Xu, S.; Zhou, J.; Sun, M.; Si, M. Fast radio map construction by using adaptive path loss model interpolation in large-scale building. *Sensors* **2019**, *19*, 712. [CrossRef]
129. Ding, G.; Tan, Z.; Wu, J.; Zhang, J. Efficient indoor fingerprinting localization technique using regional propagation model. *IEICE Trans. Commun.* **2014**, *97*, 1728–1741. [CrossRef]
130. Salamah, A.H.; Tamazin, M.; Sharkas, M.A.; Khedr, M.; Mahmoud, M. Comprehensive investigation on principle component large-scale Wi-Fi indoor localization. *Sensors* **2019**, *19*, 1678. [CrossRef] [PubMed]
131. Chawathe, S.S. Beacon placement for indoor localization using bluetooth. In Proceedings of the 2008 11th International IEEE Conference on Intelligent Transportation Systems, Beijing, China, 12–15 October 2008; pp. 980–985.
132. Chawathe, S.S. Low-latency indoor localization using bluetooth beacons. In Proceedings of the 2009 12th International IEEE Conference on Intelligent Transportation Systems, St. Louis, MO, USA, 4–7 October 2009; pp. 1–7.
133. Montaser, A.; Moselhi, O. RFID indoor location identification for construction projects. *Autom. Constr.* **2014**, *39*, 167–179. [CrossRef]
134. Li, X.; Wang, J.; Liu, C. A bluetooth/PDR integration algorithm for an indoor positioning system. *Sensors* **2015**, *15*, 24862–24885. [CrossRef] [PubMed]
135. Poulose, A.; Eyobu, O.S.; Han, D.S. A Combined PDR and Wi-Fi Trilateration Algorithm for Indoor Localization. In Proceedings of the 2019 International Conference on Artificial Intelligence in Information and Communication (ICAIIC), Okinawa, Japan, 11–13 February 2019; pp. 72–77, [CrossRef]
136. Liu, L.; Li, B.; Yang, L.; Liu, T. Real-Time Indoor Positioning Approach Using iBeacons and Smartphone Sensors. *Appl. Sci.* **2020**, *10*, 2003. [CrossRef]
137. Hou, Y.; Yang, X.; Abbasi, Q.H. Efficient AoA-based wireless indoor localization for hospital outpatients using mobile devices. *Sensors* **2018**, *18*, 3698. [CrossRef]
138. Zhang, L.; Du, T.; Jiang, C. Detection of an unknown radio transmitter using an enhanced K-nearest neighbor algorithm based on virtual reference point and RSSD information. *EURASIP J. Wirel. Commun. Netw.* **2019**, *2019*, 71. [CrossRef]

139. Kim, J.; Ji, M.; Jeon, J.i.; Park, S.; Cho, Y. K-NN based positioning performance estimation for fingerprinting localization. In Proceedings of the 2016 Eighth International Conference on Ubiquitous and Future Networks (ICUFN), Vienna, Austria, 5–8 July 2016; pp. 468–470.
140. Wang, H.; Wang, X.; Xue, Y.; Jiang, Y. UWB-based Indoor Localization Using a Hybrid WKNN-LSTM Algorithm. In Proceedings of the 2020 IEEE 4th Information Technology, Networking, Electronic and Automation Control Conference (ITNEC), Chongqing, China, 12–14 June 2020; Volume 1, pp. 1720–1725, [CrossRef]
141. Poulose, A.; Han, D.S. Performance Analysis of Fingerprint Matching Algorithms for Indoor Localization. In Proceedings of the 2020 International Conference on Artificial Intelligence in Information and Communication (ICAIIC), Fukuoka, Japan, 19–21 February 2020; pp. 661–665, [CrossRef]
142. Youssef, M.; Agrawala, A. The Horus WLAN location determination system. In Proceedings of the 3rd International Conference on Mobile Systems, Applications, and Services, Seattle, WA, USA, 6–8 June 2005; pp. 205–218.
143. Taboga, M. *"Normal distribution—Maximum Likelihood Estimation", Lectures on Probability Theory and Mathematical Statistics*, 3rd ed.; Kindle Direct Publishing: Seattle, WA, USA, 2017.
144. Zhang, W.; Liu, K.; Zhang, W.; Zhang, Y.; Gu, J. Deep neural networks for wireless localization in indoor and outdoor environments. *Neurocomputing* **2016**, *194*, 279–287. [CrossRef]
145. Bernas, M.; Płaczek, B. Fully connected neural networks ensemble with signal strength clustering for indoor localization in wireless sensor networks. *Int. J. Distrib. Sens. Netw.* **2015**, *11*, 403242. [CrossRef]
146. Kriz, P.; Maly, F.; Kozel, T. Improving indoor localization using bluetooth low energy beacons. *Mob. Inf. Syst.* **2016**, *2016*. [CrossRef]
147. Ma, R.; Guo, Q.; Hu, C.; Xue, J. An improved WiFi indoor positioning algorithm by weighted fusion. *Sensors* **2015**, *15*, 21824–21843. [CrossRef]
148. Li, C.; Qiu, Z.; Liu, C. An improved weighted k-nearest neighbor algorithm for indoor positioning. *Wirel. Pers. Commun.* **2017**, *96*, 2239–2251. [CrossRef]
149. Kumar, S.; Hegde, R.M.; Trigoni, N. Gaussian process regression for fingerprinting based localization. *Ad Hoc Netw.* **2016**, *51*, 1–10. [CrossRef]
150. Tian, Z.; Tang, X.; Zhou, M.; Tan, Z. Fingerprint indoor positioning algorithm based on affinity propagation clustering. *EURASIP J. Wirel. Commun. Netw.* **2013**, *2013*, 272. [CrossRef]
151. Subedi, S.; Pyun, J.Y. Lightweight workload fingerprinting localization using affinity propagation clustering and Gaussian process regression. *Sensors* **2018**, *18*, 4267. [CrossRef]
152. Liu, K.; Meng, Z.; Own, C.M. Gaussian Process Regression Plus Method for Localization Reliability Improvement. *Sensors* **2016**, *16*, 1193. [CrossRef]
153. Rehim, A.A.Y.A.; Amin, M. HORUS: A WLAN-Based Indoor Location Determination System. Ph.D. Thesis, University of Maryland, College Park, MD, USA, 2004.
154. Zhu, C.; Byrd, R.; Lu, P.; Nocedal, J. L_BFGS_B Fortran Subroutines for Large Scale Bound Constrained Optimization, L BFGS B_ FORTRAN sub routines for large scale bound constrained optimization. In *Technol Report Department of Electrical Engineering and Computer Science*; Northwestern University: Evanston, IL, USA, 1995.
155. Hernández, N.; Ocaña, M.; Alonso, J.; Kim, E. Continuous space estimation: Increasing WiFi-based indoor localization resolution without increasing the site-survey effort. *Sensors* **2017**, *17*, 147. [CrossRef]
156. Zhao, Y.; Zhang, Z.; Feng, T.; Wong, W.; Garg, H.K. GraphIPS: Calibration-free and Map-free Indoor Positioning using Smartphone Crowdsourced Data. *IEEE Internet Things J.* **2020**, [CrossRef]
157. Li, T.; Han, D.; Chen, Y.; Zhang, R.; Zhang, Y.; Hedgpeth, T. IndoorWaze: A Crowdsourcing-Based Context-Aware Indoor Navigation System. *IEEE Trans. Wirel. Commun.* **2020**, *19*, 5461–5472, [CrossRef]
158. Wu, C.; Yang, Z.; Liu, Y. Smartphones based crowdsourcing for indoor localization. *IEEE Trans. Mob. Comput.* **2014**, *14*, 444–457. [CrossRef]
159. Tong, X.; Liu, K.; Tian, X.; Fu, L.; Wang, X. FineLoc: A Fine-grained Self-calibrating Wireless Indoor Localization System. *IEEE Trans. Mob. Comput.* **2018**. [CrossRef]
160. Yoo, J.; Johansson, K.H.; Kim, H.J. Indoor localization without a prior map by trajectory learning from crowdsourced measurements. *IEEE Trans. Instrum. Meas.* **2017**, *66*, 2825–2835. [CrossRef]
161. Zhuang, Y.; Syed, Z.; Li, Y.; El-Sheimy, N. Evaluation of two WiFi positioning systems based on autonomous crowdsourcing of handheld devices for indoor navigation. *IEEE Trans. Mob. Comput.* **2015**, *15*, 1982–1995. [CrossRef]

162. Zang, H.; Baccelli, F.; Bolot, J. Bayesian inference for localization in cellular networks. In Proceedings of the 2010 Proceedings IEEE INFOCOM, San Diego, CA, USA, 14–19 March 2010; pp. 1–9.
163. Kanaris, L.; Kokkinis, A.; Liotta, A.; Stavrou, S. Fusing bluetooth beacon data with Wi-Fi radiomaps for improved indoor localization. *Sensors* **2017**, *17*, 812. [CrossRef] [PubMed]
164. Hartigan, J.A.; Wong, M.A. Algorithm AS 136: A k-means clustering algorithm. *J. R. Stat. Soc. Ser. C Appl. Stat.* **1979**, *28*, 100–108. [CrossRef]
165. Gokcay, E.; Principe, J.C. Information theoretic clustering. *IEEE Trans. Pattern Anal. Mach. Intell.* **2002**, *24*, 158–171. [CrossRef]
166. Dueck, D. Affinity Propagation: Clustering Data by Passing Messages. Ph.D. Thesis, University of Toronto, Toronto, ON, Canada, 2009.
167. Caso, G.; De Nardis, L.; Di Benedetto, M.G. A mixed approach to similarity metric selection in affinity propagation-based WiFi fingerprinting indoor positioning. *Sensors* **2015**, *15*, 27692–27720. [CrossRef]
168. Mézard, M. Passing messages between disciplines. *Science* **2003**, *301*, 1685–1686. [CrossRef]
169. Jain, A.K. Data clustering: 50 years beyond K-means. *Pattern Recognit. Lett.* **2010**, *31*, 651–666. [CrossRef]
170. Bezdek, J.C.; Ehrlich, R.; Full, W. FCM: The fuzzy c-means clustering algorithm. *Comput. Geosci.* **1984**, *10*, 191–203. [CrossRef]
171. Sadhukhan, P. Performance analysis of clustering-based fingerprinting localization systems. *Wirel. Netw.* **2018** *25*, 2497–2510. [CrossRef]
172. Luo, J.; Fu, L. A smartphone indoor localization algorithm based on WLAN location fingerprinting with feature extraction and clustering. *Sensors* **2017**, *17*, 1339.
173. Karegar, P.A. Wireless fingerprinting indoor positioning using affinity propagation clustering methods. *Wirel. Netw.* **2018**, *24*, 2825–2833. [CrossRef]
174. Youssef, M.; Agrawala, A. The Horus location determination system. *Wirel. Netw.* **2008**, *14*, 357–374. [CrossRef]
175. Bahl, P.; Padmanabhan, V.N.; Bahl, V.; Padmanabhan, V. RADAR: An in-building RF-based user location and tracking system. In Proceedings of the IEEE INFOCOM 2000. Conference on Computer Communications. Nineteenth Annual Joint Conference of the IEEE Computer and Communications Societies (Cat. No.00CH37064), Tel Aviv, Israel, 26–30 March 2000.
176. Ladd, A.M.; Bekris, K.E.; Marceau, G.; Rudys, A.; Wallach, D.S.; Kavraki, L.E. Using wireless ethernet for localization. In Proceedings of the IEEE/RSJ International Conference on Intelligent Robots and Systems, Osaka, Japan, 30 September–4 October 2002; Volume 1, pp. 402–408.
177. Ekahau Positioning Engine. 2020. Available online: https://www.ekahau.com (accessed on 15 January 2020).
178. Kim, Y.; Shin, H.; Chon, Y.; Cha, H. Smartphone-based Wi-Fi tracking system exploiting the RSS peak to overcome the RSS variance problem. *Pervasive Mob. Comput.* **2013**, *9*, 406–420. [CrossRef]
179. Lee, S.; Cho, B.; Koo, B.; Ryu, S.; Choi, J.; Kim, S. Kalman filter-based indoor position tracking with self-calibration for RSS variation mitigation. *Int. J. Distrib. Sens. Netw.* **2015**, *11*, 674635. [CrossRef]
180. Lee, S.H.; Lim, I.K.; Lee, J.K. Method for improving indoor positioning accuracy using extended kalman filter. *Mob. Inf. Syst.* **2016**, *2016*. [CrossRef]
181. Yadav, R.K.; Bhattarai, B.; Gang, H.S.; Pyun, J.Y. Trusted K Nearest Bayesian Estimation for Indoor Positioning System. *IEEE Access* **2019**, *7*, 51484–51498. [CrossRef]
182. Ma, Z.; Poslad, S.; Bigham, J.; Zhang, X.; Men, L. A BLE RSSI ranking based indoor positioning system for generic smartphones. In Proceedings of the 2017 Wireless Telecommunications Symposium (WTS), Chicago, IL, USA, 26–28 April 2017; pp. 1–8.

sensors

MDPI

Article

Real-Time Locating System in Production Management

András Rácz-Szabó [1,†], Tamás Ruppert [1,2,*,†], László Bántay [1], Andreas Löcklin [3], László Jakab [2] and János Abonyi [1]

1 MTA-PE Lendület Complex Systems Monitoring Research Group, Department of Process Engineering, University of Pannonia, Egyetem u., 10, POB 158, H-8200 Veszprém, Hungary; andrasraczszabo91@gmail.com (A.R.-S.); laszlo.bantay@gmail.com (L.B.); janos@abonyilab.com (J.A.)

2 Sunstone-RTLS Ltd., Kevehaza u., 1-3, H-1115 Budapest, Hungary; support@sunstone-rtls.com

3 Institute of Industrial Automation and Software Engineering, University of Stuttgart, Pfaffenwaldring 47, D-70550 Stuttgart, Germany; andreas.loecklin@ias.uni-stuttgart.de

* Correspondence: ruppert@abonyilab.com

† These authors contributed equally to this work.

Received: 15 October 2020; Accepted: 24 November 2020; Published: 26 November 2020

Abstract: Real-time monitoring and optimization of production and logistics processes significantly improve the efficiency of production systems. Advanced production management solutions require real-time information about the status of products, production, and resources. As real-time locating systems (also referred to as indoor positioning systems) can enrich the available information, these systems started to gain attention in industrial environments in recent years. This paper provides a review of the possible technologies and applications related to production control and logistics, quality management, safety, and efficiency monitoring. This work also provides a workflow to clarify the steps of a typical real-time locating system project, including the cleaning, pre-processing, and analysis of the data to provide a guideline and reference for research and development of indoor positioning-based manufacturing solutions.

Keywords: RTLS; indoor positioning system (IPS); position data; industry 4.0; traceability; product tracking

1. Introduction

Getting accurate and actual information of a process status is very important in the management and development of production systems. Information is often position located; this way, it defines the actual position of a workpiece or resource in the production area. This location based information may be suitable to connect information of resources and activities/workpieces. The purpose of this article is to introduce the potential in tools developed for indoor positioning, as well as the available technologies and the possible use of data hidden in information.

According to the ISO/IEC 24730-1:2014 standard, the real-time locating system (RTLS) is a wireless system used to locate the position of an item anywhere in a defined space at a point in time that is or is close to real-time. Indoors positioning systems (IPS) [1] locate objects in closed structures, such as office buildings, hospitals, stores, factories, and warehouses, where the GPS proves to be inaccurate [2]. In this paper we focus on how indoor positioning can be utilized in manufacturing and for simplicity, we refer to these indoor positioning systems as RTLS.

Several surveys and comparative analyses can be found on indoor tracking technology based on localization techniques [3–5]. One article has attempted to classify techniques and systems by presenting a comprehensive performance comparison of the accuracy, precision, complexity, scalability, robustness and cost [6]. Similarly, studies comparing RTLS technologies can also be found in the literature [7–13]. Furthermore, a meta-review provides a comprehensive compilation of 62 survey papers on the topic of RTLS [14], and the classification of current typical RTLS is introduced with a layered conceptual framework [15]. Many technologies are available such as infrared light, ultrasound, laser and their combinations. The reviews of these technologies focus on the technical elements and standalone applications and show that only a few specific industrial applications are available [16].

This paper aims to provide an overview of the applicability of RTLS in manufacturing to support the practical applications and provide a guideline or reference for implementation, research and development of indoor positioning and RTLS.

In order to explore the potential applications a systematic examination of literature was performed in Scopus, following the PRISMA-P protocol. The used keyword set ("real-time positioning systems" OR "indoor positioning systems") AND ("manufacturing" OR "industry") resulted near to 300 articles from which the thematic groups of the related research were identified.

Positioning data in the production system is the key information for traceability [17], and digitalization [18]. We overview the potential technologies and the possible traceability levels in Section 2. The levels represent the identification unit from the transportation unit (highest level—trucks, ships) to item unit (lowest level–raw material). Determination of the traceability level depends on more factors such as the complexity of the production process, the number of raw material types and the conditions of the information system and infrastructure.

We discuss the potential manufacturing applications according to the tasks depicted in Figure 1. The figure describes how positioning-based information allows for continuous improvement to other parts of the manufacturing environment, such as production control, logistics, applications in quality management, safety and RTLS-based efficiency monitoring. These applications and the required data analysis tasks are discussed in Section 3.

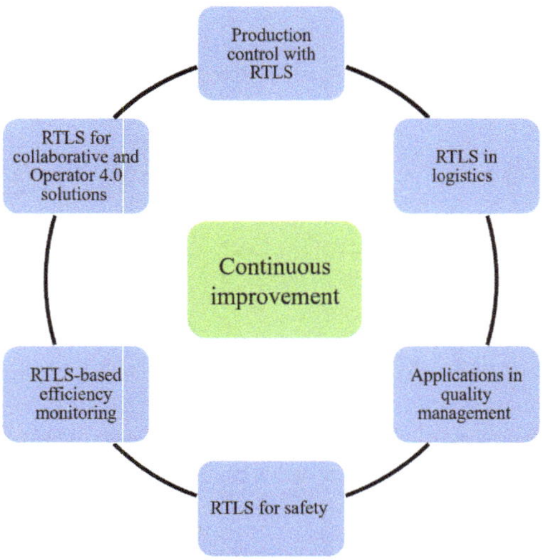

Figure 1. Use of real-time locating system (RTLS)-based positioning information by the different parts of the manufacturing environment. Continuous improvement is a central element of an RTLS project.

Section 4 describes a workflow to implement an RTLS-based digitalization project such as installation with the necessary hardware elements and data processing to allow the data integration. Finally, the applicability of the RTLS in manufacturing is illustrated by a case study presented in Section 5.

2. Levels of Location Information in Manufacturing Industries

To determine the appropriate tracking technology, we need to know the identification levels with the associated technologies. In terms of integration into our system, it is important to see the relevant characteristics for the selection process of the particular RTLS technology. In the next subsection, we present the criteria for choosing a method for a tracking solution that arises in a production system. Based on this information in the second subsection, we show recommended indoor positioning-based traceability technologies. Please note that a detailed description of the technologies is not the purpose of this article.

2.1. Identification Levels and Technology Solutions

Different applications require different types of tracking systems. Figure 2 shows the identification layers based on the possible available levels.

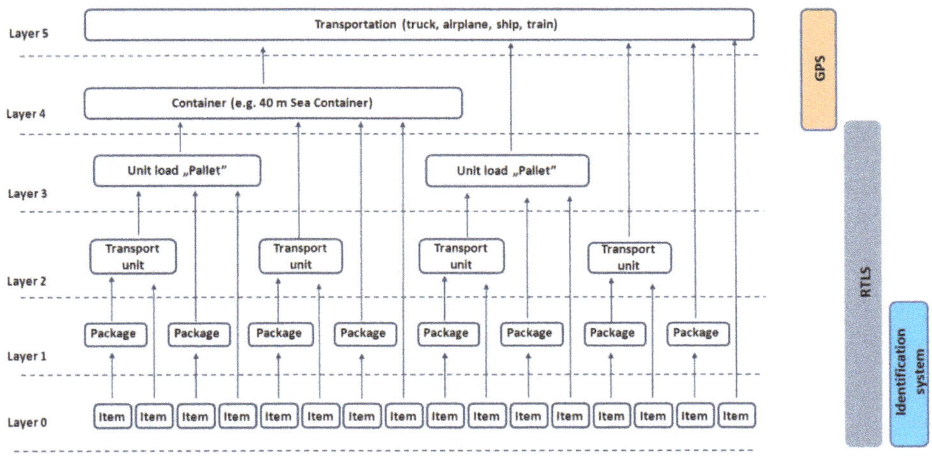

Figure 2. Identification levels in a production system. Layers define the logistic units from raw material (items) to trucks (transportation).

GPS is used for tracking containers and transportation equipment. At the lower level, where we consider the intralogistics, GPS is not accurate, or in many cases, it is unusable and not suitable for general asset tracking due to energy consumption. RTLS can handle indoor container identification as well as the unit load, transportation unit and package, that is, the third, second and first layers. RFID and barcode technologies are possible solutions for item identification, but it is essential to consider whether using a particular technology is appropriate or if it is worthwhile to combine the available technologies, such as UWB technology (which is not recommended for raw material tracking). This approach is called hybrid traceability technology in the literature [19]. However, with RFID tags, unit identification can be achieved with a lower cost [7].

Table 1 shows solutions for these different identification levels (see in Figure 2) with the advantages and disadvantages. We discuss four different traceability solutions to support the technology chosen. A decision-making model for selection is proposed in Reference [16], where UWB, RFID, Wifi, Zigbee and BLE (Bluetooth Low Energy) are compared with several aspects based on the developed methodology. The steps of this methodology are RTLS definition, market analyses, weights of criteria, ranking. They made a comparison between the technologies with the many parameters (coverage area, accuracy, room level usefulness, RF interference potential, bit rate, complexity, initial cost, security and privacy, health concern). We focus on application-oriented parameters in Table 2.

Table 1. The most widely applied solutions for traceability. The advantages and disadvantages help us to choose the right solution for the right layer (see in Figure 2).

Solution		Advantages	Disadvantages
Identification system	Barcode	Cost effective, spread technology	The potential for human error is high
			Additional installation cost per every layout changing
		The cost per item is low	The cost of printed labels can be relevant in the case of enormous stock
	RFiD	Cost effective, spread technology	Every new station needs new hardware and installation
		The cost per item is low	Additional installation cost per every layout changing
		The human error can be minimized	
GPS		Spread technology	The accuracy is not enough in the case of indoor positioning
		Many devices are already compatible	The reliability is low in the case of indoor positioning
		It is highly scalable	
RTLS		The error of human is excluded	The cost depends on the number of tracked items
		The traceability is available at the covered area	
		The system is fully flexible, any layout changing can be handle in the software application	Any new item can be added at any time to the system (highly scalable)

2.2. Structuring of Indoor Positioning Systems and Potential Traceability Technologies

Table 1 helps us to choose the right technology in the case of layer 0 and 1. Now we focus on the applicability of the RTLS. We found a multilevel selection criteria [20], where the three levels are Economic, Technical and Implementation. We made an overview of the most relevant RTLS technologies (excluding the no-radio based technologies) in Table 2. It summarizes these technologies with critical performance criteria, including accuracy, power consumption and costs.

Figure 3 gives us a classification of RTLS. We considered the left side of the figure because there are very few examples of no radio-based technologies. In the production environment, ultrasound signal transmission is also accompanied by a radio frequency (RF) pulse to combine the high accuracy of ultrasound with the high communications capacity of RF, which enables tracking of hundreds of simultaneously moving tags [21]. The technology is not used independently in the manufacturing environment because of the communications capacity, and environmental noise can degrade the localization accuracy [10]. One standalone application of RTLS is tracking the locations of construction resources such as labor, materials, machinery, and vehicles [22]. This application uses WiFi-based RTLS because GPS is limited in indoor environments, such as tunnels and buildings under construction. Another example is the development of a self-governing mobile robot navigation system for indoor construction applications [23]. Several navigation strategies with a mobile robot were tested with various combinations of localization sensors, including wheel encoders, sonar/infrared/thermal proximity sensors, motion sensors, a digital compass, and ultra-wideband (UWB) technology. The findings can be adapted to several potential construction or manufacturing applications such as robotic material delivery, inspection, and onsite security. Two RTLS applications of UWB and ultrasound technology have been tested in the SmartFactory KL [24].

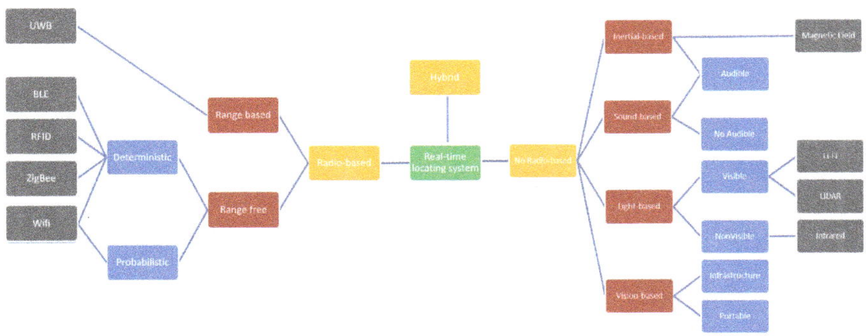

Figure 3. Classification of RTLS [25].

Compared to other technologies, Zigbee has not spread substantially in industry [26]. However, industrial applications in the literature, such as the Zigbee positioning system for coal miners [27], have also been studied [28]. Laser-based systems are also used for navigation [29] and production tracking [30]. RFID is used in the production independently as an identification system [31] because only the presence of tags, such as barcodes, can be accounted for at the RFID reader. Other technologies must be used for real-time location [32]. An RFID-based RTLS solution exists, but this solution is less widely used because it is more expensive and inaccurate than UWB [33]. RTLS must be able to locate, track and identify objects in an indoor environment; therefore, RFID technology is not appropriate for RTLS. The literature in this area is controversial, so we consider various types of technology, such as BLuetooth, WiFi, Zigbee and UWB, as having the ability to support RTLS. Apple proposed the iBeacon protocol [34] in 2013 as another tool [35]. The new iPhone 11 from Apple already includes UWB beacons, which may be suitable for indoor positioning [36]. Moreover, it is essential to mention 5G technology [37], which could be crucial for future smart manufacturing, including highly accurate indoor localization. Because of the large-signal bandwidth and beamforming capabilities, localization and tracking could be more robust and efficient [38]. The technique is only now beginning to spread in industry; currently, it can only be applied in test and development environments [39]. The different indoor positioning-based traceability technologies can be combined in RTLS (already mentioned hybrid technology) to take advantage of different solutions in one system; for example, the ZigBee and UWB technologies or RSS measurements and a fingerprinting location algorithm usage for better position estimation [40]. To use hybrid technology, a platform is also needed. In general, roughly five layers are worth defining. A hardware layer, where position data are generated; a processing layer, where the position is calculated and filtered; a data layer, where location data is stored; a service layer, where we can optimize the system; a visualization layer, where we can analyze and monitor real-time location data.

In the next section, we describe potential industrial applications, while in Section 4 we propose a workflow of setting up an RTLS-based manufacturing support system. Finally, we describe a use case to illustrate the applicability of RTLS.

Table 2. Review of indoor positioning-based traceability technologies

Techn. Scale	Tag Cost [41–43] L:<3$ M:<10$ H:>20$	Module Cost [44] L:<10$ M:<40$ H:>70$	Accuracy L:>1 m M:10 cm H:<10 cm	Space dim. 2D/3D	Power cons. [44,45] L:<100 mA H:>200 mA
Zigbee [46]	M	M	M	2D	L
RFID [47]	L	H	L/M	2D/3D	L
BLE [48]	L	L	L/M	2D	L
Wifi [49]	H	H	L/M	2D	H
UWB [50]	H	H	H	2D/3D	H

3. Industrial Applications of RTLS

An approach to RTLS selection is reviewed [16] and the RTLS based articles and the main advantages are summarized within Table 3 with the related fields of application in production and logistics categorized according to areas of use such as Quality Management, Safety and Efficiency Monitoring. Within the topics, reference is made to existing solutions, but possible directions for development are also presented in the following subsections.

Table 3. Industrial applications of RTLS technologies.

Application type	Application	Technology
Production	Cycle time optimization	UWB [51]
Production	Position data-based decision making	UWB [52]; RFID [53,54]
Production	Activity-Time monitoring in production line	UWB [55]
Production	Digital Facility Layout Planning	Independent [56]
Logistics	Logitics management	RFID [57]; Hybrid [21]; Independent [20]
Logistics	Warehouse management	RFID [58]; WiFi [59]
Logistics	Pallet management	RFID [31]
Logistics	Material/component and production tracking	WiFi [60]; UWB [31,61]; RFID [31,62–65]; Hybrid [66]; Laser [30]; Barcode [67]
Logistics	Assets tracking	Bluetooth [68]; RFID [69–72]; Hybrid [73]; UWB [74,75]; Laser [29]; Barcode [67]; ZigBee [46]
Quality	Weak spot analyzis in production	UWB [76]
Safety	Safety management	RFID [33]
Safety	Collision avoidance	UWB [77]

Table 3. *Cont.*

Application type	Application	Technology
Safety	Personal protective equipment monitoring	Hybrid [78]
Safety	Person tracking	ZigBee [27]; RFID [79]; UWB [80]
Safety	Contact tracking	Independent [81]
Efficiency monitoring	Performance of manufacturing process	RFID [82]
Efficiency monitoring	Lean manufacturing	UWB [83]; BLE [48]
Efficiency monitoring	Human resource monitoring	RFID [84]

Different application possibilities exist in manufacturing departments with RTLS. Table 4 presents various types of applications, where the RTLS-provided information is defined for every industrial application. The possible benefits define how we can validate the efficiency of the RTLS project. The applied positioning system provides real-time information about where equipment, semi-finished or finished products and specified logistic vehicles or workers are located in the manufacturing area [63].

Table 4. Application of RTLS in manufacturing, the useful information it provides and possible benefits.

Application Name	Information Provided by RTLS	Possible Benefits
Production control with RTLS (Section 3.1)	Footprint of semi-finished products and cycle time control	More efficient production planning
RTLS in logistics (Section 3.2)	Tracking of logistical assets in the production system	More cost-effective logistics process planning
Applications in quality management (Section 3.3)	Root cause analysis depends on position data	Help quality management department comply with standards and regulations
RTLS for safety (Section 3.4)	Human and material handling equipment tracking can help in collision detection	Reduction in occupational accidents
RTLS-based efficiency monitoring (Section 3.5)	Efficiency indicators provide a realistic picture of real-time production	Real-time efficiency monitoring assigned to machines or tools can support making better decisions
RTLS for collaborative and Operator 4.0 solutions (Section 3.6)	Precise real-time position of operators to predict the possible collaboration situations	More efficient decision making for the smart operator and collaborative system

3.1. Production Control with RTLS

Cycle time optimization is a critical task, especially in the case of modular or just-in-time (JIT) production [51]. A positive correlation exists between the potential of RTLS and JIT manufacturing. Several objects are defined as the main focus for tracking in the case of industry applications (mobile assets, workers, materials, key components, forklifts, pallets) [20].

The cycle time optimization is also possibly based on the position of products. For that, we need to know which products being produced are on the assembly line at every moment of time [51]. Pairing a semi-finished product with a tag makes inter-manufacturing tracking possible and makes the following information available:

- Time spent on the workstation for a given product;
- The production sequence;
- Which products are/have been on rework;
- Which products are/were in quality assurance;
- Average lead time for a particular product type (tact time);
- The goods in production are available with a continuous, real-time production status that supports production programming and shift design.

A position data-based decision-making approach is presented that relies on advanced data analytics for asset location systems to help production [52]. A potential use case in construction is discussed in Reference [68] and another application of material tracking in a pipe spool fabrication shop in [60].

Similar RTLS-based position data are already available in the literature, where a real-time connection between operator performance and varying product complexity was designed [55]. Another paper proposes an RTLS-based solution for a logistics problem with hybrid traceability technology (WiFi with RFID) to realize materials tracking, which can automate considerable amounts of warehouse work, such as stock-taking and storage positioning and checking [63]. Related research proposes an RFID-based intelligent decision support system architecture to handle production monitoring and scheduling in a distributed manufacturing environment [53]. Furthermore, RTLS technology can even be an element of reconfigurable facility layout planning. With its help, the processing steps of the activity and their relationships can be easily mapped and recorded in a database. More complex material flow can be provided (with information of the real flow between machines or congestion phenomena), which is not possible with a simple flowchart [56].

With the proliferation of Industry 4.0, it can be seen that there is also a strong emphasis on production monitoring. This is evidenced by the numerous references presented in the topic of production management.

3.2. RTLS in Logistics

Logistics efficiency is largely dependent on the movement of forklifts, pallet trucks and stacker trucks. For improving the logistics efficiency, we should know our current processes to see the points where lead time reductions can be achieved in the supply chain. A properly selected RTLS technology can be a tool for exploring logistics processes [61]. The following information is made available with logistical vehicles tracking:

- Routes and time spent in specific areas;
- Speed of forklift;
- Data for predictive maintenance;
- Forklift overall equipment effectiveness (OEE).

The objective of [58] is to propose an IoT and advanced data analytics-based warehouse management system (WMS) to enable smart logistics for Industry 4.0. The proposed IoT-based WMS can improve warehouse productivity, picking accuracy and efficiency, and it is robust to order variability. In [67], the authors present a sophisticated algorithm for tracking production and determining the traceability of a product. Reference [62] gives an example of how to apply RTLS across the supply chain and manage various assets within shop floors. A forklift based use-case is described in [66] where movement inside a warehouse determined by the RTLS is associated with assets that it picks up (attach through UHF-RFID reading) or puts down (detach through the loss of RFID signal).

IR-UWB-based RTLS has been deployed in an in-operation warehouse to track forklifts [75], and an RFID-enabled positioning system in AGV for a smart factory has also been presented [69]. Observations and lessons from simulation and testbed studies could be used to guide automated logistics within a smart manufacturing shop floor. The framework of an R-AGV-based material distribution scheme is proposed [73] based on an RTLS platform and electronic workshop map. The analysis and experimental results indicate that the R-AGV-based material distribution system provides new levels of process visibility and efficiency compared to traditional AGV-based distribution systems.

Tracking of transportation device is mandatory to obtain an accurate picture of intralogistics processes. A related paper describes an industrial forklift tracking problem that requires precise internal positioning [74]. It presents a study on the feasibility of meeting this challenge using UWB technology. Placing two tags on the forklift enables even more robust localization, as the measurements from the two tags are combined.

Based on the studied articles, it can be seen that much RTLS-related research is being done in logistics; there are still, of course, unexplored application possibilities.

3.3. Applications in Quality Management

It is possible to see where losses are generated with a real-time tracking system. The monitoring of the material flow with RTLS and the average duration of the processes provide information about problematic weak spots in the production process. Based on this information, possible reasons for the delay in production is discussed in Reference [76]. Root cause analysis is an essential component of quality assurance for the customers of the manufacturing company. After exploring the root cause, different action plans can be implemented, such as a review of the workflow, redesign of the workspace, education of workers and modification of work instructions.

Another possible advantage is RTLS-based dynamic work instruction. A crucial point in non-automated and human resource-required production processes is to ensure the well-supported work of the operators. One possible solution is showing just the information required to process the actual workpiece and no more. Based on the full traceability, we can show that the actual work instruction at every workstation based on the product information (from RTLS). There are relatively few practical applications on the topic, but potential development opportunities can be clearly identified such as dynamic work instruction.

3.4. RTLS for Safety

Collision avoidance is one possible improvement in the EHS (Environment, Health and Safety). Real-time alerts and notifications can be developed to prevent accidents based on the movement of vehicles and workers. A significant portion of the workplace accidents can be traced back to failure to use the required protective equipment. The monitoring of the personal protective equipment (PPE) usage is also an available function using RTLS [78]. RTLS technology can be used to control access to restricted areas for employees by sending automatic alerts whenever someone enters an unauthorized area [79]. Moreover, in the event of an emergency or natural disaster, such technology can be used to determine if everyone has already left the area or whether every worker used the designated route to leave the building [80].

The main purpose of contact-tracking solutions in industry is to help identify the contact matrix when the infection is recognized. RTLS is one possible solution to support to explore the potentially infected people [81]. Therefore, it can be used for the protection against the COVID-19 pandemic [85]. RTLS can be used in production systems also to monitor adherence to distance requirements between the operators. This is a possible useful function for a COVID-19-like epidemic situation. Due to the pandemic, the field of research for RTLS is topical, but there are also several solutions in the literature in the field of EHS (Environment, Health and Safety).

3.5. RTLS-Based Efficiency Monitoring

A related research article presents an RFID-based RTLS solution for performance metrics through RTLS data analysis to evaluate workflow performance and to obtain a lean process [82]. Spaghetti diagrams (visual representation using a continuous flow line to trace the path of an item or activity throughout a process) are time-consuming and static and, therefore, do not reflect the dynamics of logistics systems. RTLS was proposed to overcome this drawback [83].

The efficiency of a human resources personalized measurement is challenging. To achieve set goals, every organization must devise adequate, effective and efficient means of managing its HR. Related research reported on the development of an RFID and RTLS-based real-life personnel monitoring system to accurately and reliably estimate distance and coordinate the location of personnel at any instant [84]. This method can be used to measure—based on RTLS positioning data—how much time each product has spent at a particular station. This measurement can be further developed, and performance indicators can be obtained for workstations where operators work. By breaking down the overall process into sub-processes, we can refine the zones that allow for personalized performance tracking. The proposed RTLS can provide a solution to compute availability, a key parameter of OEE (Overall Equipment Efficiency) based on position data. Position data could improve the accuracy of the measurement of human resource efficiency (HRE) [86] and integrate other sensor measures for real-time activity monitoring [55]. Real-time sensor data assigned to the location of tools or machines on the shop floor allows online efficiency monitoring and supports the development and maintenance of digital twins [86] or intelligent decision-making systems.

In terms of human resources, there is still quite a bit of related research. More potential RTLS applications—like the mentioned personalized performance tracking—can help to achieve more efficient operations.

3.6. RTLS for Collaborative and Operator 4.0 Solutions

The future of manufacturing will be the personalization, and Industry 5.0 defines by the co-operation between man and machine [87]. During the current fourth industrial revolution, companies have realized they need to put humans back into industrial production with collaborative robots [88]. The workers need to be upskilled to provide value-added tasks in production to handle the mass customization and personalization for customers. This philosophy overlaps with the Operator 4.0 concept [89].

Shop floor trackers are one of the required technologies of Industry 5.0 [88]. RTLS could be a suitable solution for the full traceability on the shop floor. The smarter operator is an element of Operator 4.0 methodology [90], which is used to be the intelligent personal assistant-based solution. Real-time position data providing precise location information helps the system to make better decisions for operators and make possible the trajectory prediction of operators [91].

3.7. Analysis of Position Data and Building Data-Driven Solutions

Raw data provided by the RTLS cannot directly be utilized to support the manufacturing. The purpose of this section is to introduce data based solutions and the related data analysis techniques needed for data pre-processing and building data-driven solutions. Again, a systematic examination of literature in Scopus was done, using the keyword set ("indoor positioning" AND ("machine learning" OR "data science" OR "data mining"). The network of the mentioned keywords can be seen in Figure 4. Based on this network we can define the key thematic group of machine learning techniques and the related application areas that will also be discussed in detail in this section.

Data mining techniques are reviewed in Reference [92] to solve indoor navigation systems problems. The performance of the RTLS is shown via the integration of different features and classification algorithms, including decision tree, multi-layer perceptrons, and Bayesian networks [93]. In another article, naive Bayes theorem-based classification techniques and other classification techniques to enhance the classification accuracy are compared to identify the best location estimation algorithm [94]. K-nearest neighbor [95], support vector machine [96], decision tree [97], naıve Bayes [98] and Bayesian network methods [99] are compared and combined with ensemble learning algorithms to improve the performance, i.e., accuracy, f-score and computation time [100]. Decision tree-based classification is applied to estimate the position to improve the accuracy [97]. The clustering machine learning (ML) technique is usually used to improve RTLS accuracy, like K-means clustering backpropagation NN [101], Spatial Division Clustering (SDC) method [102], affinity propagation clustering [103]. The feeding behavior of cows is measured with RTLS in [104]. The presence at the feeder (feeding probability) of the cows was calculated using the logistic regression model. Support Vector Regression (SVR) is used to calculate the efficient RTLS [105]. Based on the aforementioned literature study, the common combinations of ML techniques are presented in Table 5 and RTLS technologies can be seen in Figure 5. Based on the qualitative analysis of the literature it can be highlighted that many researchers have successfully applied NNs to the indoor positioning problem via convolutional neural networks (CNNs) [106]. A ZigBee [107] indoor positioning research scheme based on the location fingerprinting approach uses an NN locating model. This model, with the signal-index-pair data pre-processing method, is used to increase positioning precision [108]. Related research uses a particle swarm optimization-based backpropagation (PSO-BP) NN to determine the relationship between RFID signals and the position of a tag for an RFID-based positioning system [109].

Furthermore, to improve the quality of training samples, the experimental data are pre-processed via Gauss filtering.

The following section is devoted to show how RTLS systems and the presented models can be integrated into one system.

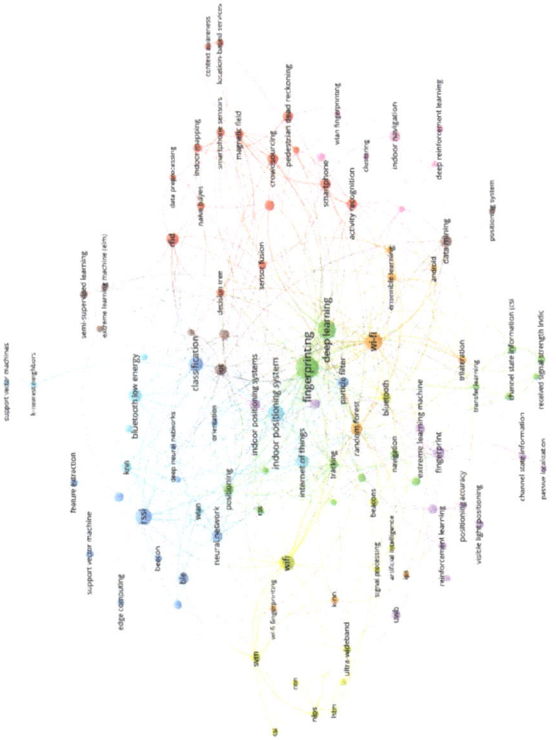

Figure 4. Network of keywords based on Scopus database.

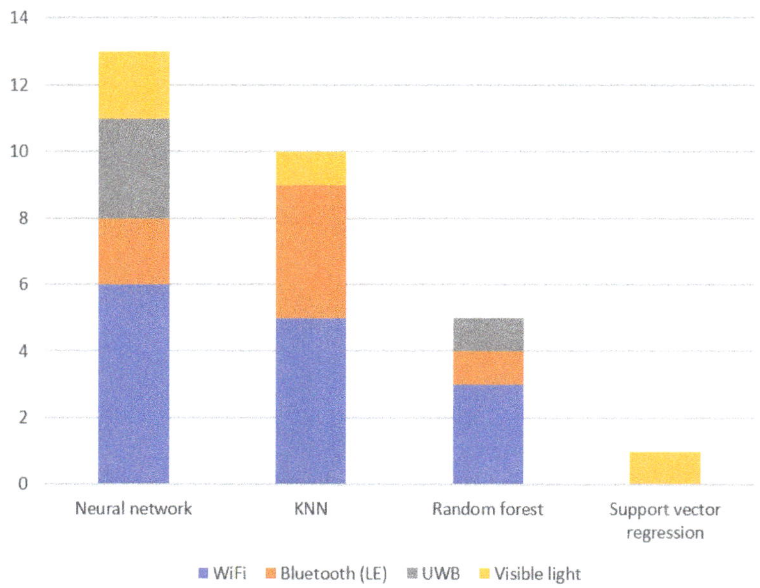

Figure 5. Concurrence of machine learning (ML) techniques and RTLS technologies in articles.

Table 5. Data mining techniques and areas of RTLS-based application

Method	Definition	Data Analytic Techniques	Application Areas	RTLS-Based Applications
Classification	Discriminating data into different labeled subsets pursuant to class attribute. Retrieving important and relevant information about data and metadata.	Neural network Support vector machine (SVM) Decision tree k-Nearest neighbor Bayesian network Genetic algorithm	Pre-defined distribution (e.g., identification of differences) Fault detection Anomaly detection problems	For intralogistics navigation problems [92], shows the performance of RTLS [93], find the best location estimation algorithm [94]
Clustering	Grouping the database according to their similarities. Discovering similarities and dissimilarities between the data.	Partition based algorithms (e.g., K-means, fuzzy c-means) Hierarchical clustering (e.g., dendrogram) Density-based method Grid-based methods Model-based methods	Data segmentation (division into homogeneous sets) Identification of typical prototypes (e.g., simultaneous identification of time-homogeneous periods and their averages/trends)	Improve RTLS accuracy [101–103] Pedestrian motion learning [110]
Regression analysis	Identifying and analyzing the relationship between variables. Predicting and forecasting the process or dependent variables.	Multivariate linear regression Neural network Regression tree	Creating a model that predicts time (e.g., creating a model for predicting temperature)	Used to calculate the efficient RTLS [105] feeding behavior of cows [104]

4. Steps of setting up an RTLS for Manufacturing Support

An installation of RTLS is described in this section with a proposed workflow to illustrate the difficulties of RTLS projects. Figure 6 shows the necessary steps for RTLS-based process analysis. In general, the first step of an RTLS based digitization project is the identification of the requirements, where we define the physical area on the shop floor and the possible applications. The next step is the installation of the sensor network. After the system is running and the position information is being successfully gathered, the accuracy of the system should be validated. The multi-tag concept substantially improves the object detection probability and makes the system more robust [111,112]. Generally, the position engines of RTLS apply filtering methods to pre-process the position data (e.g., a Kalman-filter in the case of GPS) [113]. Several accuracy improvement solutions are available based on RTLS, including regression [114] and k-nearest neighbor classification [95].

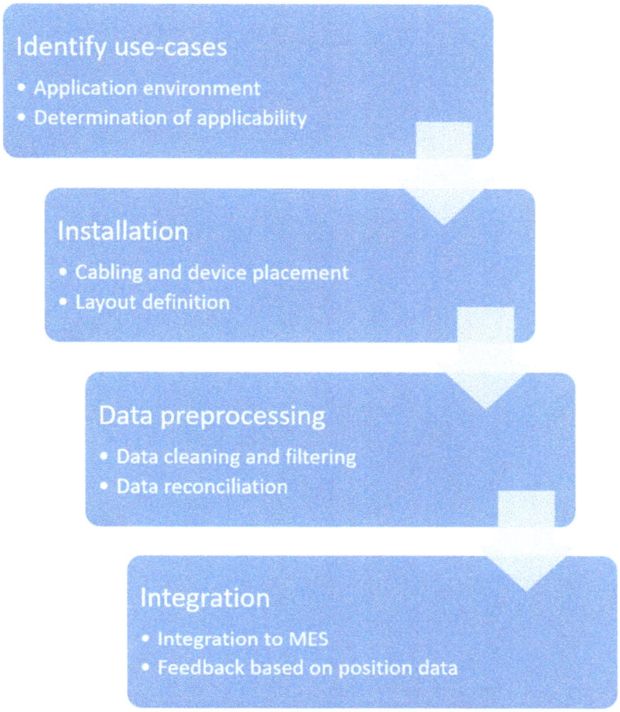

Figure 6. The full setup of RTLS in manufacturing. After the physical system installation, the layout and zone definition is necessary for system integration into the Manufacturing Execution System.

With the spread of RTLS, position data pre-processing and cleaning methods have become an important research topic, based mainly on pedestrian dead reckoning [115] and wireless signal positioning methods [116]. The cleared and filtered position data provides more accurate information to the production system.

The integration of position data into the Manufacturing Execution System (MES) is a crucial element of the implementation.

To obtain usable data from the position information, zones of the manufacturing process should be defined. A zone (Station ID) represents a workstation or storage space, and the RTLS can obtain zone information from every tag in real-time. Figure 7 shows the connections among the RTLS, MES and production. At the beginning of production, the operator pairs a product with an RTLS tag (with a barcode scanner or manually at a PC) to identify the actual product ID in the system. The RTLS provides the zone information (based on the position data of tags and zones definition) with a timestamp to the MES with an application programming interface (API). The MES changes the status of the actual product ID based on the information from RTLS (e.g., the product is tested at a testing station). If a digital interface is provided at the workstation, then the MES can show the work instruction for the actual product or can set the optimal cycle time based on the product content [51].

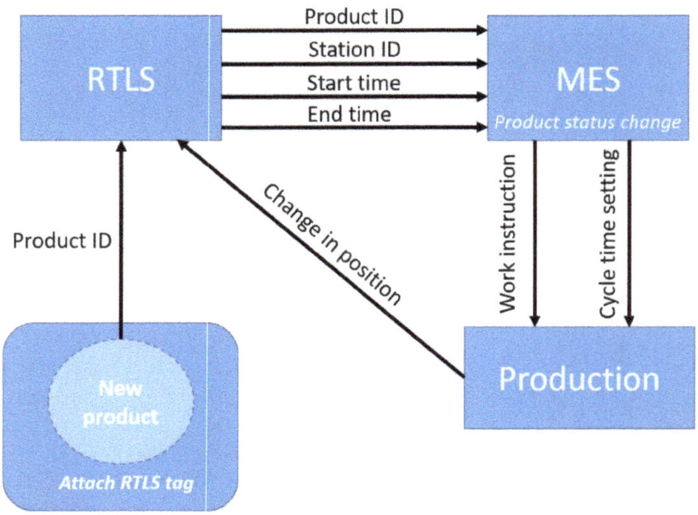

Figure 7. The real-time connection between the Manufacturing Execution System (MES) and production is available based on the RTLS.

5. Analysis Based on Position Data—A Case Study

Position data-based production tracking has considerable potential to optimize production processes. In this section, we present a case study based on an implemented RTLS in a manufacturing environment. The purpose of the position data in production is to transform to relevant information, in the interest of comparing the defined production zones with the position based clusters. Hidden information can be extracted from the position data for production management.

Our use case is an anonymized example from a Tier 1 supplier company from the automotive sector. The production company used the Sunstone-RTLS Ltd. system, which is accurate to 50 cm with eight anchors per every 2000–3000 m^2. The system architecture is shown in Figure 8. There are seven workstations that are used to produce a small wire harness. The zones define the workstations, storage units and routes. The workstations are Tubing station I., Tubing station II., Channeling station, Test station, Screwdriver station, Packaging station and Quality check. The operator attaches the RTLS tag to the product at the first station (Tubing station I/II), and the final station is the Packaging.

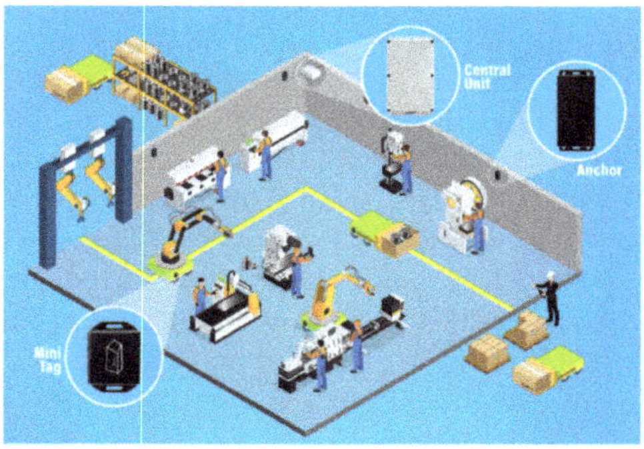

Figure 8. The infrastructure of the Sunstone-RTLS. Every central unit (CU) has eight anchors (which collect data from tags), and the CUs can be connected to create a cascade installation.

The goal of the position-based zone identification is to determine the temporary storage at the temporary station area: a K-means algorithm is applied for position data classification. Figure 9 shows the classified position data where the algorithm detects the three small storage areas over the pre-defined zones. These three undefined or unplanned stations could be the cause of several losses.

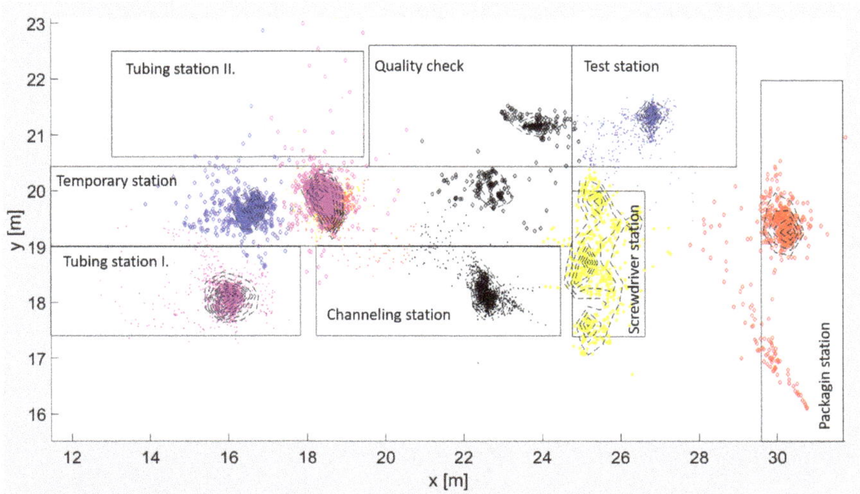

Figure 9. The production layout with seven pre-defined workstations. The classified and pre-defined (rectangles) zones are shown. The algorithm detects three small areas behind the Temporary station and indicates that Tubing station II was not used in this period.

Thanks to the position data, the full traceability of products is available. The cycle time of workstations can be measured based on the classified zone data. The operator scans the product identification label at the first workstation, where the system paired it to the current tag ID (ID is also scanned at the station). The spent time of the actual p product (T_p^z) in the actual zone (z) is the difference between the last timestamp ($T_p^z(l)$ of the actual product (p) position data in the actual zone and the first one ($T_p^z(f)$).

Figure 10 shows the cycle time deviation of every workstation. The boxplot shows the distribution of the times related to the production of more than 150 products (in one shift). We can notice that the packaging station has less cycle time and also the smallest deviation. The reason is, we cannot identify the real finish time at this station due to the fact that they collect the tags traced by the RTLS at this station after the production, and instead, the operator scans the test label during the packaging. The figure shows the Tubing station I. is the bottleneck, but we can see in Figure 9, there is a second station (Tubing station II.), which is a spare workstation. In the current situation, there are not enough resources to operate both workstations, but it could be the solution to improve the process.

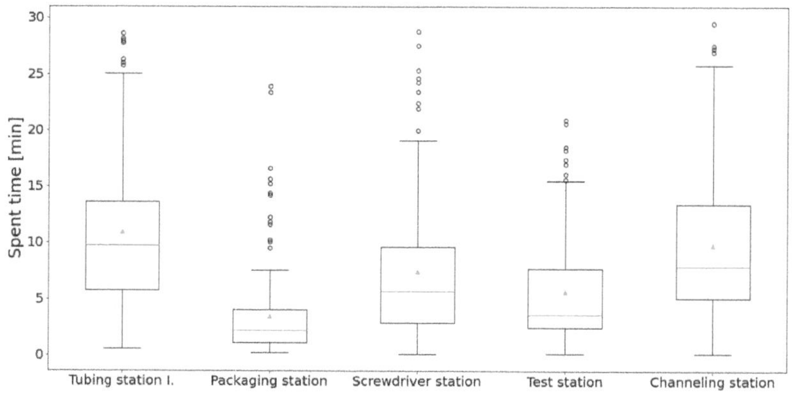

Figure 10. Boxplots of the cycle times measured at different zones of the production process. (Green triangles represent the averages, while red lines the medians).

As this example highlighted, the RTLS can provide accurate and real-time information about the current status of the production process that could be utilised in the development of production processes.

6. Conclusions, Limits and Future Direction of Research

The purpose of this article was to provide a comprehensive overview of the application and development possibilities of RTLS in the manufacturing field. The overview of the solutions determined the value of the positioning data and specified which traceability technologies are suitable for real-time locating in different situations to ensure traceability. Our research explored the possible applications in the production and logistics process. Finally, the implementation of RTLS and a data cleaning method are represented. The end of the article presented a case study, in which we demonstrated what kind of information an RTLS system can provide.

The research pointed out that information extracted from RTLS is highly applicable for performance monitoring. Based on this fact, RTLS supported LEAN projects are very important research topics of the future. This article also introduced that machine learning and state estimation techniques are getting used more and more widely in the development of position data based models. Another conclusion of the case study is that typical states of the production process can be easily determined based on clustering algorithms. Analyzing the sequence of these means a significant increase in the understanding of the processes and in support of process models. According to this we believe that process mining is the most relevant research topic for the future.

It is also important to highlight that an installed RTLS makes the integration of more sensor data available, and in this way the quick implementation of IoT solutions. We believe that this opportunity is beneficial primarily in the development of existing processes, like the introduction of brown field Industry 4.0 solutions.

The advantage of these solutions can be exploited well if the production system is supported by a Manufacturing Execution System (MES), in which a system ensures that information derived from position data can be used in production process optimizations. In line with this, an RTLS project should be connected with an MES development. This MES development process can be supported by semantic models, which are helping to structure sensor and production data. The application of these models is also a research and development topic for the future.

Author Contributions: Writing—original draft preparation, András Rácz-Szabó and Tamás Ruppert; formal analysis, László Bántay; validation, Andreas Löcklin; writing—review and editing, László Bántay and László Jakab; supervision, János Abonyi. All authors have read and agreed to the published version of the manuscript.

Funding: This research received no external funding.

Acknowledgments: This work was supported by the TKP2020-NKA-10 project financed under the 2020-4.1.1-TKP2020 Thematic Excellence Programme by the National Research, Development and Innovation Fund of Hungary. The research was also supported by the "Intelligent production optimization Industry 4.0 in existing plants" 2018-1.3.1-VKE-2018-00048 project.

Conflicts of Interest: The authors declare no conflict of interest.

References

1. Brena, R.F.; García-Vázquez, J.P.; Galván-Tejada, C.E.; Muñoz-Rodriguez, D.; Vargas-Rosales, C.; Fangmeyer, J. Evolution of indoor positioning technologies: A survey. *J. Sens.* **2017**, *2017*, 2630413.
2. Saab, S.S.; Nakad, Z.S. A standalone RFID indoor positioning system using passive tags. *IEEE Trans. Ind. Electron.* **2010**, *58*, 1961–1970.
3. Ang, J.L.F.; Lee, W.K.; Ooi, B.Y. GreyZone: A Novel Method for Measuring and Comparing Various Indoor Positioning Systems. In Proceedings of the 2019 International Conference on Green and Human Information Technology (ICGHIT), Kuala Lumpur, Malaysia, 15–17 January 2019; pp. 30–35.
4. Chen, Y.; Tang, J.; Jiang, C.; Zhu, L.; Lehtomäki, M.; Kaartinen, H.; Kaijaluoto, R.; Wang, Y.; Hyyppä, J.; Hyyppä, H.; et al. The accuracy comparison of three simultaneous localization and mapping (SLAM)-based indoor mapping technologies. *Sensors* **2018**, *18*, 3228.
5. Hameed, A.; Ahmed, H.A. Survey on indoor positioning applications based on different technologies. In Proceedings of the 2018 12th International Conference on Mathematics, Actuarial Science, Computer Science and Statistics (MACS), Karachi, Pakistan, 24–25 November 2018; pp. 1–5.
6. Liu, H.; Darabi, H.; Banerjee, P.; Liu, J. Survey of wireless indoor positioning techniques and systems. *IEEE Trans. Syst. Man, Cybern. Part C (Appl. Rev.)* **2007**, *37*, 1067–1080.
7. Batistić, L.; Tomic, M. Overview of indoor positioning system technologies. In Proceedings of the 2018 41st International Convention on Information and Communication Technology, Electronics and Microelectronics (MIPRO), Opatija, Croatia, 21–25 May 2018; pp. 0473–0478.
8. Ali, W.H.; Kareem, A.A.; Jasim, M. Survey on Wireless Indoor Positioning Systems. *Cihan Univ.-Erbil Sci. J.* **2019**, *3*, 42–47.
9. Alarifi, A.; Al-Salman, A.; Alsaleh, M.; Alnafessah, A.; Al-Hadhrami, S.; Al-Ammar, M.A.; Al-Khalifa, H.S. Ultra wideband indoor positioning technologies: Analysis and recent advances. *Sensors* **2016**, *16*, 707.
10. Zafari, F.; Gkelias, A.; Leung, K.K. A survey of indoor localization systems and technologies. *IEEE Commun. Surv. Tutor.* **2019**, *21*, 2568–2599.
11. Sakpere, W.; Adeyeye-Oshin, M.; Mlitwa, N.B. A state-of-the-art survey of indoor positioning and navigation systems and technologies. *S. Afr. Comput. J.* **2017**, *29*, 145–197.
12. Shi, G.; Ming, Y. Survey of indoor positioning systems based on ultra-wideband (UWB) technology. In *Wireless Communications, Networking and Applications*; Springer: Berlin/Heidelberg, Germany, 2016; pp. 1269–1278.
13. Huang, G.Q.; Zhang, Y.; Jiang, P. RFID-based wireless manufacturing for walking-worker assembly islands with fixed-position layouts. *Robot. Comput.-Integr. Manuf.* **2007**, *23*, 469–477.
14. Mendoza-Silva, G.M.; Torres-Sospedra, J.; Huerta, J. A meta-review of indoor positioning systems. *Sensors* **2019**, *19*, 4507.
15. Shang, J.; Yu, S.; Zhu, L. Location-aware systems for short-range wireless networks. In Proceedings of the 2009 International Symposium on Computer Network and Multimedia Technology, Wuhan, China, 18–20 January 2009; pp. 1–5.
16. Gladysz, B.; Santarek, K. An approach to RTLS selection. In *DEStech Transactions on Engineering and Technology Research, Proceedings of the 24th International Conference on Production Research (ICPR 2017), Poznan, Poland, 30 July–3 August 2017*; 2017.
17. Saez, M.; Maturana, F.P.; Barton, K.; Tilbury, D.M. Real-time manufacturing machine and system performance monitoring using internet of things. *IEEE Trans. Autom. Sci. Eng.* **2018**, *15*, 1735–1748.

18. Macagnano, D.; Destino, G.; Abreu, G. Indoor positioning: A key enabling technology for IoT applications. In Proceedings of the 2014 IEEE World Forum on Internet of Things (WF-IoT), Seoul, Korea, 6–8 March 2014; pp. 117–118.
19. Li, S.; Rashidzadeh, R. Hybrid indoor location positioning system. *IET Wirel. Sens. Syst.* **2019**, *9*, 257–264.
20. Hammerin, K.; Streitenberger, R. *RTLS—The missing link to optimizing Logistics Management?* In *Jönköping University, Production systems, Jönköping, Sweden, 20 May 2019*; 2019.
21. Holm, S. Hybrid ultrasound-RFID indoor positioning: Combining the best of both worlds. In Proceedings of the 2009 IEEE International Conference on RFID, Orlando, FL, USA, 27–28 April 2009; pp. 155–162.
22. Woo, S.; Jeong, S.; Mok, E.; Xia, L.; Choi, C.; Pyeon, M.; Heo, J. Application of WiFi-based indoor positioning system for labor tracking at construction sites: A case study in Guangzhou MTR. *Autom. Constr.* **2011**, *20*, 3–13.
23. Park, J.; Cho, Y.K.; Martinez, D. A BIM and UWB integrated mobile robot navigation system for indoor position tracking applications. *J. Constr. Eng. Proj. Manag.* **2016**, *6*, 30–39.
24. Stephan, P.; Heck, I.; Krau, P.; Frey, G. Evaluation of Indoor Positioning Technologies under industrial application conditions in the SmartFactoryKL based on EN ISO 9283. *IFAC Proc. Vol.* **2009**, *42*, 870–875.
25. Simões, W.C.; Machado, G.S.; Sales, A.; de Lucena, M.M.; Jazdi, N.; de Lucena, V.F. A Review of Technologies and Techniques for Indoor Navigation Systems for the Visually Impaired. *Sensors* **2020**, *20*, 3935.
26. Lennvall, T.; Svensson, S.; Hekland, F. A comparison of WirelessHART and ZigBee for industrial applications. In Proceedings of the 2008 IEEE International Workshop on Factory Communication Systems, Dresden, Germany, 21–23 May 2008; pp. 85–88.
27. Longkang, W.; Baisheng, N.; Ruming, Z.; Shengrui, Z.; Hailong, L. ZigBee-based positioning system for coal miners. *Procedia Eng.* **2011**, *26*, 2406–2414.
28. Uradzinski, M.; Guo, H.; Liu, X.; Yu, M. Advanced indoor positioning using zigbee wireless technology. *Wirel. Pers. Commun.* **2017**, *97*, 6509–6518.
29. Shi, J. Laser guided four-wheel drive AGV trolley. In *AIP Conference Proceedings, Wuhan, Chine*; AIP Publishing LLC: 2019; Volume 2073, p. 020060.
30. Alici, G.; Shirinzadeh, B. A systematic technique to estimate positioning errors for robot accuracy improvement using laser interferometry based sensing. *Mech. Mach. Theory* **2005**, *40*, 879–906.
31. Kirch, M.; Poenicke, O.; Richter, K. RFID in logistics and production–Applications, research and visions for smart logistics zones. *Procedia Eng.* **2017**, *178*, 526–533.
32. Zhai, C.; Zou, Z.; Zhou, Q.; Mao, J.; Chen, Q.; Tenhunen, H.; Zheng, L.; Xu, L. A 2.4-GHz ISM RF and UWB hybrid RFID real-time locating system for industrial enterprise Internet of Things. *Enterp. Inf. Syst.* **2017**, *11*, 909–926.
33. Lee, H.S.; Lee, K.P.; Park, M.; Baek, Y.; Lee, S. RFID-based real-time locating system for construction safety management. *J. Comput. Civ. Eng.* **2012**, *26*, 366–377.
34. Martin, P.; Ho, B.J.; Grupen, N.; Muñoz, S.; Srivastava, M. An iBeacon primer for indoor localization: Demo abstract. In Proceedings of the Proceedings of the 1st ACM Conference on Embedded Systems for Energy-Efficient Buildings, Memphis, USA, 5–6 November 2014; 2014; pp. 190–191.
35. Yue, H.; Zheng, X.; Wang, J.; Zhu, L.; Zeng, C.; Liu, C.; Liu, M. Research and Implementation of Indoor Positioning Algorithm for Personnel Based on Deep Learning. International Conference on Emerging Internetworking, Data & Web Technologies, Tirana, Albania, 15–17 March 2018; pp. 782–791.
36. Otim, T.; Díez, L.E.; Bahillo, A.; Lopez-Iturri, P.; Falcone, F. Effects of the body wearable sensor position on the UWB localization accuracy. *Electronics* **2019**, *8*, 1351.
37. Gohil, A.; Modi, H.; Patel, S.K. 5G technology of mobile communication: A survey. In Proceedings of the 2013 international conference on intelligent systems and signal processing (ISSP), Gujarat, India, 1–2 March 2013; pp. 288–292.
38. Witrisal, K.; Hinteregger, S.; Kulmer, J.; Leitinger, E.; Meissner, P. High-accuracy positioning for indoor applications: RFID, UWB, 5G, and beyond. In Proceedings of the 2016 IEEE International Conference on RFID (RFID), Orlando, FL, USA, 3–5 May 2016; pp. 1–7.
39. Horsmanheimo, S.; Lembo, S.; Tuomimaki, L.; Huilla, S.; Honkamaa, P.; Laukkanen, M.; Kemppi, P. Indoor positioning platform to support 5G location based services. In Proceedings of the 2019 IEEE International Conference on Communications Workshops (ICC Workshops), Shanghai, China, 20–24 May 2019; pp. 1–6.
40. Rodas, J.; Barral, V.; Escudero, C.J. Architecture for multi-technology real-time location systems. *Sensors* **2013**, *13*, 2220–2253.

41. Lin, J.R.; Talty, T.; Tonguz, O.K. On the potential of bluetooth low energy technology for vehicular applications. *IEEE Commun. Mag.* **2015**, *53*, 267–275.
42. Jiang, X.; Chen, S. Design of Wireless Point of Sale Based on ZigBee Technology. *Sens. Transducers* **2014**, *164*, 120.
43. Abdullah, A.; Ismael, A.; Rashid, A.; Abou-ElNour, A.; Tarique, M. Real time wireless health monitoring application using mobile devices. *Int. J. Comput. Networks Commun. (IJCNC)* **2015**, *7*, 13–30.
44. Lee, J.S.; Su, Y.W.; Shen, C.C. A comparative study of wireless protocols: Bluetooth, UWB, ZigBee, and Wi-Fi. In Proceedings of the IECON 2007-33rd Annual Conference of the IEEE Industrial Electronics Society, Taipei, Taiwan, 5–8 November 2007; pp. 46–51.
45. Kazeem, O.O.; Akintade, O.O.; Kehinde, L.O. Comparative Study of Communication Interfaces for Sensors and Actuators in the Cloud of Internet of Things. *Int. J. Internet Things* **2017**, *6*, 9–13.
46. Cui, Y.; Zhao, J. Real-time location system and applied research report. In Proceedings of the International Conference on Internet of vehicles, Chengdu, China, 19–21 December 2015; pp. 49–57.
47. Astafiev, A.; Zhiznyakov, A.; Privezentsev, D. Development of Indoor Positioning Algorithm Based on Bluetooth Low Energy beacons for Building RTLS-Systems. In Proceedings of the 2019 International Russian Automation Conference (RusAutoCon), Sochi, Russia, 8–14 September 2019; pp. 1–5.
48. Nowotarski, P.; Pasławski, J.; Skrzypczak, M.; Krygier, R. RTLS systems as a Lean Management tool for productivity improvement. In Proceedings of the ISARC—International Symposium on Automation and Robotics in Construction, Taipei, Taiwan, 28 June–1 July 2017; Volume 34.
49. Moreira, A.; Nicolau, M.J.; Meneses, F.; Costa, A. Wi-Fi fingerprinting in the real world-RTLS@ UM at the EvAAL competition. In Proceedings of the 2015 International Conference on Indoor Positioning and Indoor Navigation (IPIN), Banff, AB, Canada, 13–16 October 2015; pp. 1–10.
50. Mazhar, F.; Khan, M.G.; Sällberg, B. Precise indoor positioning using UWB: A review of methods, algorithms and implementations. *Wirel. Pers. Commun.* **2017**, *97*, 4467–4491.
51. Ruppert, T.; Abonyi, J. Industrial Internet of Things based cycle time control of assembly lines. In Proceedings of the 2018 IEEE International Conference on Future IoT Technologies (Future IoT), Eger, Hungary, 18–19 January 2018; pp. 1–4.
52. Gyulai, D.; Pfeiffer, A.; Bergmann, J. Analysis of asset location data to support decisions in production management and control. *Procedia CIRP* **2020**, *88*, 197–202.
53. Guo, Z.; Ngai, E.; Yang, C.; Liang, X. An RFID-based intelligent decision support system architecture for production monitoring and scheduling in a distributed manufacturing environment. *Int. J. Prod. Econ.* **2015**, *159*, 16–28.
54. Gallimore, J.J.; Quill, L.; Cagle, R.; Gruenke, J.; Hosman, C.; Matthews, E.; Faas, P.; Seyba, J.; Young, I. *User Feedback on RFID and Integrated Flightline Data for Maintenance Decisions;* Technical Report; University of Dayton Research Institute: Dayton, OH, USA, 2006.
55. Ruppert, T.; Abonyi, J. Software Sensor for Activity-Time Monitoring and Fault Detection in Production Lines. *Sensors* **2018**, *18*, 2346.
56. Peron, M.; Fragapane, G.; Sgarbossa, F.; Kay, M. Digital Facility Layout Planning. *Sustainability* **2020**, *12*, 3349.
57. Zang, Y.; Wu, L. Application of RFID and RTLS technology in supply chain enterprise. In Proceedings of the 2010 6th International Conference on Wireless Communications Networking and Mobile Computing (WiCOM), Chengdu, China, 23–25 September 2010; pp. 1–4.
58. Lee, C.; Lv, Y.; Ng, K.; Ho, W.; Choy, K. Design and application of Internet of things-based warehouse management system for smart logistics. *Int. J. Prod. Res.* **2018**, *56*, 2753–2768.
59. Ma, X.; Liu, T. The application of Wi-Fi RTLS in automatic warehouse management system. In Proceedings of the 2011 IEEE International conference on automation and logistics (ICAL), Chongqing, China, 15–16 August 2011; pp. 64–69.
60. Soleimanifar, M.; Lu, M. Streamlining an indoor positioning architecture based on field testing in pipe spool fabrication shop. In Proceedings of the Winter Simulation Conference 2014, Savanah, GA, USA, 7–10 December 2014; pp. 3260–3271.
61. Silvia, Z.; Martina, C.; Fabio, S.; Alessandro, P. Ultra Wide Band Indoor Positioning System: Analysis and testing of an IPS technology. *IFAC-PapersOnLine* **2018**, *51*, 1488–1492.

62. Hegedűs, C.; Frankó, A.; Varga, P. Asset and Production Tracking through Value Chains for Industry 4.0 using the Arrowhead Framework. In Proceedings of the 2019 IEEE International Conference on Industrial Cyber Physical Systems (ICPS), Taipei, Taiwan, 6–9 May 2019; pp. 655–660.
63. Ding, B.; Chen, L.; Chen, D.; Yuan, H. Application of RTLS in warehouse management based on RFID and Wi-Fi. In Proceedings of the 2008 4th International Conference on Wireless Communications, Networking and Mobile Computing, Dalian, China, 12–14 October 2008; pp. 1–5.
64. Akanmu, A.A.; Anumba, C.J.; Messner, J.I. An RTLS-based approach to cyber-physical systems integration in design and construction. *Int. J. Distrib. Sens. Networks* **2012**, *8*, 596845.
65. Akanmu, A.; Rasheed, S.H.; Qader, I.A. Spatial mapping approach to component tracking using RTLS system. In *AEI 2013: Building Solutions for Architectural Engineering, Proceedings of the Architectural Engineering Conference 2013, State College, PA, USA, 3–5 April 2013*; ASCE: Reston, VA, USA, 2013; pp. 364–376.
66. Frankó, A.; Vida, G.; Varga, P. Reliable Identification Schemes for Asset and Production Tracking in Industry 4.0. *Sensors* **2020**, *20*, 3709.
67. Sima, G.; Lile, R.; Dan, G.; Muncut, E. Management, traceability and control of industrial processes. In *Soft Computing Applications*; Springer: Berlin/Heidelberg, Germany, 2016; pp. 971–980.
68. Badihi, B.; Zhao, J.; Zhuang, S.; Seppänen, O.; Jäntti, R. Intelligent Construction Site: On Low Cost Automated Indoor Localization Using Bluetooth Low Energy Beacons. In Proceedings of the 2019 IEEE Conference on Wireless Sensors (ICWiSe), Pulau Pinang, Malaysia, 19–21 November 2019; pp. 29–35.
69. Lu, S.; Xu, C.; Zhong, R.Y.; Wang, L. A RFID-enabled positioning system in automated guided vehicle for smart factories. *J. Manuf. Syst.* **2017**, *44*, 179–190.
70. Kelepouris, T.; McFarlane, D. Determining the value of asset location information systems in a manufacturing environment. *Int. J. Prod. Econ.* **2010**, *126*, 324–334.
71. Choi, H.; Baek, Y.; Lee, B. Design and implementation of practical asset tracking system in container terminals. *Int. J. Precis. Eng. Manuf.* **2012**, *13*, 1955–1964.
72. Ghosh, S.; Anurag, D.; Bandyopadhyay, S. Use of wireless mesh network to improve mobile asset utilization in manufacturing industries. *Procedia Comput. Sci.* **2011**, *5*, 66–73.
73. Jiang, J.; Guo, Y.; Liao, W. Research on AGV guided by real-time locating system (RTLS) for material distribution. *Int. J. Control Autom.* **2015**, *8*, 213–226.
74. Barral, V.; Suárez-Casal, P.; Escudero, C.J.; García-Naya, J.A. Multi-sensor accurate forklift location and tracking simulation in industrial indoor environments. *Electronics* **2019**, *8*, 1152.
75. Li, H.B.; Miura, R.; Nishikawa, H.; Kagawa, T.; Kojima, F. Tracking of warehouse forklifts using an indoor positioning system based on IR-UWB. In Proceedings of the International Conference on indoor positioning and navigation (IPIN 2016), Madrid, Spain, 4–7 Octeber 2016.
76. Slovák, J.; Vašek, P.; Šimovec, M.; Melicher, M.; Šišmišová, D. RTLS tracking of material flow in order to reveal weak spots in production process. In Proceedings of the 2019 22nd International Conference on Process Control (PC19), Strbske Pleso, Slovakia, 11–14 June 2019; pp. 234–238.
77. Zhang, C.; Hammad, A. Multiagent approach for real-time collision avoidance and path replanning for cranes. *J. Comput. Civ. Eng.* **2012**, *26*, 782–794.
78. Barro-Torres, S.; Fernández-Caramés, T.M.; Pérez-Iglesias, H.J.; Escudero, C.J. Real-time personal protective equipment monitoring system. *Comput. Commun.* **2012**, *36*, 42–50.
79. Ulku, A. The Next Generation In Personnel\/People Tracking: Active RFID Technology Has Allowed for Enhanced Security and Safety. *IEEE Consum. Electron. Mag.* **2017**, *6*, 122–124.
80. Köppe, E.; Bartholmai, M.; Liers, A.; Schiller, J. Radio-based multi-sensor system for person tracking and indoor positioning. In Proceedings of the 2012 9th Workshop on Positioning, Navigation and Communication, Dresden, Germany, 15–16 March 2012; pp. 180–186.
81. Hellmich, T.R.; Clements, C.M.; El-Sherif, N.; Pasupathy, K.S.; Nestler, D.M.; Boggust, A.; Ernste, V.K.; Marisamy, G.; Koenig, K.R.; Hallbeck, M.S. Contact tracing with a real-time location system: A case study of increasing relative effectiveness in an emergency department. *Am. J. Infect. Control* **2017**, *45*, 1308–1311.
82. Arkan, I.; Van Landeghem, H. Evaluating the performance of a discrete manufacturing process using RFID: A case study. *Robot. Comput.-Integr. Manuf.* **2013**, *29*, 502–512.
83. Gladysz, B.; Santarek, K.; Lysiak, C. Dynamic spaghetti diagrams. A case study of pilot RTLS implementation. In *International Conference on Intelligent Systems in Production Engineering and Maintenance*; Springer: Berlin/Heidelberg, Germany, 2017, pp. 238–248.

84. Awolusi, A.B.; Akinyokun, O.C.; Iwasokun, G.B. RFID and RTLS-Based Human Resource Monitoring System. *J. Adv. Math. Comput. Sci.* **2016**, *14*, pp. 1–14.
85. Ho, H.J.; Zhang, Z.X.; Huang, Z.; Aung, A.H.; Lim, W.Y.; Chow, A. Use of a real-time locating system for contact tracing of health care workers during the COVID-19 pandemic at an infectious disease center in Singapore: Validation study. *J. Med Internet Res.* **2020**, *22*, e19437.
86. Ruppert, T.; Abonyi, J. Integration of real-time locating systems into digital twins. *J. Ind. Inf. Integr.* **2020**, *20*, 100174.
87. Demir, K.A.; Döven, G.; Sezen, B. Industry 5.0 and human-robot co-working. *Procedia Comput. Sci.* **2019**, *158*, 688–695.
88. Nahavandi, S. Industry 5.0—A human-centric solution. *Sustainability* **2019**, *11*, 4371.
89. Peruzzini, M.; Grandi, F.; Pellicciari, M. Exploring the potential of Operator 4.0 interface and monitoring. *Comput. Ind. Eng.* **2020**, *139*, 105600.
90. Ruppert, T.; Jaskó, S.; Holczinger, T.; Abonyi, J. Enabling technologies for operator 4.0: A survey. *Appl. Sci.* **2018**, *8*, 1650.
91. Löcklin, A.; Ruppert, T.; Jakab, L.; Libert, R.; Jazdi, N.; Weyrich, M. Trajectory Prediction of Humans in Factories and Warehouses with Real-Time Locating Systems. In Proceedings of the 2020 25th IEEE International Conference on Emerging Technologies and Factory Automation (ETFA), Vienna, Austria, 8–11 September 2020; Voume 1, pp. 1317–1320.
92. Sattarian, M.; Rezazadeh, J.; Farahbakhsh, R.; Bagheri, A. Indoor navigation systems based on data mining techniques in internet of things: A survey. *Wirel. Netw.* **2019**, *25*, 1385–1402.
93. Yazıcı, A.; Keser, S.B.; Gunal, S. Integration of classification algorithms for indoor positioning system. In Proceedings of the 2017 International Conference on Computer Science and Engineering (UBMK), Antalya, Turkey, 5–8 October 2017; pp. 267–270.
94. Zia, K.; Iram, H.; Aziz-ul Haq, M.; Zia, A. Comparative study of classification techniques for indoor localization of mobile devices. In Proceedings of the 2018 28th International Telecommunication Networks and Applications Conference (ITNAC), Sydney, NSW, Australia, 21–23 November 2018; pp. 1–5.
95. Xu, H.; Ding, Y.; Li, P.; Wang, R.; Li, Y. An RFID indoor positioning algorithm based on Bayesian probability and K-nearest neighbor. *Sensors* **2017**, *17*, 1806.
96. Adege, A.B.; Lin, H.P.; Tarekegn, G.B.; Munaye, Y.Y.; Yen, L. An indoor and outdoor positioning using a hybrid of support vector machine and deep neural network algorithms. *J. Sensors* **2018**, *2018*, 1253752.
97. Chanama, L.; Wongwirat, O. A comparison of decision tree based techniques for indoor positioning system. In Proceedings of the 2018 International Conference on Information Networking (ICOIN), Chiang Mai, Thailand, 10–12 January 2018; pp. 732–737.
98. Wu, Z.; Xu, Q.; Li, J.; Fu, C.; Xuan, Q.; Xiang, Y. Passive indoor localization based on csi and naive bayes classification. *IEEE Trans. Syst. Man, Cybern. Syst.* **2017**, *48*, 1566–1577.
99. Alhammadi, A.; Alraih, S.; Hashim, F.; Rasid, M.F.A. Robust 3D Indoor Positioning System Based on Radio Map Using Bayesian Network. In Proceedings of the 2019 IEEE 5th World Forum on Internet of Things (WF-IoT), Limerick, Ireland, 15–18 April 2019; pp. 107–110.
100. Yazıcı, A.; Keser, S.B.; Günal, S.; Yayan, U. A Multi-Criteria Decision Strategy to Select a Machine Learning Method for Indoor Positioning System. *Int. J. Artif. Intell. Tools* **2018**, *27*, 1850018.
101. Lian, Z.K.; Yuan, F.; Qi, W. Improved K-means Clustering BP Neural Network UWB Indoor Location Method. *Mod. Comput.* **2017**, *2017*, 3.
102. Mo, Y.; Zhang, Z.; Meng, W.; Ma, L.; Wang, Y. A spatial division clustering method and low dimensional feature extraction technique based indoor positioning system. *Sensors* **2014**, *14*, 1850–1876.
103. Subedi, S.; Gang, H.S.; Ko, N.Y.; Hwang, S.S.; Pyun, J.Y. Improving indoor fingerprinting positioning with affinity propagation clustering and weighted centroid fingerprint. *IEEE Access* **2019**, *7*, 31738–31750.
104. Pastell, M.; Frondelius, L. A hidden Markov model to estimate the time dairy cows spend in feeder based on indoor positioning data. *Comput. Electron. Agric.* **2018**, *152*, 182–185.
105. Abdou, A.S.; Aziem, M.A.; Aboshosha, A. An efficient indoor localization system based on Affinity Propagation and Support Vector Regression. In Proceedings of the 2016 Sixth International Conference on Digital Information Processing and Communications (ICDIPC), Beirut, Lebanon, 21–23 April 2016; pp. 1–7.

106. Lu, E.H.C.; Chang, K.H.; Ciou, J.M. Data Pre-processing Based on Convolutional Neural Network for Improving Precision of Indoor Positioning. In Proceedings of the Asian Conference on Intelligent Information and Database Systems, Phuket, Thailand, 23–26 March 2020; pp. 545–552.
107. Gislason, D. *Zigbee Wireless Networking*; Elsevier, Newnes, Australia, 2008.
108. Hung, M.H.; Lin, S.S.; Cheng, J.Y.; Chien, W.L. A ZigBee indoor positioning scheme using signal-index-pair data preprocess method to enhance precision. In Proceedings of the 2010 IEEE International Conference on Robotics and Automation, Anchorage, AK, USA, 3–7 May 2010; pp. 548–553.
109. Wang, C.; Wu, F.; Shi, Z.; Zhang, D. Indoor positioning technique by combining RFID and particle swarm optimization-based back propagation neural network. *Optik* **2016**, *127*, 6839–6849.
110. Yang, X.; Wang, Y.; Zhou, M.; Liu, Y. Pedestrian motion learning based indoor WLAN localization via spatial clustering. *Wirel. Commun. Mob. Comput.* **2018**, *2018*, 2571671.
111. Kuhn, M.J.; Mahfouz, M.R.; Turnmire, J.; Wang, Y.; Fathy, A.E. A multi-tag access scheme for indoor UWB localization systems used in medical environments. In Proceedings of the 2011 IEEE Topical Conference on Biomedical Wireless Technologies, Networks, and Sensing Systems, Phoenix, AZ, USA, 16–19 January 2011; pp. 75–78.
112. Gao, Z.; Ma, Y.; Liu, K.; Miao, X.; Zhao, Y. An indoor multi-tag cooperative localization algorithm based on NMDS for RFID. *IEEE Sens. J.* **2017**, *17*, 2120–2128.
113. Jamil, F.; Iqbal, N.; Ahmad, S.; Kim, D.H. Toward Accurate Position Estimation Using Learning to Prediction Algorithm in Indoor Navigation. *Sensors* **2020**, *20*, 4410.
114. Xu, H.; Wu, M.; Li, P.; Zhu, F.; Wang, R. An RFID indoor positioning algorithm based on support vector regression. *Sensors* **2018**, *18*, 1504.
115. Moder, T.; Hafner, P.; Wisiol, K.; Wieser, M. 3d indoor positioning with pedestrian dead reckoning and activity recognition based on bayes filtering. In Proceedings of the 2014 International Conference on Indoor Positioning and Indoor Navigation (IPIN), Busan, Korea, 27–30 October 2014; pp. 717–720.
116. He, S.; Chan, S.H.G. Wi-Fi fingerprint-based indoor positioning: Recent advances and comparisons. *IEEE Commun. Surv. Tutor.* **2015**, *18*, 466–490.

sensors

MDPI

Letter

Cooperative Localization and Time Synchronization Based on M-VMP Method

Zhongliang Deng, Shihao Tang *, Buyun Jia, Hanhua Wang, Xiwen Deng and Xinyu Zheng

School of Electronic Engineering, Beijing University of Posts and Telecommunications, Beijing 100876, China; dengzhl@bupt.edu.cn (Z.D.); jiabuyun@bupt.edu.cn (B.J.); whh0710@bupt.edu.cn (H.W.); dengxiwen@bupt.edu.cn (X.D.); buptzxy@bupt.edu.cn (X.Z.)

* Correspondence: swift71116@bupt.edu.cn; Tel.: +86-010-6119-8509

Received: 28 September 2020; Accepted: 4 November 2020; Published: 5 November 2020

Abstract: Localization estimation and clock synchronization are important research directions in the application of wireless sensor networks. Aiming at the problems of low positioning accuracy and slow convergence speed in localization estimation methods based on message passing, this paper proposes a low-complexity distributed cooperative joint estimation method suitable for dynamic networks called multi-Gaussian variational message passing (M-VMP). The proposed method constrains the message to be a multi-Gaussian function superposition form to reduce the information loss in the variational message passing algorithm (VMP). Only the mean, covariance and weight of each message need to be transmitted in the network, which reduces the computational complexity while ensuring the information completeness. The simulation results show that the proposed method is superior to the VMP algorithm in terms of position accuracy and convergence speed and is close to the sum-product algorithm over a wireless network (SPAWN) based on non-parametric belief propagation, but the computational complexity and communication load are significantly reduced.

Keywords: multi-variational message passing (M-VMP); factor graph (FG); second-order Taylor expansion; cooperative localization; joint estimation of position and clock

1. Introduction

In recent years, wireless sensor network has been widely used in agriculture, warehousing, production safety, emergency rescue and other fields [1]. The information which sensors collected and transmitted is valuable when combined with the sensors' location [2,3]. Therefore, location awareness of wireless sensor networks has become one of the most important directions in the development of wireless sensor networks. Global Navigation Satellite System (GNSS) is able to provide location information for sensor nodes, but it is difficult to apply on sensor networks due to high power consumption and high cost [2–4]. Furthermore, the poor signal penetration capabilities of GNSS lead to inadequate location information in indoor scenes. Cooperative localization can overcome these problems through ranging and exchanging locations between neighbor nodes [5]. Over the past decade, cooperative localization in wireless sensor networks has drawn considerable attention.

In the recent ten years, cooperative positioning in wireless sensor networks has been widely concerned. The multidimensional scaling (MDS) algorithm [6,7] calculates the shortest path distance between nodes according to the connectivity of the network or distance measurement. Then, the relative coordinate diagram of all nodes is constructed by using the multidimension scale algorithm. Finally, the relative coordinate diagram is transformed into an absolute coordinate graph according to the coordinate of the anchor node. The semi-definite programming (SDP) algorithm [8,9] represents geometric constraints among nodes as a set of linear matrix inequalities, then combines the inequality into a semi-definite programming problem and obtains the location of each node by global optimization. The distance-vector (DV) hop algorithm [10,11] firstly measures the minimum number of hops between the nodes to be located and each anchor node. Then, the average distance of each hop is determined according to the number and coordinate between anchor nodes. After the jump number is converted into distance, the position of the node to be located is obtained according to the trilateral measurement method. The Approximate Perfect Point-In-Triangulation (APIT) algorithm [12,13] firstly selects three anchor nodes connected with the node to be located, and then judges whether they are in the triangle composed of anchor nodes according to the pit test, and then repeats the pit test with a different anchor node combination. Finally, the center of mass position of these triangles is taken as the coordinates of the nodes to be located. The SPAWN algorithm [14] decomposes the posterior probability density function of position variables and represents it as a factor graph. Then, the message is transmitted by particles on the factor graph to calculate the probability of edge distribution of each variable. Finally, position estimation is carried out according to minimum mean square error (MMSE) or map criterion. Although the above cooperative positioning algorithm has high positioning accuracy, it has high computational complexity and high communication cost, which seriously restricts the practical application.

For the above problems, a collaborative location algorithm based on factor graph and VMP is proposed in Reference [15], which has simple message form and small computation. However, Kullback-Leibler (KL) divergence is used to Gauss the non-Gaussian confidence of the nonlinear range measurement model. The first kind of convergence hypergeometric function minimization problem is introduced, which makes the calculation complexity very high. In Reference [16], a hard decision-based cooperative algorithm is proposed to alleviate the effect of the outliers. The geometric relationship between agent position and distance is used to avoid a large deviation coursed by geometric constraints. The authors of Reference [17] propose the distributed particle filtering evolved variational message passing (DPF-E-VMP) algorithm, which improves the convergence speed of positioning estimation by using distributed particle filtering (DPF), but this performance improvement is often accompanied by greater computational consumption. By combining the average consensus method and VMP algorithm, a joint self-localization tracking algorithm called cooperative localization with outlier constraints (CLOC) is proposed in Reference [18], which has better location performance than the separate self-localization algorithm. In Reference [19], KL divergence is minimized by the Newton conjugate gradient method, but the computational complexity is still high. But, the information loss caused by VMP parameterization will have a certain influence on the positioning accuracy. At the same time, the algorithm has been reduced in accuracy and convergence speed due to the clock synchronization between the nodes to be located.

In this paper, a VMP distributed cooperative localization algorithm based on multi-Gaussian is proposed. Based on the non-line of sight (NLOS) environment ranging model, the VMP message passing strategy based on multi-Gaussian is innovated, and the second-order Taylor expansion form of position and time synchronization joint estimation is derived. Its computational complexity is far less than the approximate solution based on KL divergence. The time complexity and communication cost are connected with the traditional VMP algorithm. However, the positioning accuracy and iteration speed have been greatly improved.

The remainder of this paper will be organized as follows: A two-dimensional (2D) wireless network is first established as a system model. Then, a traditional cooperative localization algorithm is introduced and leads to the method proposed in this paper. After that, simulation performance of the proposed method is investigated. Finally, concluding remarks are presented. A list of symbols that are used in the paper is given in Table 1.

Table 1. List of symbols.

Symbol	Meaning	Symbol	Meaning
$S_{i,n}$	Set of neighbor anchor nodes of node i at time n	$C_{i,n}$	Set of neighbor nodes to be located of node i at time n
$\boldsymbol{x}_{i,n}$	Position vector of node i at time n	$\tilde{t}_{i,n}$	Measurement value of local clock of note i at time n
T_n	Real time value at time n	$a_{i,n}$	Slope of local clock of node i at time n
$a_{ij,n}$	Relative slope of local clock offset between nodes i, j	$N_{i,n}$	Set of neighbor nodes of node i at time n
$\tilde{\rho}_{ij,n}$	Range measurement between node i,j at time n	$\omega_{ij,n}$	Measurement noise of $\tilde{\rho}_{ij,n}$
σ_d^2	Variance of $\omega_{ij,n}$	$\zeta_{ij,n}$	NLOS error of $\tilde{\rho}_{ij,n}$
Ψ_n	Set of all communicable node pairs (i, j) at time n	Θ_n	Set of all node pairs (i, j) with NLOS error at time n
$\boldsymbol{x}_n$	Position vector of all nodes at time n	$\boldsymbol{a}_n$	Clock offset of all nodes at time n
$v_{i,n}$	Average velocity of node i from time $n-1$ to time n	$\omega_{i,n}$	Measurement noise of $v_{i,n}$
$\sigma_{i,n}^2$	Variance of $\omega_{i,n}$	$\boldsymbol{z}_{i,n}$	Vector to be estimated of note i at time n
$\hat{\boldsymbol{z}}_{i,n}$	Estimation result of node i at time n	$\tilde{\rho}_n$	Range measurement in all node pairs $(i, j) \in V_n$ at time n
ζ_n	NLOS error in all node pairs $(i, j) \in \Theta_n$ at time n	$\mathcal{X}_{1:n}, \mathcal{A}_{1:n}, \mathcal{P}_{1:n}, \mathcal{B}_{1:n}$	Vector sets of $\boldsymbol{x}_n, \boldsymbol{a}_n, \tilde{\rho}_n, \zeta_n$ from time 1 to time n
$\mu_{f_{ij,n}\to\boldsymbol{z}_{i,n}}(\boldsymbol{z}_{i,n})$	Message send from $f_{ij,n}$ to $\boldsymbol{z}_{i,n}$	$b(\boldsymbol{z}_{i,n})$	Belief of variable $\boldsymbol{z}_{i,n}$
$F(\cdot)$	Confluent Hypergeometric Function of the First Type	$E_{\boldsymbol{z}_{i,n}}$	Mean of belief $b(\boldsymbol{z}_{i,n})$
$V_{\boldsymbol{z}_{i,n}}$	Covariance of belief $b(\boldsymbol{z}_{i,n})$	$\eta_{M,ij}$	Weight of the M-th Gaussian distribution
$F(\boldsymbol{z}_{i,n})$	Fisher Information Matrix (FIM) of $\boldsymbol{z}_{i,n}$	$\mathrm{CRLB}(\boldsymbol{z}_{i,n})$	Cramer-Rao Lower Bound of $\boldsymbol{z}_{i,n}$

2. System Model

The anchor nodes are always deployed at the same height, and high vertical dilution of precision means the system cannot provide reliable vertical positioning results [20], which is usually obtained by other sensors [21]. So, in this paper, a 2D dynamic wireless network is considered, which includes the anchor node with known location and synchronized local time, and the node with inaccurate location information and local time out of sync. The position vector of node i at time n is $\boldsymbol{x}_{i,n} = [x_{i,n}, y_{i,n}]^T$, and the measured value of local clock $\tilde{t}_{i,n} = t_i(T_n)$. The slope of local clock at time n between node i and the external standard clock is $a_{i,n} = (\tilde{t}_{i,n} - \tilde{t}_{i,n-1})/(T_n - T_{n-1})$. The local clock of all anchor nodes is synchronized with the external reference clock, i.e., $a_{i,n} = 1 \forall i \in S$. At time n, node i has neighbor nodes set as $N_{i,n}$, where neighbor anchor node set is $S_{i,n}$, node set to be located is $C_{i,n}$, and all communicable node pairs (i, j) constitute communicable node set Ψ_n.

Considering the local clock drift, node i measures time of arrival (TOA) from neighbor node j at time n as follows:

$$\tilde{\rho}_{ij,n} = \parallel \boldsymbol{x}_{i,n} - \boldsymbol{x}_{j,n} \parallel + \zeta_{ij,n} + cTa_{ij,n} + \omega_{ij,n}, \tag{1}$$

where $\parallel \boldsymbol{x}_{i,n} - \boldsymbol{x}_{j,n} \parallel$ is the Euclidean distance, $d_{ij,n}$, between nodes i, j, $T = T_n - T_{n-1}$, $\omega_{ij,n}$ is the observed measurement noise, assuming that it obeys the Gaussian distribution, i.e., $\omega_{ij,n} \sim \mathcal{N}(0, \sigma_d^2)$, and $\zeta_{ij,n}$ is NLOS error, which is expressed as follows:

$$\zeta_{ij,n} = \begin{cases} 0, (i, j) \notin \Theta_n \\ \lambda e^{-\lambda b_{ij,n}}, (i, j) \in \Theta_n \end{cases}, \tag{2}$$

where $b_{ij,n} > 0$ [22], λ is a constant and Θ_n is the set of all node pairs (i,j) with NLOS error at time n. Where, $a_{ij,n}$ is the relative slope of local clock offset between nodes i, j, defined as follows:

$$a_{ij,n} \triangleq \begin{cases} \frac{a_{i,n}}{a_{j,n}}, t_{i,n} > t_{j,n} \\ \frac{a_{j,n}}{a_{i,n}}, t_{i,n} > t_{j,n} \end{cases}, \tag{3}$$

Define $\boldsymbol{x}_n \triangleq [\boldsymbol{x}_{1,n}^T, \boldsymbol{x}_{2,n}^T, \ldots, \boldsymbol{x}_{S+C,n}^T]^T$ as the position vector of all nodes, $\boldsymbol{a}_n \triangleq [a_{1,n}, a_{2,n}, \ldots, a_{S+C,n}]^{\mathrm{T}}$ as the clock offset slope of all nodes, $\tilde{\rho}_n \triangleq [\ldots, \rho_{ij,n}, \ldots]^T$ as the range measurement in all node pairs $(i,j) \in \Psi_n$, and $\boldsymbol{\zeta}_n \triangleq [\ldots, \zeta_{ij,n}, \ldots]^T$ as the NLOS error in all node pairs $(i,j) \in \Theta_n$. Moreover, all the vector sets from time 1 to time n are defined as follows: $\boldsymbol{\mathcal{X}}_{1:n} \triangleq \{\boldsymbol{x}_1, \boldsymbol{x}_2, \ldots, \boldsymbol{x}_n\}$, $\boldsymbol{\mathcal{A}}_{1:n} \triangleq \{\boldsymbol{a}_1, \boldsymbol{a}_2, \ldots, \boldsymbol{a}_n\}$, $\boldsymbol{\mathcal{P}}_{1:n} \triangleq \{\rho_1, \rho_2, \ldots, \rho_n\}$, $\boldsymbol{\mathcal{B}}_{1:n} \triangleq \{\zeta_1, \zeta_2, \ldots, \zeta_n\}$. The goal of the algorithm is to estimate the accurate node location vector $\boldsymbol{x}_n$ and the clock slope $\boldsymbol{a}_n$ by measuring $\boldsymbol{\mathcal{P}}_{1:n}$ and the information transmitted between nodes.

3. M-VMP Joint Estimation Algorithm

According to whether the location information is considered as a random variable, the cooperative localization algorithms are divided into two categories: non-Bayesian estimation and Bayesian estimation [5]. In non-Bayesian estimation, the location information is estimated by deterministic methods. The typical algorithms are least square (LS) [23,24] and the maximum likelihood method (ML) [25]. Bayesian estimation is based on the probability model of location information. The typical algorithms include the maximum posterior probability (MAP) estimation [26] and the MMSE method [6]. Location information node i estimates by MMSE criteria. The expression is as follows:

$$\hat{\boldsymbol{z}}_{i,n} = \int \boldsymbol{z}_{i,n} p(\boldsymbol{z}_{i,n} | \boldsymbol{\mathcal{P}}_{1:n}) d\boldsymbol{z}_{i,n} \tag{4}$$

where $\hat{\boldsymbol{z}}_{i,n}$ is the estimation result and $\boldsymbol{z}_{i,n} = [x_{i,n}, y_{i,n}, a_{i,n}]^T$ is the vector to be estimated.

3.1. Probability Model

Assume $\boldsymbol{x}_{i,n}$ evolve according to a memory-less Gauss–Markov process, then, there are:

$$\boldsymbol{x}_{i,n} = \boldsymbol{x}_{i,n-1} + \boldsymbol{v}_{i,n} T + \omega_{i,n}, \tag{5}$$

where $\boldsymbol{v}_{i,n}$ is the average velocity of node i from time $n-1$ to time n, which is measured by the sensor inside nodes. $\omega_{i,n}$ is Gaussian white noise, and its covariance matrix is $diag\{\sigma_{i,n}^2, \sigma_{i,n}^2\}$. Since the motions of all nodes are independent of each other, there are:

$$p(\boldsymbol{x}_n | \boldsymbol{x}_{n-1}) = \prod_i p(\boldsymbol{x}_{i,n} | \boldsymbol{x}_{i,n-1}) \tag{6}$$

$$p(a_n | a_{n-1}) = \prod_i p(a_{i,n} | a_{i,n-1}), \tag{7}$$

$$p(\boldsymbol{\mathcal{X}}_{1:n}) = p(\boldsymbol{x}_0) \prod_n p(\boldsymbol{x}_n | \boldsymbol{x}_{n-1}), \tag{8}$$

$$p(\boldsymbol{\mathcal{A}}_{1:n}) = p(a_0) \prod_i p(a_n | a_{n-1}), \tag{9}$$

where $p(\boldsymbol{x}_0)$ is the prior distribution of all nodes' location information at time 0, which is obtained by GNSS, base station or in other ways [27]. According to the Bayesian rule, the edge probability function in (3) is calculated by the following formula:

$$p(\boldsymbol{\mathcal{X}}_{1:n}, \boldsymbol{\mathcal{A}}_{1:n} | \boldsymbol{\mathcal{P}}_{1:n}) \propto p(\boldsymbol{\mathcal{P}}_{1:n} | \boldsymbol{\mathcal{X}}_{1:n}, \boldsymbol{\mathcal{A}}_{1:n}) p(\boldsymbol{\mathcal{X}}_{1:n}, \boldsymbol{\mathcal{A}}_{1:n}) \tag{10}$$

The likelihood function $p(\mathcal{P}_{1:n}|\mathcal{X}_{1:n}, \mathcal{A}_{1:n})$ in (10) is decomposed into the following formula with the independent observations at different times $\rho_{ij,n}$:

$$p(\mathcal{P}_{1:n}|\mathcal{X}_{1:n}, \mathcal{A}_{1:n}) \propto \prod_n \prod_{(i,j)\notin\Theta_n} p_{LOS}(\rho_{ij,n}) \prod_{(i,j)\in\Theta_n} p_{NLOS}(\rho_{ij,n}), \tag{11}$$

$$p_{LOS}(\rho_{ij,n}) \propto \exp\left[-\frac{\left(\rho_{ij,n} - \| \boldsymbol{x}_{i,n} - \boldsymbol{x}_{j,n} \| - cTa_{ij,n}\right)^2}{2\sigma_d^2}\right], \tag{12}$$

$$p_{NLOS}(\rho_{ij,n}) \propto \exp\left[-\frac{\left(\rho_{ij,n} - \| \boldsymbol{x}_{i,n} - \boldsymbol{x}_{j,n} \| - cTa_{ij,n} - \zeta_{ij,n}\right)^2}{2\sigma_d^2}\right], \tag{13}$$

According to the above derivation, (10) is expanded to the following formula:

$$\begin{aligned} p(\mathcal{X}_{1:n}, \mathcal{A}\mathrm{R}_{1:n}|\mathcal{P}_{1:n}) \\ \propto \prod_n \prod_{(i,j)\notin\Theta_n} p_{LOS}(\rho_{ij,n}) \prod_{(i,j)\in\Theta_n} p_{NLOS}(\rho_{ij,n}) \\ \cdot \prod_n p(x_{0,n}) p(a_{0,n}) \prod_i p(x_{i,n}|x_{i,n-1}) p(a_{i,n}|a_{i,n-1}), \end{aligned} \tag{14}$$

3.2. Factor Graph Model

Factor graphs intuitively reflect the spatiotemporal relationship between variables [5]. The factor graph is a dichotomous graph, which contains two kinds of nodes: factor node and variable node. Variable node represents the information to be evaluated, and factor node represents the messages passed between variable nodes. The schematic diagram of cooperative localization is shown in Figure 1.

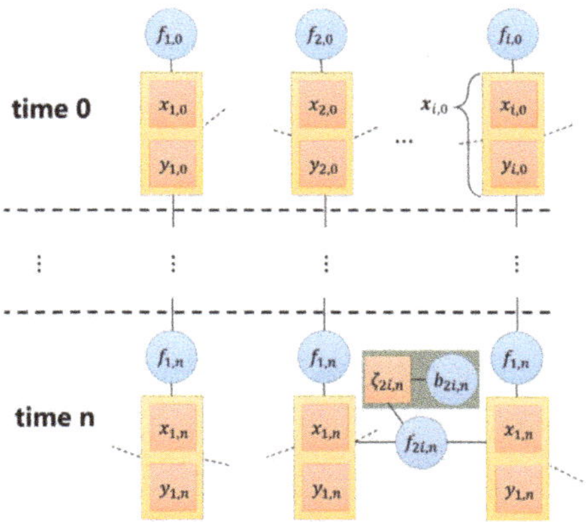

Figure 1. Factor graph model of variational message passing (VMP)-based cooperative localization problem. The circle represents the factor node and the square represents the variable node. $x_{i,n} = [x_{i,n}\ y_{i,n}]^T$ represents the localization variable of the node to be located, $f_{i,n}$ represents the message that transmits between different times, $f_{ij,n}$ represents the distance information that the nodes transmit, $\zeta_{ij,n}$ represents the non-sight distance parameter that affects the information between nodes and $b_{ij,n}$ represents the non-sight distance error probability function that affects the non-sight distance parameter.

Factor nodes are further assumed as follows:

$$f_{i,0} = p(\boldsymbol{z}_{i,0}), f_{i,n} = p(\boldsymbol{z}_{i,n}|\boldsymbol{z}_{i,n-1}), \tag{15}$$

$$f_{ij,n} = \begin{cases} p_{LOS}(\rho_{ij,n}), (i,j) \notin \Theta_n \\ p_{NLOS}(\rho_{ij,n}), (i,j) \in \Theta_n \end{cases}, \tag{16}$$

$$\zeta_{ij,n} = p(b_{ij,n}), \tag{17}$$

According to the above definition, edge probability of localization is clearly calculated in the factor graph. Next, a localization algorithm based on the VMP will be introduced.

3.3. VMP-Based Localization Algorithm

VMP uses the exponential model to deliver messages, which greatly reduces communication consumption. According to the VMP message rules based on the factor graph proposed in Reference [28] and the assumptions in Section 3.2, the message delivered from factor node to parameter node at time n is as follows:

$$\mu_{f_{ij,n}\to \boldsymbol{z}_{i,n}}(\boldsymbol{z}_{i,n}) = \exp\left\{\int \mu_{b_{ij,n}\to f_{ij,n}}(b_{ij,n})\mu_{\boldsymbol{z}_{j,n}\to f_{ij,n}}(\boldsymbol{z}_{j,n}) \ln f_{ij,n}(\boldsymbol{z}_{i,n}, \boldsymbol{z}_{j,n}) d\boldsymbol{z}_{j,n}\right\}, \tag{18}$$

$$\mu_{f_{i,n}\to \boldsymbol{z}_{i,n}}(\boldsymbol{z}_{i,n}) = p(\boldsymbol{z}_{i,n}|\boldsymbol{z}_{i,n-1}), \tag{19}$$

$$\mu_{f_{ij,n}\to b_{ij,n}}(b_{ij,n}) = \exp\left\{\int \mu_{\boldsymbol{z}_{i,n}\to f_{ij,n}}(\boldsymbol{z}_{i,n})\mu_{\boldsymbol{z}_{j,n}\to f_{ij,n}}(\boldsymbol{z}_{j,n}) \ln f_{ij,n}(\boldsymbol{z}_{i,n}, \boldsymbol{z}_{j,n}) d\boldsymbol{z}_{j,n}\right\}, \tag{20}$$

where the cooperative messages $\mu_{\boldsymbol{z}_{j,n}\to f_{ij,n}}(\boldsymbol{z}_{j,n})$ are divided into two categories: one is related to the neighbor anchor node $j \in S_{i,n}$, and the other is related to the neighbor node $k \in C_{i,n}$ to be located. The calculations are performed below:

$$\mu_{f_{ij,n}\to \boldsymbol{z}_{i,n}}(\boldsymbol{z}_{i,n}) \propto \mathcal{N}(\rho_{ij,n} \| \| x_{i,n} - x_{j,n} \|, \sigma_d^2) \tag{21}$$

$$\mu_{f_{ij,n}\to \boldsymbol{z}_{i,n}}(\boldsymbol{z}_{i,n}) = \exp\{\mathbb{E}_{b(x_k)}[\mathcal{N}(\rho_{ik,n} \| \| x_{i,n} - x_{k,n} \|, \sigma_d^2)]\}, \tag{22}$$

where $\mathbb{E}_{f(\cdot)}[g(\cdot)]$ means the mean of $g(\cdot)$ with respect to $f(\cdot)$. The belief of $\boldsymbol{z}_{i,n}$ is obtained as follows:

$$\begin{aligned} b(\boldsymbol{z}_{i,n}) &= \tfrac{1}{Z}\mu_{f_{i,n}\to \boldsymbol{z}_{i,n}}(\boldsymbol{z}_{i,n}) \prod_{j\in S_{i,n}} \mu_{f_{ij,n}\to \boldsymbol{z}_{i,n}}(\boldsymbol{z}_{i,n}) \prod_{k\in C_{i,n}} \mu_{f_{ik,n}\to \boldsymbol{z}_{i,n}}(\boldsymbol{z}_{i,n}) \\ &\propto \tfrac{1}{Z}\mathcal{N}(\boldsymbol{z}_{i,n}|\boldsymbol{z}_{i,n-1} + v_{i,n}T, \sigma_{i,n}^2) \prod_{j\in S_{i,n}} \mathcal{N}(\rho_{ij,n} \| \| x_{i,n} - x_{j,n} \|, \sigma_d^2) \\ &\times \prod_{k\in C_{i,n}} \exp\{\mathbb{E}_{b(x_k)}[\mathcal{N}(\rho_{ik,n} \| \| x_{i,n} - x_{k,n} \|, \sigma_d^2)]\} \end{aligned} \tag{23}$$

where Z is the normalization constant. Obviously, it can be seen that $b(\boldsymbol{z}_{i,n})$ is not a Gaussian function about $\boldsymbol{z}_{i,n}$, and it is difficult to transmit directly between nodes. To reduce communication overhead, Reference [19] approximates $b(\boldsymbol{z}_{i,n})$ by minimizing KL divergence: where $\mathcal{G}$ represents a Gaussian function, and $\mathrm{KLD}(q(\boldsymbol{z}_{i,n}) \| b(\boldsymbol{z}_{i,n}))$ is the KL divergence between Gaussian function $q(\boldsymbol{z}_{i,n})$ and belief $b(\boldsymbol{z}_{i,n})$, the calculation formula is as follows:

$$b(\boldsymbol{z}_{i,n}) = \underset{q(\boldsymbol{z}_{i,n})\in\mathcal{G}}{\mathbf{argmin}} \mathrm{KLD}(q^{(m)}(\boldsymbol{z}_{i,n}) | b^{(m)}(\boldsymbol{z}_{i,n})) \tag{24}$$

$$\begin{aligned}
&\mathrm{KLD}(q(\boldsymbol{z}_{i,n})|b(\boldsymbol{z}_{i,n})) \\
&= \frac{\|\boldsymbol{x}_{i,n}-\boldsymbol{x}_{i,0}\|^2+2\omega_{i,n}^2}{2\omega_{i,0}^2} - \ln \omega_{i,n}^2 \\
&+ \sum\nolimits_{j\in S_{i,n}} \ln \frac{-2\left(\tilde{\rho}_{ij,n}-cTa_{ij,n}\right)\sqrt{\frac{\pi}{2}-{}_{i,n}^{2}}F\left(-\frac{1}{2},1,\frac{\|\boldsymbol{x}_{i,n}-\boldsymbol{x}_{i,0}\|^2}{2\sigma_d^2}\right)+\|\boldsymbol{x}_{i,n}-\boldsymbol{x}_{j,n}\|+2\sigma_d^2}{2\sigma_d^2} \\
&+ \sum\nolimits_{k\in C_{i,n}} \ln \frac{-2\left(\tilde{\rho}_{ik,n}-cTa_{ik,n}\right)\sqrt{\frac{\pi}{2}-{}_{i,n}^{2}}F\left(-\frac{1}{2},1,\frac{\|\boldsymbol{x}_{i,n}-\boldsymbol{x}_{i,0}\|^2}{2\sigma_d^2}\right)+\|x_{i,n}-x_{k,n}\|+2\sigma_d^2}{2\sigma_d^2}
\end{aligned} \tag{25}$$

3.4. M-VMP-Based Joint Estimation Algorithm

In Section 3.3, the clock drift of node or model loss with single Gaussian distribution are not considered, which affects localization accuracy [19]. In this paper, a joint estimation algorithm of time synchronization and localization based on multi-Gaussian distribution VMP is proposed, that the belief $b(\boldsymbol{z}_{i,n})$ of the variable $\boldsymbol{z}_{i,n}$ to be estimated is approximately a multi-Gaussian function $\hat{b}(\boldsymbol{z}_{i,n}) \propto \sum \eta_{M,ij}\mathcal{N}(\boldsymbol{z}_{k.i,n}|E_{\boldsymbol{z}_{i,n}}, V_{\boldsymbol{z}_{i,n}})$, and the belief of $\zeta_{ij,n}$ is $\hat{b}(\zeta_{ij,n}) \propto \mathcal{N}(\zeta_{ij,n}|\tilde{\rho}_{ij,n} - cTa_{ij,n} - \| \boldsymbol{x}_{i,n} - \boldsymbol{x}_{j,n} \|, \sigma_d^2)$. The factor graph between nodes i and j at the time n is shown in Figure 2.

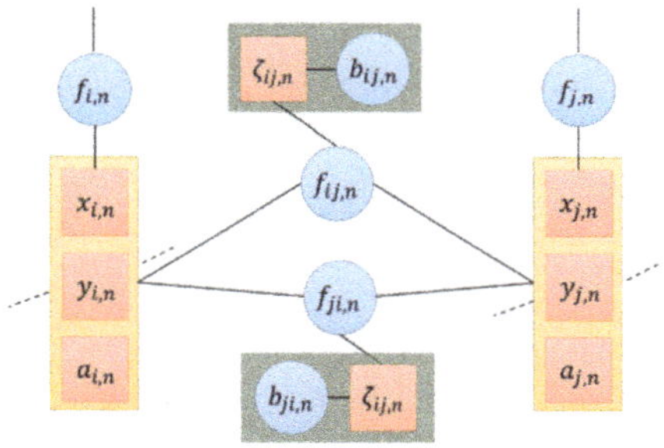

Figure 2. FG of a node pair (i, j) at time n. $x_{i,n}, y_{i,n}, a_{i,n}$ represent the localization and time synchronization parameter of the node to be estimated, $f_{i,n}$ represents the message that transmits between different times, $f_{ij,n}$ represents the distance information that the nodes transmit, $b_{ij,n}$ represents the non-sight distance parameter that affects the information between nodes and $\zeta_{ij,n}$ represents the non-sight distance error probability function that affects the non-sight distance parameter.

In order to approximate $b(\boldsymbol{z}_{i,n})$ to a Gaussian function, Formula (23) is rewritten as follows:

$$b(\boldsymbol{z}_{i,n}) \propto \left[f_{i,n}(\boldsymbol{z}_{i,n}|\boldsymbol{z}_{i,n-1}) + \sum_{j\in S_{i,n}} f_j(\boldsymbol{z}_{i,n}) + \sum_{k\in C_{i,n}} f_k(\boldsymbol{z}_{i,n}) \right], \tag{26}$$

where:

$$f_{i,n}(\boldsymbol{z}_{i,n}|\boldsymbol{z}_{i,n-1}) \triangleq \sum_{M} \frac{-\eta_{M,i}}{2}\left(\boldsymbol{z}_{i,n} - E_{\boldsymbol{z}_{i,n-1,M}}\right)^T V_{i,n-1,M}^{-1}\left(\boldsymbol{z}_{i,n} - E_{\boldsymbol{z}_{i,n-1,M}}\right) \tag{27}$$

$$f_j(\boldsymbol{z}_{i,n}) \triangleq \sum_{M} -\eta_{M,ij} \frac{\left(\tilde{\rho}_{ij,n} - \| \boldsymbol{x}_{i,n} - \boldsymbol{x}_j \| - cTE_{a_{ij,n-1,M}} - b_{ij,n}\right)^2}{2\sigma^2_{\boldsymbol{z}_{ij,n}}}, \tag{28}$$

$$f_k(\boldsymbol{z}_{i,n}) \triangleq \sum_{M} -\eta_{M,ik} \int b\left(\boldsymbol{z}_{k,n}\right) \frac{\left(\tilde{\rho}_{ij,n} - \| \boldsymbol{x}_{i,n} - \boldsymbol{x}_{k,n} \| - cTE_{a_{ik,n-1,M}} - b_{ik,n}\right)^2}{2\sigma^2_{\boldsymbol{z}_{ik,n}}} d\boldsymbol{z}_{k,n}, \tag{29}$$

Because the existence of two non-linear terms, $\| \boldsymbol{x}_{i,n} - \boldsymbol{x}_j \|$ in $f_j(\boldsymbol{z}_{i,n})$ and $\| \boldsymbol{x}_{i,n} - E_{\boldsymbol{x}_{k,n-1,M}} \|$ in $f_k(\boldsymbol{z}_{i,n})$, results in non-Gaussian of $b(\boldsymbol{z}_{i,n})$, the two non-linear terms are expanded by a second-order Taylor expansion to obtain the mean and covariance of $b(\mathbf{z}_{i,n})$. For discussion purposes, define $g_j(\mathbf{z}_{i,n}) \triangleq \| \mathbf{x}_{i,n} - \mathbf{x}_j \| - cTE_{a_{ij,n-1,M}}$ and $g_k(\mathbf{z}_{i,n}, \mathbf{z}_{k,n}) \triangleq \| \mathbf{x}_{i,n} - \mathbf{x}_{k,n} \| - cTE_{a_{ik,n-1,M}}$.

a. Second-order Taylor series expansion for nonlinear term $g_j(\mathbf{z}_{i,n})$:$g_j(\mathbf{z}_{i,n})$ is expanded at $\mathbf{z}_{i,n} = E_{\mathbf{z}_{i,n-1}}$ by a second-order Taylor to get:

$$\begin{aligned} g_j(\mathbf{z}_{i,n}) &= \| \mathbf{x}_{i,n} - \mathbf{x}_j \| - cTE_{a_{ij,n-1,M}} \\ &\approx g_j\left(E_{\mathbf{z}_{i,n-1}}\right) + \nabla^T g_j\left(E_{\mathbf{z}_{i,n-1}}\right)\left(\mathbf{z}_{i,n} - E_{\mathbf{z}_{i,n-1}}\right) \\ &+\tfrac{1}{2}\left(\mathbf{z}_{i,n} - E_{\mathbf{z}_{i,n-1}}\right)^T \nabla^2 g_j\left(E_{\mathbf{z}_{i,n-1}}\right)\left(\mathbf{z}_{i,n} - E_{\mathbf{z}_{i,n-1}}\right), \end{aligned} \tag{30}$$

where $\nabla g_j\left(E_{\mathbf{z}_{i,n-1}}\right)$ and $\nabla^2 g_j\left(E_{\mathbf{z}_{i,n-1}}\right)$ are the first and second steps of $g_j(\mathbf{z}_{i,n})$ at $E_{\mathbf{z}_{i,n-1}}$.

b. Second-order Taylor series expansion for nonlinear term $g_k\left(\mathbf{z}_{i,n}, \mathbf{z}_{k,n}\right)$: $g_k\left(\mathbf{z}_{i,n}, \mathbf{z}_{k,n}\right)$ is expanded at $\mathbf{z}_{i,n} = E_{\mathbf{z}_{i,n-1}}$ and $\mathbf{z}_{k,n} = E_{\mathbf{z}_{k,n-1}}$ by a second-order Taylor to get:

$$\begin{aligned} g_k\left(\mathbf{z}_{i,n}, \mathbf{z}_{k,n}\right) &= \| \mathbf{x}_{i,n} - \mathbf{x}_{k,n} \| - cTE_{a_{ik,n-1,M}} \\ &\approx g_k\left(E_{\mathbf{z}_{i,n-1}}, E_{\mathbf{z}_{k,n-1}}\right) + \frac{\partial g_k\left(E_{\mathbf{z}_{i,n-1},n}\mathbf{z}_{k,n-1}\right)}{\partial \mathbf{z}_{i,n}}\left(\mathbf{z}_{i,n} - E_{\mathbf{z}_{i,n-1}}\right) \\ &+\frac{\partial^T g_k\left(E_{\mathbf{z}_{i,n-1}}, E_{\mathbf{z}_{k,n-1}}\right)}{\partial \mathbf{z}_{k,n}}\left(\mathbf{z}_{k,n} - E_{\mathbf{z}_{k,n-1}}\right) \\ &+\frac{1}{2}\begin{pmatrix} \mathbf{z}_{i,n} - E_{\mathbf{z}_{i,n-1}} \\ \mathbf{z}_{k,n} - E_{\mathbf{z}_{k,n-1}} \end{pmatrix}^T \nabla^2 g_k\left(E_{\mathbf{z}_{i,n-1}}, E_{\mathbf{z}_{k,n-1}}\right)\begin{pmatrix} \mathbf{z}_{i,n} - E_{\mathbf{z}_{i,n-1}} \\ \mathbf{z}_{k,n} - E_{\mathbf{z}_{k,n-1}} \end{pmatrix}, \end{aligned} \tag{31}$$

where $\frac{\partial g_k\left(E_{\mathbf{z}_{i,n-1}}, E_{\mathbf{z}_{k,n-1}}\right)}{\partial \mathbf{z}_{i,n}}$ and $\frac{\partial g_k\left(E_{\mathbf{z}_{i,n-1}}, E_{\mathbf{z}_{k,n-1}}\right)}{\partial \mathbf{z}_{k,n}}$ are the first-order partial derivatives of $g_k\left(\mathbf{z}_{i,n}, \mathbf{z}_{k,n}\right)$ with respect to $\mathbf{z}_{i,n}$ and $\mathbf{z}_{k,n}$ at $\mathbf{z}_{i,n} = E_{\mathbf{z}_{i,n-1}}$, and $\mathbf{z}_{k,n} = E_{\mathbf{z}_{k,n-1}}$, $\nabla^2 g_k\left(E_{\mathbf{z}_{i,n-1}}, E_{\mathbf{z}_{k,n-1}}\right)$ is the Hessian matrix of $g_k\left(\mathbf{z}_{i,n}, \mathbf{z}_{k,n}\right)$ at $\mathbf{z}_{i,n} = E_{\mathbf{z}_{i,n-1}}$ and $\mathbf{z}_{k,n} = E_{\mathbf{z}_{k,n-1}}$.

After calculation, we can get the mean $E_{\mathbf{z}_{i,n}}$ and covariance $V_{\mathbf{z}_{i,n}}$ of $b(\mathbf{z}_{i,n})$:

$$\begin{aligned} V_{z_{i,n}} = \Big[V_{z_{i,n-1}}^{-1} &+ \sum\nolimits_{j\in S_{i,n}} \sum\nolimits_{M_1} \frac{\eta_{M_1,ij}}{\sigma_d^2}\left(I - \left(\tilde{\rho}_{ij,n} - cTE_{a_{ij,n-1}} - \zeta_{ij,n}\right)\nabla^2 g_j\left(E_{z_{i,n-1}}\right)\right) \\ &+ \sum\nolimits_{k\in C_{i,n}} \sum\nolimits_{M_2} \frac{\eta_{M_2,ik}}{\sigma_d^2}\big(I \\ &-\left(\tilde{\rho}_{ik,n} - cTE_{a_{ik,n-1}} - \zeta_{ik,n}\right)\nabla^2 g_k\left(E_{z_{i,n-1}}, E_{z_{k,n-1}}\right)\big)\Big]^{-1}, \end{aligned} \tag{32}$$

$$\begin{aligned} \hat{E}_{z_{i,n}} = V_{z_{i,n}}\Big\{ &V_{z_{i,n-1}}^{-1} E_{z_{i,n-1}} \\ &+\sum\nolimits_{j\in S_{i,n}} \sum\nolimits_{M_1} \frac{\eta_{M_1,ij}}{\sigma_d^2}\big(\mathbf{x}_j \\ &-\left(\tilde{\rho}_{ij,n} - cTE_{a_{ij,n-1}} - \zeta_{ij,n}\right)\left(\nabla g_j\left(E_{z_{i,n-1}}\right) - \nabla^2 g_j\left(E_{z_{i,n-1}}\right)E_{z_{i,n-1}}\right)\big) \\ &+\sum\nolimits_{k\in C_{i,n}} \sum\nolimits_{M_2} \frac{\eta_{M_2,ik}}{\sigma_d^2}\big(\mathbf{x}_k \\ &-\left(\tilde{\rho}_{ik,n} - cTE_{a_{ik,n-1}} - \zeta_{ik,n}\right)\big(\nabla g_j\left(E_{z_{i,n-1}}\right) \\ &-\nabla^2 g_k\left(E_{z_{i,n-1}}, E_{z_{k,n-1}}\right)E_{z_{i,n-1}}\big)\big)\Big\}, \end{aligned} \tag{33}$$

where $\eta_{M,ij} = Z/\left(\rho_{ij}\cdot tr\left(V_{\boldsymbol{z}_{j,n-1}}\right)\right)$, and Z is the normalization constant. Compared with the approximation method based on KL divergence minimization, the proposed method based on multi-Gaussian approximation with second-order Taylor series expansion greatly reduces the complexity of approximate calculation. When $b(\boldsymbol{z}_{i,n})$ is approximated to a Gaussian function $\hat{b}(\boldsymbol{z}_{i,n})$, each node only needs to send its own position vector and covariance matrix to its neighbor nodes, and the communication overhead is much lower than that of the particle message-based method. In addition, since each node has three parameters to be estimated, the network is required to include at least three anchor nodes. The flow of the M-VMP localization algorithm (Algorithm 1) is as follows:

Algorithm 1: M-VMP joint estimation algorithm

Initialization:
  Initialize node location distribution
  $f_{i,0}p\left(\boldsymbol{z}_{i,0}\right)$
Location estimation:
for n = 1 to N (time index) do
  Nodes $i \in C$
    1) Compute covariance $V_{\boldsymbol{z}_{i,n}}$ of belief $\hat{b}\left(\boldsymbol{z}_{i,n}\right)$ according to (32)
    2) Compute mean $\hat{E}_{z_{i,n}}$ of belief $\hat{b}\left(\boldsymbol{z}_{i,n}\right)$ according to (33)
    3) Broadcast means and variances of belief $\hat{b}\left(\boldsymbol{z}_{i,n}\right)$, and receive the mean and covariance matrix of location variables of neighbor nodes
    4) Estimate location information of node i according to MMSE principle
    5) Update location information and form messages $\mu_{\boldsymbol{z}_{i,n}\to f_{i,n+1}}$
  end parallel
end for

4. Simulation Analysis and Test Results

In this section, Cramer-Rao Lower Bound (CRLB) of $\boldsymbol{z}_{i,n}$ is derived first, then the performance of the proposed joint algorithms is analyzed through simulation.

4.1. CRLB Lower Bound of $z_{i,n}$

For node i, the Fisher Information Matrix $\boldsymbol{F}(\boldsymbol{z}_{i,n})$ can be derived as:

$$\boldsymbol{F}(\boldsymbol{z}_{i,n}) = \mathbf{A}_i diag\left(\ldots,\sigma_d^2,\ldots\right)\mathbf{A}_i^{\mathrm{T}}, \tag{34}$$

with

$$\mathbf{A}_{i,n} = \left[\ldots,\boldsymbol{u}_{ij,n},\ldots\right], j \in \mathbf{N}_{i,n} \tag{35}$$

$$\boldsymbol{u}_{ij,n} = \frac{1}{\rho_{ij,n}}\begin{bmatrix} x_{i,n}-x_{j,n} \\ y_{i,n}-y_{j,n} \\ cTa_{ij,n} \end{bmatrix}, \tag{36}$$

and CRLB of $\boldsymbol{z}_{i,n}$ is $\mathrm{CRLB}(\boldsymbol{z}_{i,n}) = \boldsymbol{F}^{-1}(\boldsymbol{z}_{i,n})$.

4.2. Simulation Scenario and Result Analysis

To analyze the performance of the proposed method, we have built a simulation scenario according to the real scene of the zone 1, underground parking lot, Beijing University of Posts and Telecommunications. The real scene is shown in Figure 3a and the top view is shown in Figure 3b.

According to the actual size of zone 1, the simulation scene is a rectangular area of 20 × 24 m. By default, the number of anchor nodes is 4, the number of nodes to be located is 20 and the initial position distribution of nodes in the site conforms to the uniform distribution. In order to limit the communication area of nodes to less than 1/2 of the whole simulation area, the maximum communication distance is set to 10 m. The performance of the proposed method is obtained in the line of sight (LOS) environment, except the last simulation, which shows localization accuracy with different NLOS probability. All simulation results are the average of 1000 independent runs.

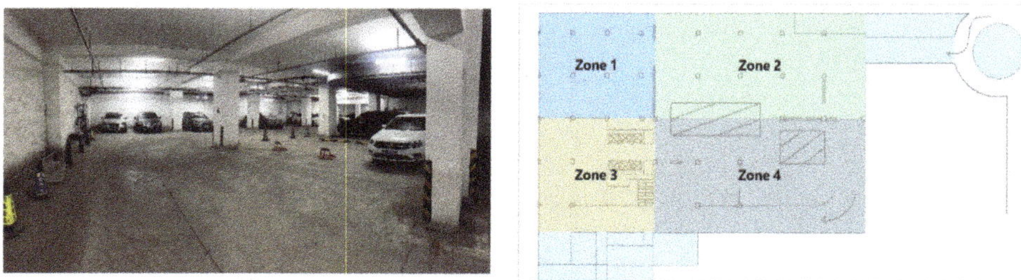

(a) (b)

Figure 3. (**a**) The real scene of the underground parking lot. (**b**) Simulation scenario. Zone 1 is the simulation area.

The initial position measurement error of the anchor node conforms to the Gaussian distribution with the standard deviation of 0.1 m. The initial position measurement error distribution of the node to be located is composed of Gaussian distributions with the standard deviation of 10 m. According to the performance of the crystal oscillator used in the hardware (TG5032CFN), the clock drift of the node to be located conforms to uniform distribution from [−1 ppm, 1 ppm], that is, the maximum distance measurement error caused by the clock drift between adjacent time (1 s) is 300 m. In the process of simulation, anchor nodes remain stationary, and the velocity of the node to be located is consistent with the uniform distribution of the maximum value of 3 m/s, and the direction is random every time step. The position movement measurement of the node conforms to the Gaussian distribution with the standard deviation of 1 m. The default process is 15 s, 10 iterations per second. Root mean square error (RMSE) and cumulative distribution functions (CDF) are used to measure the performance of algorithms.

In Figure 4a, the relationship between localization RMSE and initial location error is shown. The CLOC method [16] treats the uncertainties of nodes' positions and clock offsets as measurement noise and thus suffers performance degradation. The SPAWN [26] method is implemented by using 4000 particles to represent the messages on FG. The localization accuracy of the VMP method [17] and the proposed M-VMP method are better than the CLOC method, but slightly worse than the SPAWN method.

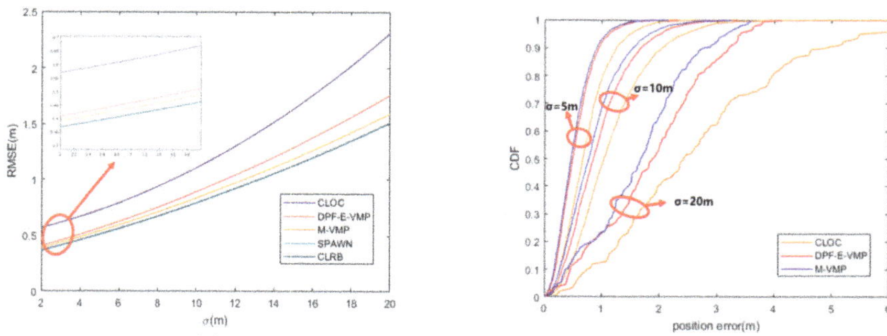

(a) (b)

Figure 4. (**a**) Root mean square error (RMSE) of position error with different initial location error. (**b**) Cumulative distribution function (CDF) of position error with different initial location error.

The position error of VMP and the proposed M-VMP method are compared in Figure 4b. The cumulative distribution function of the VMP method and the proposed method under different initial position errors is given in Figure 4b. It is found that the probability of position error less than 1.6 m of the two methods under different initial position errors is basically the same, while the proposed method has a better effect of restraining position error. The result is consistent with the conclusion in Figure 4a.

Figure 5 shows the relationship among the clock drift slope, RMSE of localization accuracy and the number of iterations, respectively. It is seen that three methods converge in the finite iterations, which proves that the algorithms are feasible. Compared with the VMP method [17], the proposed method improves the convergence speed of localization and time synchronization by adding time synchronization parameters into the localization process.

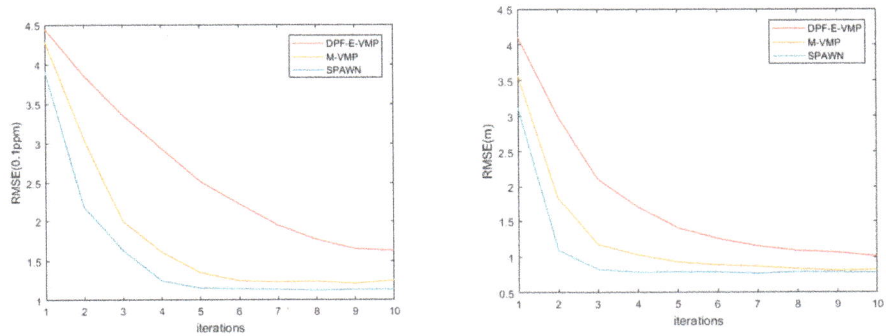

(**a**) (**b**)

Figure 5. (**a**) RMSE of the clock drift slope versus iterations and (**b**) RMSE of position error versus iterations.

In Figure 6, the CDF of the proposed method under different values of the number of Gaussian distribution M are compared. It is seen that the localization accuracy (3σ) of the proposed method is better than that of the VMP method when M is greater than 2. With the increase of M, the localization accuracy is also improved.

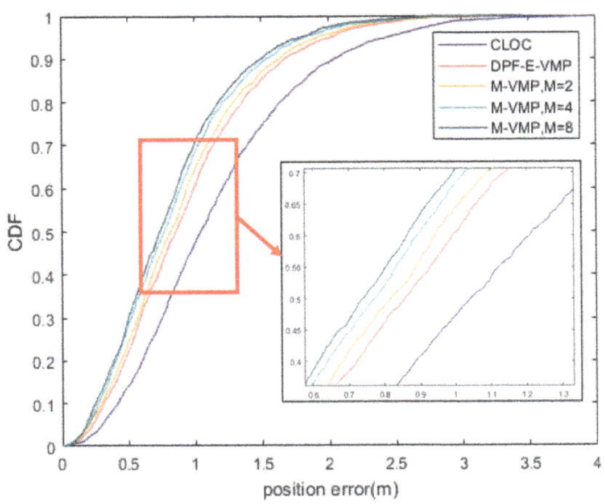

Figure 6. CDF of position error with different values of the number of Gaussian distribution M in line of sight (LOS) environment.

Figure 7 reflects the influence of communication distance on the localization accuracy of the proposed method under the same node density. Obviously, localization accuracy increases with the increase of communication distance, which increases the connectivity of nodes and the redundancy of information. At the same time, it can be seen that the improvement of localization accuracy is not obvious when the communication distance is 20 and 30 m. The difference of node connectivity between those two conditions is little, so the localization accuracy cannot be greatly improved. Because the communication distance is proportional to the transmission power of the node, it is necessary to determine the appropriate communication distance in practical application by considering the localization accuracy, node power consumption and other indicators.

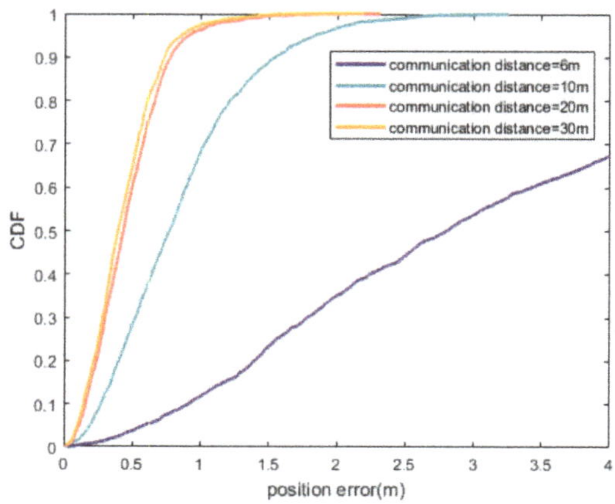

Figure 7. CDF of position error with different communication distance.

In the simulation of Figure 8, in order to amplify the influence of node density on positioning accuracy, we reset some of the simulation parameters, and set the communication distance to 15 m and the variance of the distance observation and the initial position error of the node to be located to 20 m^2. It can be seen from Figure 8 that as the density of nodes increases, the overall positioning accuracy is decreasing. However, when the density of nodes to be located is too large, the positioning error will increase. At the same time, when there are more nodes to be located, the speed at which the positioning error decreases with more anchor nodes also decreases. This is due to the increase in the density of nodes to be located, and the overall weight of the information contained in neighboring nodes to be located in the proposed method becomes larger. When the number of anchor nodes is insufficient, the increase of neighboring nodes to be located helps to improve the positioning accuracy, but the effect is limited.

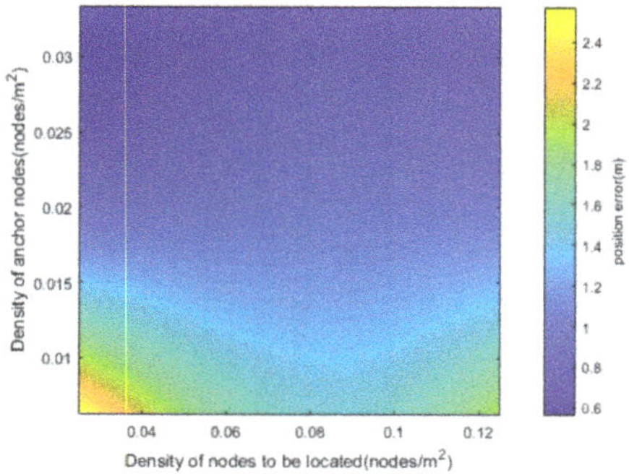

Figure 8. Relationship between position error and nodes' density.

In Figure 9, the relationship between the RMSE of localization accuracy and NLOS occurrence probability of the VMP method and the proposed method is compared. Furthermore, it compares with the method that only deals with LOS. It is seen that both the VMP and the proposed method effectively suppress NLOS error. Because the two-way NLOS parameters between nodes are added in the proposed method, the NLOS error is better suppressed.

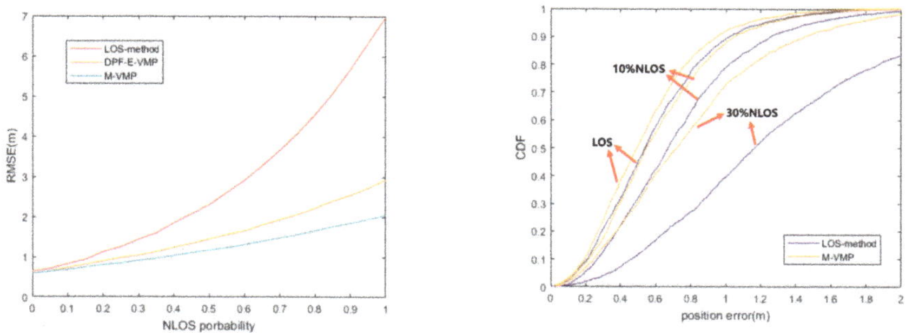

(a) **(b)**

Figure 9. (**a**) RMSE of position accuracy with different NLOS probability. (**b**) CDF of position error with different NLOS probability.

4.3. Computational Complexity and Communication Overhead Analysis

Because the proposed method in this paper and the comparison methods are distributed algorithms, the computation and communication are done by the nodes independently, so only one node's time complexity and communication overhead need to be considered in the analysis.

The time complexity is evaluated by the number of operations in local computing, and the communication cost is evaluated by the number of information parameters broadcast by nodes. The time complexity of the CLOC method is related to its neighbor nodes number N_{inb}, and its communication cost is $3{\cdot}O(1)$. The SPAWN method is based on particles. When the number of particles in each message is N_p, its time complexity is $O\left(N_p + N_p^2{\cdot}N_{inb}\right)$, and its communication cost is $O\left(N_p\right) + 2{\cdot}O(1)$. Compared with the VMP method, the number of Gaussian messages in the proposed method is M times, and every message from neighbor nodes needs to be processed once. The time complexity of dimension reduction depends on N_{inb}, so its time complexity is $O(M{\cdot}N_{inb}) + O(N_{inb})$. In the proposed method, nodes transmit a localization vector including $x_{i,n}, y_{i,n}, a_{i,n}$ and a covariance matrix, so the communication cost is $3O(M) + O(1)$. The computational complexity, run-time and communication Overhead of the three algorithms is shown in Table 2.

Table 2. Comparisons of different methods for each node at each iteration.

Method	Computational Complexity	Run-Time	Communication Overhead
M-VMP (proposed method)	$O(M{\cdot}N_{inb}) + O(N_{inb})$	84.449 ms	$3{\cdot}O(M) + O(1)$
CLOC	$O(N_{inb}^2)$	288.500 ms	$3{\cdot}O(1)$
SPAWN	$O(N_p + N_p^2{\cdot}N_{inb})$	5469.089 ms	$O(N_p) + 2{\cdot}O(1)$

4.4. Future Research Directions

In the next stage of research, we will mainly focus on the following aspects: First, analyze the performance of the positioning and simultaneous joint estimation problem and its influencing factors in principle. Secondly, expand the positioning scene from 2 dimensions to 3 dimensions, and reduce the problem of positioning accuracy degradation in dense scenes with nodes to be located through methods such as signal quality screening. Finally, the proposed algorithm will be implemented based on the hardware platform, and the measured results will be compared with the simulation results to further improve the performance of the proposed method.

5. Conclusions

This paper presented a M-VMP-based TOA localization and time synchronization joint estimation algorithm for mixed LOS and NLOS environments. Firstly, according to the VMP method, a message propagation model based on a factor graph was constructed for localization and time synchronization. Owing to the existence of nonlinear terms, it is difficult to represent the message in a closed form, so Taylor expansion was used for the linearization of nonlinear terms. Moreover, to reduce the communication cost, all messages were expressed in the form of multi-Gaussian distribution, and only the mean value, covariance and weight of each Gaussian distribution need to be transferred in the message transmission. The simulation results showed that the accuracy and convergence speed of the proposed method were close to that of SPAWN method, and the time complexity and communication cost were greatly reduced: the run-time was only 1.5% that of SPAWN.

Author Contributions: Conceptualization, S.T.; Data curation, S.T.; Formal analysis, S.T., B.J. and H.W.; Funding acquisition, Z.D. and S.T.; Investigation, S.T., B.J., H.W. and X.D.; Project administration, Z.D., S.T., B.J., H.W. and X.Z.; Resources, Z.D. and S.T.; Software, S.T., B.J., H.W. and X.D.; Supervision, B.J., H.W. and X.Z.; Validation, S.T. and X.D.; Visualization, S.T.; Writing—original draft, S.T.; Writing—review and editing, S.T., B.J., H.W., X.D. and X.Z. All authors have read and agreed to the published version of the manuscript.

Funding: This research was funded by the National Key Research & Development Program: 2016YFB0502003.

Conflicts of Interest: The authors declare no conflict of interest.

References

1. Amin, R.; Biswas, G. A secure light weight scheme for user authentication and key agreement in multi-gateway based wireless sensor networks. *Ad Hoc Netw.* **2016**, *36*, 58–80. [CrossRef]
2. Yassin, A.; Nasser, Y.; Awad, M.; Al-Dubai, A.; Liu, R.; Yuen, C.; Raulefs, R.; Aboutanios, E. Recent advances in indoor localization: A survey on theoretical approaches and applications. *IEEE Commun. Surv. Tutor.* **2016**, *19*, 1327–1346. [CrossRef]
3. Han, G.; Xu, H.; Duong, T.Q.; Jiang, J.; Hara, T. Localization algorithms of wireless sensor networks: A survey. *Telecommun. Syst.* **2013**, *52*, 2419–2436. [CrossRef]
4. Han, G.; Jiang, J.; Zhang, C.; Duong, T.Q.; Guizani, M.; Karagiannidis, G.K. A survey on mobile anchor node assisted localization in wireless sensor networks. *IEEE Commun. Surv. Tutor.* **2016**, *18*, 2220–2243. [CrossRef]
5. Wymeersch, H.; Lien, J.; Win, M.Z. Cooperative localization in wireless networks. *Proc. IEEE* **2009**, *97*, 427–450. [CrossRef]
6. Shang, Y.; Rumi, W.; Zhang, Y.; Fromherz, M. Localization from connectivity in sensor networks. *IEEE Trans. Parallel Distrib. Syst.* **2004**, *15*, 961–974. [CrossRef]
7. Sert, S.A.; Bagci, H.; Yazici, A. MOFCA: Multi-objective fuzzy clustering algorithm for wireless sensor networks. *Appl. Soft Comput.* **2015**, *30*, 151–165. [CrossRef]
8. Ghari, P.M.; Shahbazian, R.; Ghorashi, S.A. Wireless sensor network localization in harsh environments using SDP relaxation. *IEEE Commun. Lett.* **2015**, *20*, 137–140. [CrossRef]
9. Doherty, L.; El Ghaoui, L. Convex position estimation in wireless sensor networks. In Proceedings of the IEEE INFOCOM 2001. Conference on Computer Communications. Twentieth Annual Joint Conference of the IEEE Computer and Communications Society, Anchorage, AK, USA, 22–26 April 2001.
10. Niculescu, D.; Nath, B. Ad hoc positioning system (APS). In Proceedings of the GLOBECOM'01. IEEE Global Telecommunications Conference, San Antonio, TX, USA, 25–29 November 2001.
11. Shi, W.; Jia, C.; Liang, H. An Improved DV-Hop Localization Algorithm for Wireless Sensor Networks. *Chin. J. Sens. Actuators* **2011**, *24*, 83–87.
12. He, T.; Huang, C.; Blum, B.M.; Stankovic, J.A.; Abdelzaher, T. Range-free localization schemes for large scale sensor networks. In Proceedings of the 9th Annual International Conference on Mobile Computing and Networking, San Diego, CA, USA, 16–18 September 2003.

13. Liu, Z.; Fang, Z.; Ren, N.; Zhao, Y. A new range-free localization algorithm based on Annulus Intersection and Grid Scan in wireless sensor networks. *J. Inf. Comput. Sci.* **2012**, *9*, 831–841.
14. Das, K.; Wymeersch, H. Censoring for Bayesian cooperative positioning in dense wireless networks. *IEEE J. Sel. Areas Commun.* **2012**, *30*, 1835–1842. [CrossRef]
15. Pedersen, C.; Pedersen, T.; Fleury, B.H. A variational message passing algorithm for sensor self-localization in wireless networks. In Proceedings of the 2011 IEEE International Symposium on Information Theory Proceedings, St. Petersburg, Russia, 31 July–5 August 2011.
16. Wang, Z.; Jin, X.; Wang, X.; Xu, J.; Bai, Y. Hard decision-based cooperative localization for wireless sensor networks. *Sensors* **2019**, *19*, 4665. [CrossRef] [PubMed]
17. Hehdly, K.; Laaraiedh, M.; Abdelkefi, F.; Siala, M. Cooperative localization and tracking in wireless sensor networks. *Int. J. Commun. Syst.* **2019**, *32*, e3842. [CrossRef]
18. Zhang, J.; Cui, J.; Wang, Z.; Ding, Y.; Xia, Y. Distributed Joint Cooperative Self-Localization and Target Tracking Algorithm for Mobile Networks. *Sensors* **2019**, *19*, 3829. [CrossRef] [PubMed]
19. Cakmak, B.; Urup, D.N.; Meyer, F.; Pedersen, T.; Fleury, B.H.; Hlawatsch, F. Cooperative localization for mobile networks: A distributed belief propagation–mean field message passing algorithm. *IEEE Signal Process. Lett.* **2016**, *23*, 828–832. [CrossRef]
20. Liu, Y.; Tian, S. Vertical Positioning Technologies and its Application of Pseudolites Augmentation. In Proceedings of the 4th International Conference on Wireless Communications, Networking and Mobile Computing, Dalian, China, 12–14 October 2008.
21. Peral-Rosado, J.A.D.; Bavaro, M.; Lopez-Salcedo, J.A.; Seco-Granados, G.; Chawdhry, P.; Fortuny-Guasch, J.; Crosta, P.; Zanier, F.; Crisci, M. Floor Detection with indoor vertical positioning in LTE femtocell networks. In Proceedings of the Globecom Workshops, San Diego, CA, USA, 6–10 December 2010.
22. Gezici, S.; Sahinoglu, Z. UWB Geolocation Techniques for IEEE 802.15.4a Personal Area Networks. Available online: https://merl.com/publications/docs/TR2004-110.pdf (accessed on 5 November 2020).
23. Nguyen, T.V.; Jeong, Y.; Shin, H.; Win, M.Z. Least Square Cooperative Localization. *IEEE Trans. Veh. Technol.* **2015**, *64*, 1318–1330. [CrossRef]
24. Alsindi, N.A.; Pahlavan, K.; Alavi, B.; Li, X. A novel cooperative localization algorithm for indoor sensor networks. In Proceedings of the IEEE International Symposium on Personal, Helsinki, Finland, 11–14 November 2006.
25. Zheng, J.; Wu, Y.-C. Joint time synchronization and localization of an unknown node in wireless sensor networks. *IEEE Trans. Signal Process.* **2009**, *58*, 1309–1320. [CrossRef]
26. Wymeersch, H.; Ferner, U.; Win, M. Cooperative Bayesian Self-Tracking for Wireless Networks. *IEEE Commun. Lett.* **2008**, *12*, 505–507. [CrossRef]
27. Sottile, F.; Wymeersch, H.; Caceres, M.A.; Spirito, M.A. Hybrid GNSS-terrestrial cooperative positioning based on particle filter. In Proceedings of the Global Telecommunications Conference, Kathmandu, Nepal, 5–9 December 2011.
28. Yuan, W.; Wu, N.; Etzlinger, B.; Wang, H.; Kuang, J.-M. Cooperative Joint Localization and Clock Synchronization Based on Gaussian Message Passing in Asynchronous Wireless Networks. *IEEE Trans. Veh. Technol.* **2016**, *65*, 1. [CrossRef]

sensors

MDPI

Article

Deep Learning-Based Positioning of Visually Impaired People in Indoor Environments

Payal Mahida, Seyed Shahrestani and Hon Cheung

School of Computer, Data and Mathematical Sciences, Western Sydney University, Second Ave, Kingswood 2747, Australia; S.Shahrestani@westernsydney.edu.au (S.S.); H.Cheung@westernsydney.edu.au (H.C.)
* Correspondence: P.Mahida@westernsydney.edu.au

Received: 14 September 2020; Accepted: 29 October 2020; Published: 31 October 2020

Abstract: Wayfinding and navigation can present substantial challenges to visually impaired (VI) people. Some of the significant aspects of these challenges arise from the difficulty of knowing the location of a moving person with enough accuracy. Positioning and localization in indoor environments require unique solutions. Furthermore, positioning is one of the critical aspects of any navigation system that can assist a VI person with their independent movement. The other essential features of a typical indoor navigation system include pathfinding, obstacle avoidance, and capabilities for user interaction. This work focuses on the positioning of a VI person with enough precision for their use in indoor navigation. We aim to achieve this by utilizing only the capabilities of a typical smartphone. More specifically, our proposed approach is based on the use of the accelerometer, gyroscope, and magnetometer of a smartphone. We consider the indoor environment to be divided into microcells, with the vertex of each microcell being assigned two-dimensional local coordinates. A regression-based analysis is used to train a multilayer perceptron neural network to map the inertial sensor measurements to the coordinates of the vertex of the microcell corresponding to the position of the smartphone. In order to test our proposed solution, we used IPIN2016, a publicly-available multivariate dataset that divides the indoor environment into cells tagged with the inertial sensor data of a smartphone, in order to generate the training and validating sets. Our experiments show that our proposed approach can achieve a remarkable prediction accuracy of more than 94%, with a 0.65 m positioning error.

Keywords: indoor; positioning; visually impaired; deep learning; multi-layered perceptron; inertial sensor; smartphone

1. Introduction

Accessible location-based information for navigating in a complex indoor environment is a need of every individual [1]. Navigation in complex infrastructures, such as shopping malls, airports, and hospitals, is aided by the proliferation of visual maps, digital maps, and kiosks. However, visually impaired (VI people) can find it hard to use such aids effectively. Globally, 285 million people are estimated to be visually impaired. Amongst them, 39 million are legally blind. Mobility and being able to move around independently can pose significant challenges for a VI person [2].

When travelling in a new environment or public buildings, VI people may require directional assistance or some form of navigation aid. A robust outdoor navigation solution is provided by the Global Positioning System (GPS). However, the use of GPS in an indoor environment is not always possible, as the satellite signals that they rely on cannot penetrate most walls [3]. Special technologies—such as raised line maps, i.e., tactile maps and signage information in Braille—can be of assistance to a VI person in a complex environment [4]. However, not all VI individuals can read and understand the tactile maps [5]. Apart from that, the tactile maps have limitations, including static information about the changing surroundings, and the difficulty of a VI person to position themselves [6]. A white cane is a luminous physical aid that allows a VI person to scan their surrounding for obstacles. However, it fails to identify the location of the individual. The development of innovative modern technologies, such as the Internet of Things and artificial intelligence, has opened up possibilities for providing an interactive system to assist a VI person to independently navigate in an indoor environment.

The lack of a robust technology hinders the navigation of a VI person due to several issues regarding the layout complexity, accessibility, connectivity, and temporal changes of the environment [7]. Technologies need to ease the processes of VI people's navigation by solving challenging issues like the provision of suitable indoor positioning, the tracking of the moving users, obstacle avoidance, and pathfinding [8]. Currently, a variety of wireless technologies are available for indoor positioning and navigation, relying on ZigBee, Radio Frequency Identification (RFID), Beacon, Bluetooth, ultra-wideband (UWB) radio, magnetic fields, and pedestrian dead reckoning (PDR) [8,9].

In our previous works, we have reported on the provision of solutions for the movements of VI people in a smart environment using interconnected IoT devices [10]. A robust framework is the utilization of Bluetooth low energy beacon sensors in the building in order to help a VI person navigate indoors. A developed algorithm, DynaPATH, generates VI-friendly safe routes to a destination, considering significant vision constraints, such as walking along the walls, and the creation of a straight path with minimal turns. Unlike solutions that choose the shorter path [11], DynaPATH proposes a safe path considering the limitations of the VI people [12]. However, a VI person may find difficulty in positioning him/herself in an open space—such as a big hallway—due to the unavailability of external physical devices. Due to the possible loss of external signals, the system needs to maintain the position of the VI people when other external devices are out of range. This paper addresses this positioning issue and investigates the use of inertial sensors to provide a complementary solution that can be integrated into our work. The contribution of this article is to provide a self-directed, accurate, and audio-aided standalone positioning system considering the constraints of a VI person. The main idea behind the work is to demonstrate the minimum infrastructure usage that can help VI people to overcome the challenge of positioning themselves independently between the landmarks.

This paper proposes a deep learning approach to the positioning of a VI user in an indoor environment with a smartphone as inertial guidance. Each room is given a room identifier. The indoor area is divided into microcells, each of which is assigned with a unique region/place identifier that acts as a recognition layer. The vertex of each microcell has 2D (x, y) local coordinates. Figure 1 shows the representation of a sample floor plan that has undivided and divided areas. The solid black lines represent the walls of the indoor environment, and the obstacles in the rooms are represented as solid filled rectangles. The indoor space is divided into grids of cells, as depicted in the lower part of the Figure. The shaded grey rectangle is a unique microcell with 4 vertices. The local coordinates are allocated manually and are stored for each vertex in the building, resembling the latitude and longitude used in a GPS.

In this paper, we propose to map the inertial sensor measurements of a smartphone into position coordinates using the regression-based training of a deep neural network (DNN). We report the results of various experiments in order to check the suitability of our proposed approach. Our experiments used a publicly available dataset that contains records that resemble the walking and movement data of a VI person, e.g., walking straight along the walls. The novelty of our reported approach stems from the use of regression-based multilayer perceptron (MLP) neural network training to accurately find the position of the VI person in a building.

After the validating experiments, we developed an application for the indoor positioning of a VI person. Figure 2 shows the interactions between the application residing in a smartphone and the pre-trained model established by a deep neural network. The inertial sensors of a smartphone provide the inputs to the deep neural network. The MLP will then use these measurements to estimate the corresponding position. The app informs the user of their relative position through an audio interface.

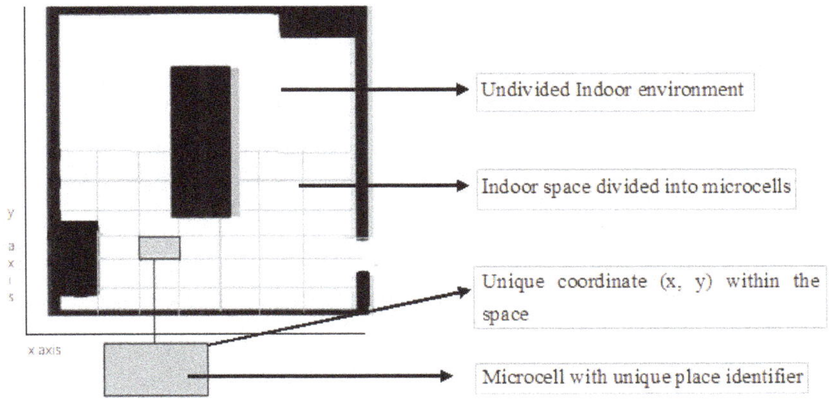

Figure 1. Grid distribution of an indoor environment.

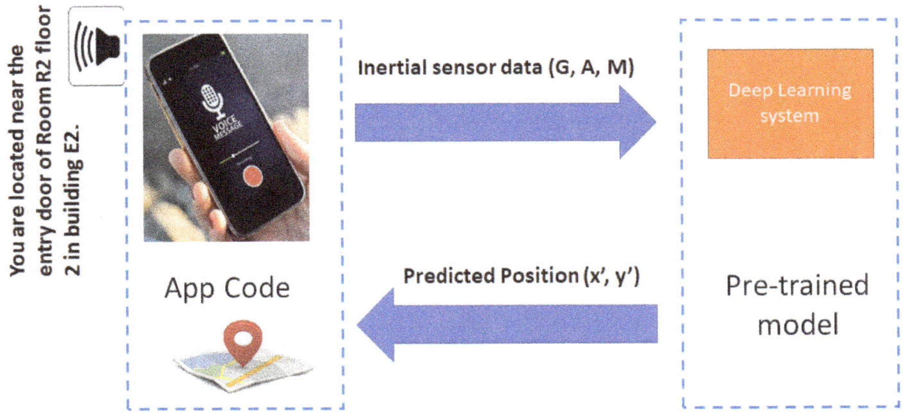

Figure 2. Interaction between the smartphone app and the pre-trained model.

Our contributions in the article, summarized below, aim to provide a robust independent inertial guidance tool to position a VI person in an indoor environment:

- The work proposes an audio assistant app, to be developed and deployed on a smartphone, that helps VI people move independently in a complex building.
- To the best of the author's knowledge, this work is the first to propose and recommend regression-based neural network training for the estimation of the position of a VI person moving in an indoor environment with a smartphone.
- We experimented with a deep neural network model to predict the position of an indoor user as a complementary system to our existing navigation framework using external sensors [10,12].

The remainder of this paper is organized as follows. Section 2 discusses the related research work that sought to solve the indoor positioning problem. The multivariate IPIN2016 dataset, with its usability to evaluate the 2D position of an indoor user, is introduced in Section 3. Furthermore, in Section 4, different versions of the regression-based deep neural network trainings are presented for experimentation. It discusses the deep network structure with the hyperparameters used in the experiments. Section 5 discusses the experimental platform and selection of deep network architecture with suitable hyperparameters, considering the prediction accuracy and the localization error. Finally, conclusions are drawn in Section 6, which also presents the limitations.

2. Background and Motivations for This Work

For a VI person, the localization error needs to be within a few centimetres in order to locate a user in the right room within a building. The system should also be able to estimate and update the location of the moving user quickly. The literature review of this work focuses on prediction accuracy, positioning error, and the usage of technologies that can use minimal resources.

The existing indoor navigation and positioning technologies for VI people can be categorized as being vison based, non-vision based, and based on IoT devices [8]. However, their popularity differs for inaccuracies due to indoor disturbances, availability, energy consumption, the cost of installation, and being high maintenance [13]. Considering these challenges, the performance of indoor location-based services highly depends on the appropriate choice of technology and approaches. Vison-based positioning technologies require the receivers and the moving object or person to be in the line of sight (LOS) to estimate the position measurements [14]. This category involves a vision-based camera and infrared ultrasonic systems [15]. Guerrero [16] suggested a micro navigation system using an infrared camera, Wiimotes, and an augmented white cane to detect the user's position and movement. This system requires massive resources, and the computation operation needed to map and position the user is too great.

Non-vision-based positioning technology includes narrow and wideband wireless radio frequency and magnetic field-based technologies [14]. Indoor positioning has been attempted using WiFi (Wireless Fidelity), infrared, RFID, ultrasound, Bluetooth, and a combination of the technologies [17,18]. A Radial based network including Infrared and RFID has acceptable localization errors. However, it suffers from high costs, as it requires additional hardware and needs offensive calibration processes [15]. Ultrasound waves are used to estimate and track the position of a user in ultrasound-based systems. However, the blockage of the line of sight might result in incorrect measurements [19]. SUGAR [20] uses multiple UWB tags that achieve suitable localization errors for VI people; up to 38 cm. However, installing the UWB system is expensive, and the positioning is purely based on a UWB tag. Nakajima has proposed the use of visible light communication (VLC) and geomagnetic sensor to position and localize the user in an indoor environment [21]. The system provides localization errors up to 1 to 2 m, which are not sufficient for VI people.

Several attempts have been made to develop indoor navigation systems; however, not many of them are successfully deployed. NavCog is a smartphone-based turn-by-turn navigation system for blind users using a network of Bluetooth low energy (BLE) beacons, which uses a K-nearest neighbour (KNN) algorithm approach [22]. The system achieves precise localization information. However, the solution demands a rerouted path to the destination due to missed turns. LowViz [23] is the latest mobile application to assist the visually impaired in indoor navigation. The system uses a wide range of technologies—including sensors, WiFi and Bluetooth low-energy beacons—to guarantee a low localization error. However, context-aware real-time pathfinding is yet to be included in the system. The app may fail when the signals from external devices fail. A variety of newly-developed technologies are being generated and tested. Still, the designs suffer from limitations in their localization error, hardware cost, availability, and lack of additivity.

Recently, there has been considerable new interest in indoor localization techniques, driven by the proliferation of smartphones and other mobile devices. Traditional approaches, such as WiFi-based fingerprinting or distance-based methods, have low prediction accuracy due to shallow learning [24]. In order to handle the shallow learning problem, the deep neural network (DNN) is implemented for the self-extraction of appropriate low and high-level features of given raw data [25–27]. DNN approaches have shown good performance against signal fluctuations, noise effects, and time-consuming manual tuning [28]. The deep networks dynamically learn from the environment by mapping noisy and complex input data to the corresponding output [29]. To the best of the authors' knowledge, not enough work has been done to provide deep learning-based positioning for VI people. Due to limitations in the positioning system for VI people, we have reviewed indirectly applicable positioning using WiFi, inertial sensors, and Channel State Information (CSI).

A novel indoor classification approach [24] was proposed with WiFi fingerprints to predict the correct floor and locations using a deep neural network. The work in [30] used heterogeneous network data, including WiFi and cellular networks with recurrent neural network algorithms, with a high average error of 9.19 m. The positioning error is approximately 9 m, which is not suitable for a VI person. A recurrent neural network (RNN) based indoor positioning solution [31] was applied to RSS data, exploiting the sequential correlation of RSS data. The work achieved an average localization error of 0.75 m, with 80% of the errors being below 1 m. The integration of Linear discriminate analysis (LDA) and MLP based on RSS was proposed in [28]. The approach has a 99.15% prediction accuracy, with a 0.98 m positioning error. RSS-based approaches have high variability at a fixed position in each time. Furthermore, RSS-based localization systems have coarse information due to multipath channels from different antennas. RSS-based approaches usually have 1–3 m of localization error, which it is difficult to further improve [32].

A localization technique based on CSI fingerprints collected using a single access point was proposed in [33]. It used a principal component analysis (PCA) feature extraction technique with different positioning errors in different rooms, varying between 0.6 m and 1.08 m. In [26], the authors compared the results of positioning using MLP and a convolutional network with RSS and CSI data. The RSS data could achieve an average of a 0.92 m localization error, with the highest error as 9 m. The results with the CSI data achieved a 0.92 m positioning error, with a maximal localization of 1.92 m. Besides WiFi network information, the magnetic field signals captured from the magnetometer are similar to the earth's non-constant magnetic field [34]. Each building has its unique magnetic field, with some local anomalies. Thus, the static magnetic field can be utilized in indoor localization and navigation systems [34–36]. An RNN deep neural network [36] approach applied to magnetic signals indoors achieved a positioning localization error of 1.062 m, compared to an average error of 3.14 m with BLE fingerprinting results.

Despite extensive research, the algorithms and technologies mentioned above are still facing issues related to accuracies, infrastructure, and computational complexity. Most of the indoor solutions focus on the use of additional high-computing devices, including beacons and RFIDs. The decrease of the cost and size of the sensors and spurring technologies have resulted in smartphones as a useful and popular IoT device. Modern smartphones have several such sensors, including accelerometers, gyroscopes, magnetometers, GPS, gravity sensors, barometers, and ambient light sensors [37]. Considering the need of a VI person for indoor navigation, we have experimented with the use of the smartphone as navigation assistance in an indoor space.

From the related works studied, most of the positioning systems have focused on solving the underlying issue as a classification problem using WiFi signals by providing room-specific information. We aim to mitigate the infrastructure dependency to position a VI person by proposing the use of a commonly-carried device: a smartphone.

3. Characteristics of the Used Dataset

The use of an appropriate dataset for the training and testing of the model is an essential step in a deep neural network. Despite many works trying to solve the indoor localization issue, there is a lack of public datasets with inertial sensor data for a controlled environment. With the limited number of datasets, we have used a multivariate IPIN2016 dataset [38] in our work to test the proposed approach. Though the pedestrian collecting the inertial sensor data is not visually impaired, the movement of the user has similar steps, including walking along the wall and walking at the same pace.

Like our design, the dataset splits the indoor environment into cells mapped with the inertial sensor data of a smartphone. This Section discusses the dataset and its usability for a controlled indoor environment for VI people. The dataset has different types of movement fingerprints, including magnetic readings from smartphone/smartwatches in the divided spaces. Magnetic readings are data captured by the magnetometer, accelerometer, and gyroscope of a smartphone/smartwatch. The multivariate IPIN2016 dataset has captured the records of the moving user in 325 different places [38]. This dataset includes 36,795 continuous samples over two scenarios of one hour at 10 Hz, which resulted in 6500 discrete samples in 325 places.

The dataset was created on the first floor of the Institute of Information Science and Technologies (ISTL), inside the Italian National Council (CNR) building. The dataset covered movements on a surface measuring 185.12 m^2. Figure 3 depicts the overall map of the building with the top view and trajectory path [38]. The top left corner of the Figure is the top view of the floorplan. The middle portion of the Figure is the highlighted corridor of the given floor.

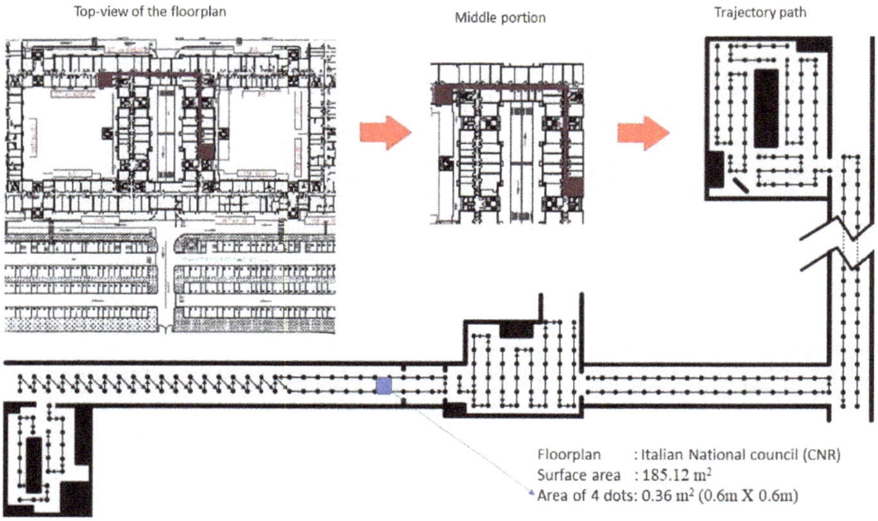

Figure 3. Indoor Floorplan with the top view and trajectory path from iPIN2016 [38].

Furthermore, the trajectory path followed by the users is shown in the bottom part with dots. Each dot in the map corresponds to a detection point, and each dot is 0.6 m from another. The dots represent the different locations at which two users acquired inertial sensor data on their smart devices. As such, the combination of each four dots occupies an area of 0.6 m × 0.6 m. Due to the fixed size of the microcell in the given dataset, our experiments use the same grid size. However, there is a further scope to observe the effects of different grid sizes on the results.

The dataset consists of two scenarios with a combination of zigzag and straight path trajectories performed by two different users holding a smartphone, in order to cover the entire target area. The walking speed of each user was 0.6 m/s on average. Each sample was collected roughly every 100 ms, and the collection time was short. The dataset is a unique combination of both WiFi signals and the inertial sensor data of both a smartphone and smartwatch. This study does not consider data from the WiFi access points and the smartwatch data. During the acquisition, the smartphone was kept at the chest level, with the screen facing up. Every time the user was at a specific location, the device recorded the following data at each dot location.

The recorded data includes the following readings at each dot, represented as a PlaceID with their local coordinates (x, y) at a given timestamp:

- X, Y, and Z-axis values of the accelerometer sensor;
- X, Y, and Z-axis values of the magnetometer;
- X, Y, and Z-axis values of the gyroscope;
- Roll, pitch, and azimuth values of the inertial sensor.

Our work focuses on the magnetic field signals of the accelerometer and gyroscope of the smartphone. Figure 4 represents the graphical representation of values from the x, y and z-axis of a magnetometer. The normalized magnitude M_{mag} of the magnetometer is calculated by Equation (1).

$$M_{mag} = \sqrt{M_x{}^2 + M_y{}^2 + M_z{}^2} \quad (1)$$

where M_{mag} is the normalized magnitude of the magnetometer. M_i is the value of the ith axis of the 3-axis accelerometer.

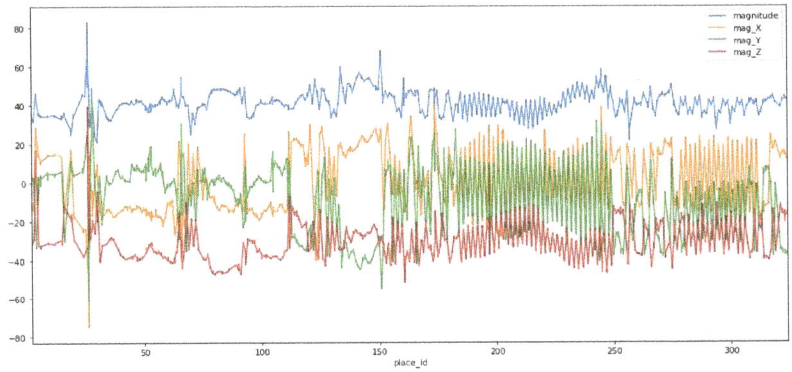

Figure 4. Graphical representation of the x, y, and z coordinates and magnitude of the magnetometer readings.

Figure 5 shows the magnetic field heatmap in each location on the trajectory of the corridor, followed by user 1. The number represented in each cell of the grid on the heatmap is the normalized magnitude, as evaluated in Equation (1). The value of the magnitude varies from 21 to 68 for the given dataset. The indoor magnetic field may be distorted over time locally because of the steel-reinforced concrete in the structures. However, the study in [39] reveals that the magnetic field's distortion pattern remains static.

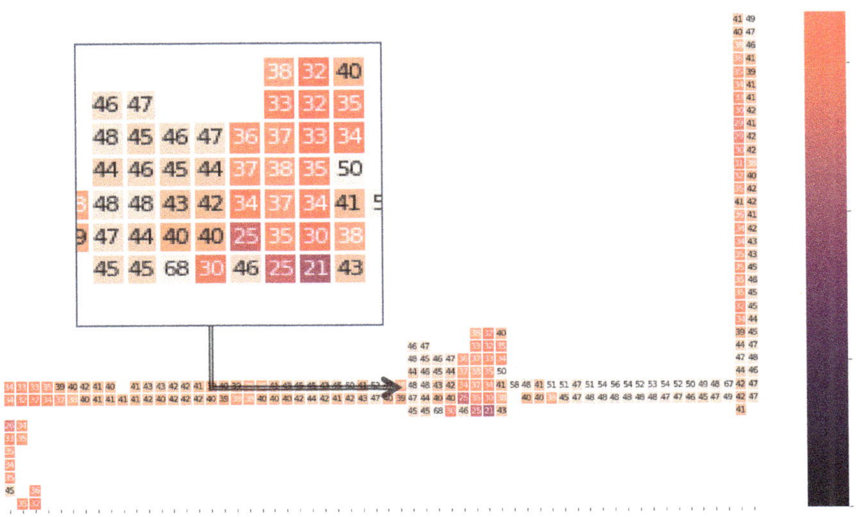

Figure 5. Heatmap of the magnitude of the magnetic field in each place in the building.

4. Deep Learning-Based Positioning

This Section discusses the detailed deep neural network model proposed to predict and evaluate the position of a moving user in a controlled environment. The MLP is characterized as fully connected layers, where each perceptron relates to every other perceptron. The MLP model is a class of feedforward artificial network that defines a mapping function, as shown in Equation (2) [29]:

$$y = \psi\left(\sum_{i=1}^{n} \omega_i x_i + b\right) = \psi\left(w^T x + b\right) \tag{2}$$

where y is the target, w denotes the vector of the weights, x is the vector of the inputs, b is the bias, and ψ is a non-linear activation function.

In this work, we propose to use a regression-based training algorithm to generate the MLP weights, mapping the inertial sensor data of a smartphone into the coordinates of the phone.

In this case, the inputs of the MLP correspond to the 3-axis inertial sensor measurements. The output layer delivers the coordinates of a point in two-dimensional space: x and y. Figure 6 represents the MLP-based DNN with three hidden layers consisting of 128, 64 and 128, neurons used in this work.

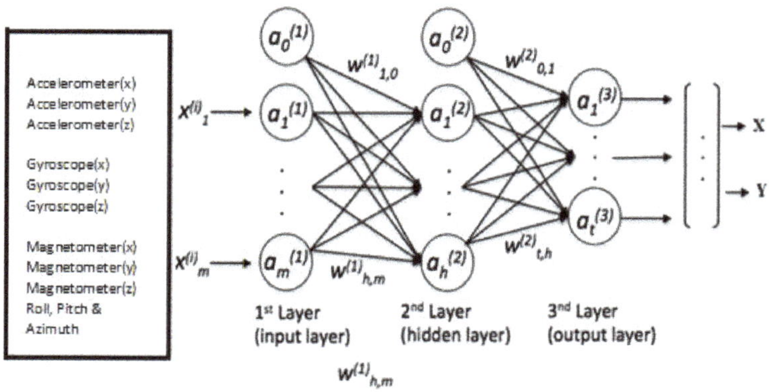

Figure 6. The MLPv1 network structure.

In order to mitigate the effect of unstable gradients with the given neural network, an additional batch normalization layer was introduced to perform an optimization on the input layers. The batch normalization layer works by performing a series of operations on the incoming input data [29]. For equal distribution amongst the input of the hidden layers and faster convergence, we adopted the batch normalization layer between the hidden layers. The weights of the hidden layer are updated by a reduction in the loss function L, as expressed in Equation (3) using the back-propagation algorithm.

$$L = \frac{1}{m}\sum_{i=1}^{m}(yi - f(xi))^2 \tag{3}$$

where m represents the number of samples of input features, and yi represents the actual coordinates of the ith sample. $f(xi)$ is a function to predict the position from the ith sample of the input features.

The data is split into two subsets to train the MLP and to validate the learning. The testing of the performance involves the use of an independent dataset that was not used for the training of the model. The size of the dataset in our simulation is comparatively small and intricate. Therefore, we used the K-Fold Cross-validation technique. The K-fold method is a resampling procedure used to evaluate a deep learning model based on a limited number of data samples [40]. It is popular because it is a less biased or a less optimistic estimate of the model than a simple train/test split. This technique involves the random division of a dataset into K groups, or folds, of approximately equal size. The first blue fold is treated as validation data, and the model is trained on the remaining K-1 training data, as depicted in the first of the K iterations in Figure 7. A validation fold is used to monitor the performance during training and is not used in training the model. In the second iteration, the second fold is used as validation data, while the rest are used in the training process, and so on.

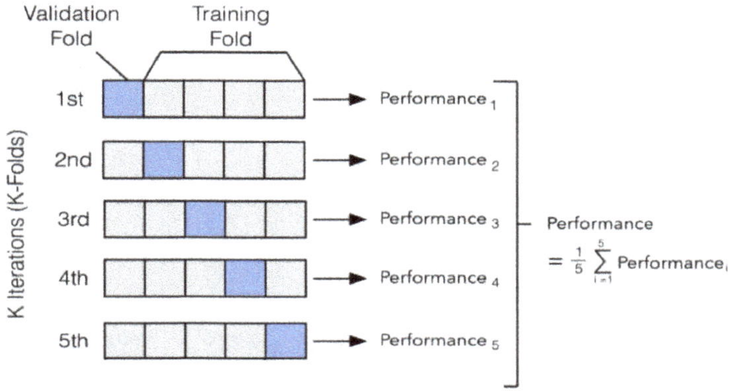

Figure 7. k-Fold Cross-validation technique.

The distribution of the training and validation data in the experiments is with k = 5 in the K-fold cross-validation technique. The dataset is equally distributed in five parts, including the first 7359 records as the validation data, and the remaining 29,436 as training data in the first iteration. Five iterations are performed over the total samples each time, in which 7359 data samples are treated as validation, and the remaining are treated as training data.

The training data is used in each iteration with fixed hyperparameters. Hyperparameters are the higher-level properties of the data model that improves the performance of the model and conveys the capacity of the model to learn the complexity of the data [41]. In order to improve the performance of the model, we involved hyperparameters, including several layers, epochs, a mini-batch size, an activation function, a dropout, regularization, and optimizers [42]. The experimental model was implemented in python, with Keras and TensorFlow libraries with different settings and hyperparameters, as listed in Table 1.

Table 1. The experimental model with the hyperparameter values.

Parameter	Hyperparameter Values in Proposed Deep MLP
Software	Python, Keras, TensorFlow
Training data	29,436
Validation data	7359
Epochs	60 to 140
Batch size	20, 40, 60, 80
Layers with Hidden neurons (with batch normalization)	3 layers—128, 64 and 128 neurons 5 layers—256, 128, 64 and 128 and 256 neurons 7 layers—512, 256, 64, 128, 256 neurons
Drop out rate	0.2 to 0.8
Activation	Selu, elu, softplus, relu
Optimizer	Adam, adamax, rmsprop, adagrad
Loss function	MAE, MSE, RMSE

The experiments were performed with different hyperparameters settings making a different version of the MLP model. Furthermore, the training data and labels are tuned to select the final model. The training data is tuned with the best hyperparameters and learning algorithms. The next Section discusses the experimental results of the different settings with the best-suited hyperparameters.

5. Setup of the Experiments and Analysis of the Results

The experimental platform used to the test the performance of the proposed neural network is presented in Section 5.1, and the performance metrics and evaluation results are discussed in Section 5.2.

5.1. Experimental Platform

Figure 8 represents the experimental platform in the context of the estimation of the position of the user using a deep MLP algorithm.

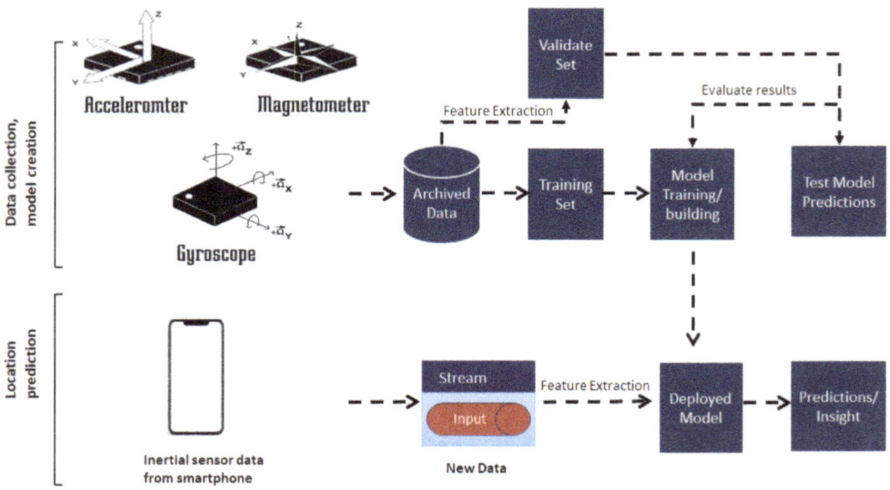

Figure 8. Experimental platform for the proposed model.

The experimental platform expects the sequence of inertial sensor values, including the accelerometer, gyroscope, and magnetometer from the dataset. The collected values are archived and passed as the training set to build the model. The model expects live stream data from a smartphone. Furthermore, the features are extracted and passed to the model for the prediction of the position.

In this work, a sequence of inertial sensor sample values from a dataset—including those from accelerometer, gyroscope, and magnetometer for a moving user—are collected and fed as an input to the neural network as training data. We train the neural network as a regression problem in order to learn the 2-dimensional location of the user based on the input information. After the training, a model is established, and it can be used to estimate a user's location based on real-time sensor data. In order to test the performance of the model, a test set based on the K-fold technique is used to evaluate the prediction accuracy of the proposed model.

5.2. Performance Metrics and Evaluation

The deep learning model is implemented using python, with libraries such as TensorFlow and Scikit-learn. The performance is measured using the mean squared distance (MAE), root mean squared error (RMSE), and mean squared error (MSE) between the ground truth and the predicted location. The model's evaluation process is the assessment of the localization error and the prediction accuracy of a model on the multivariate IPIN2016 dataset, as described in Section 3. The MAE is the mean of the absolute value of the errors, as shown in Equation (4).

$$MAE = \frac{1}{n}\sum_{i=1}^{n}|PS_i - SA_i| \tag{4}$$

The mean squared error (MSE) measures the average of the squares of the errors. It is the average squared difference between the actual value and the estimated value, as shown in Equation (5).

$$MSE = \frac{1}{n}\sum_{i=1}^{\Pi}(PS_i - SA_i)^2 \tag{5}$$

The RMSE is a measure of the average deviation of the predicted values from the actual values. It is used to measure the difference between the values predicted by a model and the values observed from the modelled environment. The average localization error of the calculated distance travelled of the proposed approach can be evaluated by calculating the root mean squared error (RMSE) as the square root of the residuals with Equation (6):

$$RMSE = \sqrt{\frac{\sum_{i=1}^{n}(PS_i - SA_i)^2}{n}} \tag{6}$$

where n represents the walking experiments conducted by each user along the given path. PS_i denotes the final location predicted by the proposed algorithm, and SA_i denotes the actual final location in the ith experiment.

We implemented three batch normalized versions of the MLP algorithm MLPv1, MLPv2, and MLPv3 with 3, 5, and 7 layers. All three versions were implemented in order to evaluate the performance of the best fit model, and to investigate the effects of a different number of hidden layers. The performance metrics, including MAE, MSE and RMSE, are shown in Figure 9a. The prediction accuracy of the model was evaluated by calculating the ratio of the number of correct prediction occurrences to the total number of predictions. The prediction accuracy is shown in Figure 9b.

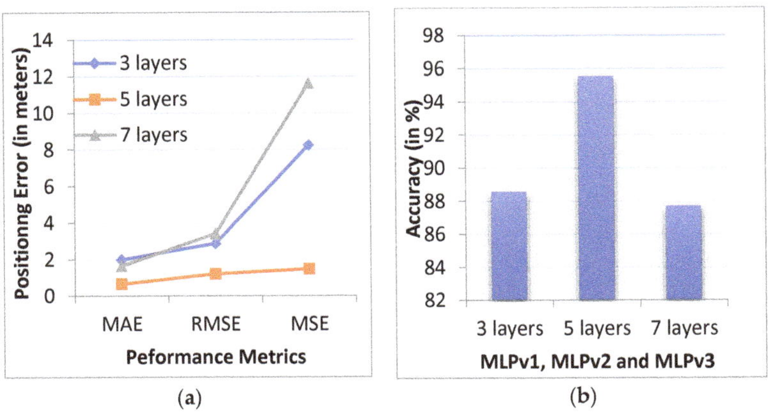

Figure 9. Comparison of: (**a**) the positioning error (MAE, MSE and RMSE) and (**b**) the accuracy for the number of the layers of the model.

The MAE positioning error for MLPv1 with three layers is 1.99 m, with a prediction accuracy of 88.57%. When the number of layers is increased to seven, the positioning error is reduced to 1.64 m, with a prediction accuracy of 87.71%. The performance of the MLPv2 with five hidden layers has a positioning error of 0.66 m, with a prediction accuracy of 95.54%. The results show that the average positioning error and prediction accuracy of the five layered network is better than the other two networks. As such, we use the network with five hidden layers in the rest of this work.

A different permutation of the hyperparameter was further implemented with a 5-layered network. As the input data is continuous and differentiating, we tested the model with non-linear (tanh, selu, relu and softmax and elu) activation functions. In the tuning process, we applied optimizers, including adam, adamax, rmsprop and adagrad. Table 2 shows the positioning errors for the variously-implemented optimizers and activation functions. For each optimizer, an appropriate learning rate was proposed. As shown in Table 1 and Figure 10a, the implementation of the adam optimizer provides the best performance, with an average of 0.71 m MAE error and 95.5% prediction accuracy. As such, we continued to keep the adaptive moment estimation (adam) with $\beta1 = 0.9$, for $\beta2 = 0.999$, and $\epsilon = (10 \times \exp(-8))$.

Table 2. Positioning error (in meters (m)) with different optimizers and activation functions.

Optimizer	MAE (m)	RMSE (m)	MSE (m)
Adam	0.71	1.30	1.70
Adamax	0.84	1.35	1.81
Rmsprop	1.04	1.84	3.39
Adagrad	5.59	3.25	3.62
Activation	**MAE (m)**	**RMSE (m)**	**MSE (m)**
relu	1.24	2.61	6.84
softplus	1.35	2.45	6.01
elu	0.92	1.85	3.45
selu	0.65	1.29	1.67

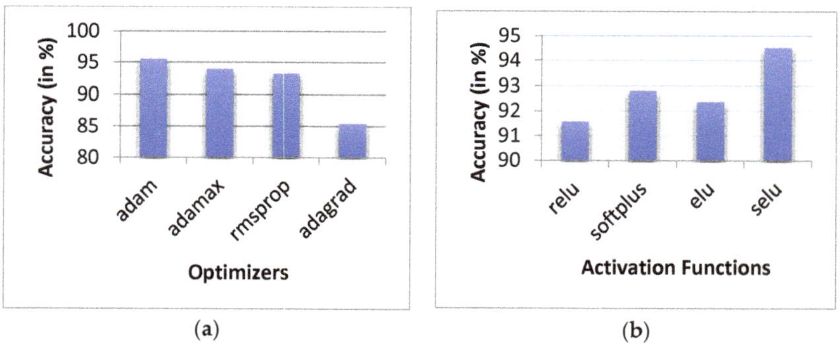

Figure 10. Prediction error with: (**a**) different optimizers and (**b**) different activation functions.

The positioning accuracies using activation functions such as relu, softplus, elu and selu are shown in Table 2. All of the activation functions were performed with adam optimizer. It is evident from the results that the selu activation function outperforms the other activation functions, with a positioning error of 0.65 m and a 94.51% prediction accuracy. Table 2 shows that the minimum positioning error is found with the selu activation function, with 0.65 m as the MAE, 1.67 m as MSE and 1.29 m as RMSE.

From Figure 10b and Table 2, we can conclude that the selu activation function outperforms the relu, softplus and elu activation functions.

The loss function declining curve with different epochs on the training and validation dataset is plotted in Figure 11a,b. It clearly shows that the curve becomes stable after 60 epochs, but we continued the observation to the 140th epoch.

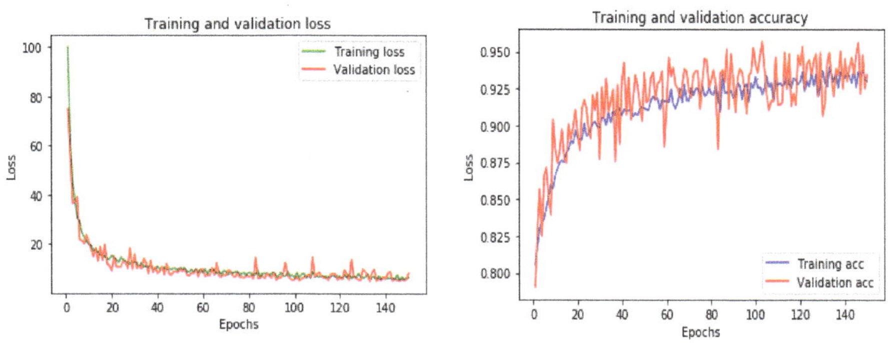

(a) (b)

Figure 11. Training and validation (**a**) loss and (**b**) accuracy for the deep MLP model.

Figure 12 represents the best-suited regression-based deep neural network model used for the predictions. Figure 13 represents traces of the actual position (x, y) from the IPIN2016 dataset for user 1. Figure 14 shows the calculated predicted positions (x', y') using the best-suited regression MLP model.

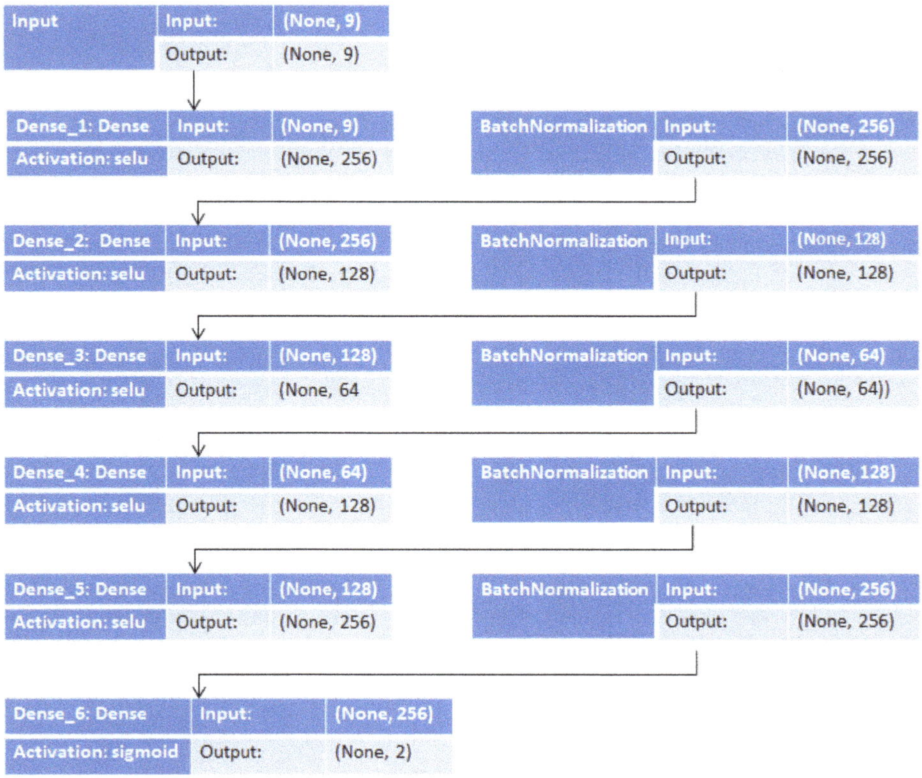

Figure 12. The best-suited regression-based deep neural network MLP model.

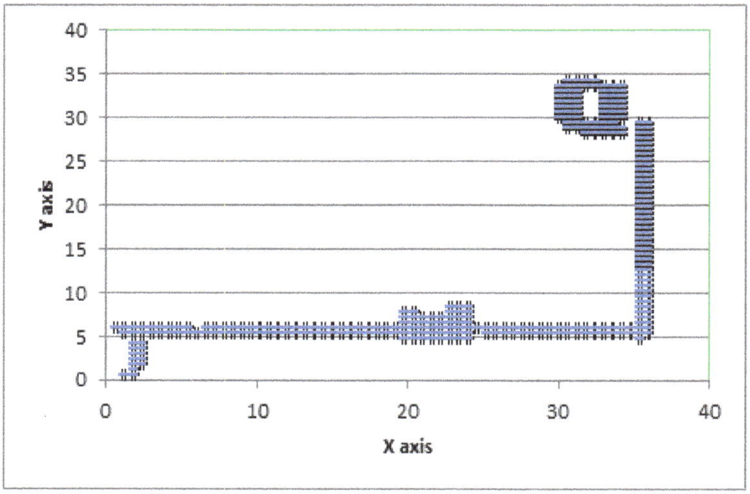

Figure 13. Actual (x, y) position based on the dataset.

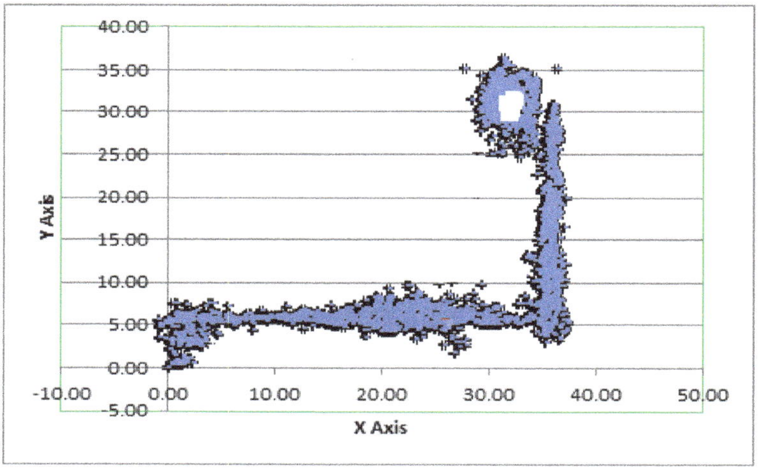

Figure 14. Predicted (x', y') position based on the deep MLP model.

The results demonstrate that the proposed model achieves a considerable prediction accuracy of 94.51%, with a 0.65 m positioning error. The highest positioning error is not higher than 0.89 m. The training time for the given model is approximately 16 s, and the prediction time for a given sample, once trained, is 5 ms. Our previous work based on an improved positioning algorithm [43], when applied on the dataset, provides an almost equal prediction accuracy of 95%. However, the positioning error was evaluated as 1.5 to 2 m. The method proposed in [10] needs an absolute position from additional devices, such as a those that are capable of generating beacon signals. The variation of the error is too high compared to the proposed model. Moreover, the computation time is almost doubled, to 10 to 12 ms.

6. Conclusions

This paper proposed a novel approach to achieve the positioning of a moving VI person as part of an indoor navigation system. The approach is based on feeding the data from the inertial sensors of a typical smartphone to a trained MLP that will map them into the 2D local coordinates of the microcell corresponding to the position of the person holding the phone. The proposed approach was tested with the data from a publicly available multivariant dataset, IPIN2016. The dataset contains data from movements that resemble the walking of a VI person. The performed experiments show that the proposed approach is capable of achieving a positioning accuracy that is close to the step size of a typical user: around 0.65 m. We performed our experiments on a grid size of 0.6 by 0.6 m. Our future work will investigate the impact of different grid sizes on the positioning error and prediction accuracy. We also intend to test our approach using a much larger, temporal dataset to observe the impact of magnetic intensity variations on the prediction error. It is noted that the proposed approach requires Internet connectivity, as it relies on the receipt of the position estimates from the trained model residing in the cloud. In our future works, we intend to explore whether this shortcoming can be addressed by using pre-trained models in a smartphone app. We also aim to complement our proposed positioning approach with other navigation components in order to facilitate easy indoor movements by a VI person.

Author Contributions: Conceptualization, Data curation, Formal Analysis, Methodology, Writing—Original Draft, P.M.; Project Administration, Supervision, Fund Acquisition, Validation, Visualization, Writing—Review and Editing, S.S.; Investigation, Methodology, Writing—Review and Editing, H.C. All authors have read and agreed to the published version of the manuscript.

Funding: This research was funded by the Australian Govt. through RTP scholarship scheme.

Conflicts of Interest: The authors declare no conflict of interest.

References

1. Davidson, P.; Piche, R. A Survey of Selected Indoor Positioning Methods for Smartphones. *IEEE Commun. Surv. Tutor.* **2016**, *19*, 1347–1370. [CrossRef]
2. Pissaloux, E.; Velázquez, R. *Mobility of Visually Impaired People: Fundamentals and ICT Assistive Technologies*; Springer International Publishing: Berlin/Heidelberg, Germany, 2017.
3. Farid, Z.; Nordin, R.; Ismail, M. Recent Advances in Wireless Indoor Localization Techniques and System. *J. Comput. Netw. Commun.* **2013**, *2013*, 185138. [CrossRef]
4. Brock, A.; Jouffrais, C. Interactive audio-tactile maps for visually impaired people. *ACM SIGACCESS Access. Comput.* **2015**, 3–12. [CrossRef]
5. Legge, G.E.; Beckmann, P.J.; Tjan, B.S.; Havey, G.; Kramer, K.; Rolkosky, D.; Gage, R.; Chen, M.; Puchakayala, S.; Rangarajan, A. Indoor Navigation by People with Visual Impairment Using a Digital Sign System. *PLoS ONE* **2013**, *8*, e76783. [CrossRef]
6. Papadopoulos, K.; Barouti, M.; Charitakis, K. A University Indoors Audio-Tactile Mobility Aid for Individuals with Blindness. In *Lecture Notes in Computer Science*; Springer: Cham, Switzerland, 2014; Volume 8548, pp. 108–115. [CrossRef]
7. Cecílio, J.; Duarte, K.; Furtado, P. BlindeDroid: An Information Tracking System for Real-time Guiding of Blind People. *Procedia Comput. Sci.* **2015**, *52*, 113–120. [CrossRef]
8. Mahida, P.T.; Shahrestani, S.; Cheung, H.; Mahida, P.T. Localization techniques in indoor navigation system for visually impaired people. In Proceedings of the 2017 17th International Symposium on Communications and Information Technologies (ISCIT), Cairns, Australia, 25–27 September 2017; pp. 1–6. [CrossRef]
9. Wei, D.; Wei, D.; Lai, Q.; Li, W.; Yuan, H. A Context-Recognition-Aided PDR Localization Method Based on the Hidden Markov Model. *Sensors* **2016**, *16*, 2030. [CrossRef]
10. Mahida, P.T.; Shahrestani, S.; Cheung, H. Indoor positioning framework for visually impaired people using Internet of Things. In Proceedings of the 2019 13th International Conference on Sensing Technology (ICST), Sydney, Australia, 2–4 December 2019; Institute of Electrical and Electronics Engineers (IEEE): Piscataway, NJ, USA, 2019; pp. 1–6. [CrossRef]
11. Mahida, P.T.; Shahrestani, S.; Cheung, H. Comparision of pathfinding algorithms for visually impaired people in IoT based smart buildings. In Proceedings of the 2018 28th International Telecommunication Networks and Applications Conference (ITNAC), Sydney, Australia, 21–23 November 2018; Institute of Electrical and Electronics Engineers (IEEE): Piscataway, NJ, USA, 2018; pp. 1–3. [CrossRef]

12. Mahida, P.T.; Shahrestani, S.; Cheung, H. DynaPATH: Dynamic Learning Based Indoor Navigation for VIP in IoT Based Environments. In Proceedings of the 2018 International Conference on Machine Learning and Data Engineering (iCMLDE), Sydney, Australia, 3–7 December 2018; Institute of Electrical and Electronics Engineers (IEEE): Piscataway, NJ, USA, 2018; pp. 8–13. [CrossRef]
13. Van Haute, T.; De Poorter, E.; Crombez, P.; Lemic, F.; Handziski, V.; Wirström, N.; Wolisz, A.; Voigt, T.; Moerman, I. Performance analysis of multiple Indoor Positioning Systems in a healthcare environment. *Int. J. Health Geogr.* **2016**, *15*, 1–15. [CrossRef]
14. Xu, Z.; Zheng, H.; Pang, M.; Su, X.; Zhou, G.; Fang, L. Utilizing high-level visual feature for indoor shopping mall localization. In Proceedings of the 2017 IEEE Global Conference on Signal and Information Processing (GlobalSIP), Montreal, QC, Canada, 14–16 November 2017; pp. 1378–1382. [CrossRef]
15. Won, S.-H.P.; Melek, W.W.; Golnaraghi, F. Remote Sensing Technologies for Indoor Applications. In *Handbook of Position Location: Theory, Practice and Advance*; Wiley: Hoboken, NJ, USA, 2011. [CrossRef]
16. Guerrero, L.A.; Vasquez, F.; Ochoa, S.F. An Indoor Navigation System for the Visually Impaired. *Sensors* **2012**, *12*, 8236–8258. [CrossRef]
17. Brena, R.F.; García-Vázquez, J.P.; Galván-Tejada, C.E.; Muñoz-Rodriguez, D.; Vargas-Rosales, C.; Fangmeyer, J. Evolution of Indoor Positioning Technologies: A Survey. *J. Sens.* **2017**, *2017*, 1–21. [CrossRef]
18. Tsirmpas, C.; Rompas, A.; Fokou, O.; Koutsouris, D. An indoor navigation system for visually impaired and elderly people based on Radio Frequency Identification (RFID). *Inf. Sci.* **2015**, *320*, 288–305. [CrossRef]
19. Chen, C.-C.; Chang, C.-Y.; Li, Y.-N. Range-Free Localization Scheme in Wireless Sensor Networks Based on Bilateration. *Int. J. Distrib. Sens. Netw.* **2013**, *9*, 620248. [CrossRef]
20. Martinez-Sala, A.S.; Losilla, F.; Sánchez-Aarnoutse, J.C.; Garcia-Haro, J. Design, Implementation and Evaluation of an Indoor Navigation System for Visually Impaired People. *Sensors* **2015**, *15*, 32168–32187. [CrossRef] [PubMed]
21. Nakajima, M.; Haruyama, S. New indoor navigation system for visually impaired people using visible light communication. *EURASIP J. Wirel. Commun. Netw.* **2013**, *2013*, 37. [CrossRef]
22. Ahmetovic, D.; Gleason, C.; Kitani, K.M.; Takagi, H.; Asakawa, C. NavCog: Turn-by-turn smartphone navigation assistant for people with visual impairments or blindness. In Proceedings of the 13th Web for All Conference on—W4A'16, Montreal, QC, Canada, 11–13 April 2016; p. 9. [CrossRef]
23. Stein, N. LowViz Guide Launched to Help Visually Impaired People Navigate. May 2015. Available online: https://indoo.rs/lowviz-guide-launched/ (accessed on 30 October 2020).
24. Nowicki, M.; Wietrzykowski, J. Low-Effort Place Recognition with WiFi Fingerprints Using Deep Learning. In *Advances in Intelligent Systems and Computing*; Springer Science and Business Media LLC: Berlin/Heidelberg, Germany, 2017; Volume 550, pp. 575–584.
25. Mehmood, H.; Tripathi, N.K. Optimizing artificial neural network-based indoor positioning system using genetic algorithm. *Int. J. Digit. Earth* **2013**, *6*, 158–184. [CrossRef]
26. Hsieh, C.-H.; Chen, J.-Y.; Nien, B.-H. Deep Learning-Based Indoor Localization Using Received Signal Strength and Channel State Information. *IEEE Access* **2019**, *7*, 33256–33267. [CrossRef]
27. Fawaz, H.I.; Forestier, G.; Weber, J.; Idoumghar, L.; Muller, P.-A. Deep learning for time series classification: A review. *Data Min. Knowl. Discov.* **2019**, *33*, 917–963. [CrossRef]
28. Adege, A.B.; Lin, H.-P.; Tarekegn, G.B.; Jeng, S.-S. Applying Deep Neural Network (DNN) for Robust Indoor Localization in Multi-Building Environment. *Appl. Sci.* **2018**, *8*, 1062. [CrossRef]
29. Alom, Z.; Taha, T.M.; Yakopcic, C.; Westberg, S.; Sidike, P.; Nasrin, M.S.; Hasan, M.; Van Essen, B.C.; Awwal, A.A.S.; Asari, V.K. A State-of-the-Art Survey on Deep Learning Theory and Architectures. *Electronics* **2019**, *8*, 292. [CrossRef]
30. Wu, L.; Chen, C.-H.; Zhang, Q. A Mobile Positioning Method Based on Deep Learning Techniques. *Electronics* **2019**, *8*, 59. [CrossRef]
31. Hoang, M.T.; Yuen, B.; Dong, X.; Lu, T.; Westendorp, R.; Reddy, K. Recurrent Neural Networks for Accurate RSSI Indoor Localization. *IEEE Internet Things J.* **2019**, *6*, 10639–10651. [CrossRef]
32. Liu, W.; Chen, H.; Deng, Z.; Zheng, X.; Fu, X.; Cheng, Q. LC-DNN: Local Connection Based Deep Neural Network for Indoor Localization with CSI. *IEEE Access* **2020**, *8*, 108720–108730. [CrossRef]
33. Zhang, L.; Ding, E.; Hu, Y.; Liu, Y. A novel CSI-based fingerprinting for localization with a single AP. *EURASIP J. Wirel. Commun. Netw.* **2019**, *2019*, 51. [CrossRef]

34. Subbu, K.P.; Gozick, B.; Dantu, R. LocateMe: Magnetic-fields-based indoor localization using smartphones. *ACM Trans. Intell. Syst. Technol.* **2013**, *4*, 1–27. [CrossRef]
35. Shu, Y.; Shin, K.G.; He, T.; Chen, J. Last-Mile Navigation Using Smartphones. In Proceedings of the 21st Annual International Conference on Mobile Computing and Networking—MobiCom '15, Paris, France, 7–11 September 2015; pp. 512–524. [CrossRef]
36. Jang, H.J.; Shin, J.M.; Choi, L. Geomagnetic Field Based Indoor Localization Using Recurrent Neural Networks. In Proceedings of the GLOBECOM 2017—2017 IEEE Global Communications Conference, Singapore, 4–8 December 2017.
37. Zeinalipour-Yazti, D.; Laoudias, C.; Georgiou, K.; Chatzimilioudis, G. Internet-Based Indoor Navigation Services. *IEEE Internet Comput.* **2017**, *21*, 54–63. [CrossRef]
38. Barsocchi, P.; Crivello, A.; La Rosa, D.; Palumbo, F. A multisource and multivariate dataset for indoor localization methods based on WLAN and geo-magnetic field fingerprinting. In Proceedings of the 2016 International Conference on Indoor Positioning and Indoor Navigation (IPIN), Alcala de Henares, Spain, 4–7 October 2016; pp. 1–8. [CrossRef]
39. Shao, W.; Zhao, F.; Wang, C.; Luo, H.; Zahid, T.M.; Wang, Q.; Li, D. Location Fingerprint Extraction for Magnetic Field Magnitude Based Indoor Positioning. *J. Sens.* **2016**, *2016*, 1945695. [CrossRef]
40. A Gentle Introduction to k-fold Cross-Validation. Available online: https://machinelearningmastery.com/k-fold-cross-validation/ (accessed on 21 January 2020).
41. Mashlakov, A.; Tikka, V.; Lensu, L.; Romanenko, A.; Honkapuro, S. Hyper-parameter Optimization of Multi-attention Recurrent Neural Network for Battery State-of-Charge Forecasting. In *Lecture Notes in Computer Science*; Springer: Cham, Switzerland, 2019; Volume 11804, pp. 482–494.
42. Nwankpa, C.; Ijomah, W.; Gachagan, A.; Marshall, S. Activation Functions: Comparison of trends in Practice and Research for Deep Learning. *arXiv* **2018**, arXiv:1811.03378.
43. Mahida, P.T.; Shahrestani, S.; Cheung, H. An improved positioning method in a smart building for Visually Impaired Users. In Proceedings of the Submitted to International Conference on Internet of Things Research and Practice (iCIOTRP2019), Sydney, Australia, 24–26 November 2019. [CrossRef]

sensors

MDPI

Article

ARBIN: Augmented Reality Based Indoor Navigation System

Bo-Chen Huang [1], Jiun Hsu [2], Edward T.-H. Chu [1,*] and Hui-Mei Wu [2]

1 Department of Computer Science and Information Engineering, National Yunlin University of Science and Technology, Yunlin 64002, Taiwan; m10717004@yuntech.edu.tw

2 National Taiwan University Hospital YunLin Branch, Yunlin 640203, Taiwan; Y00051@ms1.ylh.gov.tw (J.H.); Y01614@ms1.ylh.gov.tw (H.-M.W.)

* Correspondence: edwardchu@yuntech.edu.tw; Tel.: +886-05-534-2601-4519

Received: 27 August 2020; Accepted: 12 October 2020; Published: 17 October 2020

Abstract: Due to the popularity of indoor positioning technology, indoor navigation applications have been deployed in large buildings, such as hospitals, airports, and train stations, to guide visitors to their destinations. A commonly-used user interface, shown on smartphones, is a 2D floor map with a route to the destination. The navigation instructions, such as turn left, turn right, and go straight, pop up on the screen when users come to an intersection. However, owing to the restrictions of a 2D navigation map, users may face mental pressure and get confused while they are making a connection between the real environment and the 2D navigation map before moving forward. For this reason, we developed ARBIN, an augmented reality-based navigation system, which posts navigation instructions on the screen of real-world environments for ease of use. Thus, there is no need for users to make a connection between the navigation instructions and the real-world environment. In order to evaluate the applicability of ARBIN, a series of experiments were conducted in the outpatient area of the National Taiwan University Hospital YunLin Branch, which is nearly 1800 m^2, with 35 destinations and points of interests, such as a cardiovascular clinic, x-ray examination room, pharmacy, and so on. Four different types of smartphone were adopted for evaluation. Our results show that ARBIN can achieve 3 to 5 m accuracy, and provide users with correct instructions on their way to the destinations. ARBIN proved to be a practical solution for indoor navigation, especially for large buildings.

Keywords: augmented reality; Bluetooth; indoor positioning system; indoor navigation system; smart hospital

1. Introduction

Due to the advance of the internet of things and business opportunities, indoor navigation systems have been deployed in many large buildings, such as big train stations, shopping malls, hospitals, and government buildings. After installing a navigation mobile app, users can select a point of interest on a menu list. Then, the app will determine a route to the destination, which is usually the shortest path. Nowadays, the most commonly used user interface (UI) of navigation applications is a 2D map with a route. Users are provided with navigation instructions, such as turn left, turn right, and go straight, when they are close to an intersection. However, due to the limitations of a 2D navigation map, it could add an additional cognitive load for users to construct the relationship between the 2D navigation map and the real environment. Extra mental pressure may also be induced and make users confused [1]. Therefore, eliminating possible user confusion is important for navigator UI design.

In order to create a good user experience, several research efforts have been devoted to developing an indoor navigation system by utilizing augmented reality (AR) technology. A. Mulloni et al. [2,3] and L. C. Huey et al. [4] deployed markers as location anchors in the environment. A user can know their location by matching the markers with the associated location information stored at a remote server or on a user's phone. However, the angle of the camera must be in proper alignment with markers before the matching process can start. In addition, the markers could get dirty easily and become unrecognizable, therefore increasing maintenance costs. S. Kasprzak et al. [5], J. Kim et al. [6] first performed an image search for pre-tagged objects, such as billboards and trademarks, in the environment, and then determined the user's location based on the obtained objects. However, the more complicated the environment is, the more difficult it will be to identify pre-tagged objects. The image matching processing becomes even challenging when the layout and decoration of different parts of the space are similar. Feature matching is another method to determine user's location [1]. However, constructing point clouds of a real indoor environment is time consuming and costly, especially for a large building.

In this paper, we designed ARBIN, an augmented reality-based navigation system, by extending our previous work, WPIN [7]. WPIN utilized Bluetooth low energy (BLE) beacons, named Lbeacons version 1 (BiDaE Technology, Taipei, Taiwan), deployed at each intersection and point of interest (POIs), to get the coordinates of the current position. 2D images, such as turn left, turn right, and go straight, were provided to users as direction indicators along the route to the destination. Unlike WPIN, ARBIN uses AR technology that combines virtual objects and the real world. Navigation instructions, as well as AR 3D models, are posted on the screen on the surrounding environment through the smartphone camera. Therefore, there is no need for users to make a connection between the navigation instructions and the real-world environment. In our implementation, Google ARCore (Google Inc., Mountain View, California, United States) [8] is adopted to create AR 3D models, obtain gyroscope sensor readings, and determine where to put the models. Accuracy is the key factor for the success of an AR-based indoor navigator. The difficulties in achieving accuracy of indoor positioning, and that of AR 3D model placement are described as follows.

In WPIN [7], Lbeacons were adopted at waypoints to periodically broadcast their own coordinates to smartphones nearby. A waypoint can be an intersection, a point of interest (POI), or the middle of a corridor. After receiving a broadcast message sent from a Lbeacon, the positioning module, running on the user's smartphone, starts to estimate the distance between itself and the Lbeacon according to a RSSI (received signal strength indicator) distance model. The stronger the received signal is, the closer the user is to the Lbeacon. When the user and the Lbeacon are close enough, for example less than 5 m, a new direction indicator will pop up to guide the user to the next waypoint. The above process continues until the user arrives the destination. However, because of the machine cutting error, the size of the antenna board of each Lbeacon may not be identical, which could affect its capability for transmitting and receiving signals. Furthermore, the characteristics of the RF (Radiofrequency) circuit of each Lbeacon may also be different due to the nature of an analogy circuit. Therefore, the RSSI distance model of each Lbeacon is not exactly identical according to our experience. In our previous work, to achieve the required positioning accuracy, we constructed a RSSI model for each Lbeacon, which was time consuming and unscalable. To overcome this unavoidable and challenging hardware problem, a novel RSSI modeling method was developed to overcome the problem of the heterogeneous issues of Lbeacons, which is given in Section 3.2.

The AR 3D models, such as a left arrow or a right arrow, should be placed properly in a real-world environment to avoid possible user confusion. An inaccurate placement of the 3D model may make users confused, and avoid using it. For example, displaying a 3D model in the wrong orientation, an incorrect elevation angle, or an incorrect depression angle. Many parameters should be carefully considered before having the correct placement of a 3D model, such as the face orientation of a user, the location, and orientation of the smartphone. Constructing a relationship between these parameters and the coordinates of a 3D model is challenging. The detailed method is presented in Section 3.4 to Section 3.5.

In order to investigate the applicability of ARBIN, we first evaluated the responsiveness of the positioning module of ARBIN. We then set up a field trial in a hospital. For the former, we conducted a series of experiments in the engineering building No. 5 of the National Yunlin University of Science and Technology, crossing three floors with a total area of around 250 m^2. The experiment results showed that the adopted RSSI (received signal strength indicator) model could accurately determine the distance between a Lbeacon and a smartphone. Thus, the AR models could be displayed correctly on the smartphone screen. Furthermore, a field trial was conducted at the outpatient area of the National Taiwan University Hospital YunLin Branch, which is nearly 1800 m^2, with 35 destinations and point of interests, such as a cardiovascular clinic, x-ray examination room, pharmacy, and so on. Four different types of smartphone were adopted for evaluation. Our results show that ARBIN can achieve 3 to 5 m accuracy and give users correct instructions on their ways to the destinations. ARBIN proved to be a practical solution for indoor navigation, especially for large buildings.

2. Related Work

Due to the promise of providing a user-friendly interface to users, several researchers have utilized AR technologies to develop indoor navigation applications. Based on the positioning technologies they used, the existing AR-based navigations can be classified into three types: marker-based method, 2D image recognition-based method, and 3D space recognition-based method. Each of them is described as follows.

2.1. Marker-Based Methods

For marker-based methods, markers are first deployed in an indoor environment, and function as location anchors. A marker can be a QR-Code or a specially-designed pattern. The universally unique identifier (UUID) and coordinate of each marker are pre-stored in either the local storage of a smartphone or a remote database for future queries. When finding a marker, the user points the smartphone camera at the marker and scans it. The scanned image is then used for determining the user's location and the place to put a 3D AR model. A. Mulloni et al. [2,3], L. C. Huey et al. [4], G. Reitmayr et al. [9], F. Sato [10], and C. Feng et al. [11] used highly recognizable pictures as markers. Each marker is regarded as a node of the navigation route. When a user comes to a marker and aims the camera at it, a 3D arrow model will be shown on the screen of smartphone to guide the user to the next node. Although markers are easy to deploy, extra user training may be needed to make a marker-based navigation system successful. For example, users should be able to identify markers in the surrounding environment before getting location information and moving forward. It is particularly difficult for a user who has no idea of what a marker looks like. In addition, the camera should be aligned with the marker to ensure the correctness of marker recognition, which makes it user-unfriendly. Maintenance could also be a critical issue when markers get dirty and become unrecognizable. Owing to the above limitations, marker-based navigation is not considered in our work.

2.2. 2D Image Recognition-Based Methods

Unlike marker-based methods, 2D image recognition-based methods search for pre-annotated objects, such as billboards, trademarks, and signs in the environment, and then determine the user's location based on the obtained objects [5,6,12–15]. With the development of image recognition technology, there is a trend for replacing marker-based navigation with image recognition-based navigation. Although the image recognition-based methods may not have the problem of maintenance, they could fail when two objects are similar and undifferentiable. For example, in a large shopping mall, chairs, signboards, and decorations are usually designed in a similar way, which makes it difficult for image recognition. Many other factors can also affect the accuracy of image recognition, such as the view angle of the camera, the distance between the objects and the user, the number of moving objects in the surrounding environment, and so on. Since all these issues should be well addressed before an indoor navigation system can well function in a large and crowded building, we decided not to use an image recognition-based method. Both G. H. Nam et al. [16] and J. Wu et al. [17] adopted image recognition technology for indoor navigation. However, many factors can also affect the accuracy of image recognition, such as the view angle of camera, the distance between the objects and the user, the number of moving objects in the surrounding environment, and so on.

2.3. 3D Space Recognition-Based Methods

Compared to 2D image recognition-based methods, 3D space recognition-based methods collect features of the entire space rather than the features of 2D objects. During runtime, a feature matching mechanism is first used to determine a user's location, and then an AR engine is adopted to draw a 3D model for direction indication. For the implementation, G. Gerstweiler et al. [18], U. Rehman et al. [1] used Metaio SDK (Metaio, Munich, Germany) [19] to create a 3D point cloud for indoor positioning. T. Rustagi et al. [20] used the MapBox API (Mapbox, San Francisco, California, United States) [21] to collect vector data of the indoor environment, and create a corresponding 3D model. The Unity (Unity Technologies, San Francisco, California, United States) [22] game engine was adopted in their system to show AR models. Although 3D could deliver better indoor positioning accuracy, construing 3D models is costly and time consuming, especially for large buildings. In addition, re-modeling is required when indoor layout is changed, which usually happens to shopping malls and exhibition halls. Furthermore, moving objects inside the space could significantly reduce positioning accuracy. The more crowded the environment is, the lower positional accuracy will be. A. Koc et al. [23] adopted ARKit (Apple Inc., Cupertino, California, United States) [24] to build an AR-based indoor navigation. Similar to ARCore, ARKit constructs an AR world with associated 3D information. However, the proposed method requires the user to create coordinate information of all corners in the indoor environment, which makes it time consuming and error prone. In addition, their experimental results showed that the accumulated error becomes significant when the place is large. H. Choi et al. [25] proposed a virtual anchor (VA) point selection method for AR-assisted sensor positioning. According to their definition, a VA is a positioning reference point used by a UWB (Ultra-Wideband) positioning mechanism. The more VAs are selected, the more time is required to determine the user's location. Since UWB devices are required, the issues of hardware cost and energy consumption should also be addressed before a deployment.

Unlike existing feature-matching methods, ARBIN utilized Lbeacons deployed at each intersection and POIs to get the coordinates of the current position. ARBIN performs well in crowded spaces due to the advantage of direction antenna built into Lbeacons, which can adjust transmission power and beam width to properly cover navigation areas. Finally, ARBIN is easy to configure and maintain.

3. Methodology

3.1. System Overview

As Figure 1 shows, ARBIN consists of four modules: indoor positioning, route planning, motion tracking, and AR 3D model placement. At the beginning, the destination selected by the user is sent to the route planning module for determining of a route to the destination (Step 1). The underneath indoor positioning module continuously updates the user's location based on the received BLE advertisement messages and the associated RSSI (Step 2). When the user comes to a waypoint, the route planning module sends a message including the expected face orientation and directional indicator to the AR placement module (Step 3). The AR placement relies on the motion tracking module to obtain the direction (azimuth) and the pitch of the smartphone from the IMU (Inertial Measurement Unit) (Step 4). Based on the collected information, the placement module overlays a 3D arrow model, such as turn left or turn right, on the real-world image (Step 5).

Figure 2 lists the user interface of the ARBIN App. Frequently asked destinations are shown on the main page (Figure 2a). After a user selects a destination on the list (Figure 2b), ARBIN determines the user's current location and a route with the shortest distance to the destination. At the beginning, the user is asked to face a specific direction before the navigation service starts (Figure 2c). In other words, the navigation service will not start until the user faces the expected orientation. On the way to the destination, a 3D indicator will be placed in the real-world environment when the use approaches an intersection or a point of interest, such as stairs or elevators (Figure 2d–g). The navigation service stops when the user arrives at the destination. Finally, a message pops up to reminder the user that the navigation service is finished (Figure 2h).

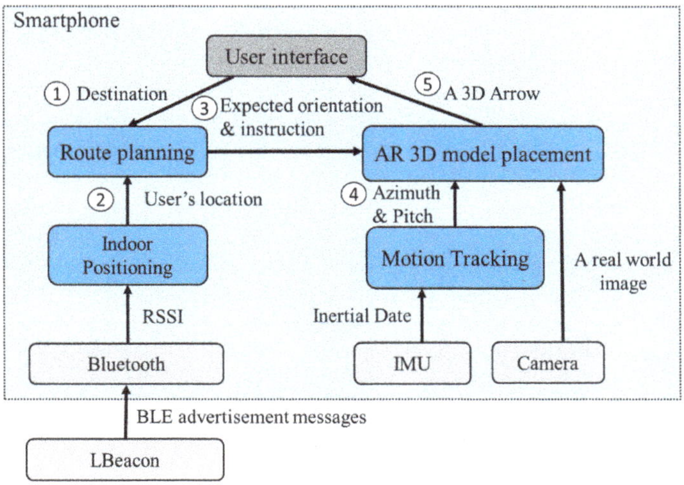

Figure 1. System architecture of ARBIN.

Figure 2. The user interface of ARBIN. (**a**) Main page of ARBIN; (**b**) Destination list; (**c**) Start navigation service; (**d–g**) 3D indicator of navigation instruction; (**h**) Arrival message.

3.2. Indoor Positioning Module

The purpose of the positioning module is to determine the user's location. As Figure 3a shows, Lbeacons are deployed at waypoints. In this work, a waypoint is defined as an intersection, a point of interest, or the middle of a corridor. Each Lbeacon periodically broadcasts its coordinate information to smartphones nearby. From the view point of the user, his or her smartphone continuously receives the coordinate information sent by Lbeacons nearby, while determining how far the smartphone is from the closest Lbeacon. If the distance between the smartphone and a Lbeacon is close enough, for example 3 m, the navigation app provides the user with a directional indicator to guide him or her to the next waypoint. An illustrated example is shown in Figure 3b, the route starts from waypoint A and ends at waypoint C. The user first receives a "go straight" command when entering the area of waypoint A, and then a "turn left" command at waypoint B. The coverage size of a waypoint depends on the size of the intersection or the point of interest. The larger the coverage area is, the larger the range of a waypoint is. In our implementation, the coverage size of a waypoint is a 3-m, 5-m, or 7-m radius circle. The key factor for waypoint-based navigation success is accurately determining the distance between the user and the Lbeacons. For this, in our previous work [7], RSSI distance models stored on the smartphone were adopted to estimate the distance. However, because of the machine cutting error and the characteristics of the RF circuit, the RSSI distance model of each Lbeacon is not identical. To achieve the required positioning accuracy, we constructed a RSSI model for each Lbeacon, but it was time consuming and unscalable.

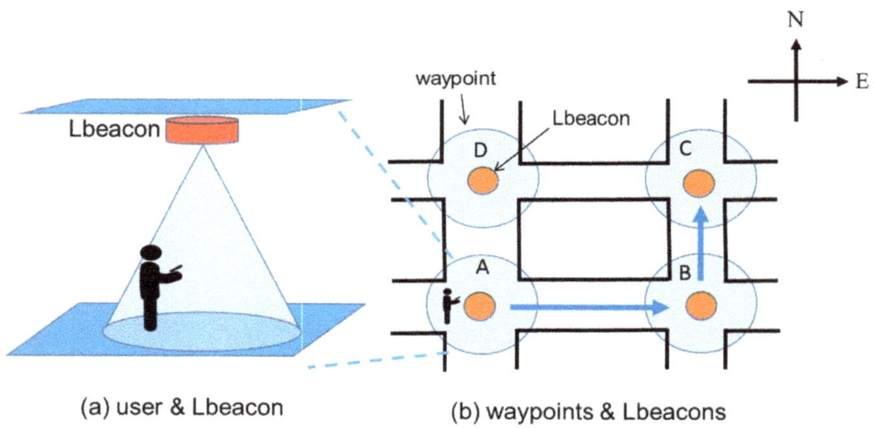

Figure 3. The positioning method of waypoint-based navigation [7]. (**a**) User and Lbeacon; (**b**) waypoint and Lbeacons.

To overcome this problem, in this work we first analyzed the characteristics of RSSI models of about 24 randomly selected Lbeacons, from 70 Lbeacons. We then classified them into four types. For each type of Lbeacon, only one RSSI model was used. Because the navigator must give a user a directional indicator when he/she enters the coverage of a waypoint, we mainly focused on the behavior of the RSSI curve in the range of 0 to 3 m, 3 to 5 m, and 5 to 7 m. As Figure 4 shows, we measured the RSSI values at the locations 1 m to 7 m away from a Lbeacon. For each location, we collected one-minute of RSSI samples (i.e., 240 samples) and took the average as the result. The measurement stops when the seven locations have been measured.

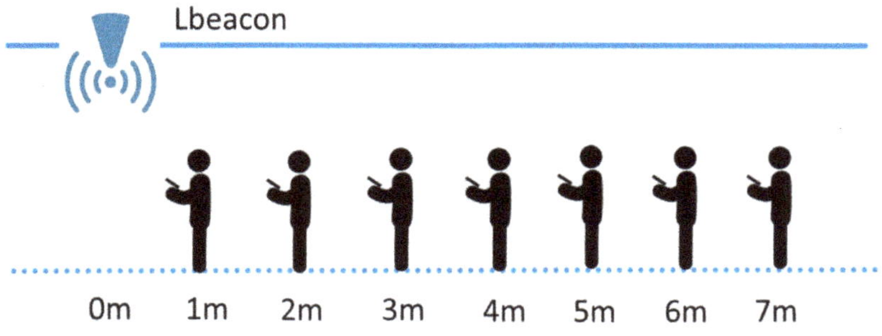

Figure 4. Collecting received signal strength indicator (RSSI) samples at different distances.

As shown in Figure 5a, four Lbeacons, numbered A1, A2, A3, and A4, were classified as type 1, in which the RSSI values drops inversely to the distance, in the range of 0–3 m and 5–7 m. Therefore, Type 1 Lbeacons are suitable to cover a waypoint with a radius of 3 m and 7 m. Similarly, Type 2 Lbeacon are only suitable to cover a waypoint with a radius of 3 m. Meanwhile, Type 3 Lbeacons are suitable for a waypoint with radius of 3, 5, or 7 m since the RSSI values drops inversely to all the distances we measured. Type 4 Lbeacons are suitable for a waypoint with a radius of 5 m. Based on the measurement, for each type of Lbeacon, we adopted a polynomial function as a regression model to represent the relationship between the distance and the RSSI values. Results are shown in Figure 6. Given a new and unknown type of Lbeacon, we first classified it into one of the four types based on the characteristic of its RSSI curve. A RSSI model was then picked from the RSSI models shown in Figure 6. In Section 5, the correctness of the proposed models shown in Figure 6 is evaluated.

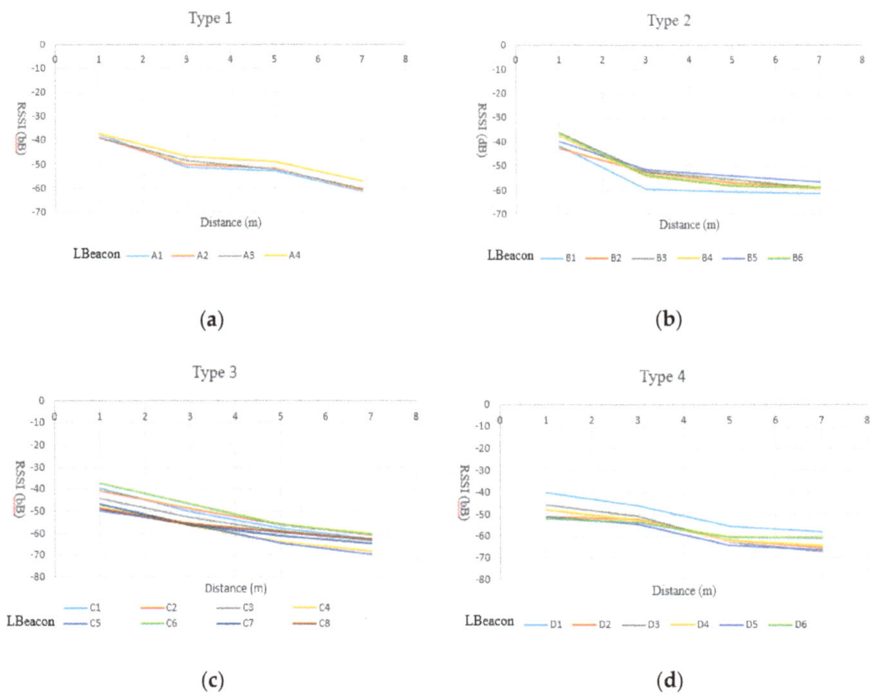

Figure 5. Four types of RSSI distance models. (**a**) Type 1; (**b**) Type 2; (**c**) Type 3; (**d**) Type 4.

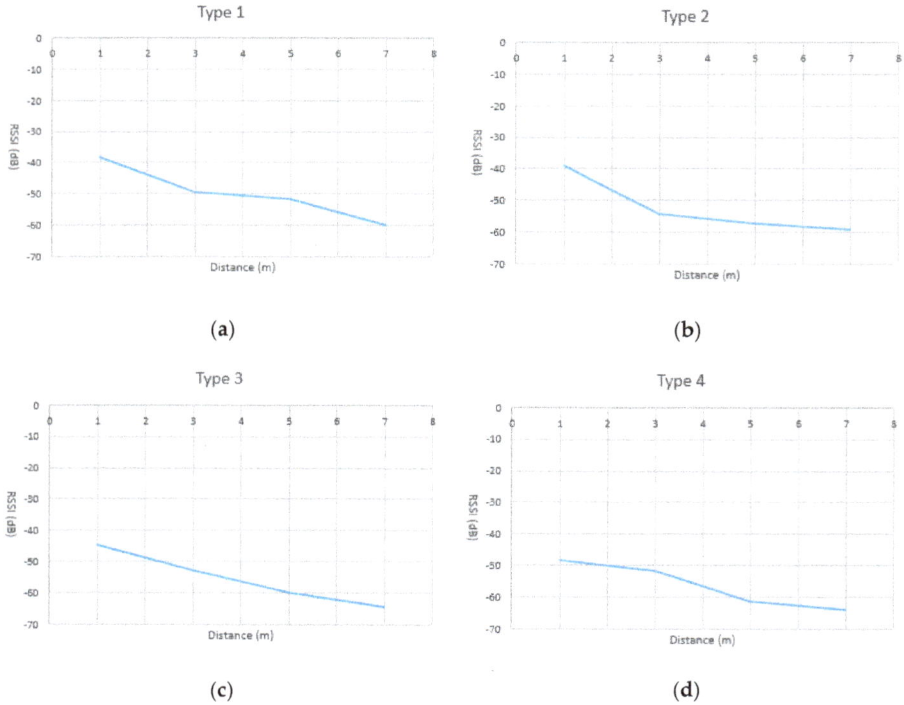

Figure 6. The regression model of the distance and the RSSI values. (**a**) The regression model of type 1 Lbeacon; (**b**) The regression model of type 2 Lbeacon; (**c**) The regression model of type 3 Lbeacon; (**d**) The regression model of type 4 Lbeacon;.

To tolerate the variability of RSSI values, we considered the RSSI values of Lbeacons nearby. Let S_i and S_j represent the highest and the second-highest RSSI detected by the smartphone. The S_i is the RSSI of waypoint *i* and the S_j is that of waypoint, *j*. Since S_j is the highest, the user is regarded as being at waypoint *i*. Based on the RSSI models, we can obtain the theoretical value of S_i and S_j at waypoint i; that is, $\acute{S}_i$ and $\acute{S}_j$. If $S_i - S_j >= \acute{S}_i - \acute{S}_j$, the user's location is updated to waypoint *i*. On the other hand, if $S_i - S_j < \acute{S}_i - \acute{S}_j$, the S_i is considered as a signal surge and will be filtered out.

All Lbeacons were classified into four types. As Figure 6a,d show, each type has its own RSSI model. The RSSI model used to determine the distance between a user's smartphone and a Lbeacon depended on the type of the Lbeacon. When getting close to a Lbeacon, the smartphone uses received UUID to look up the type of Lbeacon and its associated RSSI model, pre-stored in the smartphone.

3.3. Route Planning Module

After receiving the information of user's location and destination, shown in Figure 1, the route planning (RP) module determines a route to the destination by the well-known Dijkstra's shortest path algorithm. Based on the route, the RP module updates the AR model placement module with a direction indicator and an expected face orientation when the user comes to a waypoint. The two pieces of information are then used for placing a 3D model on the real-world environment. For example, as Figure 3b shows, the user starts at waypoint A and moves to waypoint B. When the user enters the coverage of waypoint B, the expected face orientation is east. After the user turns left and moves forward, his/her expected face orientation at waypoint C is north. For ARBIN, at each waypoint, if the user's orientation is not the same as the expected face orientation, the associated directional indicator will not show in the real-world environment. A warning message will pop up to remind the user, when needed. If this happens, possible reasons are that the user is going the wrong way, or that the user does not face to the expected orientation. The route will be recalculated if the user is found at an unexpected waypoint. In our implementation, the orientation is obtained by IMU (inertial measurement unit) sensors of the smartphone. ARBIN uses the getOrientation() of Android Sensor Manager [26] to obtain the orientation. In the above-mentioned example, if B and C are not detected when the user arrives at D, ARBIN will recalculate the route. Then D will be a new starting point.

Let R denote the expected orientation at a waypoint. The R is an integer between 0 to 7, each of which represents a type of orientation, shown in Figure 7a. For example, $R = 1$ is northeast while $R = 2$ is east. After the user passes through a waypoint, the R is updated by how many degrees the user turns to the new orientation. For example, for a turn right instruction, the R is updated by adding 90°. Additionally, for a turn left instruction, the R is updated by adding 270°. Since there are only 8 types of directional indicator in our implementation, we use L, an integer between 0 to 7, to represent the turning angle of a directional indicator. The definition of each value of L is given in Figure 7b. When the user enters a waypoint, R is updated by $(R + L)$ mod 8. For example, in Figure 3b, at the beginning, the user faces to the east and R is 2. When the user comes to waypoint B, the expected face orientation is east. After the user turns left and moves forward, the expected face orientation R at the waypoint C is updated to 0 (= (2 + 6) mod 8), which is north.

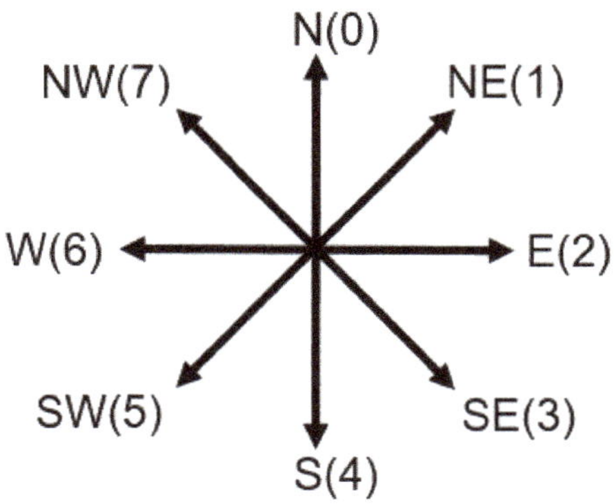

(**a**)

Type of directional indicator	Turning angle	L
Go straight	0°	0
Right front	45°	1
Right	90°	2
Right rear	135°	3
Turn around	180°	4
Left rear	225°	5
Left	270°	6
Lest front	315°	7

(**b**)

Figure 7. The definition of orientation and directional indicators. (**a**) R values; (**b**) L values.

3.4. Motion Tracking Module

The motion tracking module aims to determine the direction (azimuth) and the pitch of the smartphone based on the magnetic sensor and the acceleration sensor of a smartphone. Since the coordinate system of the smartphone and earth are different, transformation is needed before the sensor readings can be used. As shown in Figure 8a, in our usage scenario, the smartphone should be kept upright so that a 3D model can be properly put onto a real environment. If the smartphone is laid flat, shown in Figure 8b, a warning message will be provided to remind the user. Let vector V be the heading direction of the smartphone. As Figure 8a shows, V is a vector on the X-Z plane. ARBIN uses the orientation of V as the expected face orientation. Moreover, the pitch of the smartphone should be greater than 80° before a 3D model can be displayed. The definition of pitch is shown in Figure 8c.

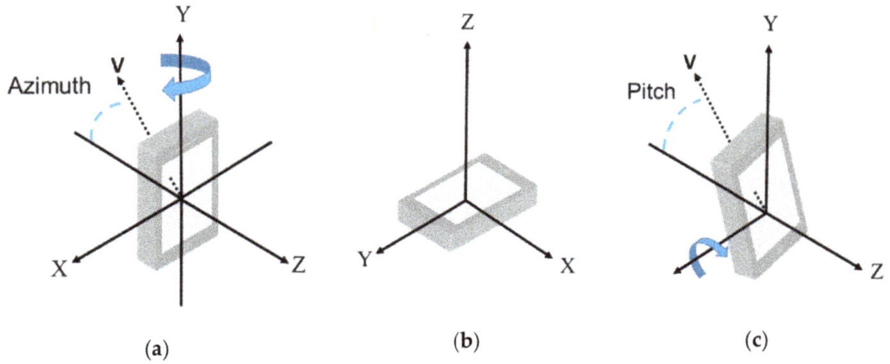

Figure 8. The orientation of a smartphone. (**a**) Phone is keeping upright; (**b**) Phone is lying flat; (**c**) The definition of pitch.

Let X, Y, and Z represent the three axes of the smartphone coordinate system S. In addition, $\acute{X}$, $\acute{Y}$, and $\acute{Z}$ represent those of the earth coordinate system G. The θ_{ij} is the angle between the i axis of S and the j axis of G, in which i = X, Y, or Z, and $j = \acute{X}, \acute{Y}$, or $\acute{Z}$. The angles can be obtained by the IMU sensor built into a smartphone. Let $(x,\ y,\ z)$ be a point on S and its associated coordinate in G be $(\acute{x},\ \acute{y},\ \acute{z})$. Therefore, we have

$$\begin{aligned}
\acute{x} &= x\cos\theta_{x\acute{x}} + y\cos\theta_{x\acute{y}} + z\cos\theta_{x\acute{z}},\\
\acute{y} &= x\cos\theta_{y\acute{x}} + y\cos\theta_{y\acute{y}} + z\cos\theta_{y\acute{z}},\\
\acute{z} &= x\cos\theta_{z\acute{x}} + y\cos\theta_{z\acute{y}} + z\cos\theta_{z\acute{z}}.
\end{aligned}$$

The $(\acute{x},\ \acute{y},\ \acute{z})$ can be represented by $(\acute{x},\ \acute{y},\ \acute{z})^T = R(x, y, z)$. The R the rotation matrix which is:

$$R = \begin{pmatrix} \cos\theta_{x\acute{x}} & \cos\theta_{x\acute{y}} & \cos\theta_{x\acute{z}} \\ \cos\theta_{y\acute{x}} & \cos\theta_{y\acute{y}} & \cos\theta_{y\acute{z}} \\ \cos\theta_{z\acute{x}} & \cos\theta_{z\acute{y}} & \cos\theta_{z\acute{z}} \end{pmatrix}.$$

Therefore, when the smartphone has a rotation around different axis, we can have a different rotation matrix [27]. They are:

$$R_P = \begin{pmatrix} 1 & 0 & 0 \\ 0 & \cos\theta_{y\acute{y}} & \cos\theta_{y\acute{z}} \\ 0 & \cos\theta_{z\acute{y}} & \cos\theta_{z\acute{z}} \end{pmatrix} = \begin{pmatrix} 1 & 0 & 0 \\ 0 & \cos P & \sin P \\ 0 & -\sin P & \cos P \end{pmatrix},$$

$$R_A = \begin{pmatrix} \cos\theta_{x\acute{x}} & 0 & \cos\theta_{x\acute{z}} \\ 0 & 1 & 0 \\ \cos\theta_{z\acute{x}} & 0 & \cos\theta_{z\acute{z}} \end{pmatrix} = \begin{pmatrix} \cos A & 0 & -\sin A \\ 0 & 1 & 0 \\ \sin A & 0 & \cos A \end{pmatrix},$$

$$R_O = \begin{pmatrix} \cos\theta_{x\acute{x}} & \cos\theta_{x\acute{y}} & 0 \\ \cos\theta_{y\acute{x}} & \cos\theta_{y\acute{y}} & 0 \\ 0 & 0 & 1 \end{pmatrix} = \begin{pmatrix} \cos 0 & \sin 0 & 0 \\ -\sin 0 & \cos 0 & 0 \\ 0 & 0 & 1 \end{pmatrix},$$

in which R_P is the rotation matrix when the smartphone has a rotation around X axis, and P is the pitch angular. In addition, R_A is the rotation matrix when the smartphone has a rotation around Y xis, and A is the azimuth angular. Furthermore, R_O is the rotation matrix when the smartphone has a rotation around the Z axis and O is the roll angular. By using the rotation matrix and rotation angles, we can transform a coordinate between S and G.

In our implementation, the Android sensor manager (Google Inc., Mountain View, California, United States) [26] is adopted to transfer the V vector from the smartphone coordinate system, S, to the earth coordinate system, G, and obtain the pitch of the smartphone. ARBIN invokes the getRotationMatrix() function to get a rotation matrix, by feeding the sensor readings of the magnetic sensor and the acceleration sensor. The rotation matrix transformation is used to transform the vectors and coordinates from the smartphone coordinate system to the earth coordinate system. Based on the rotation matrix, ARBIN then uses getOrientation() to obtain the orientation, azimuth, and pitch of the smartphone. Environment noises could affect the correctness of the IMU of the smartphones. For this reason, ARBIN can be integrated with advanced noise filters or probability models to reduce the interference. Since sensor calibration and compensation is not the major focus of this work, for the readers who are interested in this topic, please refer to [28–30].

3.5. AR 3D Model Placement Module

The purpose of the AR 3D model placement module (APM) is to overlay a 3D model on a real-world image. The process includes three steps: pitch check, face orientation check, and placement. Each of the steps is descried as follows. First, APM checks if the smartphone is kept upright. The larger the pitch angle is, the better the camera view is. In our implementation, the pitch angle is set in the range of 80 to 90°. If the pitch angle does not meet the requirement, a warning message is displayed to remind the user to adjust the pitch angle of the smartphone. Second, APM examines whether the orientation of the smartphone is the same as the expected face orientation. If both the pitch angle and the orientation of the smartphone meet the required conditions, a 3D model is placed onto a real environment.

The 3D model placement relies on visual-inertial odometry (VIO), which first uses a camera to extract special feature points of the surrounding environment, such as the corners, boundaries, and blocks. It then continuously matches the features in the contiguous frames to estimate the movement of the camera. Based on the movement of the camera, the 3D model can then be kept at the place we expected until the 3D model is not in the field of view of a camera. In our implementation, we used ViroCore SDK (Viro Media, Inc., Seattle, Washington, United States) [31] to implement the model placement module. ViroCore is a tool package built on top of AndroidARCore (Google Inc., Mountain View, California, United States) [8]. We used getLastCameraPositionRealtime () to get the camera coordinates, and getLastCameraForwardRealtime () to get the camera shooting direction. The calibration of camera depth is done by the smartphone itself. In our configuration, the 3D model is placed at 1 m away from the camera along the camera shooting direction. To have a better view, the 3D model is further put 30 cm below the camera shooting direction. For example, as Figure 9 shows, the camera coordinate is (0, 0, 0) and the camera shooting direction vector is (0, 0, −1). Taking the above-mentioned coordinates, ARBIN determines the coordinated 3D model by (0, 0, 0) + (0, 0, −1) + (0, −0.3, 0) = (0, −0.3, −1), in which the unit is meter. The ARCore then takes (0, −0.3, −1) as input and adopts VIO technology to places the 3D model in the place we expect.

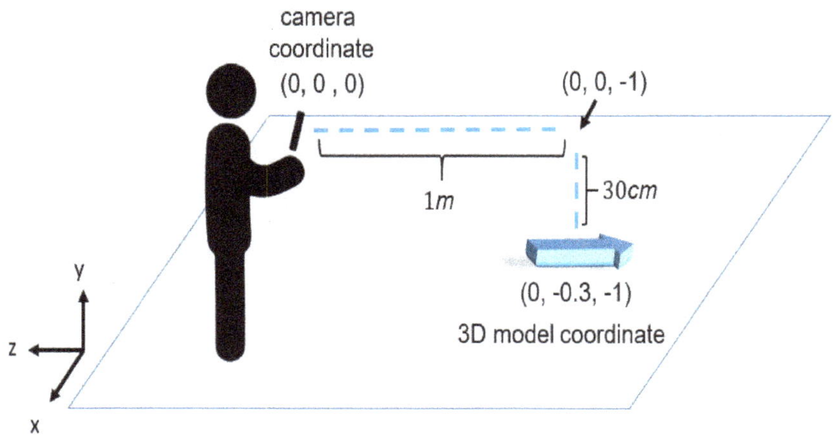

Figure 9. The coordinates of a 3D model.

4. Experiment

Our experiment included two parts: in-house experiments and a field trial. The in-house experiments were undertaken in the Engineering Building (EB) No. 5 of Yunlin University of Science and Technology (Yuntech). The purpose was to evaluate the orientation determination of a smartphone, and the correctness of the RSSI model proposed in Section 4.2. After the in-house experiments were completed, we then conducted a field trial at the National Taiwan University Hospital YunLin Branch (NTUH-Yunlin) where 35 Lbeacons were deployed in the outpatient area, over 1800 m^2, covering two floors. Volunteers were invited to evaluate the responsiveness of ARBIN.

4.1. In-House Experiments

4.1.1. Evaluation of Azimuth of Smartphones

ARBIN relies on information of face orientation to correctly place a 3D model in the real-world environment. Therefore, it is important to ensure that the azimuth value provided by a smartphone is correct. According to the Android sensor manager [26], azimuth is the angle between the smartphone's current compass direction and magnetic north. If the smartphone faces magnetic north, the azimuth is 0°; if it faces south, the azimuth is 180°. additionally, if it faces west, the azimuth is 270°, and if it faces east, the azimuth is 90°.

In our experiment, we investigated two selected smartphones, shown in Table 1, and checked if they could determine the azimuth correctly. We tested all possible orientations used by ARBIN. They are north, northeast, east, southeast, south, southwest, west, and northwest. For each direction, we kept the smartphones upright and recorded the readings of azimuth provided by the smartphone for 10 s. The average value was taken for evaluation. As Table 2 shows, both the Samsung S10e (Samsung, Seoul, South Korea) and SONY Xperia XZ premium (Sony, Tokyo, Japan) could achieve a percentage error less than 5° in determining the azimuth. The ground truth of each measurement was obtained by a real compass. For example, the azimuth error of Samsung S10e when it faces north ranged from +2.66 to –3.32°. The maximum error was 4.67° when it faced Northeast. The larger the azimuth error is, the higher is possibility it will make the user confused. According to our experience, a user's maximum tolerance level is 20°. The azimuth errors of the two smartphones were small enough to be tolerated, and will not affect the placement of a 3D model in a real-world environment. No further calibration on the azimuth angle was required.

Table 1. The smartphones used for the experiment.

Model	CPU	Memory	OS	Bluetooth
Samsung Galaxy S10e	Exynos 9820	6 GB	Android 9.0	5.0
Sony Xperia XZ Premium	Snapdragon 835	4 GB	Android 9.0	5.0

Table 2. The accuracy of azimuth.

Azimuth (Degree)	Samsung S10e		SONY Xperia XZ Premium	
	Max (Degree)	Min (Degree)	Max (Degree)	Min (Degree)
North (0)	+2.66	−3.32	+2.35	−1.46
Northeast (45)	+4.67	−1.35	+0.98	−2.18
East (90)	+3.26	−2.54	+1.23	−3.35
Southeast (135)	+2.09	−2.52	+2.33	−3.04
South (180/−180)	−0.03	+0.01	−0.01	+0.01
Southwest (−135)	+1.94	−3.48	+1.74	−3.05
West (−90)	+2.28	−3.77	+3.44	−4.31
Northwest (−45)	+2.49	−3.64	+1.28	−1.65

4.1.2. Responsiveness of ARBIN

This experiment investigated the ability of ARBIN to have a proper reaction when a user comes close to a Lbeacon. In order to provide a good user experience, we defined 3 m as the responsiveness distance. The ARBIN needs to provide the user with a directional indicator when he or she enters the area of a circle with radius of 3 m, where the Lbeacon is at the center of circle. In other words, it is meaningless to notify the user when he or she is not in the area, because the distance between the user and the Lbeacon is too far away. Similarly, notifying the user after he or she has already passed through the Lbeacon is also useless. The better the responsiveness of ARBIN, the better the user experience we create. The results of responsiveness can also represent the correctness of the four RSSI models presented in Section 4.2, because ARBIN relies on the four models to estimate the distance between the user and a Lbeacon.

The in-house experiment was conducted in EB-No.5 of Yuntech, where 10 Lbeacons were deployed in three floors, with around 250 m^2. Figure 10 shows the deployment maps. The height of the ceiling was about 3 m. When a user holds a smartphone for indoor navigation, the distance between the smartphone and the ground (i.e., 1.5 m) is almost the same as the distance between the smartphone and the ceiling. Hence, there is no significant difference in RSSI by putting Lbeacons on the ground or mounting them on the ceiling according to our experiment. To have a quick deployment without affecting the interior decoration, Lbeacons were put on the ground for the in-house experiment.

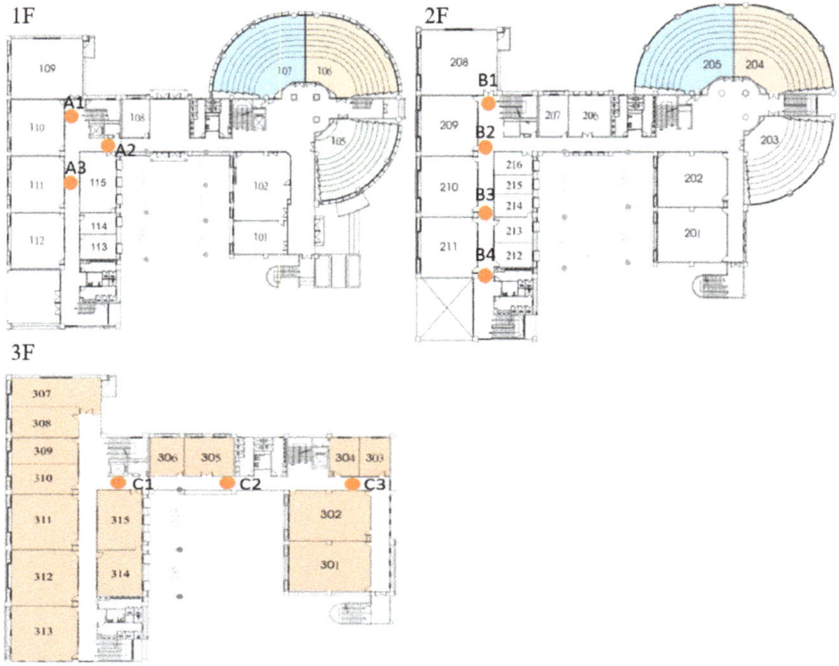

Figure 10. The deployment maps of Lbeacons in EB-No. 5 at Yuntech.

Figure 11 shows the experiment setup for measuring responsiveness. The tester walked around in the building with a normal speed of about one meter per second. As Figure 12 shows, for each Lbeacon, the tester walked back and forth five times. When receiving the directional indicator indicated by ARBIN, he stopped and measured the responsiveness distance, L, between the smartphone and the Lbeacon. As Figure 11 shows, the L is measured by an infrared rangefinder. In other words, for each Lbeacon, the tester first walked forward to the Lbeacon, then recorded the L after he was notified by ARBIN's directional instruction. Then, the tester kept walking. After leaving the range of the Lbeacon (i.e., a circle with radius of 3 m), he turned around and walked forward to the Lbeacon again from the opposite direction. The same measurement was conducted again when the tester was close to the Lbeacon. For each Lbeacon, the tester repeated the above-mentioned experiment for five rounds. The average values of L in both the forward direction and backward direction are shown in Table 3. For the Lbeacons placed on a north-south corridor, the forward direction pointed to north, and the backward direction pointed to south. In addition, for the Lbeacons deployed on an east-west corridor, the forward direction pointed to east, and the backward direction pointed to west.

In our in-house experiment, the responsiveness distance should have been less than 3 m in order to create a good user experience. As Table 3 shows, in 92.5% (=37/40) of the test cases, ARBIN could properly notify the tester when the tester was close to a Lbeacon. However, there was an exception at Lbeacon A3. When the tester held a Samsung Galaxy smartphone and moved forward to Lbeacon A3, the smartphone notified the tester earlier than was expected. In other words, the smartphone received a relatively strong signal when it was 5 m away from Lbeacon A3. A possible solution was to slightly raise the software threshold of RSSI for Lbeacon A3 to defer the notification. We also found that the responsiveness distance, L, of the same Lbeacon differed the in forward and backward directions. The possible reason is that the directional antennas of the Lbeacons may have different abilities to send out signals in different directions. Although the responsiveness distance may depend on the user's arrival direction, it does not affect the ability of ARBIN, in providing users a proper directional indicator. Furthermore, the results showed that there were no cases of notifying the tester after he had already passed the Lbeacon, which met our requirements.

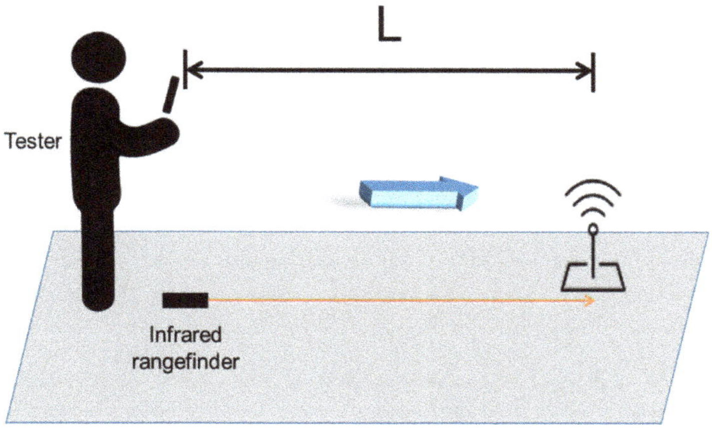

Figure 11. The experimental setup for measuring responsiveness.

Figure 12. The process of measuring the responsiveness.

Table 3. The results of responsiveness distance.

Lbeacon	Samsung Galaxy S10e		Sony Xperia XZ Premium	
	L		L	
	Forward Direction	Backward Direction	Forward Direction	Backward Direction
A1	2.0 m	0.8 m	0.2 m	0.4 m
A2	2.0 m	1.1 m	0.3 m	0.4 m
A3	3.0 m	5.3 m	1.4 m	0.4 m
B1	2.1 m	1.4 m	1.1 m	0.4 m
B2	1.0 m	3.1 m	2.2 m	0.6 m
B3	0.1 m	1.8 m	0.2 m	0.2 m
B4	2.2 m	2.0 m	1.2 m	0.2 m
C1	1.5 m	1.6 m	0.5 m	1.2 m
C2	1.3 m	3.1 m	1.1 m	1.2 m
C3	2.7 m	3.0 m	1.5 m	1.9 m
Average	1.7 m	2.3 m	1.0 m	0.7 m

The Lbeacon is equipped with a directional antenna with conical beams [32]. It can generate a 3 m range and 60° radiation pattern to provide a 3 m horizontal accuracy. However, according to our experience, the shape of the conical beams is not as perfect as is claimed. Therefore, the responsiveness distance, L, of the same Lbeacon may differ in forward and backward directions.

4.2. Field Trial

The purpose of the field trial was to evaluate the responsiveness of ARBIN in NTUH-Yunlin by collecting user's feedback. The deployment maps are shown in Figure 13, in which 35 Lbeacons were deployed in the outpatient area (i.e., B1 and 1F) of the new medical building. Figure 14 shows the installation of Lbeacons mounted on the ceiling. Each Lbeacon periodically broadcasted its UUID at 4 Hz to the smartphones nearby. The starting point was unknow to ARBIN. Based on the UUID the smartphone received, ARBIN automatically determined the starting point. We invited four volunteers, with an average age of 25, who had never either used ARBIN or been to the hospital. They were asked to judge the responsiveness of ARBIN whenever they arrived at a waypoint. The acceptable responsiveness distance was set at 5 m because the coverage area of a Lbeacon deployed in the hospital was a circle with radius of around 5 m. If the ARBIN notified a volunteer after he or she entered the coverage range of a Lbeacon, the responsiveness was moderate. Other hand, if the ARBIN notified a user before he or she entered the coverage range (i.e., the responsiveness distance was larger than 5 m), the responsiveness was fast. In addition, the responsiveness was slow when the responsiveness distance approached zero. The above standards were told to the volunteers before they started the testing. A calibration was required for a new smartphone. Users were required to stand at a specific location, for example the entrance of the building, for 5 to 10 sec. ARBIN will automatically adjust thresholds based on the received RSSI values. Since the purpose of field trial was to evaluate the user experience, volunteers judged the responsiveness visually rather than using an infrared range finder. To simulate an outpatient flow, the volunteers were asked to go to the following destinations accordingly: registration counter (A11), X-ray examination room (B3), pharmacy (A25), and exit (C1). Detailed information of each route and the Lbeacons on that route are listed in Table 4.

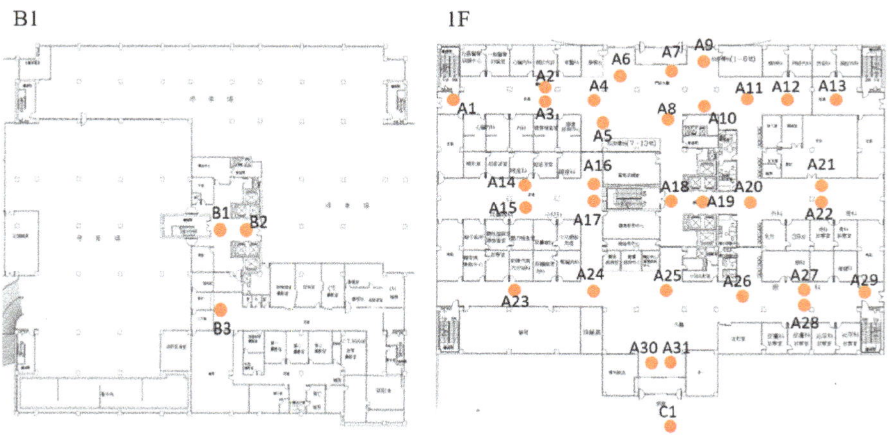

Figure 13. The deployment maps of Lbeacons in NTUH-Yunlin.

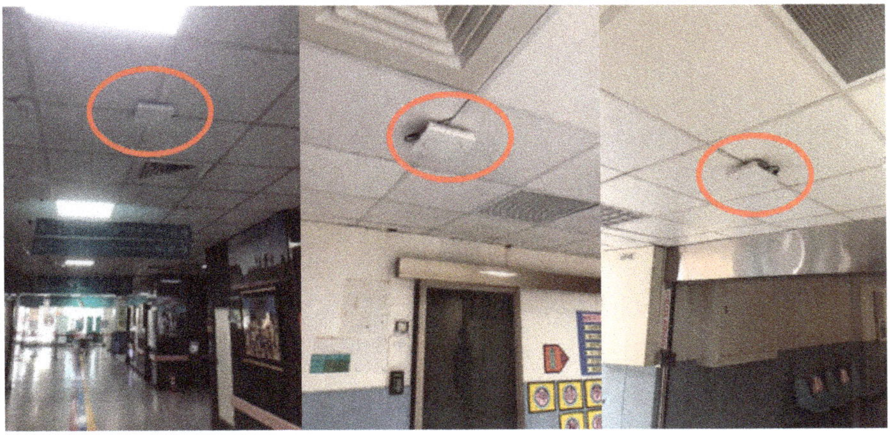

Figure 14. The installation of Lbeacons in the outpatient area of NTUH-Yunlin.

Table 4. The volunteers' feedback on responsiveness.

Route Information		Volunteer A (Samsung)	Volunteer B (Asus)	Volunteer C (Oppo)	Volunteer D (SONY)
Actions	**Lbeacons**				
Task 1: Go to The Register Counter and Then the Clinic.					
Go to the registration counter	A6–A11	Moderate	Moderate	Moderate	Moderate
Go to the clinic	A11–A20	Moderate	Moderate	Moderate	Moderate
	A20–A22	Arrive successfully	Arrive successfully	Arrive successfully	Arrive successfully
Task 2: Go to the X-ray examination room.					
Go to the stairs	A22–A20	Moderate	Moderate	Moderate	Moderate
	A20–A19	Moderate	Moderate	Moderate	Moderate
	A19–A18	Moderate	Moderate	Moderate	Moderate
Down the stairs	A18–B1	Moderate	Moderate	Moderate	Moderate
Go to the examination room	B1–B3	Arrive successfully	Arrive successfully	Arrive successfully	Arrive successfully
Task 3: Go to the pharmacy.					
Go to the stairs	B3–B1	Moderate	Moderate	Moderate	Moderate
Go up the stairs	B1–A18	Moderate	Slow	Moderate	Moderate
Go to the pharmacy	A18–A25	Arrive successfully	Arrive successfully	Arrive successfully	Arrive successfully
Task 4: Go to the exit.					
Go to the exit	A25–A31	Moderate	Moderate	Moderate	Moderate
	A31–C1	Arrive successfully	Arrive successfully	Arrive successfully	Arrive successfully

As Table 4 shows, each volunteer passed through nine Lbeacons on their way to the destinations. They judge the responsiveness whenever they reach a waypoint. The results show that 97% (35/36) of the user feedbacks were "moderate", which indicates that ARBIN can properly notify users when they approach a waypoint. Volunteer C marked A18 as slow because he experienced a deferred notification when he approached A18. A possible solution could be slightly decreasing the threshold of A18 to make ARBIN react properly. In our implementation, there was no 3D AR model placed at the destination. Therefore, we asked volunteers to check if ARBIN could guide them to the destinations successfully. The results show that ARBIN could successfully guide the volunteers to their destinations.

In addition to the user experience evaluation, we also measured the responsiveness of ARBIN in the hospital. Samsung Galaxy S10e and Sony Xperia XZ Premium were used for the evaluation. We evaluated the responsiveness of the ARBIN along the route: Entrance Hall(A6) -> Registration counter(A11) -> Outpatient clinic(A22)-> X-ray examination room (B3) -> Pharmacy(A25) -> Exit (C1). There were in total 14 waypoints on the route. As shown in Figure 15, five different walking patterns of pedestrians were evaluated. The walking speed was around one meter per second. In our experiment, the area of a waypoint was set at a radius of 5 m. We used (Slow, Moderate, Fast) to represent responsiveness distance (~0 m, 0 m~5 m, >5 m). For example, if the responsiveness distance was larger than 5 m, we marked it as fast. No signal indicates that ARBIN had no response when the user entered the waypoint. We repeated the experiment five time, and summarize the results in Table 5. In the single mode and side-by-side mode, the responsiveness distance of both smartphones was always in the range of 0 m to 5 m at every waypoint we measured. In the triangle mode, the response distance of the Sony Xperia XZ Premium became slow at waypoint B3. Additionally, in the line-up mode and stagger mode, Sony Xperia XZ Premium had no response in waypoint B3 and A10. The possible reason may have been the poor design of antenna of the Sony Xperia XZ Premium. The situation could be improved by deploying one more Lbeacon at these waypoints to reduce the possibility of no reaction or slow reaction.

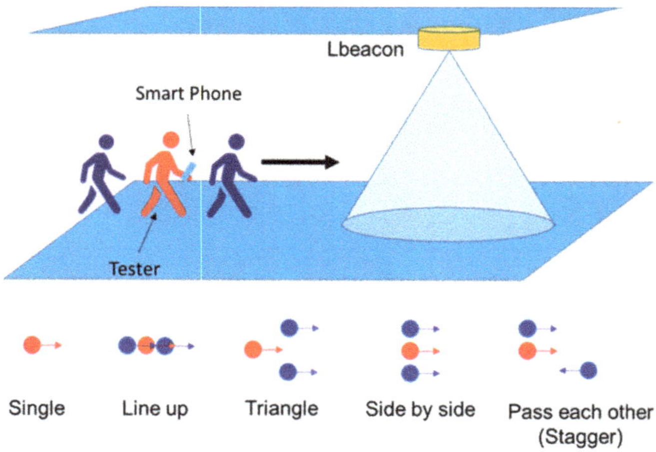

Figure 15. Different walking patterns of user.

Table 5. The volunteers' feedback on responsiveness.

Model/Working Pattern		No Signal	Slow	Moderate	Fast
Samsung Galaxy S10e	Single	0	0	14	0
	Line up	0	0	14	0
	Triangle	0	0	14	0
	Side by side	0	0	14	0
	Stagger	0	0	14	0
Sony Xperia XZ Premium	Single	0	0	14	0
	Line up	1 (B3)	0	13	0
	Triangle	0	1 (B3)	13	0
	Side by side	0	0	14	0
	Stagger	1(A10)	0	13	0

5. Conclusions

In this paper, we presented ARBIN, an augmented reality-based indoor navigation system, to guide users to their destinations in an indoor environment. When users enter the range of a waypoint, ARBIN posts a 3D directional indicator into the real-world surrounding environment. With the support of augmented reality, it is easier for users to determine their locations when walking inside a building. To address the heterogeneous problems of Lbeacons, four types of RSSI model are proposed. Experiences in correctly placing a 3D model in a real-world were also explained. Further, we conducted both in-house experiments and a field trial to verify the responsiveness and practicality of ARBIN. The in-house experiments showed that in 92.5% of the test cases, ARBIN could provide users with a proper directional indicator when they came close to a Lbeacon. For the field trial, four volunteers were invited. Of the of the user feedbacks, 97% (35/36) were moderate. Our results show that ARBIN can achieve a 3 to 5 m accuracy, and provide users with correct instructions on their way to the destinations. ARBIN proved to be a practical solution for indoor navigation, especially for large buildings. To further enhance user experience, in the future we plan to extend the capability of ARBIN by adding landmark objects into real-world environments, and showing advertisement messages provided by a surrounding information system.

Author Contributions: Conceptualization, B.-C.H., J.H., E.T.-H.C. and H.-M.W.; Data curation, B.-C.H.; Formal analysis, B.-C.H.; Funding acquisition, J.H. and E.T.-H.C.; Investigation, J.H., E.T.-H.C. and H.-M.W.; Methodology, B.-C.H., J.H. and E.T.-H.C.; Project administration, J.H., E.T.-H.C. and H.-M.W.; Resources, J.H. and E.T.-H.C.; Software, B.-C.H., J.H. and H.-M.W.; Supervision, J.H. and E.T.-H.C.; Validation, J.H. and E.T.-H.C.; Writing—original draft, B.-C.H.; Writing—review and editing, E.T.-H.C. All authors have read and agreed to the published version of the manuscript.

Funding: This work was supported by National Taiwan University Hospital YunLin Branch Project NTUHYL 109.C027 and Ministry of Science and Technology, Taiwan.

Conflicts of Interest: The authors declare no conflict of interest.

References

1. Rehman, U.; Cao, S. Augmented-Reality-Based Indoor Navigation: A Comparative Analysis of Handheld Devices Versus Google Glass. *IEEE Trans. Hum. Mach. Syst.* **2017**, *47*, 140–151. [CrossRef]
2. Mulloni, A.; Wagner, D.; Barakonyi, I.; Schmalstieg, D. Indoor positioning and navigation with camera phones. *IEEE Pervasive Comput.* **2009**, *8*, 22–31. [CrossRef]
3. Mulloni, A.; Seichter, H.; Schmalstieg, D. Handheld augmented reality indoor navigation with activity-based instructions. In Proceedings of the 13th International Conference on Human Computer Interaction with Mobile Devices and Services, Stockholm, Sweden, 30 August–2 September 2011; pp. 211–220.
4. Huey, L.C.; Sebastian, P.; Drieberg, M. Augmented reality based indoor positioning navigation tool. In Proceedings of the IEEE Conference on Open Systems, Langkawi, Malaysia, 25–28 September 2011; pp. 256–260.
5. Kasprzak, S.; Komninos, A.; Barrie, P. Feature-based indoor navigation using augmented reality. In Proceedings of the 9th International Conference on Intelligent Environments, Athens, Greece, 18–19 July 2013; pp. 100–107.
6. Kim, J.; Jun, H. Vision-based location positioning using augmented reality for indoor navigation. *IEEE Trans. Consum. Electron.* **2008**, *54*, 954–962. [CrossRef]
7. Chu, E.T.-H.; Wang, S.C.; Liu, J.W.S.; Hsu, J.; Wu, H.M. WPIN: A waypoint-based indoor navigation system. In Proceedings of the IEEE 10th International Conference on Indoor Positioning and Indoor Navigation, Pisa, Italy, 30 September–3 October 2019.
8. ARCore. Available online: https://developers.google.com/ar (accessed on 25 August 2020).
9. Reitmayr, G.; Schmalstieg, D. Location based applications for mobile augmented reality. In Proceedings of the 4th Australasian User Interface Conference, Adelaide, SA, Australia, 4–7 February 2003; pp. 65–73.
10. Sato, F. Indoor Navigation System Based on Augmented Reality Markers. In Proceedings of the 11th International Conference on Innovative Mobile and Internet Services in Ubiquitous Computing, Torino, Italy, 10–12 July 2017; pp. 266–274.
11. Feng, C.; Kamat, V.R. Augmented reality markers as spatial indices for indoor mobile AECFM applications. In Proceedings of the 12th International Conference on Construction Applications of Virtual Reality, Taipei, Taiwan, 1–2 November 2012; pp. 235–242.
12. Al Delail, B.; Weruaga, L.; Zemerly, M.J. CAViAR: Context aware visual indoor augmented reality for a university campus. In Proceedings of the IEEE/WIC/ACM International Conferences on Web Intelligence and Intelligent Agent Technology, Macau, China, 4–7 December 2012; pp. 286–290.
13. Koch, C.; Neges, M.; König, M.; Abramovici, M. Natural markers for augmented reality-based indoor navigation and facility maintenance. *Autom. Constr.* **2014**, *48*, 18–30. [CrossRef]

14. Romli, R.; Razali, A.F.; Ghazali, N.H.; Hanin, N.A.; Ibrahim, S.Z. Mobile Augmented Reality(AR) Marker-based for Indoor Library Navigation. *IOP Conf. Ser. Mater. Sci. Eng.* **2020**, *767*, 012062. [CrossRef]
15. Al Delail, B.; Weruaga, L.; Zemerly, M.J.; Ng, J.W.P. Indoor localization and navigation using smartphones augmented reality and inertial tracking. In Proceedings of the IEEE International Conference on Electronics Circuits and Systems, Abu Dhabi, UAE, 8–11 December 2013; pp. 929–932.
16. Nam, G.H.; Seo, H.S.; Kim, M.S.; Gwon, Y.K.; Lee, C.M.; Lee, D.M. AR-based Evacuation Route Guidance System in Indoor Fire Environment. In Proceedings of the 25th Asia-Pacific Conference on Communications (APCC), Ho Chi Minh City, Vietnam, 6–8 November 2019; pp. 316–319.
17. Wu, J.; Huang, C.; Huang, Z.; Chen, Y.; Chen, S. A Rapid Deployment Indoor Positioning Architecture based on Image Recognition. In Proceedings of the IEEE 7th International Conference on Industrial Engineering and Applications (ICIEA), Bangkok, Thailand, 16–21 April 2020; pp. 784–789.
18. Gerstweiler, G.; Vonach, E.; Kaufmann, H. HyMoTrack: A mobile AR navigation system for complex indoor environments. *Sensors* **2015**, *16*, 17. [CrossRef] [PubMed]
19. Metaio SDK. Available online: http://www.metaio.com/products/sdk (accessed on 25 August 2020).
20. Rustagi, T.; Yoo, K. Indoor AR navigation using tilesets. In Proceedings of the 24th ACM Symposium on Virtual Reality Software and Technology, Tokyo, Japan, 28 November–1 December 2018; pp. 1–2.
21. MapBox. Available online: https://docs.mapbox.com/api/ (accessed on 25 August 2020).
22. Unity. Available online: https://unity.com/ (accessed on 25 August 2020).
23. Koc, I.A.; Serif, T.; Gören, S.; Ghinea, G. Indoor Mapping and Positioning using Augmented Reality. In Proceedings of the 7th International Conference on Future Internet of Things and Cloud (FiCloud), Istanbul, Turkey, 26–28 August 2019; pp. 335–342.
24. ARKit. Available online: https://developer.apple.com/documentation/arkit (accessed on 25 August 2020).
25. Choi, H.; Lim, K.; Ko, Y. Improved Virtual Anchor Selection for AR-assisted Sensor Positioning in Harsh Indoor Conditions. In Proceedings of the Global Internet of Things Summit (GIoTS), Dublin, Ireland, 3 June 2020; pp. 1–6.
26. Android Sensor Manager. Available online: https://developer.android.com/reference/android/hardware/SensorManager (accessed on 25 August 2020).
27. Shuster, M.D. A survey of attitude representations. *Navigation* **1993**, *8*, 439–517.
28. Kok, M.; Hol, J.D.; Schön, T.B.; Gustafsson, F.; Luinge, H. Calibration of a magnetometer in combination with inertial sensors. In Proceedings of the 15th International Conference on Information Fusion, Singapore, 9–12 July 2012; pp. 787–793.
29. Wang, Y.; Zhang, J.Y.; Zhang, D.W. Error Analysis and Algorithm Validation of Geomagnetic Sensor. *Appl. Mech. Mater.* **2015**, *742*, 21–26. [CrossRef]
30. Zhou, Y.; Zhang, X.; Xiao, W. Calibration and compensation method of three-axis geomagnetic sensor based on pre-processing total least square iteration. *J. Instrum.* **2018**, *13*, T04006. [CrossRef]
31. ViroCore SDK. Available online: https://viromedia.com/virocore (accessed on 25 August 2020).
32. Li, C.C.; Su, J.; Chu, E.T.-H.; Liu, J.W.S. Building/environment Data/information Enabled Location Specificity and Indoor Positioning. *IEEE Internet Things J.* **2017**, *6*, 2116–2128. [CrossRef]

sensors

MDPI

Article

Position Estimation in Corridors Along the Coupled Mode of Radiating Cables

Olga Blaszkiewicz, Jaroslaw Sadowski * and Jacek Stefanski

Faculty of Electronics, Telecommunications and Informatics, Gdansk University of Technology, 80-233 Gdansk, Poland; olga.blaszkiewicz@pg.edu.pl (O.B.); jacek.stefanski@eti.pg.edu.pl (J.S.)

* Correspondence: jaroslaw.sadowski@eti.pg.edu.pl

Received: 28 July 2020; Accepted: 3 September 2020; Published: 6 September 2020

Abstract: Radiating cables are mostly used to provide radio communication in tunnels or corridors, but they can also be used to estimate the position of a mobile terminal along the cable. In this paper, a measuring receiver's position was estimated by measuring the difference in the direct signal's reception time, which was generated by a transmitter connected to one end of the radiating cable, and the delayed signal retransmitted from another end. During tests, a relatively narrowband (23 MHz) signal was used in the unlicensed band (2.4 GHz) and 50 m long coupled mode radiating cable. The cable was installed along a corridor in the office building. Measurement results used different equipment configurations (i.e., return signal only amplified or amplified and frequency-shifted), which presented possible sources of errors.

Keywords: indoor positioning; indoor navigation; radiating cable; leaky feeder

1. Introduction

One of the most dynamic developing applications of radio communication is the position estimation of people and objects using radio. In outdoor positioning and navigation, a general trend of building versatile solutions has been observed, which may fulfill expectations of different user groups. In such situations, it is not a surprise that global navigation satellite systems (GNSS) become the worldwide standard for commercial and personal positioning. However, as the availability of GNSS services is usually limited to outdoor environments, position estimation inside buildings requires different technologies. Many indoor positioning methods and systems that use different principles of radio wave propagation have already been developed [1] and there is no global agreement on the indoor position estimation technology. Thus, new solutions are still being explored [2].

Position estimation that uses radio techniques in indoor environments often suffers from insufficient accuracy caused by local anomalies in radio wave propagation [3,4]. In many indoor positioning systems that are based on signal level and/or time measurement, a multipath propagation or shadowing via walls, furniture, and humans causes variation in radio frequency (RF) fields. This introduces errors in radio-based position estimations [5–7]. Generally speaking, a longer propagation path in the indoor environment results in higher positioning errors [8–10], so in many applications it is crucial to deploy the reference nodes close to the positioning system's area of operation [11]. Unfortunately, this results in either a reduction of the network operation area or an increase in the number of reference nodes required to ensure the correct operation of the positioning system. Instead of using many reference nodes with separate antennas, a radiating cable may be used to provide a better quality positioning signal due to a reduction in propagation path length within a variable environment.

This paper is organized as follows: Section 2 reviews the state-of-the-art research on radiating cable-based positioning. Section 3 describes the principles of radiating cable positioning when measuring differences in reception time of signals travelling in a cable in opposite directions. Sections 4 and 5 present a measuring stand and the results of position estimation in an indoor building environment, respectively. Section 6 discusses position estimation errors in the proposed solution, while the last section concludes the paper.

2. Related Works

As a concept, using radiating cables for positioning is not new. For example, Nishikawa et al. [12] presented a two-dimensional (2D) position of a mobile antenna near a radiating cable and calculated it using a vector network analyzer (VNA), which measured the arrival time for two signal components received via a mobile antenna: a direct signal emitted from a radiating cable and a signal reflected from the open end of the cable. Moschevikin et al. [13] proposed a different approach—i.e., a two-dimensional position estimation of an experimental active terminal—equipped with a transmitter and receiver that simultaneously used round-trip time (RTT) and signal power measurements (e.g., RSS, received signal strength). Although this paper does not summarize position accuracy evaluation, it provide useful information regarding observed propagation of chirp sounding signals emitted by radiating cables. There have been other studies that investigated radio communication quality and RTT distance measurements in both indoor and outdoor environments using a narrowband radiated mode leaky feeder [14].

The RSS measurements were also presented by Engelbrecht et al. in [15]. This publication focused on the construction of a coaxial radiating cable optimized for a system where a cellular phone's position on the radiating cable (one dimension only) was estimated using two receivers connected to both ends of the cable. Signal's transmitted via the mobile phone were then coupled to nearby radiating cables and were received by receivers with different power levels. Coupling a signal from a cellular phone to a radiating cable influences both measured signal levels in similar way. However, different power measurement results are caused by longitudinal signal attenuation in the radiating cable which corresponds to terminal position. The same authors described [16,17] a solution based on signal transmission in opposite directions, i.e., two ends of the radiating cable connected to two wireless local area network (WLAN) access points. Signal levels were measured by the terminal located in a long hallway near the radiating cable and compared with a previously prepared radio map. It is considered a variation on the fingerprinting method. Weber et al. [18] presents results and a detailed discussion on how to improve the quality of RSS-based position estimations using a radiating cable. Further, they comment on data smoothing and Kalman filtering.

F. Pereira [19,20] described the simultaneous emission of two signals generated by two transmitters connected to both ends of a radiating cable. He further described how to map signal levels recorded along the cable installed in the tunnels. In addition, Pereira [20] considered the possibility of using a phase-difference measurement in a very high frequency (VHF) band to estimate signal propagation time in a positioning system with a leaky feeder. However, no details, results, nor estimated accuracy were discussed.

Nakamura et al. [21] presents another principal used in this system. The mobile terminal was equipped with a transceiver that amplified and filtered the test signal received from the radiating cable, performed frequency conversion, and retransmission. One end of the radiating cable was connected to the transmitter, which triggered measurements by unmodulated carrier emission. Moreover, the measuring receiver estimated the distance to the mobile device via the round-trip time measurements.

Shirai et al. [22] proposed another method of position estimation using two radiating cables. Both cable ends were connected to four-port receiver. A MUSIC algorithm was used to estimate the impulse response of a MIMO channel. The mobile transmitter was placed near cables and was estimated from delay of the MIMO components in the received signal.

Inomata et al. [23] presented interesting details on passive detection of persons using a pair of radiating cables. In this solution, a sounding signal was radiated from one leaky feeder and received by the second cable, which was parallel to the first one. Target detection was performed when extracting the signal scattered around the moving object. Propagation delay time of the scattered wave was utilized to determine location. Further, they implemented a leaky feeder perimeter intruder detection system. In contrast, the solution presented by Shah et al. [24] detected the presence of an intruder using a radiating cable-based channel state information evaluation without positioning.

Many of the positioning examples mentioned above estimated only one coordinate along the radiating cable, which may be enough for corridors and tunnels. The positioning accuracy reached 0.25 m for a two-dimensional (2D) case using time-based measurements [12], however, was not confirmed in any other publication, even those limited to one-dimensional (1D) position estimation. This may be partially explained by the inclusion of a very wide bandwidth in the sounding signal [12] that reached 1 GHz and used VNA. These factors made one-way propagation delay measurements. A drawback of this method is the necessity to connect the mobile antenna to the VNA via cable; it is not strictly a wireless system. Moreover, a measurement scenario [12] was limited to a cable length of only 5 m, while such short radiating cables were not used in indoor or tunnel radio communications. This paper presents the test results for radiating cables in time-based positioning systems with more realistic signal parameters and configurations. In our research, a general-purpose wideband radiating cable was used for communication systems. But it is worth to mention that special design of radiating cable with non-uniform deployment of slots, proposed by Hassan et al [25] may have improved position estimation quality due to side lobe reduction in cable radiation patterns modelled as linear antenna arrays.

3. Principle of Radiating Cable Positioning

A radiating cable is a transmission line (e.g., coaxial, symmetrical) designed to radiate to an external environment a controlled part of the energy of a transmitted signal. In coaxial radiating cables, the emission is caused by an imperfect shielding, i.e., a loosely woven braid or a perforated solid screen. Regarding the different geometry of slots (e.g., shape, spacing), different energy conversion principles are used to model the feeder coupling with the environment [26,27]. The first model assumes that every slot in the cable acts as an elementary magnetic dipole [28]. A resultant electromagnetic (EM) field is a superposition of radiation from every slot. This model is suitable for cables with a non-uniform slot pattern, thus optimizing them for good performance in a narrow frequency band. Such cables are often considered as the "radiating mode". The cables with closely spaced slots—i.e., the distances between them much shorter than a wavelength—are modeled as a controlled conversion of the energy between the coaxial mode inside the coaxial cable and the one-wire mode between the shielding and environment [29]. These kinds of cables are called the "coupled mode"; their performance depends on the scattering of local fields by nearby objects [30,31]. However, they are frequently used because of their wide bandwidth.

When comparing the indoor radio communication systems with antennas mounted in selected points in buildings, the radiating cable allows for us to achieve more uniform signal power distribution, as the main part of the propagation path is in the cable with predictable longitudinal attenuation. This advantage was used in the positioning method [15,16]. However, a relatively short distance between the measuring device and the nearest part of the radiating cable should also give high repeatability of propagation delay determined by constant signal velocity in the cable and device position along the cable. Assuming that both radiating cable sides are connected to transmitters Tx1 and Tx2 (Figure 1), which transmit the positioning signals at time t_1 and t_2, respectively ($t_2 - t_1 = T$ is known), the measuring receiver Rx receives both signals with a time difference as follows:

$$\Delta t = T + (d_2 - d_1) \cdot v_{prop}, \quad (1)$$

where v_{prop} is a velocity of the signal in the radiating cable. Therefore, it should be possible to estimate the position of the mobile receiver along the cable by measuring only the difference in positioning signal time of arrival (TDOA). Compared to method presented by Nishikawa et al. [12], this restricts positioning as one dimensional and assumes that signal detection time is caused mostly by emission from the nearest cable section.

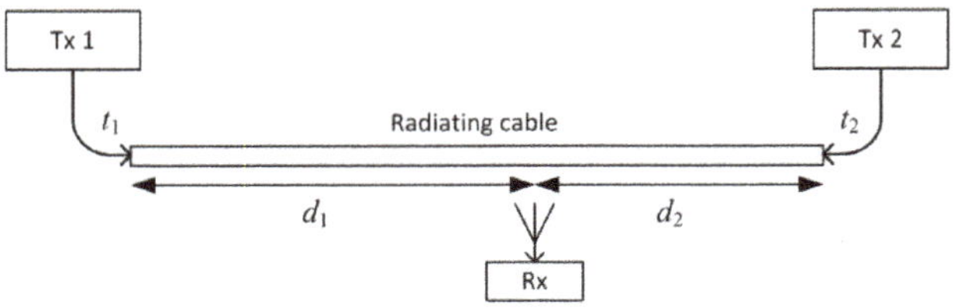

Figure 1. Principle of position estimation along the radiating cable using the time difference of arrival (TDOA) method.

The received signal is a superposition of components radiated by a long section of the cable around the nearest point. Due to signal propagation velocity differences in the cable and air, EM wave emission direction is not perpendicular to the cable axis. Thus, position estimation accuracy should depend on the distance of the receiver from the cable. However, the results of our measurements showed that, when using band limited signals, positioning errors caused by these effects may be comparable or even smaller than other error components, such as random errors caused by multipath propagation in corridors when propagation time differences for different paths are lower than signal bandwidth, or errors caused by limited time measurement resolution.

4. Measuring Stand

The possibility to estimate the measuring receiver position along the radiating cable was tested in the industrial, scientific, and medical (ISM) band at 2.45 GHz. The signal was generated by a Rohde&Schwarz SMU200 vector signal generator and was modulated using binary phase shift keying (BPSK) modulation with a pseudo random binary sequence (PRBS20) (20 MHz chip rate). The root-raised cosine filter limited bandwidth of transmitted signal to 23.3 MHz (99% of power). The signal level at the generator output was +10 dBm. Therefore, taking into account the coupling loss of the radiating cable, the emission level was far below legal limits. A relatively long PRBS sequence ($2^{20} - 1$ chips) was necessary to achieve a high processing gain during receiver correlation. This was crucial to extract the test signal from interferences from the IEEE 802.11 networks, which were present in a building where tests were performed.

Test signals were transmitted using a 50 m long RCT4-WBC-1X-RNA coupled mode coaxial radiating cable, which was on the floor of a straight corridor in a faculty building. This eight-story building had reinforced concrete ceilings and columns with brick walls. The dimension of the building was 115 × 12 m and there was a straight corridor along the entire length of the building on every floor. The cross-section of the corridor and the general view are presented in Section 4.3.

The test signals were received with a universal software radio peripheral (USRP) that had a sampling rate equal to 25 MHz and 12-bit conversion. After upsampling 10-times, a time measurement resolution, based on searching for local maxima in the cross-correlation discrete-time function of received signal and PRBS template signal, was equal to 4 ns. During the tests described below, the receiving part was placed on selected points in the corridor (stationary measurements) with an antenna 0.3 m above the floor (0.5 m from the radiating cable). The description of measurement conditions and geometry during the final campaign is presented in Section 4.3.

4.1. One Feeder

The first test verified the possibility of position estimation based on a signal transmitted from one end of the radiating cable and reflected from the open (not terminated) end. Nishikawa et al. [12] successfully presented such a scenario when conducting a test that used VNA. Unfortunately, a limited bandwidth of the pseudo random signal used in our research, made it almost impossible to distinguish between the direct and reflected signal from the radiating cable (Figure 2). Through the direct signal we understand the pseudo random signal from a vector signal generator, transmitted (and delayed) through a radiating cable and radiated into the air in the proximity of the receiver. The reflected signal is the same signal from the vector signal generator, which is transmitted through the radiating cable. It travels length-wise to its unterminated end, reflects from the open end, transmits in a backward direction, and radiates into the air in the proximity of the receiver.

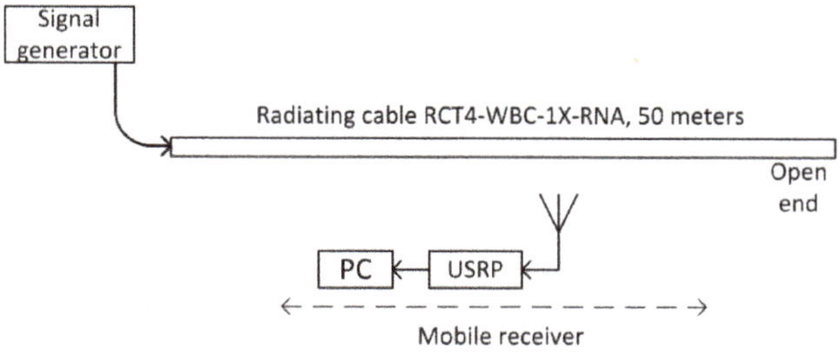

Figure 2. Measuring stand for estimation of position based on the reflection of the test signal from the open end of radiating cable.

Near the open end of the cable, the reflected signal was not visible in the correlation function (Figure 3) due to a high level of correlation side lobes. Moreover, in the area closest to the cable end with signal generator, the reflected signal power was attenuated by a long propagation path at the far end of the cable and back. Therefore, automatic detection of the reflected component was not possible. All correlation charts presented in this paper were computed separately using 52.4 ms long fragments of the recorded signals, which was the repetition time of the PRBS20 sequence clocked at a 20 MHz chip rate. No signal filtering or averaging was used, and the recorded signals contained interferences from ISM devices.

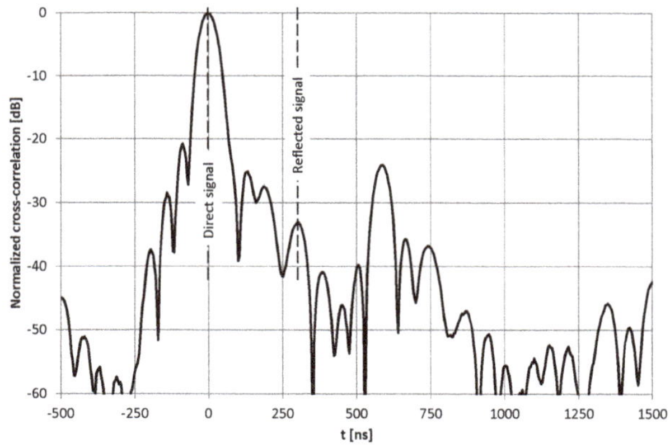

Figure 3. Example of a cross-correlation function of a signal recorded at d_1 = 13.3 m using an unterminated radiating cable.

As there is only one source of the test signal, the receiver position, d_1, related to the end of the cable connected to signal generator can be expressed by:

$$d_1 \approx D - \frac{\Delta t}{2} \cdot v_{prop}, \tag{2}$$

where Δt is the measured time difference between the direct and reflected signal in the receiver, D is the total length of radiating cable, and v_{prop} is the velocity of the signal propagation in the radiating cable. According to the datasheet, v_{prop} for cable type RCT4-WBC-1X-RNA is 0.88 c. Although v_{prop} measurements used the vector network analyzer returned value 0.89 c, we decided to use 0.88 c declared by the producer, because this observed difference in v_{prop} may cause maximal position error estimation of 0.56 m at $d_1 = 0$, which was far below the proposed method's expected accuracy. Equation (2) was accurate only when both signal components received by a measuring device were radiated perpendicular from the section of cable closest to the receiver. Therefore, the time of signal propagation in the air was equal for both components. If the receiver was located close to the cable, the real signal emission at an angle other than perpendicular ([12]) was neglected. The proposed solution was only used for 1D position estimation and in areas close to radiating cable, so possible applications are limited to corridors or tunnels. However, in the real environment, the receiver signal is a superposition of components radiated from some certain part of the leaky feeder. As long as there was sufficient fragment of the radiating cable available in both directions, resulting errors should at least partially cancel each other out. Yet when a mobile receiver was placed near the cable end, the uneven condition of reception of direct and reflected components may cause systematic position estimation errors.

4.2. Two Feeders with Amplifier

To improve quality of reception of signal reflected from the end of a cable, an amplifier was used to amplify the signal before sending it back toward the generator.

Two-directional connection of the amplifier to the same cable requires a directional coupler with separation higher than the amplifier gain. This was not available, so we used two parallel radiating cables (Figure 4). Measured coupling loss between two parallel cables, terminated with matched load, placed 20 cm apart, was below -45 dB. To ensure that no oscillations occurred, the amplifier gain was set to 30 dB and the radiating cables were spaced 30–40 cm apart. The second radiating cable was terminated by a matched load. As there was no reflection of the signal from the open end of the radiating cable, the second component of the signal recorded by the measuring receiver was called the return signal. The return signal in this scenario was the pseudo random signal from the vector signal generator connected to the first radiating cable. Then, it was transmitted (and delayed) through the whole length of the first cable, amplified by a wideband amplifier, and delayed in additional coaxial cables. Finally, it is transmitted through the second radiating cable and radiated into the air in the proximity of the receiver.

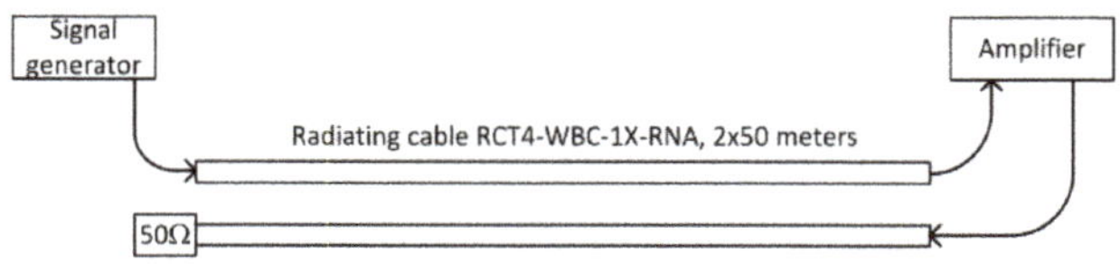

Figure 4. Measuring stand with two radiating cables for a case with return signal amplification.

In this scenario, measuring the receiver position may be estimated as:

$$d_1 \approx D - \frac{\Delta t - \tau_c - \tau_a}{2} \cdot v_{prop}, \tag{3}$$

where τ_c represents the sum of all additional signal delay in connecting cables and τ_a is a signal delay in an amplifier. All assumptions listed under Equation (2) are still valid.

The amplification and additional delay of the return signal made the detection of the main lobes of the correlation for both components (direct and return) easier. However, in some results the two highest peaks in the correlation function did not correspond to the main lobes of measurement signals.

To automatically detect both signals, it is important that near the end of the cables the return signal level may be higher than the level of the direct one (Figure 5).

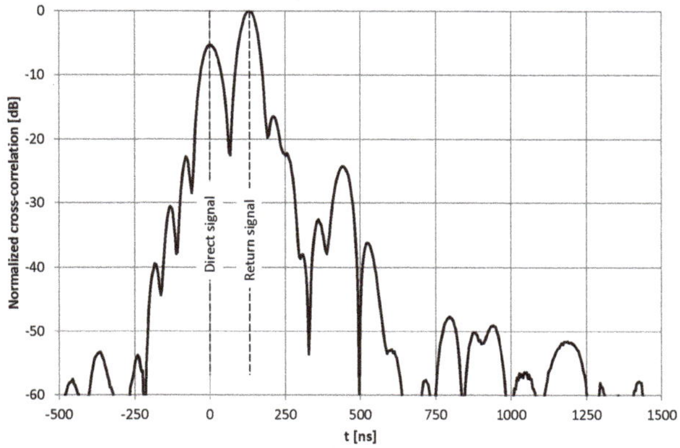

Figure 5. Example of cross-correlation function of signals recorded at d_1 = 39.3 m using two radiating cables with the amplifier.

4.3. Two Feeders with Signal Frequency Conversion

Full separation of the correlation charts for the direct and return signal may be reached with an additional signal delay before amplification using delay times longer than the duration of the unwanted components (side lobes) in the correlation function. It is also possible by modifying the return signal shape or frequency. The simplest method was a frequency conversion that used a balanced mixer and a second signal generator as a heterodyne in the measuring stand (Figure 6). The return signal in this scenario was a frequency-shifted version of the previously defined return signal. It is a pseudo random signal from generator number 1. It was transmitted and delayed in the first radiating cable. Then, it was delayed in an additional coaxial connecting cable and multiplied by a sinusoidal signal from generator number 2. Finally, the return signal was amplified and delayed, then transmitted through the second radiating cable and radiated into the air in the proximity of the measuring receiver. Therefore, both components of the test signal (direct and return) were generated by the first signal generator, but the return signal was additionally mixed with a low-frequency carrier from the second generator. The measuring receiver position was estimated using (3) and by taking into account that τ_c was the sum of the signal delay in cables connecting the first radiating cable with the mixer, the mixer with the amplifier, and the amplifier with the second radiating cable.

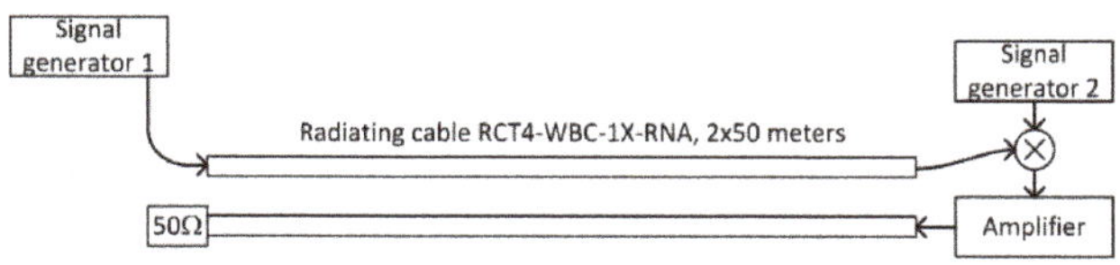

Figure 6. Measuring stand with the frequency conversion of the return signal.

To fully examine the position estimation accuracy in this scenario, measurements were taken in 12 different configurations of the radiating cables (black dots) and the receiving antenna position (squares), presented on a cross-section of the corridor in Figure 7. Firstly, the radiating cables were placed along one wall with cable no. 1 close to the wall. This cable radiated the direct signal. Next, the cables were moved nearer the opposite wall of the corridor so that cable no. 1, radiating the direct signal, was closer to the center of the corridor. The measuring receiver was mounted on a hand cart equipped with a wheel encoder for reference position measurements. Accuracy of this reference data varied from less than 5 cm at the beginning of the measurement trajectory (beginning of radiating cable) up to approximately 20 cm near the end of the cable, due to the measuring wheel slip. The receiving antenna was placed at a height of 0.3 m and 1 m above the floor. The cart speed was from 0.2 to 0.4 m/s along three parallel tracks spaced 0.5 m apart. Additional attenuation of the unconverted signal in the mixer allowed us to reduce the distance between radiating cables to 0.2 m only. In Figure 7, numbers near square marks indicate a measurement series.

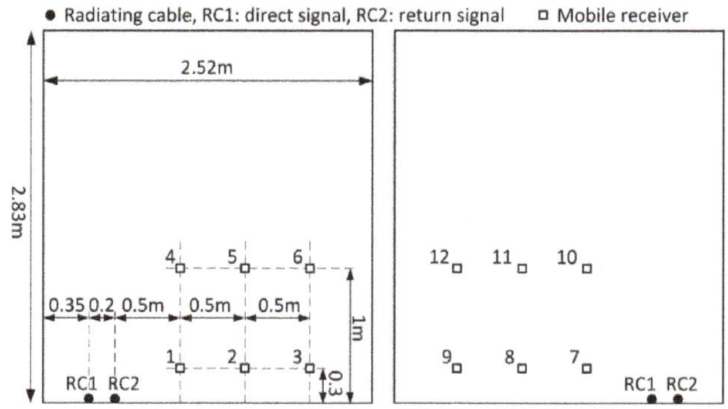

Figure 7. Position of radiating cables and receiving antenna in a corridor in the scenario with two feeders and frequency conversion.

Figure 8 shows a view of the corridor with radiating cables on the floor and the hand cart with the receiver and the wheel encoder.

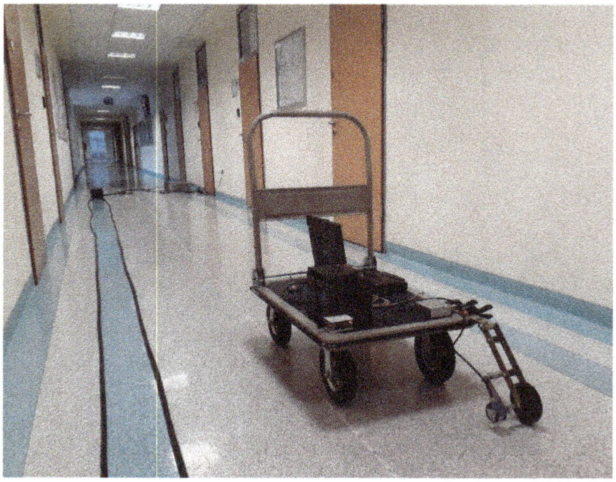

Figure 8. Measurement setup in the scenario with two feeders and frequency conversion.

The frequency of the signal from the second generator was set to 100 kHz, which was low compared to the transmitted signal's carrier frequency (2.45 GHz) and the occupied bandwidth (23.3 MHz). Yet when the pseudo random test signal was received by the correlation receiver with an integration time equal to 52.4 ms ($2^{20} - 1$ times the chip rate), even such a low frequency shift was enough to avoid spectrum despreading of unwanted recorded signal components. Relatively high signal attenuation without conversion in the balanced mixer (over 40 dB) ensured that the signal transmitted by the return cable was composed of only two components at frequencies 2.4499 GHz and

2.4501 GHz, which together occupied a bandwidth of 23.5 MHz. Therefore, the direct and return signals shared approximately the same spectrum. This method does not require wider channel bandwidth in comparison to the previous examples. For such a low-frequency shift, there is no possibility of using a diplexer to separate direct and return signals. In case of using only one radiating cable, the directional coupler is still needed.

The conversion of the return signal frequency ensured that the reception time of the direct signal was always related to the global maximum in the signal correlation function at a nominal frequency of 2.45 GHz. The reception time of the return signal was obtained from the global maximum of the signal correlation at frequencies 2.45 ± 0.0001 GHz (Figure 9). Therefore, fully automatic detection of both received signal components was trivial. However, examples of the correlation function obtained near the beginning (d_1 = 4.3 m), center (d_1 = 29.5 m), and end of the radiating cable (d_1 = 48.7 m) (Figure 9), shows that only in the center section of the cable shape of the correlation for the direct and return signals is almost the same. Distortion of the correlation function at both ends of the cable, caused by unequal conditions of emission of signals traveling in the opposite direction, may have a significant impact on position estimation accuracy.

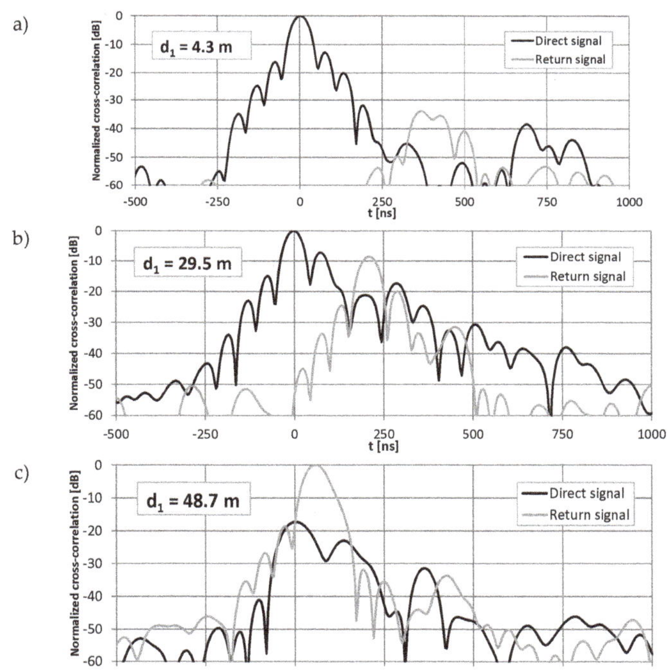

Figure 9. Examples of cross-correlation functions of direct and return signals recorded at d_1 = 4.3 m (**a**), 29.5 m (**b**) and 48.7 m (**c**) using two radiating cables with frequency conversion.

5. Results of Position Estimation

Due to the low quality of the reflected signal reception in the scenario with an unterminated radiating cable, we could not estimate the mobile receiver's position. It was caused by overlapping of the reflected signal main lobe with the higher-level side lobes from the direct signal. In this scenario, a wider bandwidth of the test signal would probably improve the discrimination of both components in the received signals. Results obtained from the two other configurations of the transmitting section are more promising.

5.1. Two Feeders with Amplifier

The results of position estimation along two radiating cables with return signal amplification (Figure 10) are median values calculated from 18 repetitions of PRBS signals received during one-second-long signal recordings by the stationary receiver. Error bars in Figure 10 represent the standard deviation of the results. Averaging the results was used to reduce random errors but also caused a reduced update rate to one result per second. If a higher update rate is needed, other methods of data filtering may be used, including the running average and Kalman filtering. Reducing random errors caused by ISM device interference may be achieved after choosing another frequency band.

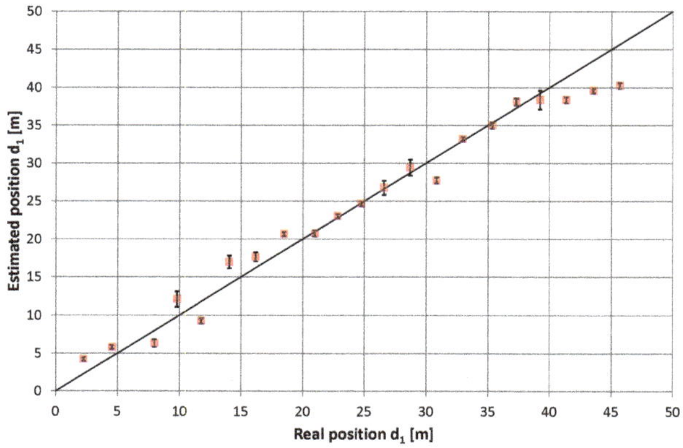

Figure 10. Results of position estimation in the scenario with return signal amplification.

In general, we saw a high level of repeatability for the obtained results. The standard deviation of position estimation at subsequent measuring points varied from 0.29 m to 1.23 m, but the average error of mean position was several times higher (from −5.41 m to +2.9 m). In addition to the random errors present during measurements, which were characterized by a standard deviation of obtained results, systematic errors were also present in many measurement points and had a higher impact on position estimation accuracy. The best accuracy was found near the center of the radiating cable. High values of mean errors occurred in certain sections of the measured area, which may suggest that it was caused by overlapping of the correlation lobes of the direct and return signals. The systematic shift, observed near both ends of the radiating cable, may be caused by uneven conditions of reception of direct and return signals travelling in opposite directions.

5.2. Two Feeders with Signal Frequency Conversion

Converting the return signal frequency should theoretically reduce the mutual impact of the overlapping lobes in the correlation function (i.e., after independent correlation of the direct and return signals). Therefore, the results of the position estimation presented in Figures 11–14 reflect the effects caused by the environment and limitations of the proposed positioning method. These measurements were taken on a different day than those presented in Figure 10 (and probably with a slightly different location of the radiating cable in the corridor). Thus, a direct comparison of both charts is not possible. However, there are similarities between results on the charts in Figures 10–12, such as systematic errors of position estimation at distances near 31–32 m.

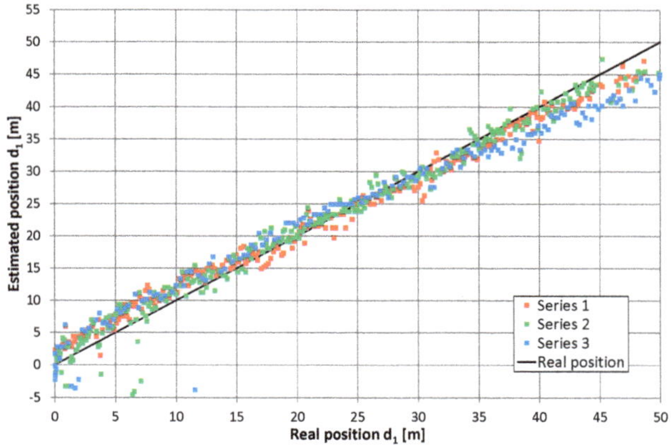

Figure 11. Results of position estimation in the case of return signal frequency conversion; the height of receiving antenna: 0.3 m, series 1–3.

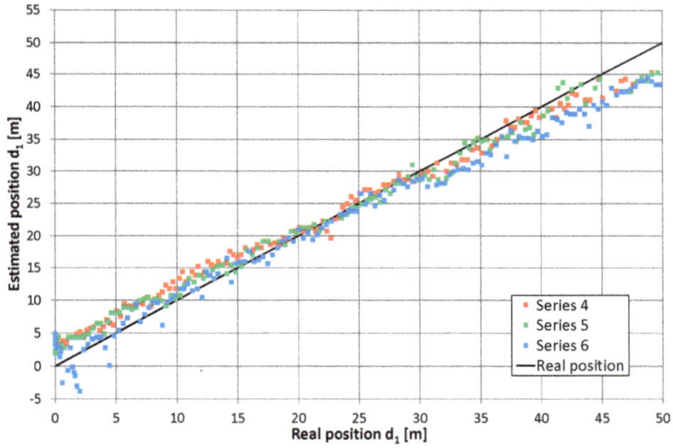

Figure 12. Results of position estimation in the case of return signal frequency conversion; the height of receiving antenna: 1 m, series 4–6.

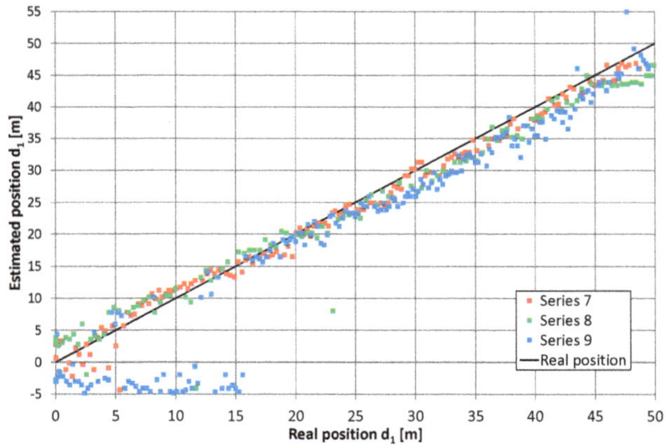

Figure 13. Results of position estimation in the case of return signal frequency conversion; the height of receiving antenna: 0.3 m, series 7–9.

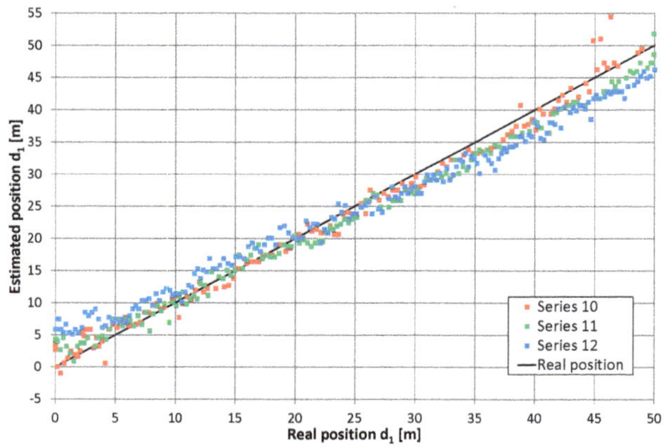

Figure 14. Results of position estimation in the case of return signal frequency conversion; the height of receiving antenna: 1 m, series 10–12.

An important difference in the method of measurements presented in this subsection was the receiver's movement. The estimated position of the receiver was a median value from 18 repetitions of correlation of PRBS signals recorded for one second by a receiver mounted on a moving hand cart. Measurements in motion allowed us to obtain more results in a limited time with an accuracy comparable to those presented in Section 5.1. However, we observed increased dispersion of 18 subsequent results for 1 s long measurements. This was not surprising, as the duration of one correlation of a PRBS20 sequence was comparable and even longer than coherence time for radiating cable communication [32]. Table 1 shows the obtained position estimation accuracy, where $\bar{\varepsilon}$ is the mean value of position error, defined as the difference between estimated and real coordinate d_1, while σ_ε is the standard deviation of errors in the final position estimate. The next two variables, included in Table 1, specify short-term data dispersion in a 1-s long measurement. $\overline{\sigma_s}$ is a mean value of the observed short-term standard deviation in the whole series. $\max(\sigma_s)$ is the maximal value of this parameter in the series. Both the mean value and standard deviation of the position estimation errors were comparable to results presented by other authors for systems with narrowband signals [17].

Table 1. Evaluation of position estimation accuracy.

Series	$\bar{\varepsilon}$ [m]	σ_ε [m]	$\overline{\sigma_s}$ [m]	$\max(\sigma_s)$ [m]
1	−0.02	1.95	1.46	3.74
2	0.05	2.17	2.3	8.25
3	−0.7	3.18	1.96	4.72
4	0.64	2.09	2.07	15.7
5	0.17	2.06	1.69	7.82
6	−1.32	2.61	2.27	5.98
7	−0.62	1.71	2.79	9.95
8	−1.02	2.93	3.47	10.7
9	−3.77	7.76	3.63	11.4
10	−0.17	1.71	2.57	7.57
11	−0.92	1.89	2.47	7.6
12	−0.39	3.23	1.73	3.48

Figure 15 presents the cumulative distribution function (CDF) of position estimation error, which is defined as the difference between the estimated and real value of coordinate d_1. In Figure 15, positive quantities indicate position estimates shifted toward the end of the cable with an amplifier, while negative results correspond to position estimates closer to the end of the cable with a signal generator. These charts show almost no difference between results obtained for the receiver antenna at height 0.3 m (series 1–3) and 1 m (series 4–6), as well as no systematic error (CDF equal 0.5 for error value close to zero) when the cable radiating direct signal was closer to the corridor wall. In the second configuration, with the cable radiating direct signal was placed closer to the center of the corridor, the mean error (0.6 m) can be observed for measurements with the receiver antenna at height 1 m (series 10–12). However, when the receiver antenna is 0.3 m above the corridor floor, several measurements returned to the incorrect position of −3.5 m, which is visible in Figure 13 for series no. 9 and in the cumulative distribution function for series 7–9. The almost equal value of position estimation in these incorrect results corresponded to the correct reception of the direct signal and incorrect reception of the return signal traveling through whole length of radiating cable number 2 and then reflected from the end of the cable which was correctly terminated with a 50-ohm load. Therefore, some impedance mismatch or "end effects" [27] probably occurred.

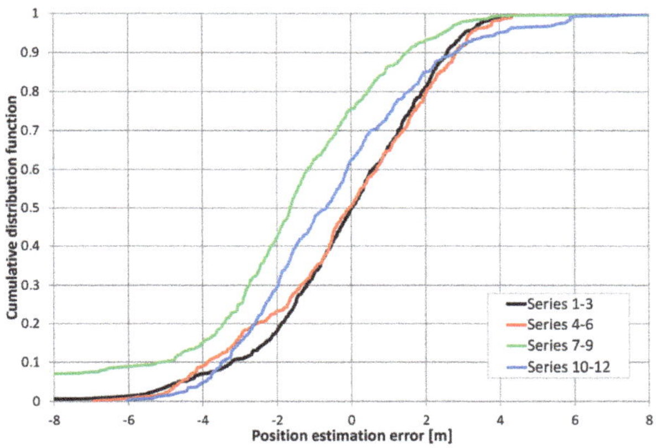

Figure 15. Cumulative distribution of position estimation errors.

Although position estimation errors in Figures 11–14 were apparently uncorrelated, they were probably caused by inhomogeneous distribution of EM fields inside the corridor. Moreover, they could be caused by a limited measurement setup (e.g., signal bandwidth, measurement resolution). Measurement repetition was performed in exactly the same conditions and showed a high level of error repeatability, which is clearly visible on exemplary charts in Figure 16. Presented results were obtained during three measurement repetitions in series number 12, with the receiver antenna placed 1 m above the corridor floor. Meanwhile, a hand cart moved along the same path with, at most, 5 cm accuracy. In all measurement repetitions, two kinds of errors were distinguished. The first was a systematic shift of position estimates near both ends of the radiating cable. Results obtained in this corridor section indicated that the receiver was closer to the center of the radiating cable; thus, it seemed that these systematic errors may be reduced after evaluating the nonlinear correction function. The second type of error was the repetitive local deviation from the general trend, which at many points exceeded 3 m. This was probably caused by an inhomogeneous building structure and a radiating cable coupling to the building structure. Compensation of these errors may be more difficult and require some kind of fingerprinting method.

We observed high values in position estimation errors for some measurement points, which cannot be explained by corresponding anomalies in the direct and return signals power levels. In general, signal levels along the radiating cables were not stable with random differences exceeding 15 dB. However, no significant changes in power level distribution were found in regions with higher position estimation errors.

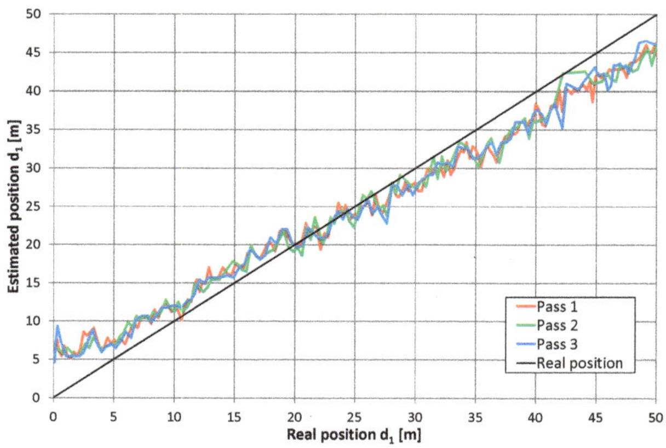

Figure 16. Results of position estimation in three measurement repetitions in series 12; the height of receiving antenna: 1 m.

In our experiment, the mobile receiver position estimation accuracy was comparable to results presented in the literature. For example, Weber et al. in [16] found position estimation using differences in signal power levels, showing an accuracy of 2.5 m in 50% of all cases and approximately 4.5 m at an 80% threshold level. Pereira et al. [19] had slightly worse results, with 20 m position estimation accuracy at 88% confidence level. They were obtained using GSM and WLAN signals, and they were not dedicated positioning signals. Nakamura et al. [21] showed variable accuracy of distance measurements, from 0.2 m to 8.1 m, with an average error value of 2.4 m. Therefore, it may be concluded that different methods of positioning with radiating cables are comparable for achieving accuracy.

6. Discussion on Position Estimation Errors

When evaluating the obtained position estimation accuracy, one should refer to sounding signal parameters, especially in terms of signal bandwidth, which is inversely related to time measurement resolution. For example, ultra-wideband (UWB) indoor positioning systems, based on IEEE 802.15.4 UWB modems using a 499.2 MHz bandwidth, allow for a ranging accuracy of several centimeters [33]. The positioning system based on Nanotron modules, which uses a chirp signal in 2.4 GHz ISM band with a bandwidth of 80 MHz, allows for 1.5 m distance measurement accuracy [34]. However, switching to a 22 MHz bandwidth results in three times worse accuracy. Therefore, the chirp-based solution with a 22 MHz bandwidth may be used as a reference to compare against the proposed solution. In case of the code-division multiple access (CDMA) signal reception in the presence of the Gaussian noise (e.g., AWGN channel, no multipath propagation), theoretical accuracy of tracking the peak of the cross-correlation function may be calculated from Equation (4) [35]:

$$\sigma = \frac{T_c}{\sqrt{2 \cdot SNR}}, \tag{4}$$

where σ is the standard deviation of peak time measurement, T_c is chip time (50 ns), and SNR is signal-to-noise ratio after spectrum despreading [35]. Exemplary charts, as presented in Section 4, show that during measurement, SNR exceeded 40 dB. Therefore, time measurement accuracy limit in the AWGN case reached 0.35 ns, which corresponded to a 0.09 m distance measurement error for radiating cable with v_{prop} = 0.88 c. However, such good accuracy was not reachable due to multipath propagation, which caused shape degradation of the correlation function's main peak, which is clearly visible on the first and third charts in Figure 9. Another reference for positioning accuracy evaluation may be the width of the main peak in the cross-correlation function of received signals. Laboratory measurements that used a cable connection between the signal generator and USRP receiver (no multipath or external interferences) gave a main peak width equal to 48 ns at −3 dB, which corresponded to a distance of 12.7 m. However, the time measurement resolution 4 ns, defined by the receiver's sampling frequency (25 MHz) and a 10-times up-sampling rate, corresponded to 1.06 m of a one-way distance measurement resolution. Therefore, we achieved position estimation accuracy close to the receiver's measurement resolution.

We mitigated time measurement uncertainty during tests by using the following tools: the accuracy of frequency standard in signal generators (at most 10^{-7}), the accuracy of a reference oscillator in the USRP receiver (at most 2×10^{-6}), the accuracy of the signal delay in the amplifier and connecting cables (±0.2 ns), and propagation speed v_{prop}. The impact of all other sources of uncertainty were several orders of magnitude smaller than the observed errors caused by inhomogeneous emissions of radio signals and cables that effect the environment.

The wide width of the main peak in the correlation function equaled 48 ns and was strictly connected to the limited possibility of separating the multipath components in the receiver. There was no possibility of investigating multipath phenomena using signals recorded by the setup presented in this paper. Signals received in the shorter period overlap, thus distorting the shape of correlation function and causing errors when detecting signals' timing based on peak tracking. However, in typical indoor or outdoor radio positioning systems, based on radio signal propagation in the air, time measurement quality may be improved with a leading-edge detector because the multipath components reach the receiving antenna after the signal travels in a straight (and short) path. However, in a positioning system based on a radiating cable, unwanted multipath components may be received before a signal radiated perpendicular to the cable, i.e., signal emissions with high power from a section of cable not close to a point near the receiver. This signal can reach the receiver after time of propagation in the air (with speed c), even if earlier than wanted component which has to travel through the radiating cable (with speed v_{prop} lower than c). Therefore, the advantage of slope detection over peak detection in positioning systems using radiating cables is questionable.

Results from all measurements clearly shows regularity. The best accuracy was available near the center of radiating cables, while position estimates obtained near both ends of the cable were systematically shifted toward the center. This effect was not caused by the wrong value of signal propagation speed in cable v_{prop}, because incorrect v_{prop} in the setup (Figures 4 and 6) would cause the best match near the end of the radiating cable and increase error in the region closer to the beginning of the cable. These systematic errors are likely caused by unequal radiation conditions when signals travel in the opposite direction in finite-length radiating cables. The direction of radiation of the EM field is skewed to the direction of signal propagation in the cable [12,29]. Additionally, the received signal is always superposition of components radiated from some section of the cable. Thus, the measurements taken near the end of the cable may correspond to different EM field distribution comparing to center section of the cable. In general, electromagnetic field emissions from radiating cables was not uniform [36–38], causing time measurements errors and a large variations of instantaneous received signal power values that exceeded 12 dB. This is visible on the power chart presented in Figure 17, as a random deviation from linear trend of power drop which was caused by leaky feeder longitudinal attenuation.

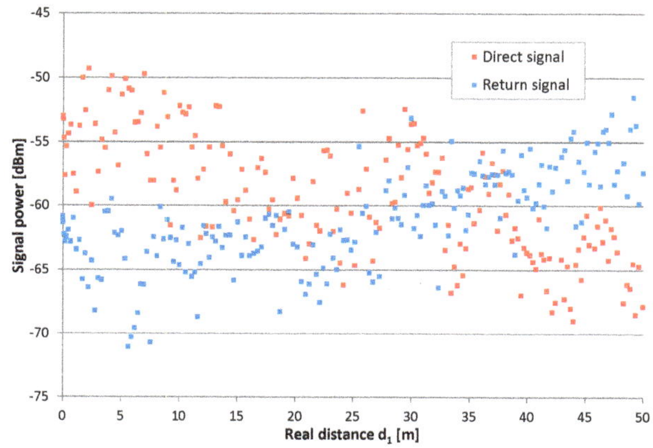

Figure 17. Exemplary values of the power level of direct and return signal, series 12.

A general trend of power level changes in Figure 17 is similar to the systematic character of the obtained position estimation error. The correction of position had a linear approximation of difference in the power of direct (P_{dir}) and return (P_{ret}) signals (in decibels) when using Equation (5):

$$d_1' = d_1 - 0.298 \cdot (P_{dir} - P_{ret}) + 0.95 \tag{5}$$

Only partially improved position estimation accuracy came from reducing the mean error $\overline{\varepsilon}$ to zero and standard deviation σ_ε from series 12 to 2.37 m (from 3.23 m without correction). It may be expected that a large dispersion of instantaneous power from the received signals was caused by a multipath fading phenomena and standing waves, which may have increased local dispersion of the position estimation results even when the systematic position shifted near both ends of the cable was reduced. It was not possible to remove the fading effect from the measured power levels without spatial data averaging over a long path, exceeding tens of wavelengths. Thus, data correction by using signal power levels is of limited use. Another method of data correction was evaluated using the least squares linear model of the position errors, which may be summarized as follows:

$$d_1' = d_1 + 0.233 \cdot d_1 - 5.28 \tag{6}$$

Such a simple correction allowed us to reduce a standard deviation to 1.46 m, giving better results than corrections based on differences in the received signal power. Both data correction methods are presented in Figure 18 using blue (correction based on the signal power levels) and green (correction based on a linear error model) lines, respectively. In general, both methods reduced systematic position shifts near the ends of the radiating cable. Both gave the mean value of the position error close to zero. Differences between them were visible not only for standard deviation value but also on local result variation (Figure 18). Maximal values of uncorrected errors in series 12 were −6.04 m and +7.29 m. Corrections based on signal power levels reduced maximal error values to −5.24 m and +6.71 m, while corrections based on the simple linear model gave maximal errors of −4.13 m and +3.75 m. Therefore, both data correction methods are able to reduce systematic position shift, which is visible in the raw data near both ends of the radiating cable. Unfortunately, reducing local anomalies in position estimates using received signal power levels was unsuccessful, as this method results in higher errors.

It is difficult to explain high values of errors visible on some charts in the 17–20 m and 30–32 m regions. The whole corridor was free from obstacles during measurements. The first region was situated near a staircase. The second region was not connected with any changes in the geometry of the corridor, but it turned out that, in this region, there was a boundary between two structural sections of the building with thick reinforced walls on both sides of the corridor instead of brick walls.

Taking into account that the radiating cable's signal emissions were connected with scattered EM fields, anomalies in position estimation may be caused by inhomogeneous geometry and building structure. Unfortunately, we could not find any area with a strictly homogeneous structure, because even in the outdoor environment, some underground infrastructure was always present (e.g., pipes, cables).

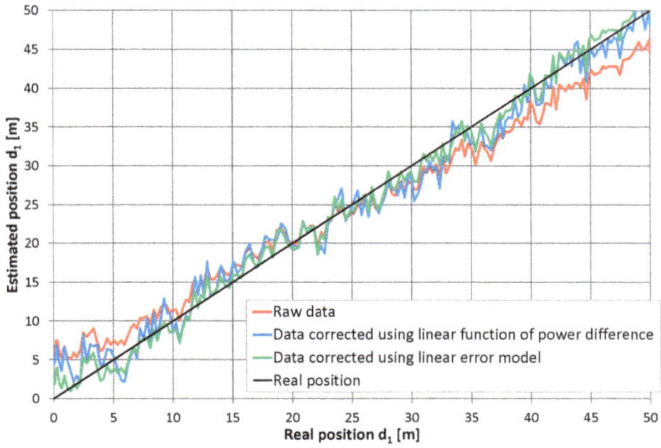

Figure 18. Results of position estimation in series 12 with correction based on the difference in direct and return signal power levels and corrections based on the linear model's position errors.

7. Conclusions

Estimating the receiver position along the radiating cable using time-difference measurements of a relatively narrowband (23 MHz) signal is possible and very promising. The obtained results can form a basis for developing radiolocation systems in corridors or tunnels where radiating cables are already installed, without the need to deploy a dense network of reference nodes required in UWB-based solutions. For a full scope of the possibilities presented by the proposed solution, extended measurements should be performed in different parts of the corridor or with different deployments of the radiating cable, such as under the ceiling and separated from any conductive elements. However, the proposed solution only allows for one dimensional position estimation in a limited area near the radiating cable, which may be assessed as the biggest disadvantage of radiating cable-based positioning.

Author Contributions: Conceptualization, J.S. (Jaroslaw Sadowski) and J.S. (Jacek Stefanski); formal analysis, O.B. and J.S. (Jaroslaw Sadowski); investigation, O.B., J.S. (Jaroslaw Sadowski), and J.S. (Jacek Stefanski); methodology, O.B. and J.S. (Jaroslaw Sadowski); supervision, J.S. (Jacek Stefanski); validation, O.B.; visualization, O.B.; writing—original draft, J.S. (Jaroslaw Sadowski). All authors have read and agreed to the published version of the manuscript.

Funding: This research received no external funding.

Acknowledgments: This work was supported under ministry subsidy for research for Gdansk University of Technology.

Conflicts of Interest: The authors declare no conflict of interest.

References

1. Liu, H.; Darabi, H.; Banerjee, P.; Liu, J. Survey of Wireless Indoor Positioning Techniques and Systems. *IEEE Trans. Syst. Man Cybern. Part C Appl. Rev.* **2007**, *37*, 1067–1080. [CrossRef]
2. Laoudias, C.; Moreira, A.J.; Kim, S.; Lee, S.; Wirola, L.; Fischione, C. A Survey of Enabling Technologies for Network Localization, Tracking, and Navigation. *IEEE Commun. Surv. Tutori.* **2018**, *20*, 3607–3644. [CrossRef]
3. Müller, P.; Raitoharju, M.; Piché, R. A Field Test of Parametric WLAN-Fingerprint-Positioning Methods. In Proceedings of the 17th International Conference on Information Fusion, Salamanca, Spain, 7–10 July 2014.

4. Wang, W.; Marelli, D.; Fu, M. Fingerprinting-Based Indoor Localization Using Interpolated Preprocessed CSI Phases and Bayesian Tracking. *Sensors* **2020**, *20*, 2854. [CrossRef] [PubMed]
5. Zayets, A.; Steinbach, E. Robust WiFi-based indoor localization using multipath component analysis. In Proceedings of the 2017 International Conference on Indoor Positioning and Indoor Navigation (IPIN), Hideo Makino, Jesús Ureña, 18–21 September 2017; pp. 1–8.
6. Xue, W.; Qiu, W.; Hua, X.; Yu, K. Improved Wi-Fi RSSI Measurement for Indoor Localization. *IEEE Sens. J.* **2017**, *17*, 2224–2230. [CrossRef]
7. Phutcharoen, K.; Chamchoy, M.; Supanakoon, P. Accuracy Study of Indoor Positioning with Bluetooth Low Energy Beacons. In Proceedings of the 2020 Joint International Conference on Digital Arts, Media and Technology with ECTI Northern Section Conference on Electrical, Electronics, Computer and Telecommunications Engineering (ECTI DAMT & NCON), Pattaya, Thailand, 11–14 March 2020; pp. 24–27.
8. Di Pietra, V.; Dabove, P.; Piras, M.; Lingua, A. Evaluation of positioning and ranging errors for UWB indoor applications. In Proceedings of the 2019 International Conference on Indoor Positioning and Indoor Navigation (IPIN), Pisa, Italy, 30 September–3 October 2019.
9. Barua, B.; Kandil, N.; Hakem, N. On performance study of TWR UWB ranging in underground mine. In Proceedings of the 2018 Sixth International Conference on Digital Information, Networking, and Wireless Communications (DINWC), Beirut, Lebanon, 14–16 November 2018; pp. 28–31.
10. Minghui, Z.; Huiqing, Z. Research on model of indoor distance measurement based on receiving signal strength. In Proceedings of the 2010 International Conference on Computer Design and Applications, Qunhuangdao, China, 25–27 June 2010.
11. Ji, M.; Kim, J.; Jeon, J.; Cho, Y.; Myungin, J. Analysis of positioning accuracy corresponding to the number of BLE beacons in indoor positioning system. In Proceedings of the 2015 17th International Conference on Advanced Communication Technology (ICACT), Seoul, Korea, 13–16 April 2015; pp. 92–95.
12. Nishikawa, K.-I.; Higashino, T.; Tsukamoto, K.; Komaki, S. A new position detection method using leaky coaxial cable. *IEICE Electron. Express* **2008**, *5*, 285–290. [CrossRef]
13. Moschevikin, A.; Serezhina, M.; Sikora, A. On the possibility to use leaky feeders for positioning in chirp spread spectrum technologies. In Proceedings of the 2014 2nd International Symposium on Wireless Systems within the Conferences on Intelligent Data Acquisition and Advanced Computing Systems, Offenburg, Germany, 11–12 September 2014; pp. 56–65.
14. Serezhina, M.; Moschevikin, A.; Evmenchikov, R.; Sikora, A. Using radiating cable for time-of-flight CSS measurements indoors and outdoors. In Proceedings of the 2015 IEEE 8th International Conference on Intelligent Data Acquisition and Advanced Computing Systems: Technology and Applications (IDAACS), Warsaw, Poland, 24–26 September 2015; pp. 91–101.
15. Engelbrecht, J.; Collmann, R.; Birkel, U.; Weber, M. Methodical leaky feeder design for indoor positioning considering multipath environments. In Proceedings of the 2010 IEEE Radio and Wireless Symposium, New Orleans, LA, USA, 10–12 October 2010; pp. 164–167.
16. Ott, A.T.; Shalaby, M.; Siart, U.; Kaliyaperumal, E.; Eibert, T.F.; Engelbrecht, J.; Collmann, R. Performance analysis of a low cost wireless indoor positioning system with distributed antennas. *Adv. Radio Sci.* **2011**, *9*, 79–84. [CrossRef]
17. Birkel, U.; Weber, M.; Collmann, R. Radiating Cable for indoor localization using UMTS. In Proceedings of the 2012 Ubiquitous Positioning, Indoor Navigation and Location Based Services, Helsinki, Finland, 3–4 October 2012; pp. 1–6.
18. Weber, M.; Birkel, U.; Collmann, R.; Engelbrecht, J. Comparison of various methods for indoor RF fingerprinting using leaky feeder cable. In Proceedings of the 2010 7th Workshop on Positioning, Navigation and Communication, Dresden, Germany, 11–12 March 2010.
19. Pereira, F.; Theis, C.; Moreira, A.; Ricardo, M. Multi-technology RF fingerprinting with leaky-feeder in underground tunnels. In Proceedings of the 2012 International Conference on Indoor Positioning and Indoor Navigation, Sydney, Australia, 13–15 November 2012; pp. 1–11.
20. Pereira, F. Positioning Systems for Underground Tunnel Environments. Ph.D. Thesis, University of Porto, Porto, Portugal, 2016.

21. Nakamura, M.; Takagi, H.; Terashima, J.; Einaga, K.; Nishikawa, T.; Moriyama, N.; Wasaki, K. Development of a Simple Multiple-Position Identifying System with a Long Range Multiband Leaky Coaxial Cable for Rescue Operations in Tunnels or Passages in Underground Facilities. In Proceedings of the Asia-Pacific Microwave Conference, Yokohama, Japan, 7–10 December 2010; pp. 163–166.
22. Shirai, K.; Higashino, T.; Okada, M. An Experimental Investigation of the MUSIC-based Wireless Position Location using LCX antenna at 5GHz band. In Proceedings of the 2019 19th International Symposium on Communications and Information Technologies, Ho Chi Minh City, Vietnam, 25–27 September 2019.
23. Inomata, K.; Hirai, T.; Yamaguchi, Y.; Yamada, H. Two-Dimensional Target Location Estimation Technique Using Leaky Coaxial Cables. *IEICE Trans. Commun.* **2008**, *91*, 878–886. [CrossRef]
24. Shah, S.I.; Shah, S.Y.; Shah, S.A. Intrusion Detection through Leaky Wave Cable in Conjunction with Channel State Information. In Proceedings of the 2019 UK/China Emerging Technologies, Glasgow, UK, 21–22 August 2019.
25. Hassan, N.; Fernando, X.N. Reduced side lobe MM-wave leaky feeder transceiver by slot space optimization. In Proceedings of the Global Symposium on Millimeter-Waves, Montreal, QC, Canada, 25–27 May 2015.
26. Wang, J.H.; Mei, K.K. Theory and Analysis of Leaky Coaxial Cables with Periodic Slots. *IEEE Trans. Antennas Propag.* **2001**, *49*, 1723–1732. [CrossRef]
27. Martin, D.J.R. A general study of the leaky-feeder principle. *Radio Electron. Eng.* **1975**, *45*, 205. [CrossRef]
28. Cao, H.; Zhang, Y.P. Radio Propagation along a Radiated Mode Leaky Coaxial Cable in Tunnels. In Proceedings of the 1999 Asia Pacific Microwave Conference, Singapore, 30 November–3 December 1999; pp. 270–272.
29. Guo, Y.C.; Zhang, Y.P. Radio Propagation along a Coupled Mode Leaky Coaxial Cable in Tunnels. In Proceedings of the 1999 Asia Pacific Microwave Conference, Singapore, 30 November–3 December 1999; pp. 273–275.
30. Aragon-Zavala, A. *Indoor Wireless Communications*; John Wiley & Sons: Hoboken, NJ, USA, 2017.
31. Delogne, P.P.; Deryck, L. Underground Use of a Coaxial Cable with Leaky Sections. *IRE Trans. Antennas Propag.* **1980**, *28*, 875–883. [CrossRef]
32. Pan, Y.-T.; Liu, X.; Zheng, G.-Z.; Guan, K. Temporal Autocorrelation of Small-Scale Fading Using Leaky Coaxial Cable in Confined Space. *IEEE Wirel. Commun. Lett.* **2018**, *7*, 1082–1085. [CrossRef]
33. Malajner, M.; Planinsic, P.; Gleich, D. UWB ranging accuracy. In Proceedings of the 2015 International Conference on Systems, Signals and Image Processing, London, UK, 10–12 September 2015; pp. 61–64.
34. Pivato, P.; Dalpez, S.; Macii, D. Performance Evaluation of Chirp Spread Spectrum Ranging for Indoor Embedded Navigation Systems. In Proceedings of the 7th IEEE International Symposium on Industrial Embedded Systems, Karlsruhe, Germany, 20–22 June 2012; pp. 307–310.
35. Yu, K.; Sharp, I.; Guo, J. *Ground-Based Wireless Positioning*; John Wiley & Sons: Chichester, UK, 2009.
36. Santos, V.; da Fonseca, F.; de Matos, L.; Meza, W.; Siqueira, G. Indoor Signal Coverage of a Leaky Feeder Cable. In Proceedings of the IEEE MTT-S International Microwave & Optoelectronics Conference, Rio de Janeiro, Brazil, 4–7 August 2013.
37. Buffi, A.; Nepa, P.; Tellini, B. Measurement System with Leaky Coaxial Cables operating as Distributed Antennas for UHF-RFID Readers. In Proceedings of the 2017 IEEE International Workshop on Measurement and Networking, Naples, Italy, 27–29 September 2017.
38. Jiang, J.; Wang, L.; Wang, G. Leaky Coaxial Cable for Near-field UHF RFID Applications. In Proceedings of the 2017 Sixth Asia-Pacific Conference on Antennas and Propagation, Xi'an, China, 16–19 October 2017.

sensors

MDPI

Article

Toward Accurate Position Estimation Using Learning to Prediction Algorithm in Indoor Navigation

Faisal Jamil, Naeem Iqbal, Shabir Ahmad and Do-Hyeun Kim *

Department of Computer Engineering, Jeju National University, Jejusi 63243, Korea; faisal@jejunu.ac.kr (F.J.); naeemiqbal@jejunu.ac.kr (N.I.); shabir@jejunu.ac.kr (S.A.)

* Correspondence: kimdh@jejunu.ac.kr

Received: 13 July 2020; Accepted: 6 August 2020; Published: 7 August 2020

Abstract: Internet of Things is advancing, and the augmented role of smart navigation in automating processes is at its vanguard. Smart navigation and location tracking systems are finding increasing use in the area of the mission-critical indoor scenario, logistics, medicine, and security. A demanding emerging area is an Indoor Localization due to the increased fascination towards location-based services. Numerous inertial assessments unit-based indoor localization mechanisms have been suggested in this regard. However, these methods have many shortcomings pertaining to accuracy and consistency. In this study, we propose a novel position estimation system based on learning to the prediction model to address the above challenges. The designed system consists of two modules; learning to prediction module and position estimation using sensor fusion in an indoor environment. The prediction algorithm is attached to the learning module. Moreover, the learning module continuously controls, observes, and enhances the efficiency of the prediction algorithm by evaluating the output and taking into account the exogenous factors that may have an impact on its outcome. On top of that, we reckon a situation where the prediction algorithm can be applied to anticipate the accurate gyroscope and accelerometer reading from the noisy sensor readings. In the designed system, we consider a scenario where the learning module, based on Artificial Neural Network, and Kalman filter are used as a prediction algorithm to predict the actual accelerometer and gyroscope reading from the noisy sensor reading. Moreover, to acquire data, we use the next-generation inertial measurement unit, which contains a 3-axis accelerometer and gyroscope data. Finally, for the performance and accuracy of the proposed system, we carried out numbers of experiments, and we observed that the proposed Kalman filter with learning module performed better than the traditional Kalman filter algorithm in terms of root mean square error metric.

Keywords: inertial navigation system; artificial neural network; motion tracking; sensor fusion; indoor navigation system

1. Introduction

Today, when most of the world is well explored, navigation resides an essential part of our society. Today's technologies enable us to use the navigation in an entirely new way than our predecessors could. After the invention of smartphones, a vast number of location-based services have been introduced. These location-based services help users to find a way to a certain point of interest. During the last two decades, after the Global Positioning System (GPS) reached it fully operational capacity, the significant of different kinds of location-based services depend on positioning and navigation capabilities have increased tremendously [1,2]. Currently, GPS is recognized as famous for calculating the user's current location using the satellite. The popular examples of navigation that using GPS are aviation, timing, agriculture, car navigation system and so forth [3]. Even though GPS is considered to be well-known technology for locating the target in an outdoor environment, but it is not feasible for an indoor navigation system as it requires a continuous connection to communicate with satellite [4]. There are many other reasons why GPS will not work in an indoor environment, for example, signal attenuation in an indoor environment because of weak GPS signal, and signal disturbs due to hurdles like steel and concrete walls. The disturbance and hurdles continuously penetrate and block the signal coming from the satellite [5]. Therefore using GPS, it is not reliable to calculate the precise user location in an indoor environment. Hence in consideration of these problems, the GPS is not reliable for indoor positioning systems (IPS) [6,7].

The IPS is a system that used certain information in order to locate the target in an indoor environment. This information includes radio waves, sensors data, WLAN nodes, magnetic field, acoustic signal and so forth [8]. Currently, significant research is being done in the area of indoor localization. However, still, there exist many problems faced by the users due to no standard solution or service for indoor positioning [1]. Nevertheless, many technologies exist that can be used to calculate the position in an indoor environment. The problem with these services is that they were created for other purposes rather than to locate persons or objects, which sometimes make them very unreliable. These issues lead to the development of many miniaturized chips specifically for determining the object or a person in an indoor environment. These chips are called inertial measurement unit (IMU) [9,10].

IMU is an electronic device that is used to measures and detects the body orientation, angular rate, and body-specific force using a combination of accelerometers, gyroscopes, and sometimes magnetometers. During the past several years many IMU has been designed in order to get the precise position estimation in indoor environment [9]. IMU provides a 3-axis sensor, that is, accelerometer, gyroscope, and magnetometer. These sensor data is used to calculate the position of the target in an indoor environment. Double integration is the popular method to calculate the position of the object using accelerometer with respect to time. Similarly, for orientation estimation, Euler angle is used, which includes the information of roll, pitch and yaw using gyroscope data. However, these sensors readings have dynamic noise and bias in their measurements; therefore, we used a different type of filter, for example, Kalman filter, and alpha-beta filter and so forth that are responsible for removing these noise from sensors readings [11,12].

Many solutions have been suggested to predict the position using machine learning (ML). These model uses historical data that reflect the behaviour of the process being modelled. Machine learning techniques for predicting accurate position estimation includes ANNs, adaptive neuro-fuzzy inference systems (ANFIS), support vector machine (SVM), and extreme learning mechanism (ELM). The ANNs method has several advantages over conventional NN as it is easy to use, fast to learn, provide good generalization results, has minimum inaccuracies in training and achieves minimum standard weights. Nowadays, deep learning methods are used in many areas for predictive purposes, such as deep neural networks, deep networks of faith, and recurring neural networks [13,14].

The main contribution of the proposed position estimation based on learning to prediction approaches are followed as:

- The main objective of the proposed system is to get an accurate position estimation by minimizing the error in IMU sensor readings using the prediction algorithm.
- The learning module is based on Artificial Neural Network, and Kalman filter are used as a prediction algorithm to predict the actual accelerometer and gyroscope reading from the noisy sensor reading.
- The learning module continuously controls, observes, and enhances the efficiency of the prediction algorithm by evaluating the output and taking into account the exogenous factors that may have an impact on its outcome.
- In position estimation module, the Kalman filter is used to fuse the IMU data to get noise and drift-free position in an indoor environment.
- Finally, for evaluating system performance, we analyzed the results using the well-known statistical measures such as RMSE, MAD, and MSE. Our proposed system experiments indicate

that learning to prediction algorithm improves the system accuracy as compared to tradition prediction algorithm.

Permitting prediction algorithms to encounter ever-changing data or varying surrounding conditions is a demanding job. In this study, we introduce a comprehensive architecture to gain the precision and execution of the prediction module by applying the learning module in indoor navigation. We have used 3-axis sensor values, that is, accelerometer, and gyroscope, which is acquired from the IMU sensor in order to calculate the orientation and position estimation. The design system is comprised of two modules, that is, learning to prediction module and position estimation using sensor fusion in indoor navigation. The learning module is based on ANN and is continuous monitors the prediction algorithm performance by analyzing the output as feedback. The learning module is also responsible for considering the external parameters, (i.e., bias and drifting error) that may affect the outcome of the prediction algorithm. Once, the learning module updates the adjustable settings or commutes the trained model of the prediction algorithm to raise its efficiency regarding prediction accuracy. Similarly, for the learning model, we have used the back-propagation neural network for predicting the accurate parameter to tune the prediction algorithm. The hidden layer comprises of ten neurons, a total of three inputs are assigned to three input layers, and the output layer contains one neuron. The linear and sigmoid functions are employed as activation functions. The structure of the rest of the paper is organized as follows: Section 3 delineates a brief overview of contemporary state-of-the-art approaches; Section 4 encompasses details about the proposed heuristic model. The empirical analysis of the experiments carried out in this study is explained in Section 4, and Section 5 concludes the paper with directions for future work.

2. Related Work

In navigation, indoor navigation and tracking is a crucial process due to the limited available resources, that is, less GPS signal and satellite availability and so forth. Over the last few years, several location estimation algorithms have been proposed to calculate the distance travelled in indoor and outdoor environments [15,16]. These algorithms are segregated into six categories, that is, fingerprinting, connectivity/neighbourhood, triangulation, inertial and motion sensor, proximity, and dead reckoning [17]. However, in this study, our primary focus is to discuss the IMU-based inertial and motion sensor applications with pros and cons in an indoor navigation system. The overviews of the approaches mentioned above are summarised in Figure 1.

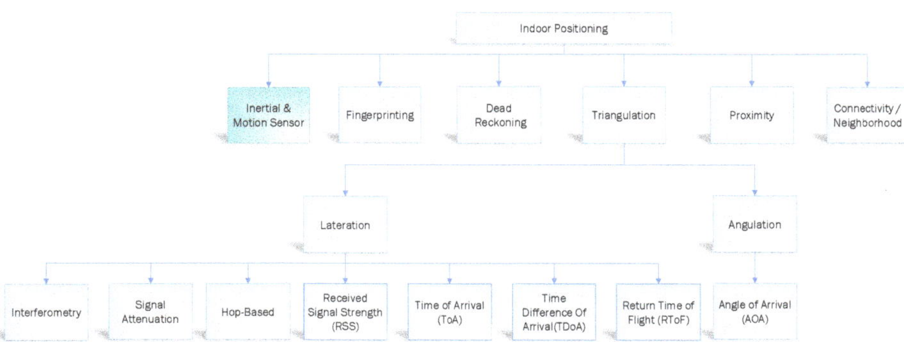

Figure 1. Taxonomy of indoor positioning system.

2.1. Inertial and Motion Sensor

Inertial and motion sensors are the types of sensors that use information, for example, acceleration, gyroscope, and magnetometer and so forth to calculate the position of the object in an indoor environment. This sensor information like accelerometer is used to calculate the position estimation using double integration method. Thus, the gyroscope tends to determine the orientation using the roll, pitch and yaw. Likewise, the magnetic field direction pertinent to the earth is calculated using the magnetometer. There are many systems proposed during the last few years that uses inertial and motion sensor. The contemporary inertial and motion sensor-based application discussed in the literature below [17].

In Reference [18], the author proposed an algorithm for calculating the orientation of the body using the MEME gyroscope. Moreover, this study also keeps track of sports activity using the IMU and improve the orientation using an extended Kalman filter by removing the uncertainty from the measurement. The performance of the system is measured using the VICON OPTICAL, and it shows that the system is accurate with less root mean square error.

The author of Reference [19] presented a model technique to improve the orientation using the reading output from the inertial measurement unit. A new sensor fusion algorithm MUSE has been implemented for orientation tracking. MUSE is a magnetometer-centric sensor fusion algorithm used orientation tracking. Moreover, this paper also proposed a new sensor fusion method to fully leverage the restriction of human arm movement by shoulder joints and elbow movement.

In Reference [20], the author presented an approach based on artificial neural network. The proposed system is adequate for combining the artificial neural network with the inertial measurement unit in order to get the accurate pedestrian positioning system. The developed system comprised of two possible states, that is, stationary state while the object is not moving regardless of its orientation, and the second is the object is moving equipped with IMU on his body. The further state can also be added to the classification results for the ANN, for example, shaking, jogging, spinning, falling, driving, and flying.

The author of Reference [21] presented an integrated navigation system using a fuzzy logic adaptive Kalman filter (FLAKF). The system is used to overcome the dynamic noise of the accelerometer output and also detect the bias in the sensor reading and resolve the error from the conventional Kalman filter. The main goal of this study is to adjust the weight of the traditional Kalman filter

In Reference [22], the author presented a distance measuring technique using two methods. The first method for measuring distance is integrating twice the acceleration to get the position estimation. However, in this method, the results are not satisfactory due to exponentially increase in error. Therefore to prevent error, the second method has been implemented, which count the number of steps and angles between legs during movement. The second method uses the accelerometer and gyroscope data to calculate the numbers of steps and angle. The main advantage of the second method is low cost and probability sensor circuit.

2.2. *Connectivity/Neighborhood*

Connectivity/Neighborhood is a method that can be used to analysis of connectivity, that is, numbers of attainable neighbours. In this approach, the numbers of reference points are defined, and they have spatially disseminated the object through which the reference point establishes a connection with the neighbour. In case of signal coverage overlap between the reference point in a suitable way, then the location of the object can be measured using the intersection of all its neighbour's coverage areas. The accuracy of the presented approach depends on the number of reference points, their distribution and coverage in terms of signal range [17,23].

2.3. *Proximity*

A proximity-based indoor positioning system aims to provide the specific point relative location information to the user whose corresponding point is close proximity. The receiver is used to determines the position of the user when the user is close to the product or an object which is directly connected to the corresponding receiver. The connection to the receiver is based on the signal strength. In case the user is beyond a single receiver then it can be linked with the receiver with the highest signal strength. The proximity-based system can be developed using cell ID (CID), radio frequency identification (RFID), Bluetooth, and infrared radiation (IR) and so forth. These technologies can be used to estimate the location of the users. In the case of CID, a unique number code is used to determine the base transceiver station (BTS). Since the CID of the BTS is received by the mobile users to which they are connected through which the location of the mobile user can be approximated to the proximity of the BTS with the CID information [17,24,25].

2.4. Triangulation

The triangulation based indoor position estimation system is based on the geometric properties, which is similar to GPS uses for an outdoor environment. The target location can be identified by several reference points using the angles. The calculation of position estimation using the above-discussed triangulation is called angulation. The common algorithm to measure the angulation is the angle of arrival (AOA). Similarly, to calculate the position using triangulation, we can also use lateration. In the case of lateration the distance of the target location is measured using several reference points. The common algorithms to calculate the lateration are interferometry, hop-based, signal attenuation, received signal strength (RSS), time difference of arrival (TDoA), time of arrival (ToA), and return time of flight (RToF) [17,26,27].

2.5. Dead Reckoning

Dead reckoning is the process of calculating the user's current position using the previously calculated position based on estimated speeds over elapsed time and course. The common example using the dead reckoning are animal navigation, marine navigation, localization of mobile sensor nodes, air navigation, automotive navigation, pedestrian dead reckoning, and directional dead reckoning and so forth [17,27].

2.6. Fingerprinting

The fingerprinting algorithm comprises of two steps the training of data and its positioning. The step in training is responsible for constructing a database which contains fingerprints, and in case of step positioning, location estimation is measured using database comparison with existing computing signal strength. In the training process, the database is constructed using the chosen reference point through which the access point signal strength is computed. Finally, each reference point is stored in the database. Similarly, in the positioning step, the signal strength of all the access points is used to measured the target and then compared it with data stored in the database by a deterministic or probabilistic approach. The output of the positioning step is the approximated position of the target. The performance of fingerprinting improves with an expansion in the number of reference points measurements and reference points. The increased number of datum points increases the accuracy as well. Though the training step for fingerprinting is a very burdensome work, and it is demanding in an active indoor condition such as an airport [17,27].

2.7. Navigation using Machine Learning Approaches

Nowadays, many machine learning algorithms are used to measure and predict body motion for wearable devices based on IMU Data. Several machine learning systems are working in diverse domain from marketing to medical services [28–34]. In Reference [35], a fingerprinting based indoor positioning uses a deep neural network to reduce the error in positioning. Similarly, in Reference [36], the author introduced a location-based car park system based on the conventional neural network. This system is used to localize and identify the car in the parking area. In another study, the two indoor localization techniques using machine learning algorithms are used to improve the indoor localization, that is, dead reckoning (DR) and data fusion [37]. In the first method, the DR technique uses an inertial sensor to improve the robustness and continuity of the indoor localization. Similarly, in the second approach, the data fusion approach integrated with machine learning model is used to predict the uncertainty in the wireless-based localization. From the last several decades, ' many indoor studies have been introduced, which uses the machine learning approaches to predict and track the location of the object in an indoor environment [38]. The contemporary indoor localization integrated with machine learning algorithms uses a different kind of input data such as, inertial sensor data [39], camera data [40], sound data [41] and LiDAR (light detection and ranging) [42]. These input data can be used for several intents, for instance, pass data as an input to the machine learning and get output. Most of the studies output three dimensional location data (x, y, z) axis, angle information [43], angle of arrival [44], distance [45], and object movement status [46].

Table 1 shows the critical analysis of indoor localization based on machine learning algorithms. We concise the comparative analysis into four categories, such as inertial measurement unit data [39,46], radio signal strength [45,47–50], channel state information [51], and angle of arrival [52].

Table 1. Comparative analysis of Indoor localization using machine learning approaches.

Approach	Reference	Input Data	Machine Learning Algorithm	Hidden Layer	Output
Inertial Measurement Unit Data	[39]	Inertial Sensor Data (acclerometer, gyroscope, magnetometer)	Artificial Neural Network	2–4	Step Length
	[46]		Recurrent Neural Network	4	Static Detection
Radio Signal Strength	[47]	WiFi Data (Access point, nodes)	Feed-Forward Neural Network	1–3	Location
	[48]		Generative Adversarial Neural Network	3	Distance
	[49]		Artificial Neural Network	1	Location
	[50]		Radial basis Function Neural Network	1	Location
	[45]		Adaptive Neural Fuzzy Inference System	3	Distance
Channel State Information	[51]	WiFi Data (Access point, nodes)	Generalized Cross-correlation	1–2	Location
Angle of Arrival	[52]	Radio, Optical or Acoustic	Convolution Neural Network	8	Location
Learning to Prediction	**Proposed Solution**	**Inertial Sensor Data (acclerometer, gyroscope, magnetometer)**	**Artificial Neural Network**	**10**	**Position**

As below mentioned, Table 1 related to indoor navigation system based on machine learning techniques have many drawbacks in terms of system accuracy and performance. These approaches directly use sensor data as an input to the machine learning algorithm in order to predict and identify the object or location. These sensor reading contains the bias and drifting error which affect the accuracy of the position estimation. However, in the proposed system, we use a prediction algorithm, which is used to minimize the noise in the sensor reading. Moreover, the prediction algorithm is to monitor and control using the artificial neural network to enhance the prediction accuracy of the design system.

As stated above, these systems are not adequately designed for indoor navigation and also have some overcoming in terms of accuracy. To the best understanding of the author, there has been no working tracking scheme for indoor navigation systems appertaining to learning prediction model created as yet.

3. Proposed Methodology

The three-axis sensor like accelerometer and gyroscope scope is acquired from next-generation IMU in order to compute the angular velocity (L_w) and linear acceleration (L_a) in the proposed position estimation system. The designed system consists of learning to prediction module and the position estimation using sensor fusion in an indoor navigation system. The first module briefly explains the step-by-step working of position estimation using sensor fusion in indoor navigation, and the second module describes the detailed learning to prediction module using ANN and Kalman filter algorithm. Next, we briefly explain the step-by-step working of these models.

3.1. Scenario of Position Estimation in Indoor Navigation

The proposed position estimation system is divided into two modules, that is, the position estimation using sensor fusion and learning to prediction module. The position estimation further divided into four sub-modules (i.e., sensor fusion based on Kalman filter algorithm, IMU acceleration, Integrator, and position estimation) as shown in Figure 2.

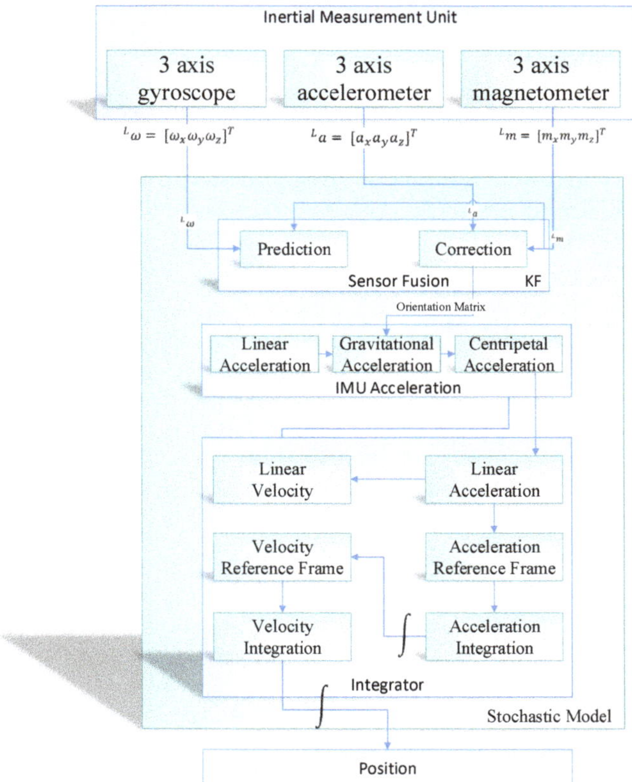

Figure 2. Position Estimation for proposed Indoor Navigation System.

The 3-axis output produced by the IMU in the form of a magnetic field vector, angular velocity, and linear acceleration is passed to the sensor fusion module. In the sensor fusion module, the output from the IMU is fuse together using the Kalman filter to get the drift-free and noise-free orientation of the object in an indoor environment. The Kalman filter in the sensor fusion module works as a prediction algorithm, which comprised of two parts, that is, prediction and correction [53,54]. The prediction is calculated by taking the integral of gyroscope measurement and then correct the prediction using the accelerometer and magnetometer readings. The fusion that happens inside the Kalman filter is a probabilistic fusion that combines and correct the inputs based on maximum likelihood. After calculation the drift and noise-free orientation from IMU, the orientation matrix is passed to the IMU acceleration module to remove the gravity and centripetal force. The processed data is further passed to the integrator module in which the velocity is calculated by taking the integral acceleration. Once getting velocity, then take the second integral of velocity to get the position of an object. The integrator is used to performs the mathematical operation of integration concerning time and calculate the velocity and position.

In an indoor environment, the position and orientation of the object are determined by the non-linear matrix. In past researches, accelerometer, magnetometer, and gyroscope's output gives the orientation estimation, which helps in finding the object orientation. In the case of gyroscope, the complete adjustment could not be calculated at once considering the tendency related to gyroscope readings, as shown in Equation (1). Integrate the angular velocity to get the orientation (roll, pitch, yaw). However, in this case, we have a drifting error.

$$L_{\omega} = L_{\omega_{true}} + b_g + v_g. \tag{1}$$

Here, L_{ω} is angular velocity vector in the local sensor frame. v_g is a frequency noise, b_g is a sensor bias a low-frequency noise we also considered bias as a constant in a small window of time and $L_{\omega_{true}}$ is a true angular velocity that we get from the sensor.

In case of acceleration, assuming there is no motion of the body, the only acceleration measure of gravity is divided by three-component and then uses some trigonometry and gets the roll and pitch with respect to the vertical axis. However, in this case, the accelerometer has a very high-frequency noise, as shown in Equation (2).

$$L_a = L_{a_{body}} + L_g + b_a + v_a, \tag{2}$$

where L_a is the acceleration vector in the local frame, $L_{m_{true}}$ represents the true acceleration due to the motion of the person or object or in other words linear acceleration, L_g is the gravitational acceleration, v_a is a frequency noise, b_a is a sensor bias a low frequency noise we also considered bias as a constant in a small window of time.

In the case of the magnetometer, it is used to calculate the yaw. Therefore, we combine both accelerometer and magnetometer to calculate orientation estimation. However, in this case, we still have noise estimation as represented in Equation (3).

$$L_m = L_{m_{true}} + L_{m_{int}} + b_m + v_m, \tag{3}$$

where b_m and v_m represent the bias and noise. $L_{m_{true}}$ represents the true magnetic field which is the earth magnetic field that is used for heading. $L_{m_{int}}$ is the magnetic disturbances. In order to solve the issue of bias and noise, the idea would be to get the best out of two kinds of estimation. The estimation, which does not have a lot of noise and at the same time does not have the drift. Therefore the best way to get the drift-free and noise-free orientation is to fuse together all the sensors, as shown in Figure 3. In the proposed system, we used the Kalman filter algorithm as a prediction algorithm in order to apply the sensor fusion. In this case, the first step is the prediction step in which we get the data from the IMU sensor and calculate the estimate through the mathematical model and then correct the estimation with measurement (i.e., correction). The Kalman filter comprised of two steps, that is, prediction and correction, in case of orientation estimation of IMU, the prediction is calculated by the integration of gyroscope reading and then correct the measurement through accelerometer and magnetometer reading. The fusion that happens inside the Kalman filter is a probabilistic fusion that combines and corrects the input based on maximum likelihood. In Figure 3, the σ_{ω} is the uncertainty in the gyroscope measurement, and angular velocity is symbolized as ω. Similarly, a is represented as acceleration and σ_a is the uncertainty in acceleration measurement. Finally, the magnetic field is denoted as m and σ_m is the uncertainty in magnetometer measurement.

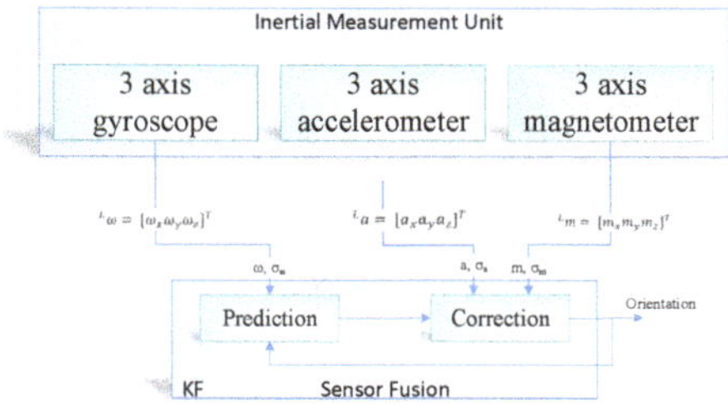

Figure 3. Orientation estimation using sensor fusion based on the Kalman filter algorithm.

3.1.1. Sensor Fusion Using Kalman Filter

In the study, we used the nonlinear version of the Kalman filter algorithm. The Kalman filter linearized the mean based on the state-space and compute the covariance-based on current state estimation. The Kalman filter is based on a discrete dynamic equation, which can be represented based on two phases; the first phase is the prediction phase, and the second one is the updated measurement phase. In the sensor fusion module of the proposed model, we have used the Kalman filter. The Kalman filter takes the linear acceleration, angular velocity, and magnetic field vector as an input from the sensor and provides the corrected predicted parameter as output. The working flow of the Kalman filter is shown in Figure 4. First, the initial estimation error covariance and initial state are to be determined and calculated using Equations (4) and (5). Besides input, measurement values, process and measurement noise covariance are also considered as an input. Equation (5) helps in finding the gain of Kalman. Both the estimation error covariance and the state estimation based on Equations (7) and (8) for their correctness.

The time update equation of the Kalman filter is:

$$\hat{x}_{\bar{k}} = \hat{x}_{k-1} + B_k \tag{4}$$

$$P_{\bar{k}} = AP_{k-1}A^T + Q. \tag{5}$$

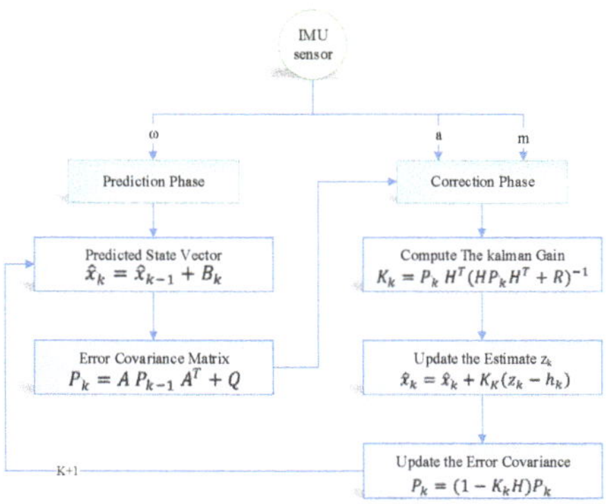

Figure 4. Configuration diagram of Kalman Filter in indoor navigation.

The state update equation of the Kalman filter is:

$$K_k = P_{\bar{k}} H^T (H P_{\bar{k}} H^T + R)^{-1} \tag{6}$$

$$\hat{x}_k = \hat{x}_{\bar{k}} + k_k (z_k - h_k) \tag{7}$$

$$P_k = (I - K_k H) P_{\bar{k}}, \tag{8}$$

where the predicted state-vector is represented as $\hat{x}_{\bar{k}}$, which includes 3-axis coordinates along with the heading using the proposed sensor fusion technique. B_k is the state space model matrix. Furthermore, $P_{\bar{k}}$ is the error covariance matrix which is comprised of two parts—(i) predicted error noise symbolized as (Q); (ii) State matrix which is denoted as A. The P_k value is modified at every iteration using three parameters. The parameters are predicted error covariance matrix ($P_{\bar{k}}$), obtained Kalman-Gain (K_k), and updates the measurements (H).

3.1.2. IMU Acceleration

The output from the IMU accelerometer is divided into three component, that is, IMU acceleration, gravity correction, centripetal force correction, as shown in Figure 5.

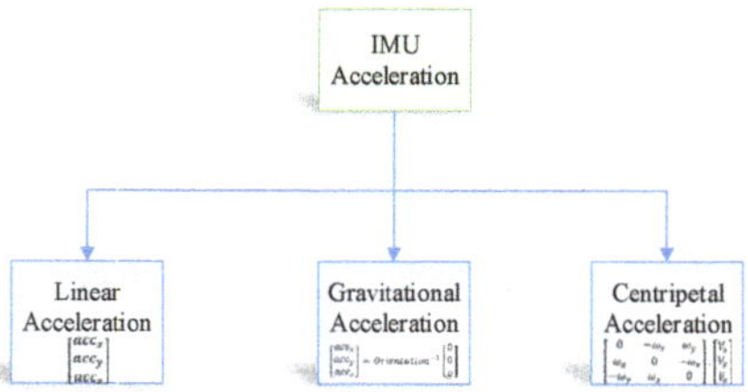

Figure 5. Inertial measurement unit (IMU) accelerometer output.

In the proposed position estimation system, we use linear acceleration in order to obtain the position in an indoor environment; therefore, it is essential to remove both centripetal and gravitational force from the sensor data. The IMU fixed frame contains gravity acceleration, which can be formulated in Equation (9)

$$\begin{bmatrix} acc_{x_{gravity}} \\ acc_{y_{gravity}} \\ acc_{z_{gravity}} \end{bmatrix} (t) = Orientation^{-1}(t) \begin{bmatrix} 0 \\ 0 \\ g \end{bmatrix}, \tag{9}$$

where gravitational acceleration magnitude is symbolized as g and acceleration vector with respect to earth fixed frame is denoted by $\begin{bmatrix} 0 \\ 0 \\ g \end{bmatrix}$. Similarly gravity acceleration vector with respect to IMU fixed frame symbolized as $\begin{bmatrix} acc_{x_{gravity}} \\ acc_{y_{gravity}} \\ acc_{z_{gravity}} \end{bmatrix}$.

The rotation of the object can take place in two ways—(i) the object rotation around a point in the space and (ii) the object rotation around itself. Therefore, in the proposed system the rotation of the object is calculated using the Equation (10):

$$\begin{bmatrix} acc_{x_{centripetal}} \\ acc_{y_{centripetal}} \\ acc_{z_{centripetal}} \end{bmatrix} = \begin{bmatrix} 0 & -\omega_z & \omega_y \\ \omega_z & 0 & -\omega_x \\ -\omega_y & \omega_x & 0 \end{bmatrix} \cdot \begin{bmatrix} V_x \\ V_y \\ V_z \end{bmatrix}. \tag{10}$$

The centripetal force is defined as the cross product of the linear velocity and angular velocity.

3.1.3. Integrator Module

Integrator is used to performs the mathematical operation of integration with respect to time and calculate the velocity and position from the linear acceleration [55]. The integrator module is based on the following steps. In the first step, we take the integral of acceleration measurement to get the velocity of the object in an indoor environment, as shown in Equation (12). In Equation (11), a denotes the acceleration measurement taken from the IMU sensor.

$$a = constant \tag{11}$$

$$v = \int adt = v_o + at. \tag{12}$$

In the second step, we take the integral of the computed velocity to get the position of the object in indoor environment. The Equations (12),(13) and (14) demonstrate the second integration, where y donates position, velocity is denoted as v_0, t represents time and a is the acceleration.

$$y = \int vdt \tag{13}$$

$$y = \int (v_o + at)dt \tag{14}$$

$$y = y_o + v_o t + \frac{1}{2}at^2. \tag{15}$$

3.2. *Learning to Prediction Model*

The proposed learning to prediction model is classified with a couple of modules, that is, the prediction algorithm and learning module. Traditionally, the historical data are used to train the prediction algorithm so that the relationship and hidden pattern can be learned among the output parameters and the input parameters. Afterwards, the output of any given input data is predicted using the trained model. The performance of the prediction algorithm depends upon a couple of things. The training data conditions are as same as the application scenario data and the input data. Nevertheless, none of the current prognostication algorithms adopts model well enough to train dynamic input states. Therefore, to vanquish the existing studies' limitations, we design a novel learning to a prediction model for the proposed indoor navigation system, as illustrated in Figure 6.

In the proposed learning to prediction model, the prediction algorithm is tuned using the learning module in order to improve the accuracy and performance of the prediction algorithm. Furthermore, the learning module continuously evaluates the performance of the prediction algorithm by receiving the output as feedback. The external parameter is also considered by the learning module that may create an impact on prediction algorithms in terms of performance. After monitoring the current output and external factors, the learning module updates the tunable parameters of the prediction algorithm or upgrade the complete trained model in the prediction algorithm to improve the prediction accuracy.

The design learning to prediction model comprised of two-part, that is, learning model and the prediction model. The learning model is based on ANN, and the prediction model uses a Kalman filter algorithm as a prediction algorithm, as demonstrated in Figures 7 and 8. In Figure 7, the Kalman filter algorithm is used to predict the actual accelerometer readings from the noisy accelerometer sensor readings, which are heavily influenced by the gyro bias. Traditionally the Kalman filter algorithm does not require historical data for prediction, but it only requires the previous state to predict the actual state of the system. Similarly, in the case of the learning module, we have used the feed-forward back propagation neural network (FFBPNN) to tune the prediction algorithm. The learning module takes three inputs, that is, accelerometer reading, gyroscope reading, and the Kalman filter predicted reading as feedback. The accelerometer sensor reading A_t is passed to Kalman filter as input at time t, and the output from the Kalman filter is the predicted accelerometer sensor reading P_a without noise. Noise in accelerometer sensor readings is due to the gyro bias, which is G_t.

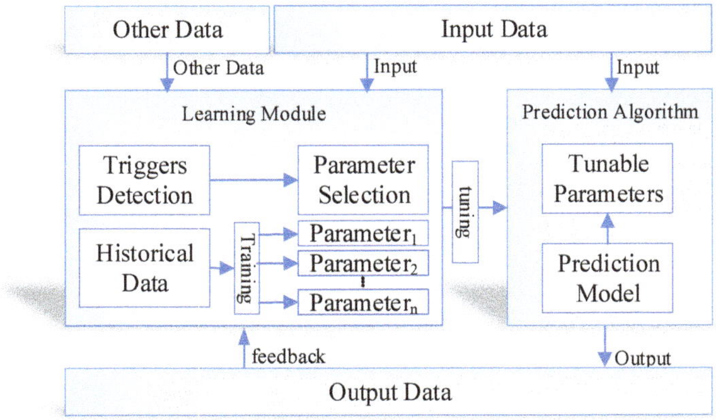

Figure 6. Conceptual view of the proposed learning to prediction model in indoor navigation system.

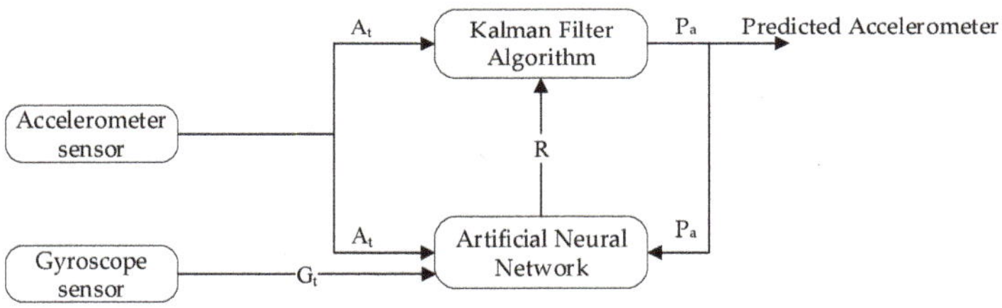

Figure 7. Block diagram of Accelerometer prediction using learning to prediction model.

Similarly, in Figure 8, the gyroscope reading G_t is predicted using the Kalman filter algorithm, and the output is predicted gyroscope sensor reading P_g without noise. Noise in the gyroscope is due to the impact of accelerometer value.

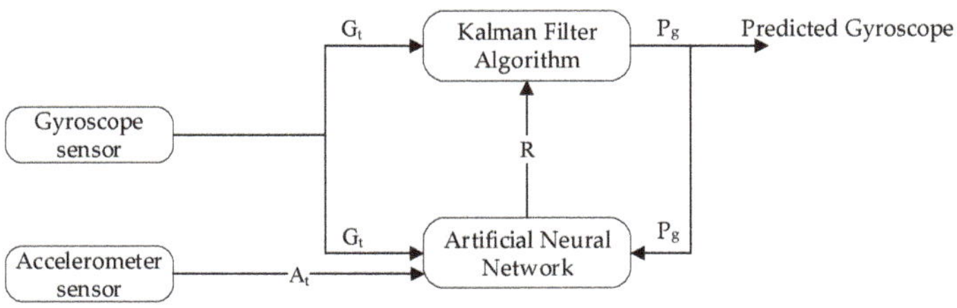

Figure 8. Block diagram of Gyroscope prediction using learning to prediction model.

In the intended learning to prediction model, the tunable parameter can control the prediction algorithm performance. , which is Kalman, gain K. To intelligently update the gain K after every iteration, the estimated error in sensors' reading R and the covariance matrix P are used. The next subsection describes the Kalman filter in detailed.

3.3. Kalman Filter Algorithm

The requirement of historical data in the Kalman filter is not necessary because of its lightweight prediction. For intelligent prediction, the Kalman filter only required previous state information in order to predict the actual state of the system. K, which is also known as Kalman gain, is the necessary parameter that is updated based on the situation to control weights given to the system's own predicted state or sensor readings. The detailed working of the Kalman filter is illustrated in Figure 9.

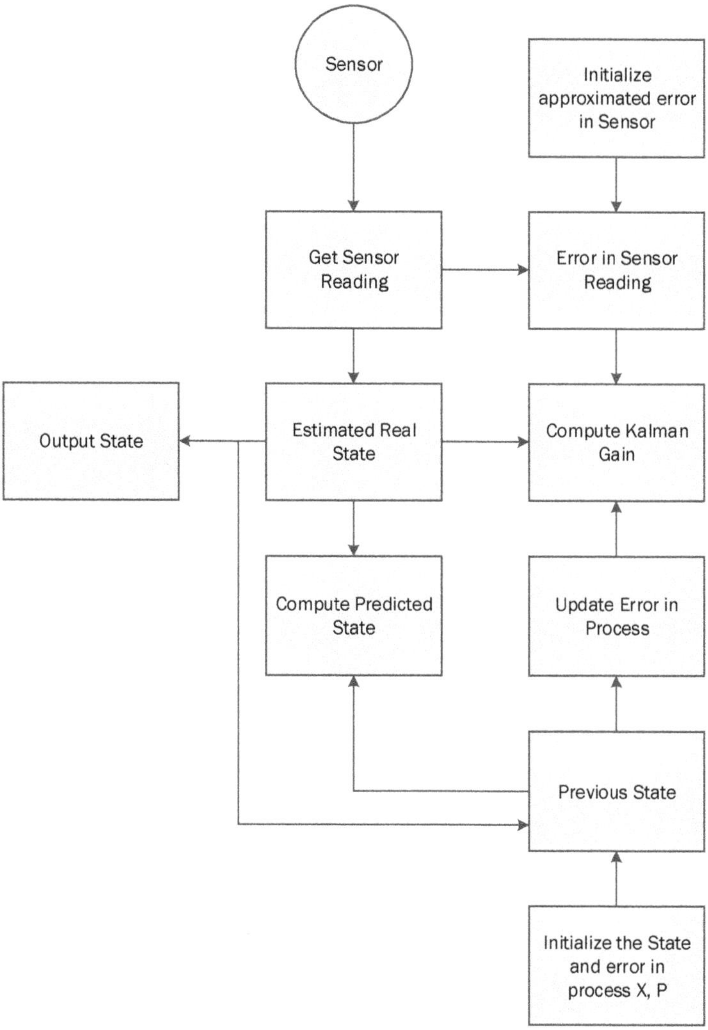

Figure 9. Working of the Kalman filter algorithm.

In every environment, there is a noise factor which creates a serious impact on sensor readings in that environment. In the proposed methodology, we considered an accelerometer and gyroscope sensor reading having noise, and let us considered P_a and P_g is the accelerometer and gyroscope at time t. The internal prediction regarding system state is based on Kalman filter, that is, estimated accelerometer P_{a+1} at time $t+1$. Afterwards, we describe the step by step process of the Kalman filter algorithm, that is how it removes the noise from sensor data.

The first step is to compute the predicted sensor reading, that is, accelerometer and gyroscope from the previously estimated value using Equation (16).

$$S_p = A.AG_{t-1} + B.u_t, \tag{16}$$

where S_p is the internally predicted sensor reading, that is, accelerometer and gyroscope, the state transition and control matrix is denoted as A and B receptively. AG_{t-1} is the predicted accelerometer and gyroscope at time $t-1$ that is previously computed and control vector denotes as u_t.

The uncertainty in the internally predicted sensor reading, that is, accelerometer and gyroscope is calculated using the covariance factor which is computed using Equation (17)

$$P_{pred} = A.P_{t-1}.A^T + Q, \tag{17}$$

where the estimated error in the process is symbolized as Q, and the old value of covariance is denoted as P_{t-1}. A is the state transition matrix, and A^T represents the transpose of the state transition matrix.

Equation (18) compute the internal system estimate of the next state and updating covariance based on Kalman gain.

$$K = \frac{P_{pred}.H^T}{H.P_{pre}.H^T + R}, \tag{18}$$

where an estimated error in the readings is denoted as R, H and H^T are the observation matrix and its transpose.

We consider a scenario in which current reading obtained form the accelerometer sensor at time t is denoted as A_t. Similarly, in case of gyroscope, current reading, which is symbolized as G_t is passed to the Kalman filter. Afterwards, the predicted accelerometer P_a and predicted gyroscope P_g given by the Kalman filter is computed using the Equation (19).

$$P_a = P_{a_{pred}} + K(G_t - H.P_{a_{pred}}) \tag{19}$$

$$P_g = P_{g_{pred}} + K(A_t - H.P_{g_{pred}}). \tag{20}$$

From Figure 10, we calculated the predicted sensor reading using Equation (21)

$$AG_t = AG_{pred} + K(S_t - H.AG_{pred}). \tag{21}$$

In Equation (22), the covariance for the next iteration is finally updated as below

$$P_t = (I - K.H)P_{pred}. \tag{22}$$

3.4. ANN-Based Learning to Prediction for Kalman Filter

The traditional Kalman filter works fine when there is no changing in the estimated error in the sensor. However, if the change in the estimated error occurred due to some external factor, then we need to update the value of R, which is the estimated error in the measurement as shown in Figure 9. In the designed system, we considered a scenario where accelerometer data is affected due to gyro bias, and similarly, accelerometer value causes the noise in the gyroscope. The traditional Kalman filter fails to predict the actual accelerometer and gyroscope under these dynamic conditions when the value of sensor reading changed due to external factors. Figure 10 shows the complete functioning layout of the intended learning to prediction model. The three inputs of the learning module are the previously predicted sensor value by Kalman filter algorithm, acceleration and gyroscope. The expected error is the output in the sensor reading, which is moreover divided by the fixed constituent denoted as F to calculate the estimated error in the sensor reading that is R. Afterwards, the Kalman filter takes the updated value of R as input and adjusting the Kalman gain K by appropriately tuning its prediction accuracy. The intended model predicts the actual accelerometer and dynamic error rate reading from the noisy sensor reading of gyroscope.

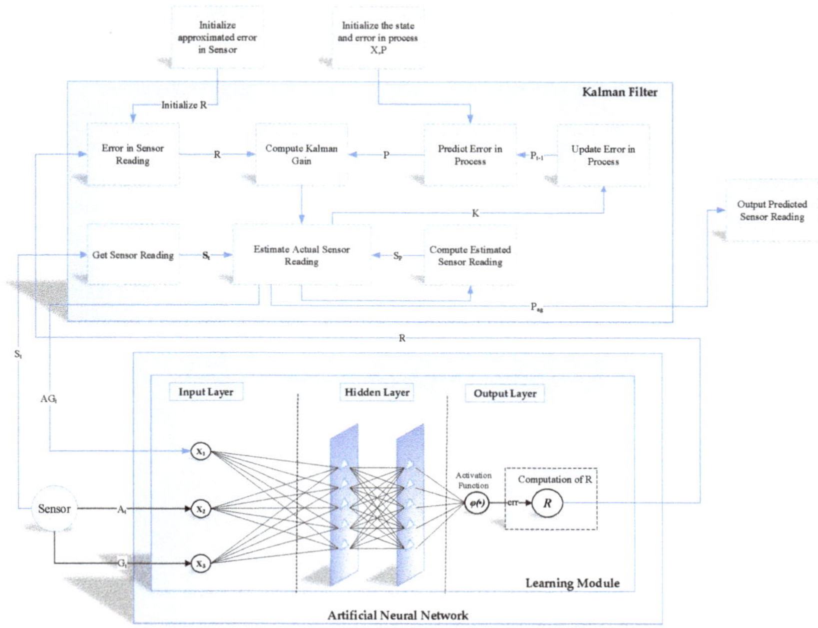

Figure 10. Position Estimation based on learning to prediction.

4. Experimental Results and Discussion

4.1. Development Environment

The development environment of the proposed work is categorized into two models, i.e., learning to prediction model and position estimation using the sensor fusion algorithm. In the case of the stochastic model, we have used the next-generation inertial measurement unit(NGIMU), the next generation inertial measurement unit designed for data acquisition with the onboard sensor and data processing algorithm. The on-board sensor includes 3-axis accelerometer, gyroscope, and magnetometer sensor, which is further used to calculate the position in an indoor environment. The characteristic of NGIMU is mentioned in Table 2.

Table 2. Characteristic of NGIMU.

Sensor	Description	
Gyroscope	Range	$\pm 2000°/s$
	Resolution	$0.06°/s$
	Sample Rate	400 Hz
Accelerometer	Range	± 16 g
	Resolution	490 μg
	Sample Rate	400 Hz
Magnetometer	Range	± 1300 μT
	Resolution	~ 0.3 μT
	Sample Rate	~ 20 Hz

The next generation inertial measurement unit sensor was used to acquired data for inertial navigation in an indoor environment. The data were taken while the object walked from corridor to room number D242 in building 4 of Jeju National University, Republic of Korea. The sample data were acquired for a time duration of approximately 1 minute in which the first starting 10 sec remained inactive so that the algorithm could congregate in a stable state. The tool and technologies for developing the proposed stochastic model are summarized in Table 3. Similarly, for the learning to prediction model, we used an ANN for tuning the prediction algorithm in order to enhance the accuracy of the prediction algorithm. The detailed summary of the development environment for learning to prediction model is mentioned in Table 4.

Table 3. Development environment for stochastic model.

Component	Description
IDE	MATLAB R2018a
Operating System	Window 10
CPU	Intel(R) Core(TM) i5-8500 CPU @ 3.00GHz
Memory	8GB
Data smoothing and prediction algorithm	Kalman Filter
API	NGIMU

Table 4. Development environment for learning to prediction model.

Component	Description
IDE	MATLAB R2018a
Operating System	Window 10
CPU	Intel(R) Core(TM) i5-8500 CPU @ 3.00GHz
Memory	8GB
Artificial Neural Network	Feed Forward Backpropagation
Hidden Layer	10
output Layer	1
Input	3
Prediction algorithm	Kalman Filter

All the implementation and experiments of the proposed system were carried out on Window 10 64 bit with 8GB memory and Intel(R) Core i5-8500 @ 3.00GHz processor. Furthermore, for development, we used MATLAB R2018a and the NGIMU application programming interface (API) to acquire the sensor data.

4.2. Implementation

The proposed system is developed in order to evaluate the performance of the prediction algorithm, that is, the Kalman filter with the learning module. The experiment was performed on the real dataset taken in the engineering building of Jeju national university. At the start, the data were loaded into the application through NGIMU API. The data has ten inputs, that is, 3-axis accelerometer, 3-axis gyroscope, 3-axis magnetometer and time at which the data is taken. Afterwards, we calculated the orientation using sensor fusion based on the Kalman filter. The orientation matrix was further processed by removing the gravitational and centripetal forces and finally applied double integration to compute the position of an object in an indoor environment. Furthermore, the RMSE value of IMU sensor reading was computed by comparing its values with the actual IMU data such as the accelerometer and gyroscope sensor values. The RMSE for the IMU sensor reading was recorded as 5.25, which is considered very high.

Moreover, we used a Kalman filter algorithm to forecast the actual sensor reading (i.e., accelerometer and gyroscope) form the noisy sensor reading. The develop interface provides manual training of the internal parameter of the Kalman filter, that is, the estimated error in measurement (R). Multiple experiments were carried out with different values of R in order to evaluate the performance of the proposed system. The RMSE of the predicted accelerometer and gyroscope was 2.30 at $R = 20$. The predicted RMSE was better than the RMSE of sensor readings, that is, a 55% reduction of error.

In the learning to prediction module, we had to use the ANN algorithm, which is used to enhance the accuracy of the prediction algorithm. The ANN algorithm was comprised of three neurons as an input layer and one neuron as a layer (i.e., accelerometer, gyroscope and Kalman filter predicted reading) and predicting the error in the sensor reading, respectively. Furthermore, we used n-fold cross-validation in order to avoid bias in the training process. For this purpose, we split the dataset into four equal subsets (i.e., 2490 samples in each subset) as shown in Figure 11. According to 4-fold validation, 75% of the dataset was used for training, and other 25% was used for testing the ANN algorithm. Furthermore, in the proposed system, we used 100 epochs which were used for training the ANN algorithm. In the ANN-based learning module, the data normalization was done using Equation (23).

$$\tilde{d_i} = \frac{d_i - d_{min}}{d_{max} - d_{min}}, \tag{23}$$

where $\widehat{d_i}$ is the normalized value for the i_{th} position of the input and output parameters, that is, accelerometer, gyroscope and predicted sensor data. The maximum and minimum value for each parameter in the dataset is denoted by d_{min} and d_{max}. Traditionally, in ANN training is done using normalized data, therefore in order to compute the predicted error we de-normalized the output data of the neural network using Equation (24).

$$er_i = \tilde{er}_i x(er_{max} - er_{min}) + er_{min} \tag{24}$$

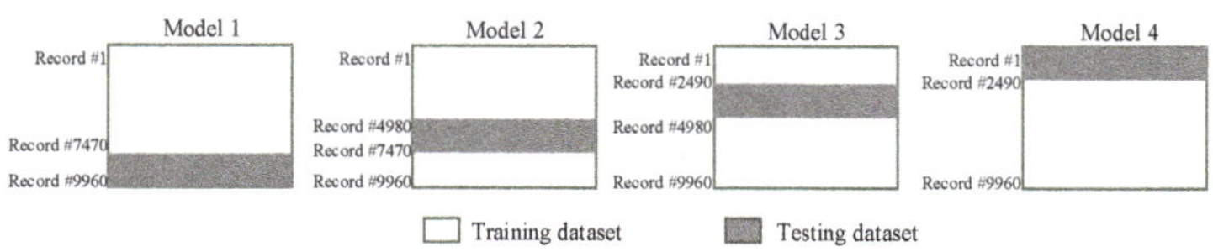

Figure 11. n-fold cross validation.

Furthermore, the proposed model accuracy was evaluated using three different matrices such as mean absolute deviation (MAD), root mean squared error (RMSE), and mean square error (MSE) as shown in Equations (25)–(27).

$$MAD = \sum_{n}^{i+1} |T_i - \hat{P}_i| \tag{25}$$

$$MAE = \sum_{n}^{i+1} (T_i - \hat{P}_i)^2 \tag{26}$$

$$RMSE = \sqrt{\frac{\sum_{i+1}^{n} (T_i - \hat{P}_i)^2}{n}}, \tag{27}$$

where total observation is denoted as n, the target value is represented as T, and $\hat{P}$ indicates the estimated value.

4.3. Results and Discussion

The open-source NGIMU was used to acquired data in order to calculate the object position in an indoor environment. Moreover, for analyzing the proposed system performance, we compared the result predicted by the conventional Kalman filter algorithm with the learning to prediction model. In Figure 12, the raw accelerometer data is shown, which was acquired from the next-generation inertial measurement unit along with the time at which the data were taken. The 3-axis representation of the accelerometer data is denoted by x, y, and z. The dotted line represents the filtered data using a Butterworth filter based on the defined cut-off frequency, and the solid black line shows the stationary data, which is the magnitude of 3-axis acceleration. The stationary data represent the state of the object if the magnitude is less than 0.05; the object state is stationary; otherwise, the object is moving.

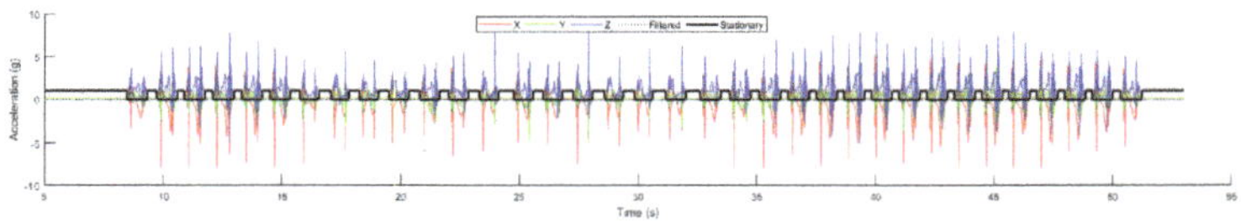

Figure 12. Acceleration.

Figure 13 shows the value of the gyroscope, through which we calculated the value of the angular velocity of the moving object. The 3-axis gyroscope is represented by x,y, and z. The first integration of the angular velocity with respect to time leads to Euler angle, which is required to define the orientation of the object. The Euler angle is usually used to calculate the Roll, Pitch, and Yaw. The angular velocity is the rate of change in prescription of the object moving over time. The formula of angular velocity is mentioned in Equation (28).

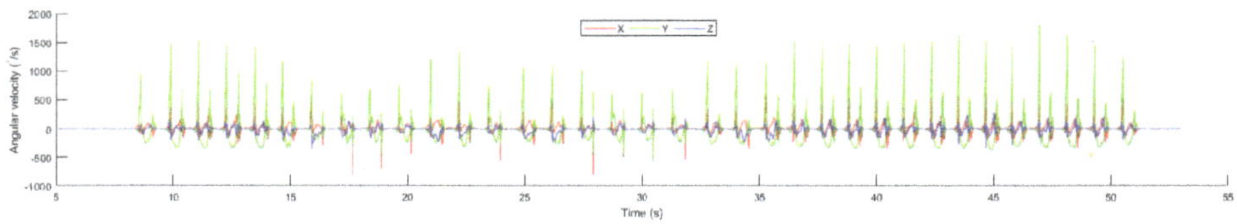

Figure 13. Gyroscope.

$$\omega = \frac{\theta_f - \theta_i}{t}, \tag{28}$$

where θ_f and θ_i denotes the final angle and the initial angle of the object. The change of angle is denoted by $\Delta\theta$, and finally t represents the time.

Figure 14 illustrated the acceleration m/s^2 of the object in an indoor environment. The acceleration m/s^2 of the object is measured as the rate of velocity over time and is calculated using the formula mentioned in Equation (29)

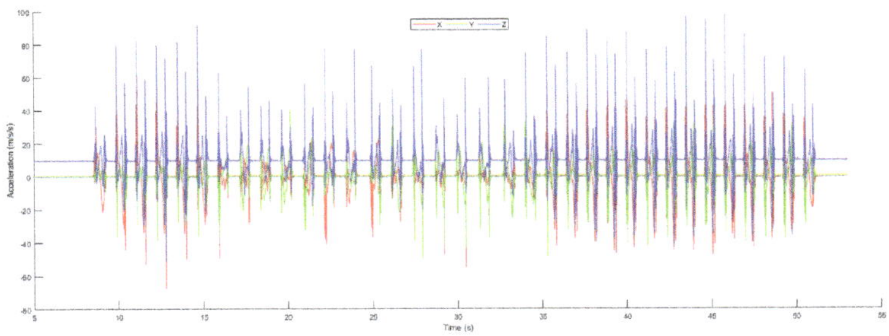

Figure 14. Acceleration m/s^2.

$$a = \frac{\Delta v}{t}, \tag{29}$$

where Δv denotes the change in velocity, a represent the acceleration in m/s^2 and time is denoted by t.

The velocity of the object was measured as to how fast the object is moving in an indoor environment. Figure 15 illustrated the 3-axis velocity of the object, which is calculated using the formula mentioned in Equation (30).

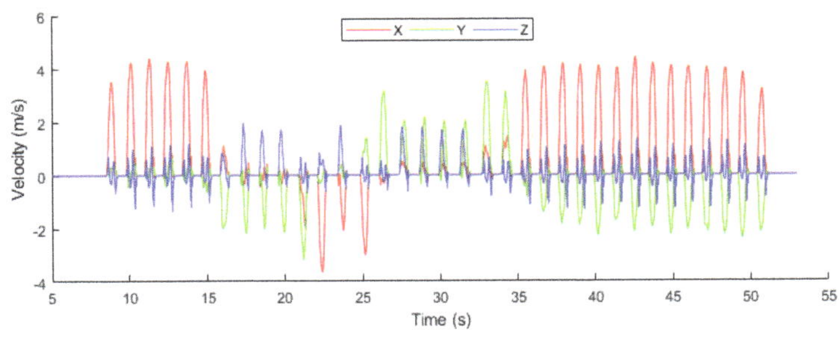

Figure 15. Velocity.

$$v = \frac{\Delta x}{t}, \tag{30}$$

where Δx represents the change in position of the object within an indoor environment, v denotes velocity and t represents the time at which the object changes its position.

In Figure 16, the 3-axis position of the object is represented using a 2-dimensional graph where x, y, and z represent the axis.

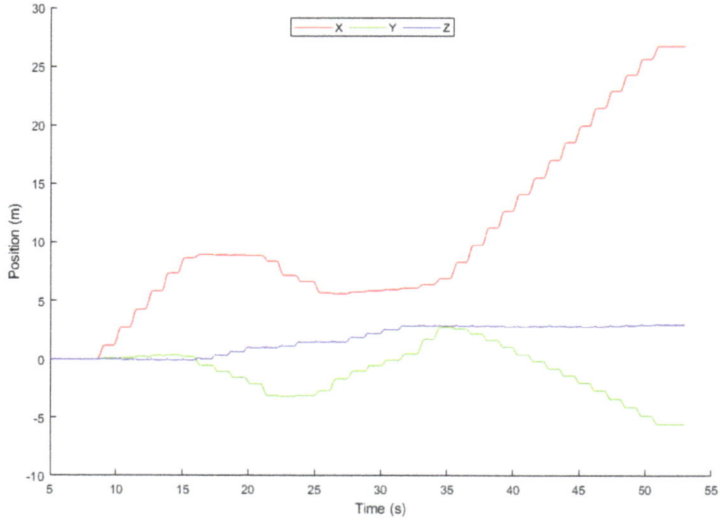

Figure 16. Position.

In Figure 17, we present the predicted result of the accelerometer sensor using the traditional Kalman filter without the learning model. We compared the original sensing data with different values of R. The optimal value of R is based on the dataset, and it is not fixed. Hence it is difficult to find the optimal value of R manually, so we considered different values of R. Also from the graph, it can be seen that the Kalman filter prediction accuracy changed with changing the values of R

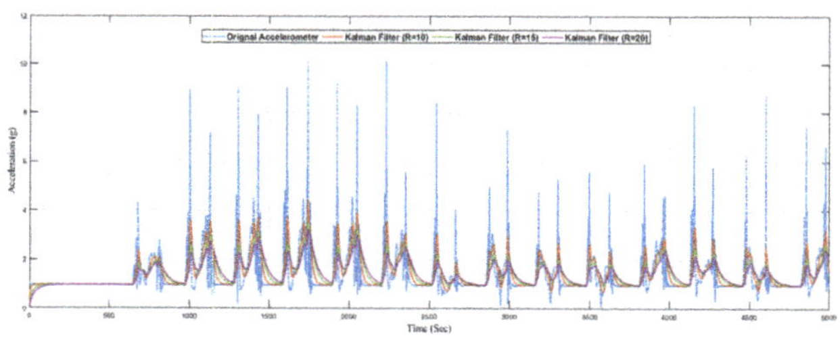

Figure 17. Position.

Similarly, Figure 18 shows the result of predicted angular velocity using a conventional Kalman filter without the learning model. The graph represents the variation in prediction results by varying the value of R in the Kalman filter configuration. The gyroscope sensor reading is predicted using three different configurations as summarized in Table 5.

Table 5. Kalman filter prediction results with and with learning module.

Metric	Kalman Filter without ANN-Based Learning Module			Learning to Prediction Model		
	R = 10	R = 15	R = 20	F = 0.01	F = 0.02	F = 0.1
RMSE	2.527	2.495	2.494	2.404	2.388	2.481
MAD	0.166	0.163	0.163	0.156	0.156	0.156
MSE	6.388	6.224	6.222	5.770	5.701	6.157

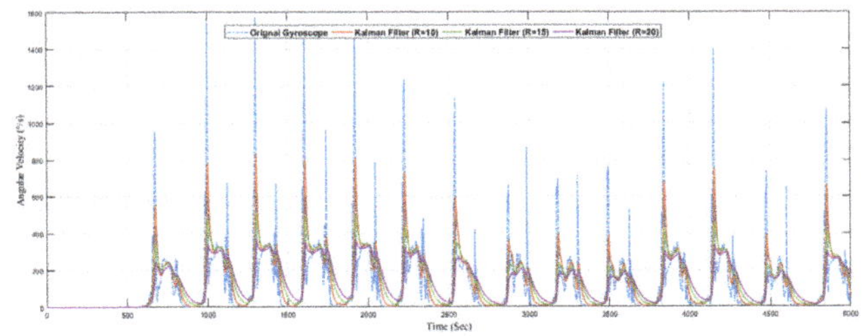

Figure 18. Position.

Next, we present the results of the proposed learning to prediction for both accelerometer and gyroscope sensors readings, as illustrated in Figures 19 and 20. We used the ANN trained model in order to improve the performance by tuning its R parameter. The predicted error rate update the R for the Kalman filter algorithm based on R; we choose the suitable value for F, also called the error factor using Equation (31).

$$R = \frac{er_i}{F}, \tag{31}$$

where proportionality constant also called as error factor denoted as F.

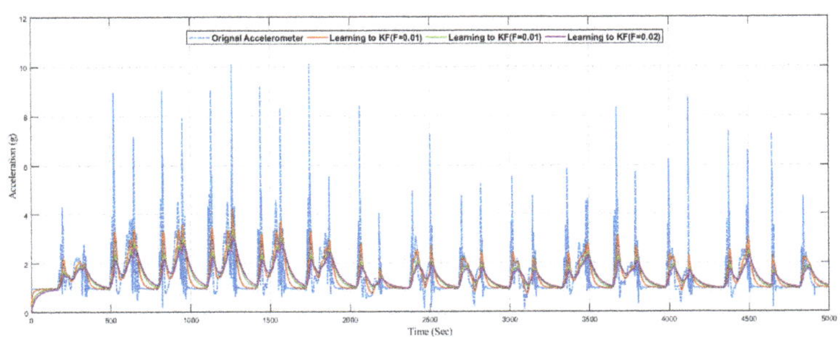

Figure 19. Prediction of acceleration using learning to prediction model with selected F values.

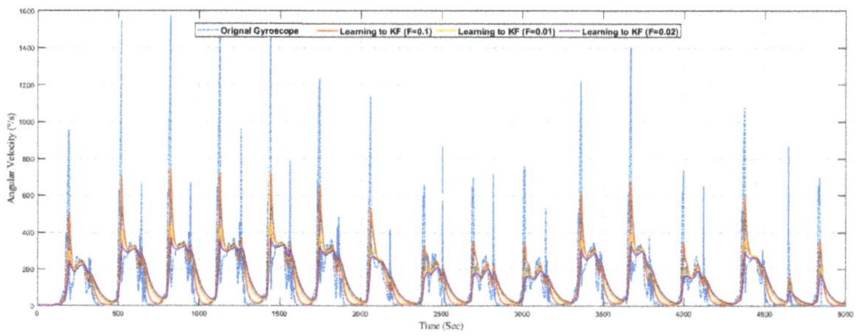

Figure 20. Prediction of gyroscope using learning to prediction model with selected F values.

The graph shows the position data for 60 s in which the first 10 s represents the stationary state. If we investigated the position plot, then we came to know that the proposed system significantly reduced the drift and error from the sensing data. Figures 21 and 22 illustrated the trajectory of a person walking in room no D242 toward the main corridor. In both scenarios, the person starts walking from the starting point where the first 10 s remained stationary so that the algorithm could converge on a stable state and stop walking at the endpoint in the main corridor. The black line shows the trajectory of the person calculated using a stochastic model, that is, position estimation using sensor fusion, whereas the red line is the predicted trajectory using the learning to prediction model. As we see in the results that the amount of drift in the sensor reading is greatly reduced after tuning prediction algorithms using artificial neural networks. In scenario 1, illustrated in Figure 21, the starting point is $(12, 2)$ and the endpoint is $(1, 24)$ computed using traditional Kalman filter with respect to defined reference point $(0, 0)$. Similarly, in the case of learning to Kalman filter model, the accuracy of the indoor system is improved in which the starting point and the ending point are $(13, 5.5)$ and $(-2, 24)$ receptively. All the computed coordinates are mapped according to the defined reference point $(0, 0)$.

Similarly in scenario 2, presented in Figure 22, we compute the result of the conventional Kalman filter with $(3, 6)$ as a starting point, and the ending point is mapped as -3 as x-coordinate and 24 as y-coordinate with $(0, 0)$ reference point. For the learning to prediction model, the accuracy is improved with respect to reference point $(0, 0)$, where the start point is mapped as $(3, 6)$ and the endpoint is $(-3, 24)$.

Table 6 presents the RMSE in position with the prediction model and the learning to prediction model. The results indicate that the error in position estimation is improved by 19% in the case of the learning to prediction model. Furthermore, the proposed model precise the sensor reading based on bias error correction, which results in improving the system accuracy.

As it was challenging to differentiate the result presented in Figures 19–22 . Thus, there is a need for several statistical methods to evaluate the above-presented results in a single quantifiable comparative analysis as mentioned in Equations (25)–(27). We have conducted multiple results for evaluating the performance of a traditional Kalman filter with different R values. Likewise, in the case of the learning to prediction module, the performance is assessed with the selected value of error factor (f). Experimental results show that Kalman filter with learning to prediction module with $F = 0.02$ performed well as compared to other statistical measures. The best outcome for the Kalman filter with no learning module where $R = 20$, as a result of 2.49 prediction accuracy in terms of RMSE. Similarly, the best case for the learning to prediction module is recorded with $F = 0.02$, with a result of 2.38 prediction accuracy in terms of RMSE. Significant enhancement in the prediction accuracy of the learning to prediction module as compared to the Kalman filter with the learning module is 0.041% for the best case and 0.11% for the worst case in terms of RMSE. The statistical summary of Kalman filter for both learning and without learning module is summarized in Table 5.

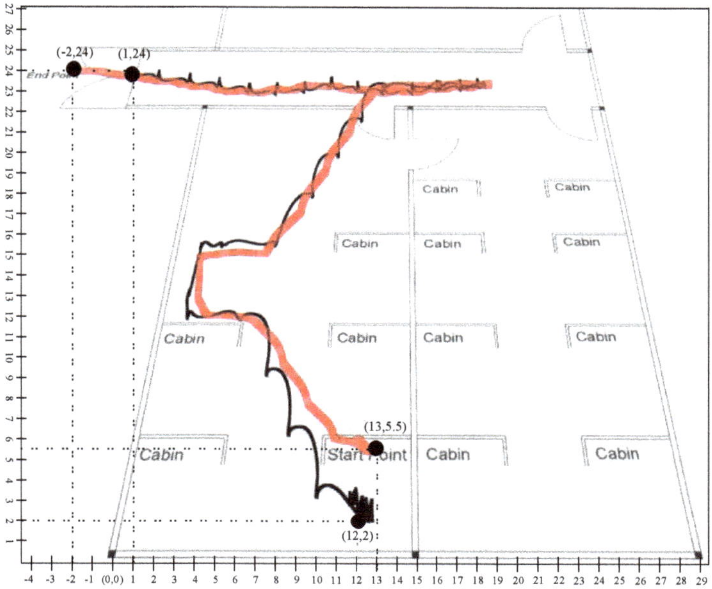

Figure 21. Scenario 1: Person tracking in indoor environment.

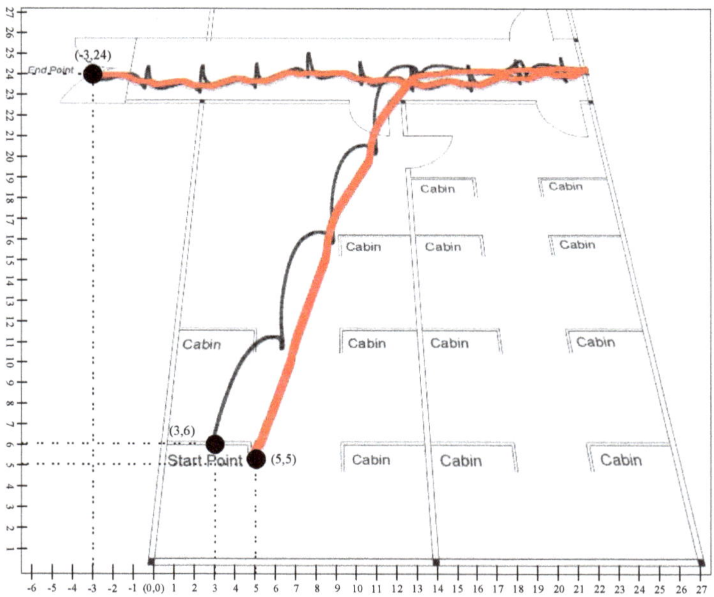

Figure 22. Scenario 2: Person tracking in indoor environment.

Table 6. Position error with the prediction model and the learning to prediction model.

Experiment ID	Position Error with Prediction Model (mm)	Position Error with Learning to Prediction Model(mm)
1	0.132	0.105
2	0.115	0.099

5. Conclusions and Future Work

In this paper, we designed a learning to prediction approach, which was used to enhance the accuracy of the prediction algorithm in an indoor environment. The proposed system is a combination of learning to prediction and position estimation using a sensor fusion algorithm. In the proposed position estimation module, we used a sensor fusion technique based on Kalman filter to fuse all three sensors' measurements in order to get the noise and drift-free orientation estimation for calculating the accurate position in indoor navigation. Likewise, we have used an ANN-based learning model in the learning to prediction module, which enhanced the accuracy of the prediction algorithm. In the designed system, we considered a scenario where the accelerometer and gyroscope sensor is affected by the external conditions, where a conventional Kalman filter failed to extract the noise-free sensor reading from the actual readings. The proposed system improved the performance of the prediction algorithm by tuning the R parameter. For comparative analysis, we analyzed the results using the well know statistical measures such as RMSE, MAD, and MSE. The comparative analysis suggested that the learning to prediction model in the indoor navigation system performed better with the error factor of 0.02, which resulted in a 2.38 RMSE prediction accuracy. The results indicated that the proposed learning to prediction model significantly improves the prediction accuracy and gives us the confidence to further explore the application to improve the performance of other prediction algorithms in indoor navigation systems

Author Contributions: Conceptualization, F.J.; Data curation, F.J.; Funding acquisition, D.-H.K.; Methodology, F.J. and N.I.; Project administration, D.-H.K.; Resources, D.-H.K.; Software, F.J. and N.I.; Supervision, D.-H.K.; Validation, N.I. and S.A.; Visualization, N.I. and S.A.; Writing—original draft, N.I.; Writing—review editing, S.A. All authors have read and agreed to the published version of the manuscript.

Funding: This research was supported by Energy Cloud R&D Program through the National Research Foundation of Korea(NRF) funded by the Ministry of Science, ICT (2019M3F2A1073387), and this research was supported by the MSIT(Ministry of Science and ICT), Korea, under the ITRC(Information Technology Research Center) support program(IITP-2020-2016-0-00313) supervised by the IITP(Institute for Information & communications Technology Planning & Evaluation). Any correspondence related to this paper should be addressed to Dohyeun Kim.

Conflicts of Interest: The authors declare no conflict of interest.

References

1. Harle, R. A survey of indoor inertial positioning systems for pedestrians. *IEEE Commun. Surv. Tutor.* **2013**, *15*, 1281–1293. [CrossRef]
2. Virrantaus, K.; Markkula, J.; Garmash, A.; Terziyan, V.; Veijalainen, J.; Katanosov, A.; Tirri, H. Developing GIS-supported location-based services. In Proceedings of the Second International Conference on Web Information Systems Engineering, Kyoto, Japan, 3–6 December 2001; pp. 66–75.
3. Grewal, M.S.; Weill, L.R.; Andrews, A.P. Global Positioning Systems, Inertial Navigation, and Integration. Available online: https://onlinelibrary.wiley.com/doi/book/10.1002/0470099720 (accessed on 6 August 2020).
4. Bill, R.; Cap, C.; Kofahl, M.; Mundt, T. Indoor and outdoor positioning in mobile environments a review and some investigations on wlan positioning. *Geogr. Inf. Sci.* **2004**, *10*, 91–98.
5. Wu, C.; Yang, Z.; Liu, Y. Wireless Indoor Localization. *IEEE Trans. Parallel Distrib. Syst.* **2012**, *24*, 839–848.
6. Bulusu, N.; Heidemann, J.; Estrin, D. GPS-less low-cost outdoor localization for very small devices. *IEEE Pers. Commun.* **2000**, *7*, 28–34. [CrossRef]
7. Liu, H.; Darabi, H.; Banerjee, P.; Liu, J. Survey of wireless indoor positioning techniques and systems. *IEEE Trans. Syst. Man Cybern. Part C Appl. Rev.* **2007**, *37*, 1067–1080. [CrossRef]

8. Al Nuaimi, K.; Kamel, H. A survey of indoor positioning systems and algorithms. In Proceedings of the 2011 International Conference on Innovations in Information Technology, Abu Dhabi, UAE, 25–27 April 2011; pp. 185–190.
9. Collin, J.; Davidson, P.; Kirkko-Jaakkola, M.; Leppäkoski, H. Inertial Sensors and Their Applications. Available online: https://link.springer.com/book/10.1007/978-3-319-91734-4 (accessed on 6 August 2020).
10. Emil, Š.; Elena, P.; Ladislav, K. Design of an Inertial Measuring Unit for Control of Robotic Devices. *Mater. Sci. Forum* **2019**, *952*, 313–322.
11. Jamil, F.; Kim, D.H. Improving Accuracy of the Alpha–Beta Filter Algorithm Using an ANN-Based Learning Mechanism in Indoor Navigation System. *Sensors* **2019**, *19*, 3946. [CrossRef]
12. Zhang, S. Exploring IMU Attitude and Position Estimation for Improved Location in Indoor Environments. Master's Thesis, Oregon State University, Corvallis, OR, USA, 12 June 2019.
13. Witt, C.; Bux, M.; Gusew, W.; Leser, U. Predictive performance modeling for distributed batch processing using black box monitoring and machine learning. *Inf. Syst.* **2019**, *82*, 33–52. [CrossRef]
14. Lewandowicz, E.; Lisowski, P.; Flisek, P. A Modified Methodology for Generating Indoor Navigation Models. *ISPRS Int. J. Geo. Inf.* **2019**, *8*, 60. [CrossRef]
15. Mortari, F.; Clementini, E.; Zlatanova, S.; Liu, L. An indoor navigation model and its network extraction. *Appl. Geomat.* **2019**, *11*, 413–427. [CrossRef]
16. Gu, Y.; Lo, A.; Niemegeers, I. A survey of indoor positioning systems for wireless personal networks. *IEEE Commun. Surv. Tutor.* **2009**, *11*, 13–32. [CrossRef]
17. Fuchs, C.; Aschenbruck, N.; Martini, P.; Wieneke, M. Indoor tracking for mission critical scenarios: A survey. *Pervasive Mob. Comput.* **2011**, *7*, 1–15. [CrossRef]
18. Yuan, Q.; Asadi, E.; Lu, Q.; Yang, G.; Chen, I.M. Uncertainty-Based IMU Orientation Tracking Algorithm for Dynamic Motions. *IEEE/ASME Trans. Mechatron.* **2019**, *24*, 872–882. [CrossRef]
19. Shen, S.; Gowda, M.; Roy Choudhury, R. Closing the gaps in inertial motion tracking. In Proceedings of the 24th Annual International Conference on Mobile Computing and Networking, New Delhi, India, 29 October–2 November, 2018; pp. 429–444.
20. Tejmlova, L.; Sebesta, J.; Zelina, P. Artificial neural networks in an inertial measurement unit. In Proceedings of the 2016 26th International Conference Radioelektronika (RADIOELEKTRONIKA), Kosice, Slovakia, 19–20 April 2016; pp. 176–180.
21. Ding, M.; Wang, Q. An integrated navigation system of NGIMU/GPS using a fuzzy logic adaptive Kalman filter. In Proceedings of the 2005 International Conference on Fuzzy Systems and Knowledge Discovery, Changsha, China, 27–29 August 2005; pp. 812–821.
22. Muset, B.; Emerich, S. Distance measuring using accelerometer and gyroscope sensors. *Carpathian J. Electron. Comput. Eng.* **2012**, *5*, 83.
23. Tenmoku, R.; Kanbara, M.; Yokoya, N. A wearable augmented reality system for navigation using positioning infrastructures and a pedometer. In Proceedings of the Second IEEE and ACM International Symposium on Mixed and Augmented Reality, Tokyo, Japan, 10 October 2003; pp. 344–345.
24. Groves, P.D. Principles of GNSS, inertial, and multisensor integrated navigation systems. *IEEE Aerosp. Electron. Syst. Mag.* **2013**, *30*, 26–27. [CrossRef]
25. Lin, C.F. Positioning and proximity warning method and system thereof for vehicle. U.S. Patent 6,480,789, 12 November 2002.
26. Koyuncu, H.; Yang, S.H. A survey of indoor positioning and object locating systems. *IJCSNS Int. J. Comput. Sci. Netw. Secur.* **2010**, *10*, 121–128.
27. Dag, T.; Arsan, T. Received signal strength based least squares lateration algorithm for indoor localization. *Comput. Electr. Eng.* **2018**, *66*, 114–126. [CrossRef]
28. Jamil, F.; Hang, L.; Kim, K.; Kim, D. A novel medical blockchain model for drug supply chain integrity management in a smart hospital. *Electronics* **2019**, *8*, 505. [CrossRef]
29. Jamil, F.; Iqbal, M.A.; Amin, R.; Kim, D. Adaptive thermal-aware routing protocol for wireless body area network. *Electronics* **2019**, *8*, 47. [CrossRef]
30. Jamil, F.; Ahmad, S.; Iqbal, N.; Kim, D.H. Towards a Remote Monitoring of Patient Vital Signs Based on IoT-Based Blockchain Integrity Management Platforms in Smart Hospitals. *Sensors* **2020**, *20*, 2195. [CrossRef]

31. Ahmad, S.; Imran; Jamil, F.; Iqbal, N.; Kim, D.; others. Optimal Route Recommendation for Waste Carrier Vehicles for Efficient Waste Collection: A Step Forward Towards Sustainable Cities. *IEEE Access* **2020**, *8*, 77875–77887. [CrossRef]
32. Iqbal, N.; Jamil, F.; Ahmad, S.; Kim, D. Toward Effective Planning and Management Using Predictive Analytics Based on Rental Book Data of Academic Libraries. *IEEE Access* **2020**, *8*, 81978–81996. [CrossRef]
33. Ahmad, S.; Jamil, F.; Khudoyberdiev, A.; Kim, D. Accident risk prediction and avoidance in intelligent semi-autonomous vehicles based on road safety data and driver biological behaviours. *J. Intell. Fuzzy Syst.* **2020**, *38*, 4591–4601. [CrossRef]
34. Khan, P.W.; Abbas, K.; Shaiba, H.; Muthanna, A.; Abuarqoub, A.; Khayyat, M. Energy Efficient Computation Offloading Mechanism in Multi-Server Mobile Edge Computing—An Integer Linear Optimization Approach. *Electronics* **2020**, *9*, 1010. [CrossRef]
35. Wang, X.; Gao, L.; Mao, S.; Pandey, S. CSI-based fingerprinting for indoor localization: A deep learning approach. *IEEE Trans. Veh. Technol.* **2016**, *66*, 763–776. [CrossRef]
36. Kumar, A.K.T.R.; Schäufele, B.; Becker, D.; Sawade, O.; Radusch, I. Indoor localization of vehicles using deep learning. In Proceedings of the 2016 IEEE 17th international symposium on a world of wireless, mobile and multimedia networks (WoWMoM), Coimbra, Portugal, 21–24 June 2016; pp. 1–6.
37. Li, Y.; Gao, Z.; He, Z.; Zhuang, Y.; Radi, A.; Chen, R.; El-Sheimy, N. Wireless fingerprinting uncertainty prediction based on machine learning. *Sensors* **2019**, *19*, 324. [CrossRef]
38. Doostdar, P.; Keighobadi, J.; Hamed, M.A. INS/GNSS integration using recurrent fuzzy wavelet neural networks. *GPS Solut.* **2020**, *24*, 29. [CrossRef]
39. Gu, F.; Khoshelham, K.; Yu, C.; Shang, J. Accurate step length estimation for pedestrian dead reckoning localization using stacked autoencoders. *IEEE Trans. Instrum. Meas.* **2018**, *68*, 2705–2713. [CrossRef]
40. Patel, M.; Emery, B.; Chen, Y.Y. Contextualnet: Exploiting contextual information using lstms to improve image-based localization. In Proceedings of the 2018 IEEE International Conference on Robotics and Automation (ICRA), Brisbane, Australia, 21–25 May 2018; pp. 1–7.
41. Zhang, X.; Sun, H.; Wang, S.; Xu, J. A new regional localization method for indoor sound source based on convolutional neural networks. *IEEE Access* **2018**, *6*, 72073–72082. [CrossRef]
42. Guan, X.; Cai, C. A new integrated navigation system for the indoor unmanned aerial vehicles (UAVs) based on the neural network predictive compensation. In Proceedings of the 2018 33rd Youth Academic Annual Conference of Chinese Association of Automation (YAC), Nanjing, China, 18–20 May 2018; pp. 575–580.
43. Valada, A.; Radwan, N.; Burgard, W. Deep auxiliary learning for visual localization and odometry. In Proceedings of the 2018 IEEE international conference on robotics and automation (ICRA), Brisbane, Australia, 21–25 May 2018; pp. 6939–6946.
44. Sun, Y.; Chen, J.; Yuen, C.; Rahardja, S. Indoor sound source localization with probabilistic neural network. *IEEE Trans. Ind. Electron.* **2017**, *65*, 6403–6413. [CrossRef]
45. Gharghan, S.K.; Nordin, R.; Jawad, A.M.; Jawad, H.M.; Ismail, M. Adaptive neural fuzzy inference system for accurate localization of wireless sensor network in outdoor and indoor cycling applications. *IEEE Access* **2018**, *6*, 38475–38489. [CrossRef]
46. Wagstaff, B.; Kelly, J. LSTM-based zero-velocity detection for robust inertial navigation. In Proceedings of the 2018 International Conference on Indoor Positioning and Indoor Navigation (IPIN), Nantes, France, 24–27 September 2018; pp. 1–8.
47. Adege, A.B.; Yen, L.; Lin, H.p.; Yayeh, Y.; Li, Y.R.; Jeng, S.S.; Berie, G. Applying Deep Neural Network (DNN) for large-scale indoor localization using feed-forward neural network (FFNN) algorithm. In Proceedings of the 2018 IEEE International Conference on Applied System Invention (ICASI), Nantes, France, 24–27 September 2018; pp. 814–817.
48. Zhu, C.; Xu, L.; Liu, X.Y.; Qian, F. Tensor-generative adversarial network with two-dimensional sparse coding: Application to real-time indoor localization. In Proceedings of the 2018 IEEE International Conference on Communications (ICC), Kansas City, MO, USA, 20–24 May 2018; pp. 1–6.
49. Aikawa, S.; Yamamoto, S.; Morimoto, M. WLAN Finger Print Localization using Deep Learning. In Proceedings of the 2018 IEEE Asia-Pacific Conference on Antennas and Propagation (APCAP), Auckland, New Zealand, 5–8 August 2018; pp. 541–542.

50. Li, J.; Wei, Y.; Wang, M.; Luo, J.; Hu, Y. Two indoor location algorithms based on sparse fingerprint library. In Proceedings of the 2018 Chinese Control And Decision Conference (CCDC), Shenyang, China, 9–11 June 2018; pp. 6753–6758.
51. Wu, G.S.; Tseng, P.H. A deep neural network-based indoor positioning method using channel state information. In Proceedings of the 2018 International Conference on Computing, Networking and Communications (ICNC), Maui, HI, USA, 5–8 March 2018; pp. 290–294.
52. Wang, X.; Wang, X.; Mao, S. Deep convolutional neural networks for indoor localization with CSI images. *IEEE Trans. Netw. Sci. Eng.* **2018**. [CrossRef]
53. Haykin, S. Kalman Filtering and Neural NetworksAvailable online: https://onlinelibrary.wiley.com/doi/book/10.1002/0471221546 (accessed on 6 August 2020)
54. Zhuang, Y.; Ma, J.; Qi, L.; Liu, X.; Yang, J. Extended kalman filter positioning method based on height constraint. U.S. Patent 16,309,939, 16 May 2019.
55. Rönnbäck, S. Developement of a INS/GPS Navigation Loop for an UAV Available online: https://www.diva-portal.org/smash/get/diva2:1029914/FULLTEXT01.pdf (accessed on 6 August 2020)

sensors

MDPI

Letter

An RSSI Classification and Tracing Algorithm to Improve Trilateration-Based Positioning

Yong Shi [1,2], Wenzhong Shi [2,*], Xintao Liu [2] and Xianjian Xiao [1]

1 School of Computer Information and Engineering, Changzhou Institute of Technology, Changzhou 21300, China; shiy@czu.cn (Y.S.); xiaoxj@czu.cn (X.X.)

2 Department of Land Surveying and Geo-Informatics, The Hong Kong Polytechnic University, Hong Kong; xintao.liu@polyu.edu.hk

* Correspondence: lswzshi@polyu.edu.hk

Received: 22 June 2020; Accepted: 29 July 2020; Published: 30 July 2020

Abstract: Received signal strength indicator (RSSI)-based positioning is suitable for large-scale applications due to its advantages of low cost and high accuracy. However, it suffers from low stability because RSSI is easily blocked and easily interfered with by objects and environmental effects. Therefore, this paper proposed a tri-partition RSSI classification and its tracing algorithm as an RSSI filter. The proposed filter shows an available feature, where small test RSSI samples gain a low deviation of less than 1 dBm from a large RSSI sample collected about 10 min, and the sub-classification RSSIs conform to normal distribution when the minimum sample count is greater than 20. The proposed filter also offers several advantages compared to the mean filter, including lower variance range with an overall range of around 1 dBm, 25.9% decreased sample variance, and 65% probability of mitigating RSSI left-skewness. We experimentally confirmed the proposed filter worked in the path-loss exponent fitting and location computing, and a 4.45-fold improvement in positioning stability based on the sample standard variance, and positioning accuracy improved by 20.5% with an overall error of less than 1.46 m.

Keywords: trilateral indoor positioning; RSSI filter; RSSI classification; stability; accuracy

1. Introduction

With widespread Wi-Fi, Bluetooth (iBeacon), and smart mobile terminal deployment, received signal strength indicator (RSSI)-based indoor positioning technology has attracted research attention due to its advantages of low complexity, low cost and high accuracy [1,2]. However, the received signal strength (RSS) is easily blocked and easily interfered with by objects and environmental effects. These influences usually increase RSSI variance; thus, RSSIs vary sharply over time even when the actual signal strength remains constant. The variation reduces accuracy and stability for an RSSI-based indoor positioning system (IPS) [3–5]. Many assisted and combined technologies were proposed to achieve higher accuracy, e.g., pedestrian dead reckoning (PDR), computer vision, space-scenario, and artificial intelligence techniques [1,4]. However, fundamentally, dealing with RSSI is a crucial step during the whole IPS process.

RSSI is environment-dependent. Therefore, it is significant to filter the raw RSSIs before substituting them into the positioning process. Many RSSI purification technologies such as the Gaussian filter [6], Kalman filter [7], and particle filter [8,9] are typically designed to mitigate either the linear or non-linear noise through smoothing. Still, they may not effectively deal with the ever-changing dynamics of the indoor environment [10] and have left-skewed distributions [11]. The mean filter [12] is widely accepted because it has similar accuracy and anti-interference performance, but has less burden in filtering computation [13,14]. Besides smoothing, RSSI screening is another effective filtering method, e.g., by selecting the max N RSSIs (N = 13 is optimal) [13], and the least variance RSSIs over time [15].

Importantly, RSSI classification is also an effective filtering method, especially in combination with clustering algorithms for RSSI filtering and singular RSSI tracing [16,17].

RSSI-based IPS is generally divided into two categories: trilateration-based IPS and fingerprint-based IPS [10,18,19]. The fingerprint-based IPS [20–22] gets more concern in terms of the number of references retrieved, where the number is 1,910,000 vs. 30,000 from Google Scholar. It also gains positive effect by using unsupervised machine-learning algorithms to reduce the data dimensionality, and fingerprint matching calculations required [23]. Meanwhile, the trilateration-based positioning system is widely applied in outdoor environments [24,25]. However, it faces challenges in indoor environments, achieving 3–5 m accuracy without assistive technology and device, and lower stability [3]. The trilateration-based IPS positioning is easy to understand and easy to construct. Usually, RSSI-based trilateration IPS is undertaken in two main steps: distance mapping and position computation [10]. The distance between the unknown position sensor and the known position sensor anchor is obtained by some RSSI propagation model [26]; its accuracy depends on signal transmission anchor, path-loss exponent in the RSSI propagation model, and RSSI sampling, etc. [10,13]. Based on this, the unknown location is obtained by trilateration methods such as least square and maximum likelihood [4,10].

This paper will propose a new filtering technique to play a role in the propagation parameter fitting and the RSSI sampling purification, to improve RSSI-based IPS accuracy and stability, particularly for trilateration-based positioning, considering that accuracy and stability are equally important. The overall accuracy may diminish due to energy consumption [27,28], and stability may significantly influence the user experience and application promotion.

The main contributions of this study are as follows:

(1) We propose a tri-partition classification of RSSIs, considering that the interference effects are finite and unambiguous, increasing and decreasing while the interference sources are multiple and fuzzy. And we also propose a clustering algorithm to trace the tri-partition classification quantitatively and seek the partition distribution centers, helping to reveal the interference and judge whether the sun-classification conforms to normal distribution.
(2) We take the proposed algorithm as an RSSI filter and discuss its work mechanism. And we infer that it is feasible by analyzing the features in terms of sample count and deviation and advantages compared to the mean filter.
(3) We verify that the proposed filter works in path-loss exponent fitting and location computing, and analyze the improvement of IPS it yields.

The remainder of this paper is arranged as follows. In Section 2, we review RSSI-based trilateration positioning technologies and RSSI filters. We detail the proposed RSSI filter based on an RSSI classification and tracing algorithm in Section 3. In Section 4, we analyze the filtering performance, including features adapted to real-time IPS and advantages over the mean filter. Section 5 introduces a test to examine the positioning performance using the proposed filter and compares its performance with the mean filter. Finally, in Section 6, we summarize and conclude the paper.

2. Literature Review

2.1. Received Signal Strength Indicator (RSSI)-Based Trilateration Indoor Positioning System (IPS)

Trilateration-based positioning technology is easily understood and widely used in the positioning of pedestrians and robots and things [1–3]. In the method we should know the position of the anchors (reference nodes) as $(x_1, y_1), (x_2, y_2), \ldots, (x_n, y_n)$, and their distances from the target node, which is calculated by the RSSI-distance mapping, $d_1, d_2, \ldots, d_3$. If we assume the target node's coordinates as

(x, y), then the essential geometric functions as follows. When the anchors are more than 4, the least square and maximum likelihood [4,10] are used to calculate the target node's optimal coordinates.

$$\begin{cases} (x_1 - x)^2 + (y_1 - y)^2 = d_1{}^2 \\ (x_2 - x)^2 + (y_2 - y)^2 = d_2{}^2 \\ \quad \cdots \\ (x_n - x)^2 + (y_n - y)^2 = d_n{}^2 \end{cases} \tag{1}$$

The positioning complexity is partly to know the anchor points' location in advance, especially when anchors change and a lot of update work required. The difficulty is that RSSI-based trilateration techniques depend on an accurate estimation of distance by RSSI-distance mapping. RSSI is a function of distance and is generally affected by the environment and any changes therein. Usually, researchers use the following simplified propagation model to measure RSSI (Formula 1) and map RSSI to distance (Formula 2) [26,29,30]:

$$\rho = \alpha - 10\beta \log(d) \tag{2}$$

$$d = 10^{((\rho-\alpha)/(10*\beta))} \tag{3}$$

where d is the distance from the current position to some beacon, ρ is the RSSI at the current position, α is the RSSI at some referenced distance (usually 1 m), β is the path-loss exponent, and the parameters α and β are obtained to adapt to different sensors and environments; the path-loss exponent β generally has a value in the range of 1.6–1.8 in an indoor environment [26,31,32]. More importantly, it is necessary to fit the path-loss exponent according to the actual environment in which RSSIs are collected [33,34]. These parameters should be calculated again when a target node moves across the boundary of two different environments [34], even be constantly updated if necessary [35]. Furthermore, the piecewise fitting and min–max method are proposed for local adaption to the real environment, 4 m and 8 m are the breakpoints, and the curvatures of different sections are noticeably different [15,34,36].

Accuracy is the most crucial performance metric of the positioning system. It is related to devices and its effective coverage. Wi-Fi and iBeacon belong to high-frequency signals, but the RSSI performance received is also different due to the difference in transmitting power and antenna angle. For example, iBeacon, particularly the loss of packets, is serious for the interval beyond 10 m [10,15,36]. The deployment density is also computing- and maintenance-cost related. When the anchors' density is 0.27 nodes/m^2, localization estimation error can be decreased to 1.5–2 m [37]. However, increasing the number of anchor nodes does not result in higher average accuracy, and with more than 50 anchors, the average accuracy declines [26]. Meanwhile, paper [38] proposes a sensor deployment method based on wireless sensor network topology optimization, and [39] suggests a novel technique related to pedestrian density, gaining an accuracy of 1.8–3.9 m.

Considering the overall accuracy may diminish due to energy consumption [27,28], that accuracy and stability are equally important, and stability may significantly influence the user experience and application promotion. This paper's core contribution is to adopt the same positioning method without increasing the calculation amount of positioning, which reflects the function advantages of proposed classification and filtering in this paper.

2.2. RSSI Filtering Technologies

Raw RSSIs measurement is related to the parameter fitting and distance mapping process, and play a decisive role for IPS performance [33]. Compared with the mean filter [11], Kalman filter [7], particle filter [8,9], least-squares estimator [40,41], and maximum-likelihood estimator [42] have advantages in terms of accuracy but are computationally expensive. They use a moment before estimation and the current observations to update the state variables' estimate, while the mean filter just takes the average. Thus, the mean filter is widely recommended because it has similar accuracy and anti-interference performance [11]. Considering their relatively small computational overheads and the fact they can be used in a real-time context, the rolling mean filter, exponential moving mean filer, and moving median filter have been discussed [12]. However, these filters are typically intended to mitigate these influences by smoothing and could suffer from left-skewed distributions caused by RSSI multipath propagation [15].

RSSI screening is another effective filtering method based on analyzing the spatial resolution of the signal strength and RSSI signal characteristics under different scenarios. Gaussian filter [6] selects the high RSSI probability RSSIs and takes the average value as filter results, lowers the influence of the small probability and interference over the measurement. An algorithm using the maximum RSSI average has been proposed and suggests that N = 13 is optimal [13]. Relative to selecting the maximum, [15] selected the least variance RSSIs over time, arguing that the normal variances are not dramatic. Based on dichotomy, Study [16] and [17] propose an RSSI classification to distinguish singular RSSIs from normal path-loss RSSIs. Paper [17] proposes a k-means clustering algorithm tracing the rating.

Applying artificial intelligence (including statistical inference methods) for RSSI processing is a new trend, e.g., using a k-means clustering algorithm for singular RSS tracing [22] and filtering, and RSSI fingerprint matching [23,24]. Moreover, support vector machine (SVM) [43], artificial neural network (ANN) [44,45] and deep learning [46] have been proposed to aid RSSI purification and high positioning accuracy. However, in indoor environments, all of the above algorithms still face challenges of spatial ambiguity, RSSI instability, and RSSI's short collecting time per location [47,48].

This paper will propose a three-classification and tracking method to explore its distribution center instead of the mean center based on the classification idea and unsupervised learning algorithm. More importantly, we will discuss the filtering performance under the conditions of small samples and small sampling time.

3. RSSI Classification and Tracing

3.1. Tri-Partition RSSI Classification

Environmental conditions, scenario changes, anchor deployment, transmission power, and interferences between anchor nodes can affect the RSSI values. Furthermore, it is challenging to determine antenna gains [10]. However, the actual effects can be summarized as increasing and decreasing, so unlike the dichotomy, we propose a tri-partition of RSSIs, as shown in Figure 1. We classify RSSI samples into three collections:

1. The decreased collection (DC) represents singular weakened RSSIs such as blocked and reflected signals.
2. The normal collection (NC) represents normal path-loss and fading RSSIs.
3. The increased collection (IC) represents singular enhanced RSSIs caused by transmitting equipment such as antenna gained power or transmitting power mutation.

The tri-partition can help with the quantitative analysis of the RSSI. Tracing the item count of sub-classification can reveal that the subsequent positioning is dependable or not, where NC RSSIs do not conform to normal distribution means more significant errors. Thus, it helps select different path-loss parameters to adapt to the environment.

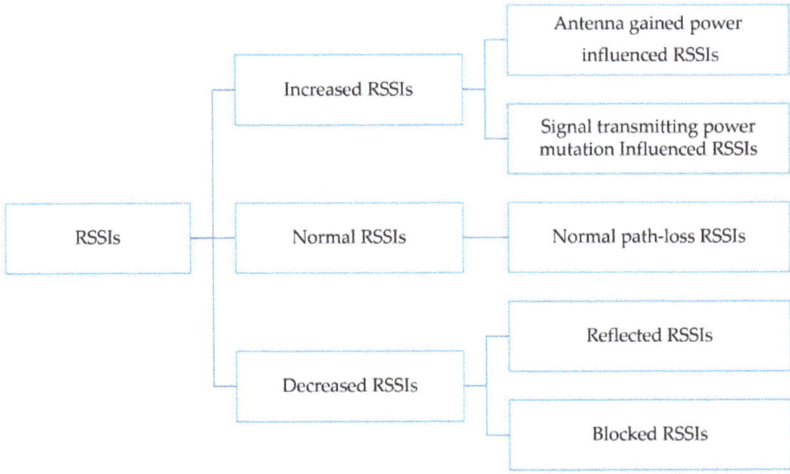

Figure 1. Received signal strength indicator (RSSI) classification based on the view of influence effects.

3.2. RSSI Tracing Algorithm

We propose an RSSI tracing algorithm based on k-means clustering to determine the partition and its distribution center. The proposed algorithm considers the RSSI sample (R) to be a one-dimensional collection. It uses the absolute value of RSSI and sub-classification center subtraction as the clustering factor, takes the maximum RSSI as the initial center of the IC, the minimum RSSI as the initial center of the DC, the mean RSSI as the initial center of the NC, and then defines the assistant function MIN(R) to obtain the minimum RSSI from R, the function MAX(R) to obtain the maximum RSSI from R, and the function AVERAGE(R) to obtain the mean value from R. The algorithm steps are as follows Algorithm 1:

Algorithm 1. Sub-Classification Tracing.

```
Input: R = {RSSI_1, RSSI_2, ... , RSSI_n}//the collected RSSI samples
Output: NC, IC, DC

Define://initial the center of IC, NC, and DC
        IV = MAX(R), NV = AVERAGE(R), and DV = MIN(R)
        IC_T = ϕ, NC_T = ϕ, and DC_T = ϕ //temporal collections
        TD = ID = ND = DD = 0.0            //temporal values.
Classify:
For each (RSSI_i ∈ R){
//calculate its distance to IV, NV, and DV
ID = abs (IV- RSSI_i)
        ND = abs (NV- RSSI_i)
        DD = abs (DV- RSSI_i)
        //select the minimum distance
        TD = MIN ({ID, ND, DD})
        //add RSSIi to sub-classification
        If (TD= =ID)
        IC = IC + { RSSI_i }
        Else if (TD= =ND)
        NC = NC+{ RSSI_i }
        Else
        DC = DC + { RSSI_i }
        }
```

```
Judge:
        //judge the change of sub-classification
        If (IC_T = = NC_T = = DC_T = = ϕ){
                IC_T = IC
                NC_T = NC
                DC_T = DC
                }
GOTO (Reset)
//if convergence, exit
        If (IC_T = = IC and NC_T = = NC and DC_T = = DC
        GOTO (Exit)
        Else {
                IC_T = IC
                NC_T = NC
                DC_T = DC
                }
Reset: //reset the sub-classification center
IV = AVERAGE (IC)
        NV = AVERAGE (NC)
        DV = AVERAGE (DC)
        GOTO (Classify)
Exit: //deal with ϕ
        If (NC = = ϕ) {
          TD = (ID + DD)/2
          NC+ = {TD}
          }
EXIT(-)
```

where NC equals ϕ in the Algorithm (1) means that sample R has a polarization distribution, and partition NC is affected by substantial RSSI deviation from reality. When we set the mean value as a new RSSI, the whole count will increase by one each time.

3.3. Apply to Trilateration-Based Positioning

Trilateration-based IPS includes many steps [10,13], here we divide these steps into two stages: offline and online. The offline stage aims to adapt the path-loss exponent, including anchor deployment, RSSI collecting, raw RSSI measurement, and path-loss exponent fitting (may fit as needed during the online stage [15,34]). The online stage aims to optimize positioning, including real-time RSSI collecting, raw RSSI measurement, distance mapping, coordinate calculation, and positioning optimization and correction. This paper adopts the proposed tracing algorithm as an RSSI filter to re-establish the raw RSSI measurement for the path-loss exponent fitting (offline) and distance calculating (online) steps, as shown in Figure 2.

Figure 2 shows the application mechanism that focuses on the sample count and the partition item count. The minimum sample count is related to real-time positioning, as RSSI's shortest collecting time per location usually less than 1 s. The partition item size is related to influences when the tri-partition counts conform to a normal distribution, indicating that the filtering result is reliable. For the filtering process, the partition center is the filtering result. Next, we will discuss the minimum sample and performance to reveal the availability of the proposed filter.

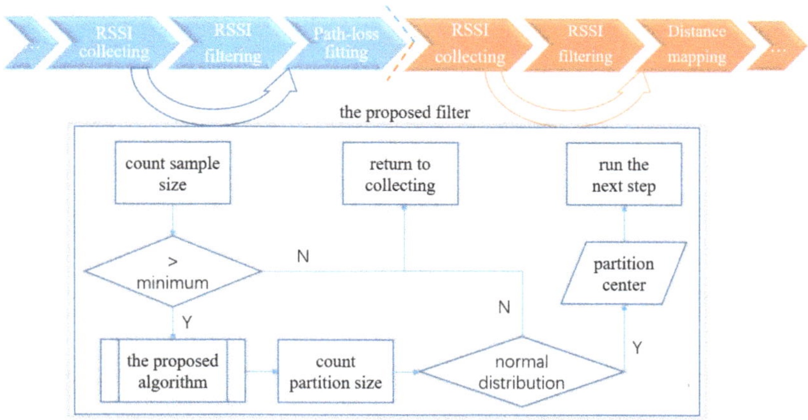

Figure 2. Application mechanism of the proposed filter for the indoor positioning system (IPS).

4. Feasibility and Performance as an RSSI Filter

To examine the proposed filter's feasibility and performance advantages, we collected a large RSSI sample (named Sample ALL) over approximately 10 min using iBeacon as the signal sender. And we defined six type sample groups, named Sample X (X = 10, 20, 30, 40, 50, 60). Each sample group had 10 sample arrays, and each array had the same count RSSIs. The sample array data structure is as follows:

$$\text{Test Sample} = \{\text{Array}_{i,j}, \text{I} = 10, 20, \ldots, 60, \text{j} = 1, 2, \ldots, 10\} \quad (4)$$

where i represents different groups, and j represents different arrays in the current group. For example, $\text{Array}_{30,1}$ represents that the collection is in group 30, and the collection's item count is 30.

In preparation, we cut the sample ALL into each test sample sequentially and continuously according to the sample type, and then filter each test sample and the sample ALL. Figure 3 shows the filtering results, where each node represents a filtering result of some Test Sample. For each type of sample, ten filtering tests are conducted in sequence, with a total of 60 filtering times. Figure 3 also shows Sample 10 has the max deviation from the Sample ALL, and with the sample count is larger than 20, the gap range gradually stabilized.

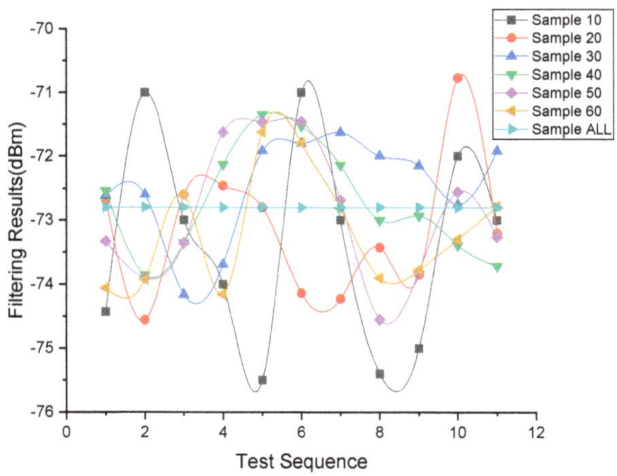

Figure 3. Filtering results of each test sample in sample group and the sample ALL.

4.1. Features of the Proposed Filter

4.1.1. Low Deviation

RSSI fluctuation is an inherent issue for wireless signals, hence reducing the variance range for each sample count is essential to improve the accuracy and stability [49,50]. Using the sample ALL as the comparative standard, we measured the deviation and variance of each test sample, which is as shown in Table 1. It summarizes the deviations from all samples, and together with Figure 3 shows that the proposed filter achieves lower variance. Sample 10 and Sample 20 have the highest variation with a maximum deviation range of less than 2.5 dBm and the mean deviation range of less than 1.44 dBm. When the test sample count was greater than 20, the maximum deviation range was less than 1.64 dBm, and the mean variation was below 0.92 dBm. As experiments have confirmed that a change of 1 dBm represents 0.08 m at some distance [13], therefore, the filter obtained a significantly lower fluctuation, and the test sample gains a similar performance to the sample. ALL, especially when the sample size is more significant than 20.

Table 1. Low deviation feature of each sample group compared to sample ALL.

Sample Type	Max Positive Deviation (dBm)	Max Negative Deviation (dBm)	Mean Absolute Deviation (dBm)
Sample 10	1.99	2.51	1.44
Sample 20	2.22	1.572	0.89
Sample 30	1.36	1.17	0.83
Sample 40	1.64	0.87	0.70
Sample 50	1.53	1.56	0.92
Sample 60	1.36	1.17	0.83

4.1.2. Minimum Sample Count

While RSSIs are Gaussian distribution and random, the time of sampling can affect the collected RSSI count and positioning quality [10,51]. Most current IPSs and RSSI analyses take a long-time interval for sample collecting, such as 1–3 or 3–6 min, but it is a gap from second-level real-time requirements.

By counting the items of each sample, as shown in Table 2, the Sample 10 had an item count of 103 when the expected count was 100, and sample 20 s item count was 201 when the expected count was 200, indicating that the NC in Sample 10 and Sample 20 generated ϕ during filtering. Further calculating the sub-classification distribution rate, as shown in Figure 4, the tri-partition collection items conform to the normal distribution when the sample count is over 20. Therefore, the proposed filter obtained a minimum sample count of 20 suitable for the positioning process.

In summary, the proposed filter has such features; the small test sample gains a similar performance to the larger sample ALL, and the sun-classification conforms to normal distribution when the sample count is larger than 20. Therefore, the proposed filter adapts to real-time positioning, for it has a lower deviation, and the minimum sample count is 20.

Table 2. Items count and rate statistics for sample groups. DC, NC, IC represnts the RSSI count of the decreased, normal, and increased collection.

Sample Type	Items Count	DC Rate	NC Rate	IC Rate
Sample 10	103	28%	32%	38%
Sample 20	201	21%	60%	19%
Sample 30	300	26%	51%	23%
Sample 40	400	25%	55%	20%
Sample 50	500	25%	50%	25%
Sample 60	600	19%	57%	23%

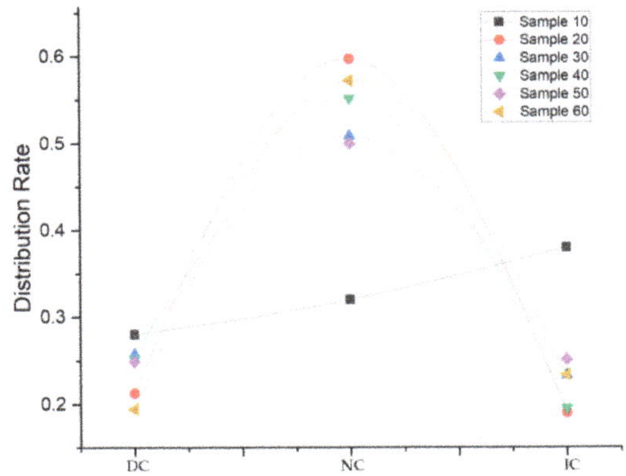

Figure 4. Distribution of tri-partition for sample groups.

4.2. *Advantages over the Mean Filter*

Using the same RSSI test sample from the upper section, Figure 5 shows the filtering results comparison by the proposed filter and the mean filter, respectively, where each node represents a filtering result of some test sample. For each type of sample, 10 filtering tests are conducted in sequence, with a total of 120 filtering computations. Based on the filtering performance, we discuss the advantages associated with reductions in variance below.

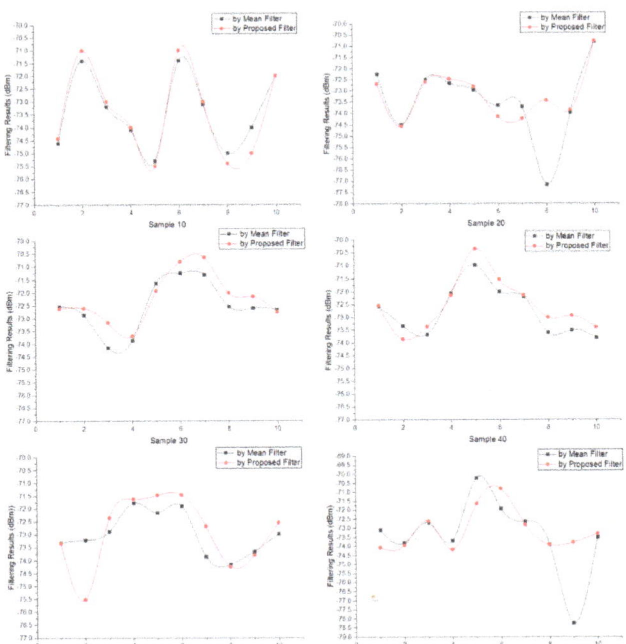

Figure 5. Filtering performance comparisons of each type sample by the proposed filter and the mean filter, respectively.

4.2.1. Reducing Variance over Time

The comparisons use variance range and sample variance (SV, calculated using the sample standard deviation function in EXCEL) to measure the proposed and mean filters' performance. As Figure 5 shows the filtering result, Table 3 shows the variance comparisons in detail. The proposed filter achieves a smaller maximum variance range than the mean filter, and a lower SV when the sample count is greater than 10.

Table 3. Comparison between the mean and proposed filters in terms of variance range and sample variance over time. SV = sample variance.

Sample Type	Mean Filter				Proposed Filter			
	Max Range	Min Range	Average Range	SV	Max Range	Min Range	Average Range	SV
Sample 10	3.2	1.0	1.96	2.28	3.9	1.96	2.25	2.68
Sample 20	3.46	0.15	1.69	2.28	3.08	0.03	1.11	1.54
Sample 30	2.23	0.06	0.66	1.03	1.77	0.01	0.69	0.95
Sample 40	1.62	0.1	0.76	0.96	1.32	0.07	0.67	0.82
Sample 50	1.96	0.3	0.82	1.03	1.25	0.01	0.65	0.83
Sample 60	5.35	0.71	2.12	2.77	2.53	0.13	1.12	1.44

Overall, the average maximum variance for the proposed filter was 2.31 dBm, while for the mean filter, this was 2.97 dBm. The average minimum deviation for the proposed filter was 0.37 dBm, while the mean filter was 0.39 dBm. The average variance for the proposed filter was 1.08 dBm, while the mean filter was 1.34 dBm, and the average SV for the proposed filter was 1.38 dBm, while the mean filter was 1.73 dBm. Thus, the proposed filter resulted in a lower and more stable variance.

4.2.2. Reducing Left-Skewness

Traditional RSSI filtering commonly produces left-skewed distributions [11,26,50], i.e., the returned value is lower than the actual value. Table 4 shows that the proposed filter mitigates this problem, achieving a 65% superior performance in terms of skewness reduction. The left 35% which did not mitigate skewness; they had an average of 96.7% deviation at less than one dBm and an average of 78.9% deviation at less than 0.5 dBm, so the proposed filter successfully reduced skewness.

Table 4. Skewness reductions by comparisons among filters.

Sample Type	Mitigating	Others Less 1 dbm	Others Less0.5 dbm
Sample 10	70%	100%	66.7%
Sample 20	50%	100%	80%
Sample 30	70%	100%	100%
Sample 40	70%	100%	66.7%
Sample 50	80%	100%	100%
Sample 60	50%	80%	60%

In summary, the proposed filter shows feasibility because it has significant low variance and a minimum sample of 20 properties and mitigating advantages over the mean filter which is vastly used.

5. Positioning Improvement

5.1. Experiment Design

Figure 6 shows the 4 × 4 m testbed used to investigate the proposed filter's positioning performance. The experiment selects (1,1), (1,3), and (3,3) as test points, deploys four iBeacon (SEEKCY s1u) sensors as RSS senders at the grid corners and selects an android mobile phone (MI max2) as the RSS receiver.

To seek reliable positioning performance under real-time, we set the minimum sample count of 30 each time. The experimental steps are as follows.

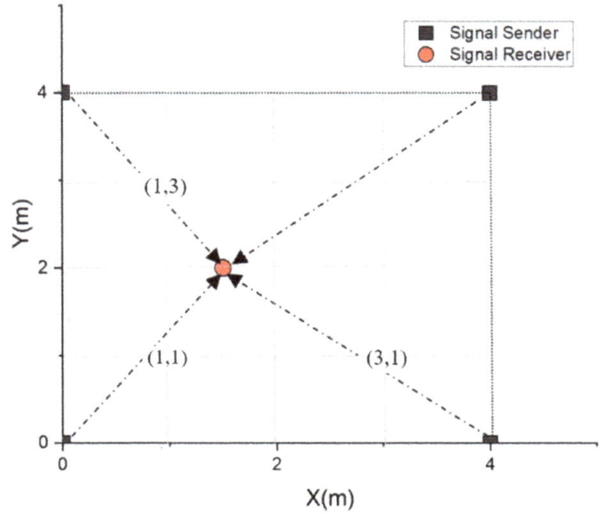

Figure 6. Testbed diagram and selected test points.

1. Fit RSSI propagation model factors,
   a. collect RSSI samples at each fixed distance (1 m, 2 m, ... , 10 m),
   b. filter each sample using the proposed and mean filters, and
   c. fit the propagation model according to Formula (2).
2. For each positioning process,
   a. filter RSSI samples using the proposed and mean filters,
   b. perform traditional trilateral positioning, and
   c. compare the positioning results.
3. Repeat step (2) for the next test process.

5.2. Experiment Results

5.2.1. Propagation Factors

Using MATLAB Fitting Tools and a similar fitting process in paper [14,15], Table 5 shows the that α and β are distinctly different. It is foreseeable that the distance mapping process will still make a difference using the factors.

Table 5. The simplified RSSI propagation model fitting.

Filter	Filtered RSSI at 1 m (α)	Fitted Path-Loss Parameter (β)
The proposed filter	–80.14 dBm	4.613
The mean filter	–83.82 dBm	2.326

The α is as RSSI sample filtering result at 1 m, and the b is as the fitting result of collected RSSI samples at each fixed distance from 1 m to 10 m, by the proposed filter and the mean filter.

Considering studies have confirmed that many packets may mass lose when the measured distance exceeds 10 m [36,52].

5.2.2. Positioning Results

Table 6 shows a total of 24 positioning coordinates. Figure 7 shows the positioning results and the relative positions of these test points. The proposed filter achieves a significantly compact location distribution at each test point, whereas the mean filter has a relatively loose distribution, with point (3,1) exhibiting the highest accuracy.

Table 6. Positioning results coordinates at point (1,1), point (1,3) and point (3,1).

Test Point	Positioning Results Using the Mean Filter		Positioning Results Using the Proposed Filter	
	Coordinate x	Coordinate y	Coordinate x	Coordinate y
(1,1)	2.82	1.02	2.69	1.79
	2.56	0.77	2.56	1.63
	4.38	2.83	2.51	1.66
	2.63	0.87	2.74	1.82
(1,3)	2.84	2.25	2.87	2.25
	4.19	2.46	2.97	2.25
	2.80	2.21	2.89	2.27
	4.38	2.10	3.06	2.09
(3,1)	2.48	0.80	2.56	1.18
	2.54	0.51	2.55	1.15
	2.54	0.56	2.54	1.14
	2.42	1.01	2.49	1.22

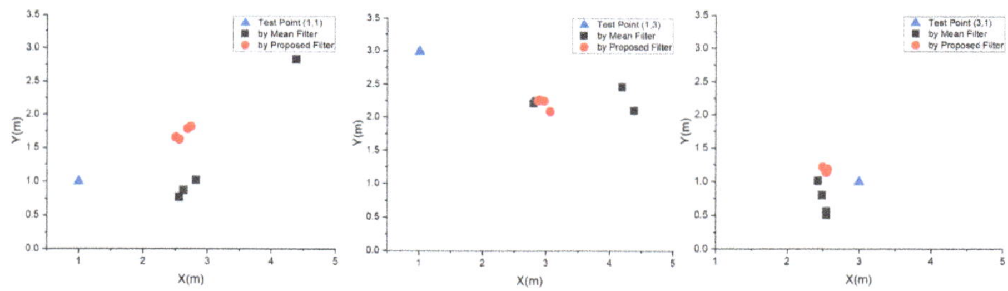

Figure 7. Positioning results distributions at point (1,1), point (1,3) and point (3,1).

5.3. *Improvement Analysis*

5.3.1. Accuracy

Figure 8 shows the calculated positioning errors in sequence. Based on the accuracy changes at each test point, errors by the proposed filter change little, while the errors by the mean filter change significantly. Unfortunately, despite the proposed filter gains in the overall improvement, it does not improve every positioning, and about 58% have higher accuracy compared to the mean filter.

In detail, Table 7 shows the comparison of the proposed filter and the mean filter in accuracy promotion. For each test point, the proposed filter achieves smaller maximum and average errors than the mean filter, and the positioning accuracy is improved by approximately 20% for each test point. The overall average error is 1.46 and 1.84 m for the proposed and mean filters, respectively, and the overall average accuracy improvement is 20.5%. Still, it can be concluded that even if the error by the proposed filter is larger, the difference between the errors by the mean filter is still small, this phenomenon is consistent with the mitigation performance analysis in chapter 4.

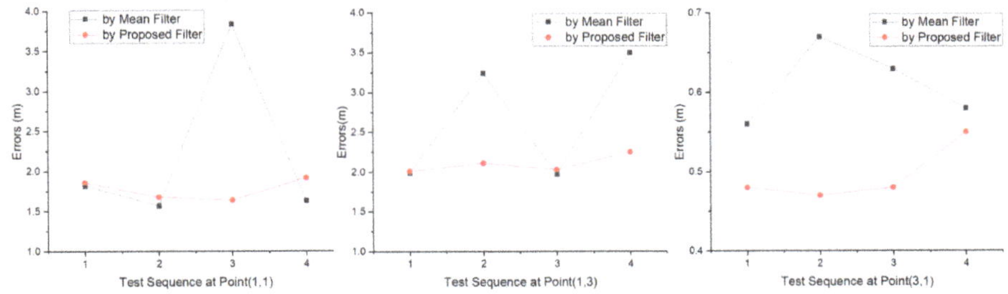

Figure 8. Positioning error comparison with the proposed filter and the mean filter at point (1,1), point (1,3) and point (3,1).

Table 7. Positioning errors and average accuracy promotion.

Test Point	Positioning Errors Using the Mean Filter			Positioning Errors Using the Proposed Filter			Average Accuracy Promotion
	Maximum	Minimum	Average	Maximum	Minimum	Average	
(1,1)	3.84	1.57	2.22	1.92	1.64	1.78	19.8%
(1,3)	3.5	1.97	2.68	2.25	2.01	2.1	21.6%
(3,1)	0.67	0.56	0.61	0.55	0.48	0.51	16.0%

5.3.2. Stability

Using sample standard deviation (SSD) to measure positioning stability, the SSD will be calculated by the following formula:

$$SSD = STDEV(R)/AVERAGE(R) \tag{5}$$

where *STDEV* (–) is the sample standard deviation function in EXCEL, and *AVERAGE* (–) is the mean function in EXCEL; *R* is the positioning error of each test point.

Table 8 shows that the proposed filter achieves smaller SV and closer average positions for each test point than the mean filter. The positioning stability is 6.45, 5.82, and 1.08-fold better for test points (1,1), (1,3), and (3,3), respectively, for the proposed filter compared with the mean filter. The overall average stability is improved 4.45-fold.

Table 8. Positioning stability and increase for the proposed filter vs. the mean filter. SSD represents the sample stand deviation and calculated by the Formula (5).

Test Point	Positioning Stability Using the Mean Filter			Positioning Stability Using the Proposed Filter			SSD Promote
	STDEV(R)	*AVERAGE(R)*	*SSD*	*STDEV(R)*	*AVERAGE(R)*	*SSD*	(–fold)
(1,1)	1.089	2.215	0.491	0.136	1.775	0.077	6.45
(1,3)	0.81	2.675	0.303	0.109	2.1	0.052	5.82
(3,1)	0.05	0.61	0.081	0.037	0.5	0.05	1.08

In summary, the proposed filter achieved a 20.5% improvement in positioning accuracy, with an overall error of less than 1.46 m and a 65% probability of higher accuracy. Significantly, the proposed filter gained a 4.45-fold improvement in positioning stability.

6. Conclusions

Considering RSSI is easily blocked and affected by things and the environment, RSSI-based IPS faces the challenge of an unfortunate application effect. To improve the accuracy and stability, particularly for trilateration-based positioning, we propose a tri-partition RSSI classification as the decreased RSSIs, normal RSSIs, and increased RSSIs, and proposed a novel RSSI tracing algorithm based on k-means clustering as an RSSI filter.

The proposed filter adapts to the real-time IPS, for it shows a characteristic achieving a lower variance (<1 dBm) when the minimum sample size is greater than 20. In contrast, traditional RSSI variation can exceed 10 dBm [26]. The proposed filter offers several advantages compared to the mean filter, including lower variance range and sample variance, and 65% probability to mitigate RSSI left-skewness. Thus, the proposed filter is feasible in real-time positioning.

We design a trilateration-based positioning test within a room. The RSSI propagation model fitting achieves a difference path-loss exponent where 4.613 by the proposed filter while 2.326 by the mean filter. Based on this, the positioning results confirm a 20.5% improvement in positioning accuracy and 4.45-fold improvement in stability for the proposed filter compared to the mean filter. Thus, the proposed filter significantly outperforms traditional mean filtering, providing an excellent option for large-scale IPS improvement.

Author Contributions: Y.S. conceived the main concept of the algorithm, designed the system and the experiments and wrote the paper. W.S. and X.L. contributed analysis tools, offered professional instructions and supervised the whole experimental process. X.X. conducted the experiments, analyzed the experimental data and wrote part of the paper. All authors have read and agreed to the published version of the manuscript.

Funding: This work was supported in part by the National Natural Science Foundation of China (Grant No. 41571382, No. 41901323), in part by the Natural Science Foundation of the Jiangsu Higher Education Institutions of China (Grant No. 15KJB170006, No.16KJB 520003, No. 17KJD520001), in part by Jiangsu industry-university-institute collaboration project (Grant No. BY2019021), in part by Jiangsu Collaborative Innovation Center for Cultural Creativity (Grant No.XYN1702, No.XYN1703), in part by Taizhou Science and technology support program of China (Grant No. TS201621), in part by Changzhou Science and technology support program of China (Grant No. CE20172023), in part by Collaborative Innovation Center of Changzhou Institute of Technology for Digital Information Technology, in part by Excellent Scientific and Technological Innovation Team of Changzhou Institute of Technology, and in part by Natural Science Foundation of Changzhou Institute of Technology (Grant No.YN1726).

Acknowledgments: The authors would like to thank the editor and the anonymous reviewers who provided insightful comments on improving this article, thank The Hong Kong Polytechnic University for providing the opportunity of a visiting scholar, and thank student Jiaxin Li, Wenlong Zhou, Zhongqi Cui for data collecting.

Conflicts of Interest: The authors declare no conflict of interest.

References

1. Ali, W.H.; Kareem, A.A.; Jasim, M. Survey on Wireless Indoor Positioning Systems. *CUESJ* **2019**, *3*, 42–47. [CrossRef]
2. Leel, K.; Naml, Y.; Min, S.D. An indoor localization solution using Bluetooth RSSI and multiple sensors on a smartphone. *Multimed. Tools Appl.* **2018**, *77*, 12635–12654. [CrossRef]
3. Oguntala, G.; Abd-Alhameed, R.A.; Jones, S.; Noras, J.; Patwary, M.; Rodriguez, J. Indoor location identification technologies for real-time IoT-based applications: An inclusive survey. *Comput. Sci. Rev.* **2018**, *30*, 55–79. [CrossRef]
4. Brena, R.F.; García-Vázquez, J.P.; Galván-Tejada, C.E.; Muñoz-Rodriguez, D.; Vargas-Rosales, C.; Fangmeyer, J. Evolution of Indoor Positioning Technologies. *Surv. J. Sens.* **2017**, *2017*, 2630413. [CrossRef]
5. Huang, H.; Gartner, G.; Krisp, J.M.; Raubal, M.; Van de Weghe, N. Location based services: Ongoing evolution and research agenda. *J. Locat. Based Serv.* **2018**, *12*, 63–93. [CrossRef]
6. Ito, K.; Xiong, K. Gaussian filters for nonlinear filtering problems. *IEEE Trans. Autom. Control* **2020**, *45*, 910–927. [CrossRef]
7. Sung, Y. RSSI-Based Distance Estimation Framework Using a Kalman Filter for Sustainable Indoor Computing Environments. *Sustainability* **2016**, *8*, 1136. [CrossRef]
8. Chao, C.H.; Chu, C.Y.; Wu, A.Y. Location-Constrained Particle Filter human positioning and tracking system. In Proceedings of the 2008 IEEE Workshop on Signal Processing Systems, Washington, DC, USA, 8–10 October 2008; pp. 73–76.

9. Liu, H.H.; Chen, W.C.; Hsu, P.C. Comparison of NVA and RLOWESS Algorithms in Indoor Positioning System. In Proceedings of the 2019 20th Asia-Pacific Network Operations and Management Symposium (APNOMS), Matsue, Japan, 18–20 September 2019; pp. 1–4.
10. Shojaifar, A. *Evaluation and Improvement of the RSSI-Based Localization Algorithm Received Signal Strength Indication (RSSI)*; Blekinge Institute of Technology: Karlskrona, Sweden, 2015.
11. Luo, J.; Zhan, X. Characterization of Smart Phone Received Signal Strength Indication for WLAN Indoor Positioning Accuracy Improvement. *J. Netw.* **2014**, *9*, 739. [CrossRef]
12. Koledoye, M.A.; De Martini, D.; Rigoni, S.; Facchinetti, T. A Comparison of RSSI Filtering Techniques for Range-based Localization. In Proceedings of the 2018 IEEE 23rd International Conference on Emerging Technologies and Factory Automation (ETFA), Turin, Italy, 4–7 September 2018; pp. 761–767.
13. Xue, W.; Qiu, W.; Hua, X.; Yu, K. Improved Wi-Fi RSSI Measurement for Indoor Localization. *IEEE Sens. J.* **2017**, *17*, 2224–2230. [CrossRef]
14. Shi, Y.; Long, Y.; Lu, F.; Xu, Z.; Xiao, X.; Shi, S. Indoor RSSI Trilateral Algorithm Considering Piecewise and Space-Scene. In Proceedings of the 2017 IEEE International Conference on Smart Cloud, New York, NY, USA, 3–5 November 2017; pp. 278–282.
15. Kaemarungsi, K.; Krishnamurthy, P. Properties of indoor received signal strength for WLAN location fingerprinting. In Proceedings of the First Annual International Conference on Mobile and Ubiquitous Systems: Networking and Services, MOBIQUITOUS 2004, Boston, MA, USA, 26 August 2004; pp. 14–23.
16. Wang, B.; Gan, X.; Liu, X.; Yu, B.; Jia, R.; Huang, L.; Jia, H. A Novel Weighted KNN Algorithm Based on RSS Similarity and Position Distance for Wi-Fi Fingerprint Positioning. *IEEE Access* **2020**, *8*, 30591–30602. [CrossRef]
17. Subedi, S.; Gang, H.S.; Ko, N.Y.; Hwang, S.S.; Pyun, J.Y. Improving Indoor Fingerprinting Positioning with Affinity Propagation Clustering and Weighted Centroid Fingerprint. *IEEE Access* **2019**, *7*, 31738–31750. [CrossRef]
18. Liu, H.H.; Lo, W.H.; Tseng, C.C.; Shin, H.Y. A WiFi-based weighted screening method for indoor positioning systems. *Wirel. Pers. Commun.* **2014**, *79*, 611–627. [CrossRef]
19. Suroso, D.J.; Cherntanomwong, P.; Sooraksa, P.; Takada, J.-I. Location fingerprint technique using fuzzy C-means clustering algorithm for indoor localization. In Proceedings of the TENCON 2011-2011 IEEE Region 10 Conference, Bali, Indonesia, 21–24 November 2011; pp. 88–92.
20. Yunfei, C.; Taihang, D.; Chundong, J.; Ling, Q.I.; Sun, S. Optimal processing of singular values of K-means clustering algorithm in indoor location. *Sci. Technol. Eng.* **2018**, *18*, 95–99.
21. Chandrawanshi, V.S.; Tripathi, R.K.; Pachauri, R. An Intelligent Energy Efficient Clustering Technique for Multiple Base Stations Positioning in a Wireless Sensor Network. *J. Intell. Fuzzy Syst.* **2019**, *36*, 2409–2418. [CrossRef]
22. Zhang, W.; Hua, X.; Yu, K.; Qiu, W.; Zhang, S.; He, X. A novel WiFi indoor positioning strategy based on weighted squared Euclidean distance and local principal gradient direction. *Sens. Rev.* **2019**, *39*, 99–106. [CrossRef]
23. Zhang, C.; Zhang, F.; Zhao, L. Research and Optimization of BLE Fingerprint Indoor Positioning Algorithm Based on Fusion Clustering. In Proceedings of the 2019 IEEE International Conference on Artificial Intelligence and Computer Applications (ICAICA), Dalian, China, 29–31 March 2019; pp. 95–100.
24. Honkavirta, V.; Perala, T.; Ali-Loytty, S.; Piché, R. A Compartive Survey of WLAN Location Fingerprinting Methods. In Proceedings of the 2009 6th Workshop on Positioning, Navigation and Communication, Hannover, Germany, 19 March 2009.
25. Teoman, E.; Ovatman, T. Trilateration in Indoor Positioning with an Uncertain Reference Point. In Proceedings of the 2019 IEEE 16th International Conference on Networking, Sensing and Control (ICNSC), Banff, AB, Canada, 9–11 May 2019; pp. 397–402.
26. Heurtefeux, K.; Valois, F. Is rssi a good choice for localization in wireless sensor network? In Proceedings of the 2012 IEEE 26th International Conference on Advanced Information Networking and Applications, Fukuoka, Japan, 26–29 March 2012; pp. 732–739.
27. Huang, K.; He, K.; Du, X. A Hybrid Method to Improve the BLE-Based Indoor Positioning in a Dense Bluetooth Environment. *Sensors* **2019**, *19*, 424. [CrossRef]
28. He, S.; Chan, S.H.G. Wi-Fi Fingerprint-Based Indoor Positioning: Recent Advances and Comparisons. *IEEE Commun. Surv. Tutor.* **2015**, *18*, 466–490. [CrossRef]

29. Zhang, W.; Dong, Y.; Wang, X. Indoor Positioning methods using Rfid and Block clustering. *Comput. Eng. Appl.* **2016**, *52*, 112–117.
30. Yang, G.L.; Kong, Y.P.; Zhong, Z.M.; Zhang, Y.L. Multimode-fingerprint matching based indoor positioning System Design and Implementation. *Comput. Eng. Des.* **2013**, *34*, 1896–1900.
31. Mehra, R.; Singh, A. Real Time RSSI Error Reduction in Distance Estimation Using RLS Algorithm. In Proceedings of the IEEE 3rd International Advance Computing Conference, IACC 2013, Ghaziabad, India, 22–23 February 2013; pp. 661–665.
32. Zhenghua, Z.; Lei, D.; Jiawen, L.; Xu, T. Research of optimizing weighted centroid distance correction localization algorithm based on RSSI. In Proceedings of the 2013 IEEE 11th International Conference on Electronic Measurement & Instruments, Harbin, China, 16–19 August 2013; pp. 552–556.
33. Zhou, C.; Yuan, J.; Liu, H.; Qiu, J. Bluetooth Indoor Positioning Based on RSSI and Kalman Filter. *Wirel. Pers. Commun.* **2017**, *96*, 4115–4130. [CrossRef]
34. Chen, Y.; Pan, Q.; Liang, Y.; Hu, Z. AWCL: Adaptive weighted centroid target localization algorithm based on RSSI in WSN. In Proceedings of the 2010 3rd International Conference on Computer Science and Information Technology, Chengdu, China, 9–11 July 2010; pp. 331–336.
35. Golestani, A.; Petreska, N.; Wilfert, D.; Zimmer, C. Improving the Precision of RSSI-based Low- Energy Localization Using Path Loss Exponent Estimation. In Proceedings of the 2014 11th Workshop on Positioning, Navigation and Communication (WPNC), Dresden, Germany, 12–13 March 2014.
36. Yong, S.H.I. Research on Indoor-Positioning Methods and Technologies Considering Indoor Space-Scene. Ph.D. Thesis, Nanjing Normal University, Nanjing, China, December 2016.
37. Sugano, M.; Kawazoe, T.; Ohta, Y.; Murata, M. Indoor localization system using RSSI measurement of wireless sensor network based on ZigBee standard. *Wirel. Opt. Commun.* **2006**, *538*, 1–6.
38. Zhou, H.; Hu, J.; Wu, H.; Guo, C. Indoor positioning research based on wireless sensor network topology optimization. In Proceedings of the 2019 Chinese Automation Congress (CAC), Hangzhou, China, 22–24 November 2019; pp. 3951–3957.
39. Chen, F. Global Optimization Placement of Bluetooth Nodes Based on Indoor Maps. Master's Thesis, Nanjing Normal University, Nanjing, China, June 2017.
40. Li, X.R. RSS-based location estimation with unknown path-loss model. *IEEE Trans. Wirel. Commun.* **2006**, *5*, 3626–3633. [CrossRef]
41. Lin, L.X.; So, H.C.; Chan, Y.T. Accurate and simple source localization using differential received signal strength. *Digit. Signal Process.* **2013**, *23*, 736–743. [CrossRef]
42. Coluccia, A.; Ricciato, F. Maximum likelihood trajectory estimation of a mobile node from RSS measurements. In Proceedings of the 2012 9th Annual Conference on Wireless On-Demand Network Systems and Services (WONS), Courmayeur, Italy, 9–11 January 2012; pp. 151–158.
43. Brunato, M.; Battiti, R. Statistical learning theory for location fingerprinting in wireless LANs. *Comput. Netw.* **2005**, *47*, 825–845. [CrossRef]
44. Chen, H.; Zhang, Y.; Li, W.; Tao, X.; Zhang, P. ConFi: Convolutional neural networks based indoor Wi-Fi localization using channel state information. *IEEE Access* **2017**, *5*, 18066–18074. [CrossRef]
45. Cui, X.; Yang, J.; Li, J.; Wu, C. Improved Genetic Algorithm to Optimize the Wi-Fi Indoor Positioning Based on Artificial Neural Network. *IEEE Access* **2020**, *8*, 74914–74921. [CrossRef]
46. Wu, L.; Chen, C.-H.; Zhang, Q. A Mobile Positioning Method Based on Deep Learning Techniques. *Electronics* **2019**, *8*, 59. [CrossRef]
47. Hoang, M.T.; Yuen, B.; Dong, X.; Lu, T.; Westendorp, R.; Reddy, K. Recurrent Neural Networks for Accurate RSSI Indoor Localization. *IEEE Internet Things J.* **2019**, *6*, 10639–10651. [CrossRef]
48. Hoang, M.T.; Zhu, Y.; Yuen, B.; Reese, T.; Dong, X.; Lu, T.; Westendorp, R.; Xie, M. A soft range limited K-nearest neighbours algorithm for indoor localization enhancement. *IEEE Sens. J.* **2018**, *18*, 10208–10216. [CrossRef]
49. Neishaboori, A.; Harras, K. Energy saving strategies in WiFi indoor localization. *Proc. Acm Mswim* **2013**, 399–404. [CrossRef]
50. Li, Y.; He, Z.; Li, Y.; Xu, H.; Pei, L.; Zhang, Y. Towards Location Enhanced IoT: Characterization of LoRa Signal for Wide Area Localization. In Proceedings of the 2018 Ubiquitous Positioning, Indoor Navigation and Location-Based Services (UPINLBS), Wuhan, China, 22–23 March 2018.

51. Xu, Y. *Autonomous Indoor Localization Using Unsupervised Wi-Fi Fingerprinting*; Kassel University Press: Kassel, Germany, 2016.
52. Jayakanth, K.; AbdelGhani, K.; Somaya, A.M.; Al-Ali, A. Indoor positioning and wayfnding systems: A survey. *Hum.Cent. Comput. Inf. Sci.* **2020**, *10*, 1–41.

sensors

MDPI

Article

A Method of Multiple Dynamic Objects Identification and Localization Based on Laser and RFID

Wenpeng Fu [1], Ran Liu [1,2,*], Heng Wang [1], Rashid Ali [1,3], Yongping He [1], Zhiqiang Cao [1] and Zhenghong Qin [1]

1 School of Information Engineering, Southwest University of Science and Technology, Mianyang 621010, China; wenpeng_fu628@mails.swust.edu.cn (W.F.); wh839@swust.edu.cn (H.W.); rashidcs@uot.edu.pk (R.A.); yongping_he@mails.swust.edu.cn (Y.H.); caozhiqiang@mails.swust.edu.cn (Z.C.); qinzhy527@mails.swust.edu.cn (Z.Q.)

2 Engineering Product Development, Singapore University of Technology and Design, Singapore 487372, Singapore

3 Department of Computer Science, University of Turbat, Balochistan 92600, Pakistan

* Correspondence: ran_liu@sutd.edu.sg; Tel.: +86-0816-608-9122

Received: 11 June 2020; Accepted: 13 July 2020; Published: 16 July 2020

Abstract: In an indoor environment, object identification and localization are paramount for human-object interaction. Visual or laser-based sensors can achieve the identification and localization of the object based on its appearance, but these approaches are computationally expensive and not robust against the environment with obstacles. Radio Frequency Identification (RFID) has a unique tag ID to identify the object, but it cannot accurately locate it. Therefore, in this paper, the data of RFID and laser range finder are fused for the better identification and localization of multiple dynamic objects in an indoor environment. The main method is to use the laser range finder to estimate the radial velocities of objects in a certain environment, and match them with the object's radial velocities estimated by the RFID phase. The method also uses a fixed time series as "sliding time window" to find the cluster with the highest similarity of each RFID tag in each window. Moreover, the Pearson correlation coefficient (PCC) is used in the update stage of the particle filter (PF) to estimate the moving path of each cluster in order to improve the accuracy in a complex environment with obstacles. The experiments were verified by a SCITOS G5 robot. The results show that this method can achieve an matching rate of 90.18% and a localization accuracy of 0.33m in an environment with the presence of obstacles. This method effectively improves the matching rate and localization accuracy of multiple objects in indoor scenes when compared to the Bray-Curtis (BC) similarity matching-based approach as well as the particle filter-based approach.

Keywords: dynamic objects identification and localization; laser cluster; radial velocity similarity; Pearson correlation coefficient; particle filter

1. Introduction

In recent years, with the development of wireless communication technology, location-based services have been widely used in search and rescue, medical services, intelligent transportation, logistics management, and other fields [1], and many positioning technologies have been investigated in the research community. Among them, it is very important to identify and localize the dynamic object in the indoor environment. This field has a wide range of applications, such as monitoring teams with different identities, and tracking designated objects, library book management, real-time control of equipment and participants in the venue, etc. [2].

Identification and localization are usually regarded as two separate tasks, which are solved by different methods. In this field, it is common to deploy cameras in the environment in order to identify and locate objects. Xu et al. [3] realized the positioning of indoor mobile robots by installing the camera above the robot head and extracting the ceiling features. Wang et al. [4] constructed a three-diensional (3D) human tracking system by fusing the information of visual and ultrasonic sensors. Liu et al. [5] analyzed target tracking under different illumination conditions in intelligent monitoring of public places. Although the vision-based method can obtain better positioning accuracy and recognition rate, however, it usually requires clear images to identify and localize, this kind of method has great limitations in practical application, because it must overcome occlusion, motion uncertainty, and environmental appearance changes, and, in some cases, this method may lead to privacy violations. Mostly, the laser range finder is used in the field of object localization due to its high precision, wide range, and fast transmission speed. However, it can only obtain sparse environmental information, and it is difficult to distinguish objects with similar appearance. It is necessary to extract effective features from the sparse information for identification, which makes the algorithm very complicated. Radio Frequency Identification (RFID) technology has a unique tag ID, which can quickly identify the object by relying on the radio frequency signal, so it can save a lot of computing resources, and can solve the problems of occlusion and other environmental factors. However, due to its hardware limitation, accurate localization cannot be achieved.

It is difficult to accurately identify and locate the object by only relying on a certain kind of sensor in the indoor environments, for the proper identification and localization it requires the fusion of multiple sensors information. Many researchers have conducted extensive research in this field. Xing et al. [6] designed a multi-sensor information fusion method based on a visual marker called ArUco. This method uses Kalman filter to fuse the information of visual marker, ultrasonic, and inertial sensors to localize micro air vehicles in indoor environments with an accuracy up to 4 cm. You et al. [7] used Unscented Kalman filter to fuse UWB and IMU data for the localization of quadrotor UAV in indoor environments. Li et al. [8] employed sliding window filter (SWF) to fuse camera and IMU data for accurate 3D motion tracking and reconstruction. Peng et al. [9] proposed a multi-sliding window classification adaptive unscented Kalman filter (MWCAUKF) method with timestamp sort updating, which could fuse multiple kinds of sensors data. Shi et al. [10] realized the positioning and navigation of the mobile robot by integrating laser and geomagnetic sensors. Zhao et al. [11] proposed a method that fuses the information obtained by 3D LiDAR and camera. The average identification accuracies of their method for cars and pedestrians are 89.04% and 78.18%, respectively. Digiampaolo et al. [12] proposed a localization system based on the passive signal phase, and used extended Kalman filter (EKF) to fuse RFID and odometer information to achieve the object localization, with an accuracy of up to 4 cm. Tian et al. [13] proposed a low-cost INS and UWB integrated pedestrian tracking system, which only uses single UWB anchor node at an unknown location, minimizing infrastructure cost and setup.

Each frame of data measured by a laser range finder can be used to represent a static environment. Therefore, if each frame of data is correlated based on time series, the localization of dynamic objects in the environment can be achieved [14,15]. Tang et al. [16] designed a real-time indoor positioning system based on laser scan matching for unmanned ground vehicles in a large area. Zhang et al. [17] designed an object localization algorithm based on laser range finder, which realized the object localization under the airborne reconnaissance platform. However, when we use the laser range finder to identify the object, as mentioned above, it is difficult to distinguish objects with similar appearance, so there is a certain degree of singularity [18]. Researchers have done a lot of research in using laser measurement for object identification. Wang et al. [19] used 3D laser sensors and sliding windows to achieve object detection and identification. Huang et al. [20] proposed using feature fusion and establishing spatio-temporal feature vectors to realize the detection and recognition of dynamic obstacles, with a recognition rate of up to 87.7%. Tong et al. [21] proposed an object recognition method by combining laser and infrared information. This method can achieve more than 90% recognition accuracy for objects with large shape differences, but the recognition accuracy for objects with similar shapes is low.

RFID is widely used in the field of object identification due to its characteristics of fast recognition speed, high recognition accuracy, and wide coverage. Liu et al. [22] proposed a method for quickly tracking dynamic targets based on the RSS measurements from a pair of RFID antennas. Fan et al. [23] proposed a method, called "M^2AI", which uses convolutional neural networks for deep learning to achieve multiple objects identification. The identification rate of this method can reach 97%, but, as the number of dynamic objects (for example three objects) increases, the recognition rate will drop to about 80%.

The unique ID of the RFID tag can be used to solve the singularity problem of laser sensor in object identification. The advantages of high measurement accuracy of laser range finder can be used to make up for the low positioning accuracy of RFID. Therefore, we fuse information from RFID and laser range finder to achieve dynamic multiple objects identification and localization. More specifically, we use the phase difference of RFID in the adjacent time to calculate the radial velocities of moving tags. Meanwhile, after clustering the laser points, we use the Pearson correlation coefficient in the update stage of particle filter to estimate the moving path and radial velocity of each cluster. Through the similarity matching algorithm based on the sliding time window, we can match the laser cluster-based radial velocities and the phase-based radial velocities in each window, and find the cluster with the highest similarity for each dynamic RFID tag, and the center coordinate of the cluster is considered as the position of the object.

We summarize the contributions of this article, as follows.

(1) We present a solution that incorporates RFID phase information and laser range measurements for multiple dynamic objects identification and localization in indoor environments.
(2) We propose incorporating the Pearson correlation coefficient into the update stage of particle filtering. This method can effectively estimate the historical trajectories of moving objects in an environment with obstacles.
(3) We set up different paths in different environments on campus, and thoroughly evaluated our method. Our method can effectively identify and locate the passing pedestrians in indoor scenes when compared to the Bray–Curtis (BC) similarity matching-based approach as well as the particle filter-based approach.

We organize the subsequent sections of this paper, as follows. The related work is discussed in Section 2. An overview of the system is described in Section 3. In Section 4, we present the details of the dynamic multi-object identification and localization method. We show the experimental results in Section 5 and conclude the paper with possible extensions in Section 6.

2. Related Work

This section gives a thorough overview of the work related to the identification and localization of dynamic objects using RFID techniques.

Dynamic objects localization approach is widely used in intelligent monitoring, human-robot interaction, virtual reality, and robot navigation. Typical methods for localizing dynamic objects using RFID systems are the LANDMARK-based approach [24], the SpotON-based approach [25], and the VIRE(VIrtual Reference Elimination)-based approach [26]. When the LANDMARC system is used for localization, there is uncertainty in the selection of the number K of neighbor labels. Besides, it is necessary to add additional reference tags in order to obtain a more accurate localization result, but too many tags will cause signal interference between tags, which will affect the positioning accuracy of the system. The SpotON method measures the RSS of a series of tags, and establishes the regression equation of the distance between RSS and the tag to the reader antenna, and then estimates the distance from the target to the antenna through RSS, and finally determines the location of the target by trigonometry using the ranging information of multiple antennas. However, RSS distance model needs to be built in advance, which is a heavy workload. The VIRE algorithm uses linear interpolation to replace the virtual position with the real position to improve the system accuracy. However, the real

RSS value changes non-linearly. There is an error between the virtual position and the real position while using linear mathematical interpolation. In addition, the VIRE algorithm does not have a good localization effect on the border area. The calculation cost will be increased if additional real labels are added to the border area.

Many researchers combine RFID with other sensors to realize the identification and localization of dynamic objects, as it is difficult to meet the requirements of high positioning accuracy and low cost by only using RFID system. Alvarado et al. [27] combined a RFID system, an omnidirectional Mobotix C25 camera, and a laser range finder to realize the localization of tour-guide robot. Choi et al. [28] combined RFID system and ultrasonic sensors to overcome uncertainties in previous RFID systems for mobile robot localization. Suparyanto et al. [29] presented a system to localize container truck while using indoor Global Positioning System (iGPS), RFID, and Inertial Measurement Unit (IMU) consisting of accelerometer, gyroscope, and magnetometer. Wang et al. [30] and Li et al. [31] combined the RFID and Kinect camera to realize the identification and localization of multiple dynamic objects, but its accuracy is low. Parr et al. [32] realized tag tracking by combining the information of RFID and IMU. Faramondi et al. [33] combined the information of IMU, RFID, and triad-magnetometer in order to realize the identification and localization of rescuers, but the accuracy is low. In this paper, we show that we can achieve the identification and localization of multiple dynamic objects with high accuracy by fusing the information from RFID and laser range finder.

3. System Overview

If an object is attached with an RFID tag, the identification of this object is known to us. However, we cannot know the exact location of the object in an environment due to the physical limitations of the RFID system. In other words, we only know that there are several objects in the environment, but we can't distinguish them. Therefore, this paper proposes a method for localizing multiple dynamic objects with known identities in an indoor environment by matching laser clusters with RFID tag IDs. The entire system uses RFID and two-dimensinal (2D) laser range finder to measure obstacles and moving objects in the environment, as shown in Figure 1. Subsequently, the algorithm uses the phases collected from the RFID to calculate the phase-based radial velocity of each RFID tag, and uses laser range finder data to calculate the laser cluster-based radial velocity of each cluster. In addition, we use Pearson correlation coefficient combined with particle filter to realize the tracking of a cluster in each sliding time window. Finally, the radial velocity similarity matching is used to locate the dynamic targets with known identities.

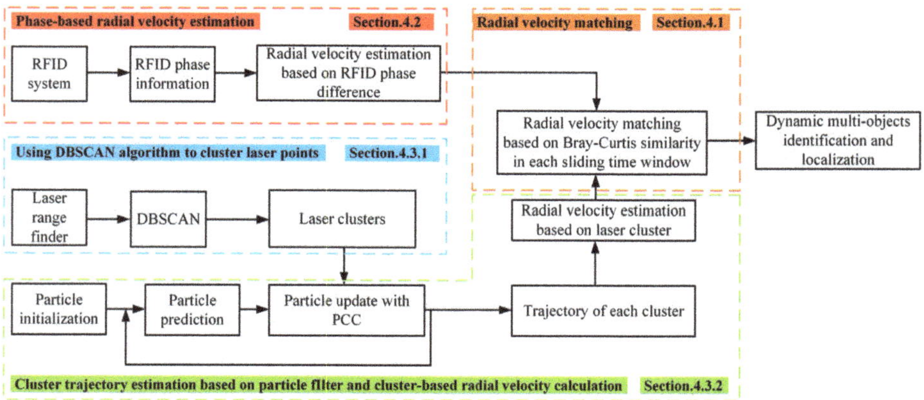

Figure 1. System overview.

More specifically, we collect the measurements through RFID reader and 2D laser range finder. The RFID reader controls the antenna to scan the tagged dynamic objects in the environment, and the

2D laser range finder is used to measure the distance and angle from the surrounding obstacles in an environment. Afterwards, the information that is detected by the two sensors is stored the database. We calculate the moving object's radial velocity based on the phase information reported from the RFID reader; meanwhile, we use the DBSCAN algorithm to cluster the laser ranging information. After obtaining the cluster information, we use the Pearson correlation coefficient and particle filter to estimate the moving path of each cluster, and estimate the radial velocities of clusters based on the distance between the two clusters at adjacent moments. Then, we define a fixed-size sliding time window w with the current time T as the end, and use the Bray–Curtis similarity algorithm to constantly matching the radial velocity estimated by two sensors in each window. According to the matching results, we select the cluster with highest similarity to realize the identification and localization of multiple dynamic objects. The specific implementation method will be described in detail in the next part of this paper.

4. Indoor Multiple Dynamic Objects Identification and Localization

4.1. Object Identification and Localization Based on Radial Velocity Matching

Restricted by the RFID physical system, when we estimate the information between the object and RFID antenna, we can only obtain the phase information, and it is difficult to estimate the exact coordinates of the object based on phase. However, we can infer the radial velocity of an object based on the phase difference. Radial velocity refers to the velocity component of the object's movement velocity in the direction of the observer's line of sight, which is, the projection of the velocity vector in the direction of the line of sight, which is often used to represent the rate of change of the distance from the object to the observation point. For the same observation point, assuming that there are multiple objects in the environment, the objects cannot be in the same position at the same time, which means that their distance and angle to the observation point are different, resulting in their radial velocity being different from each other. Therefore, we use laser sensor to estimate the radial velocity of each cluster in the environment in order to better match the object's radial velocity estimated by the RFID system. The similarity of two velocities are compared to realize the identification and localization of the moving objects. Different applications often need to use different similarity measurement methods.

Different similarity measures are often used in different applications. In this paper, it is necessary to find a reliable similarity measurement function to match the speed estimated by RFID due to the large number of clusters obtained by laser sensors. Vorst et al. [34,35] compared different similarity algorithms in order to achieve self-localization with passive RFID fingerprints. They showed that the Bray–Curtis (BC) measure gives the best performance among all of the evaluated methods. The Bray–Curtis similarity algorithm [36] is often used to calculate the similarity between different samples. Samantaray et al. [37] used BC similarity to effectively realize medical image recognition and retrieval. Therefore, we use BC similarity to match the radial velocity. Besides, in order to achieve better results, this article refers to the sliding window method commonly used in image object identification. By using sliding time windows to limit the number of laser frames, we ensure that all of the objects can be identified in each window. This paper defines a fixed-size sliding time window w with the current time T as the end, constantly matching the radial velocity estimated by two sensors in each window. The window range is denoted as $[T\text{-}w\text{+}1\text{:}T]$. The similarity of laser cluster-based radial velocity and RFID phase-based radial velocity can be computed as:

$$Sim = (1 + \frac{1}{w}\sum_{t=T}^{T-w+1} \frac{|V_t^R - V_t^{L,i}|}{|V_t^R| + |V_t^{L,i}|}), \tag{1}$$

where i represents the i-th cluster at the current time, and $V_t^{L,i}$ represents its radial velocity at time t. V_t^R represents the phase-based radial velocity at time t. In addition, if the radial velocities that are estimated by laser and RFID belong to the same object, the similarity should be high. Therefore, we choose the cluster which has the highest similarity and assigns the tag ID to it. The central coordinate of this cluster is considered as the estimated position of the moving object, thereby realizing the identification and localization of objects.

Estimating the radial velocities of moving tags based on RFID phase, and radial velocities estimation method based on laser clusters will be given in Sections 4.2 and 4.3, respectively. Table 1 lists the symbols in this paper and their meanings.

Table 1. Symbols and their meanings.

Mathematical Symbol	Meaning
Sim	The $Bray - Curtis$ similarity between phase-based radial velocity and laser cluster-based radial velocity
w	The size of each time window
V_t^R	RFID phase-based radial velocity at time t
$V_t^{L,i}$	Radial velocity of the i-th cluster at time t
θ_t	The phase of RFID signal at time t
d_t	The radial distance from the moving tag to the antenna at time t
$\Delta\theta$	Phase difference
$\Delta\theta'$	The phase differences after normalized to the main value interval of $[-\pi : \pi]$
$C_t^{n_t}$	Clusters in total at time t
$(\bar{x}_t^{(j)}, \bar{y}_t^{(j)})$	Center coordinates of the j-th cluster at time t
$v_t^{(i)}$	The radial velocity of cluster i at time t
N	Number of particles
$(x_{i,t}^{[n]}, y_{i,t}^{[n]})$	The position of particle n to track i-th cluster at time t
$\omega_{i,t}^{[n]}$	The weight of particle n to track i-th cluster at time t
$N(0, \sigma)$	The Gaussian noise with zero mean and standed deviation of σ
μ	The normalization coefficient

4.2. Estimating the Radial Velocities of Moving Tags Based on RFID Phase Difference

This paper uses the RFID phase information to estimate the radial velocities of the moving tags. The RFID phase information is a periodic function of the distance between the tag and the antenna (the period is 2π) and is given by:

$$\theta_t = 2\pi \cdot \frac{2 \cdot d_t}{\lambda} \cdot mod(2\pi), \tag{2}$$

where θ_t is the signal phase at time t, λ is the wavelength of the receiving signal, and d_t is the radial distance from the moving tag to the antenna at time t. In this paper, the phase differences at adjacent times are first processed and normalized to the main value interval of $[-\pi,\pi]$. The specific implementation is given by:

$$\Delta\theta' = \begin{cases} \Delta\theta, & -\pi < \Delta\theta < \pi \\ \Delta\theta - 2\pi, & \Delta\theta \geq \pi \\ \Delta\theta + 2\pi, & \Delta\theta \leq -\pi, \end{cases} \tag{3}$$

We may get two different values with a phase difference of π when a tag is not moving [38] due to the limitation of the RFID signal processing algorithm. Therefore, it is necessary to set an appropriate threshold to eliminate the effect of the π phase jump. This paper assumes that the phase difference of the same object should not exceed the threshold φ in the adjacent time. When the phase difference is between $-\varphi$ to φ, we use this phase difference to estimate the radial velocity, otherwise we set the radial velocity to an invalid value. The radial velocity of each tag estimated by the RFID at the current moment can be computed as:

$$V_t^R = \frac{\Delta\theta'}{4\pi\Delta t} \cdot \lambda, \tag{4}$$

4.3. Estimiating the Radial Velocities of Laser Clusters

4.3.1. Laser Clustering Based on DBSCAN

The DBSCAN algorithm is a spatial clustering algorithm based on density [39]. The remarkable advantage of this algorithm is that the algorithm is fast, and it can divide the regions with enough high density into a group, which effectively deals with noise points and quickly finds spatial clustering of arbitrary shape. The experiment divides the laser points into core points, boundary points, and noise points. For a given point, if the number of adjacent points in the neighborhood with radius ρ is greater than ξ, then this point is regarded as the core point. If it is within the ρ neighborhood of the core point, we treat it as a boundary point; otherwise, we treat it as a noise point. After clarifying the categories of the respective laser points, the core points and the boundary points are merged into one cluster. Finally, we will get n_t clusters, where $C_t = C_t^1, C_t^2, \ldots, C_t^{(n_t)}$, and we can estimate the central coordinate of each cluster as: $\bar{P}_t^{(j)} = (\bar{x}_t^{(j)}, \bar{y}_t^{(j)}), j \in [1 : n_t]$. Figure 2 shows one example of the clustering result.

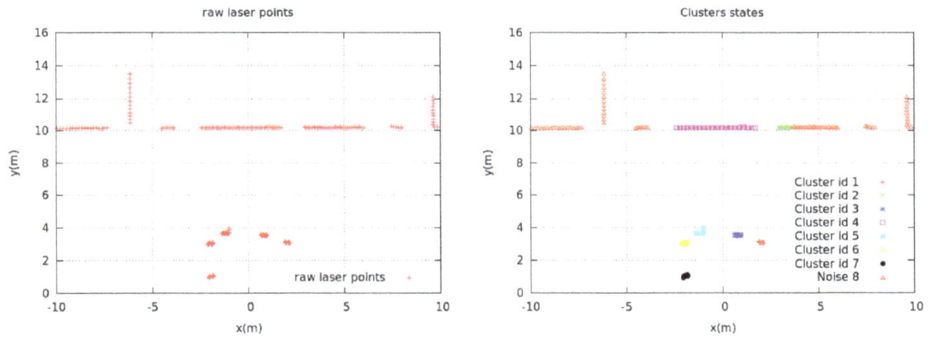

(a) (b)

Figure 2. Clustering results. (**a**) raw laser points; (**b**) DBSCAN results.

The two most important parameters in DBSCAN algorithm are the radius ρ and number of laser spots ξ in each cluster. Our previously published paper [40] showed a comparison of the clustering performance under different ρ and ξ. The parameters of DBSCAN are not discussed in the article in order to save the space of the paper. According to past experience, we set the parameters of DBSCAN as $\rho = 0.1$, $\xi = 2$.

4.3.2. Cluster Trajectory Estimation Based on Particle Filter

We need to calculate the radial velocity of each cluster in order to match the radial velocity estimated by RFID. Therefore, it is necessary to find the historical trajectory of each cluster at the previous moment. The particle filter is widely used in the field of navigation and positioning due to its advantages of high robustness, high accuracy, and excellent performance in non-linear and non-Gaussian systems. In this paper, we set up an independent particle filter to track the trajectory of each cluster. In each particle filter, the position of a moving object can be represented by a set of weighted particles: $X_{i,t} = \{X_{i,t}^{[n]}, w_{i,t}^{[n]}\}_{n=1}^{N}$, where N represents the number of particles, i represents the particle filter of the i-th object, and $i \in [1 : n_t]$. Besides, $X_{i,t}^{[n]} = x_{i,t}^{[n]}, y_{i,t}^{[n]}$ represents the two-dimensional

(2D) coordinates of each particle, and $w_{i,t}^{[n]}$ represents its weight. The particle filter will predict and update iteratively based on measurements arrival.

- Prediction

In this paper, we use the position $X_{i,t}^{[n]}$ of the particles at time t and the motion model to estimate the position $X_{i,t-1}^{[n]}$ in the previous time. The Gaussian function is chosen as the model of motion prediction and the corresponding parameters of the Gaussian function are used to adjust the particle distribution density because of the uncertainty of the direction and velocity of the moving object. The prediction is performed with the following:

$$\begin{cases} x_{i,t-1}^{[n]} = x_{i,t}^{[n]} + N(0,\sigma) \\ y_{i,t-1}^{[n]} = y_{i,t}^{[n]} + N(0,\sigma), \end{cases} \tag{5}$$

where $N(0,\sigma)$ is the Gaussian noise with zero mean and standard deviation of σ. Because the sampling time interval of the laser range finder is 0.1 s, we assume that the maximum moving distance of an object in each sampling time will not exceed 0.1 m, so we let σ = 0.1. After the prediction, we use the Pearson correlation coefficient in the update stage to match the motion model in Equation (5), to adjust the weight of the particles, thereby correcting the prediction information.

- Update

The update stage of particle filter is to use the measurement value of every moment in historical time to update the particle's weight. In the prediction stage, we generate particles with Gaussian distribution, so, in the update stage, we use the probability density function of Gaussian distribution to update the particle's weight. We calculated the distance between the cluster and the n-th particle. Therefore, the weight $w_{i,t-1}^{[n]}$ of each particle in ordinary particle filter is updated according to the following:

$$w_{i,t-1}^{[n]} = \mu \cdot w_{i,t}^{[n]} \cdot \frac{1}{n_{t-1}} \sum_{i=1}^{n_{t-1}} \frac{1}{\sqrt{2\pi\sigma}} \cdot \exp(-\frac{\Delta d^2}{2\tau^2}), \tag{6}$$

where

$$\Delta d = \sqrt{(x_{i,t}^{[n]} - \bar{x}_{t-1}^{(j)})^2 + (y_{i,t}^{[n]} - \bar{y}_{t-1}^{(j)})^2}, \tag{7}$$

where n_t indicates that there are n_t clusters in the time t, Δd represents the distance between the cluster and the particle, τ is the translational coefficient of the distance from particle to cluster, and μ is the normalization coefficient. In this paper, we assume that the radius of a human cluster is 0.1 m, therefore, we set τ = 0.1. The estimated position of each object at time t can be obtained by the weighted average of all particles.

Traditional particle filter generally uses a single feature to construct object's model, but such a model often has different disadvantages due to different features selected: for example, using color features, when the color of the object is similar to the background, the tracking process will become unstable; using the contour or texture features, if the object rotates, deforms or occludes in the process of moving, the tracking effect will become poor or even the tracking fails. This is because, in all of the above cases, when the traditional particle filter algorithm processes the weight of particles in the update stage, it cannot correctly reflect the degree of matching between the particles and the objects. In the resampling stage, after many iterations, there are too few effective particles to approximate the real state of the object, which makes the particles easily drift to the obstacles, leading to tracking failure. To solve this problem, we introduce the Pearson correlation coefficient [41]. It is usually used for the indoor localization by comparing the similarity between the location fingerprint and a known fingerprint database [42]. In addition, it is often used in the template matching stage of image processing in order to compare the similarity between the two images [43]. In this paper, we compare the i-th cluster at time t with the clusters in each historical time to obtain the Pearson correlation coefficient. Afterwards, we substitute the results into the probability density function. The updated weight of the particle is computed as:

$$w_{i,t-1}^{[n]} = \mu \cdot w_{i,t}^{[n]} \cdot \frac{1}{n_{t-1}} \sum_{i=1}^{n_{t-1}} (1-p) \cdot \exp(-\frac{\Delta d^2}{2\tau^2}), \quad (8)$$

where

$$p = \frac{\sum_{n=1}^{N} (x_{i,t}^{[n]} - \bar{x}_{t-1}^{(j)}) \cdot (y_{i,t}^{[n]} - \bar{y}_{t-1}^{(j)})}{\sqrt{\sum_{n=1}^{N} (x_{i,t}^{[n]} - \bar{x}_{t-1}^{(j)})^2} \cdot \sqrt{\sum_{n=1}^{N} (y_{i,t}^{[n]} - \bar{y}_{t-1}^{(j)})^2}}, \quad (9)$$

where $\{\bar{x}_{t-1}^{(j)}, \bar{y}_{t-1}^{(j)}\}$ represents the center coordinate of the j-th cluster at the historical time, and $j \in [i : n_t]$.

Lastly, the resampling tackles the issue of particle degradation that occurs after many iterations. We replicate a series of particle sets by eliminating particles with small weights and replicating particles with large weights.

4.3.3. Calculating Each Cluster's Radial Velocity

Now we have each cluster's historical position, then we need to calculate the clusters of the same cluster moving in the adjacent time, the radial velocity of cluster at each time can be expressed as:

$$V_t^{(i)} = \frac{\sqrt{(\bar{x}_t^{(i)})^2 + (\bar{y}_t^{(i)})^2} - \sqrt{(\bar{x}_{t-1}^{(i')})^2 + (\bar{y}_{t-1}^{(i')})^2}}{\Delta t}, \quad (10)$$

where Δ_t represents the time difference between adjacent times.

In addition, the experimental results obtained by the other two methods (namely BC-based method and PF-based method) are compared. In the first method, we only use the BC similarity algorithm to match the radial velocity estimated by two sensors in each sliding time window. Different from the algorithm proposed in this paper, the BC-based method does not use particle filter to estimate the historical information of cluster, but it finds the nearest cluster $C_{t-1}^{(i')}$ in the previous time $t-1$ for each cluster $C_t^{(i)}$ of current time t. In the BC-based method, we assume that the moving distance of the same object should be very small at the adjacent time. Therefore, the previous cluster of a moving object (denoted as i') can be determined by:

$$i' = arg \min_j \sqrt{(\bar{x}_t^{(i)} - \bar{x}_{t-1}^{(j)})^2 + (\bar{y}_t^{(i)} - \bar{y}_{t-1}^{(j)})^2}, \quad (11)$$

After finding the historical information of each cluster, the radial velocity of each cluster can be calculated by Equation (10), and matched with the object's phased-based radial velocity by Equation (1).

As for the second method, we use the traditional particle filter described above, and use the traditional method to update the weight of particles (Equation (6)). Similarly, after obtaining the historical information of each cluster, we also calculate the radial velocity of the cluster by Equation (10), and it performs similarity matching by Equation (1).

5. Experimental Results Analysis

5.1. Experimental Setups

The feasibility of the proposed method in this paper is verified by using the SCITOS G5 robot. We conducted experiments in an indoor environment that is shown in Figure 3. This robot integrates a UHF RFID reader, named Impinj Speedway Revolution R420, two circularly polarized antennas of the type Laird Technologies S9028, and a laser range finder, named SICK S300. The RFID reader provides a maximum measuring distance up to 10 m. The laser range finder provides a maximum measuring distance up to 29 m and a horizontal scanning angle of 270° with a scanning resolution of 0.5°.

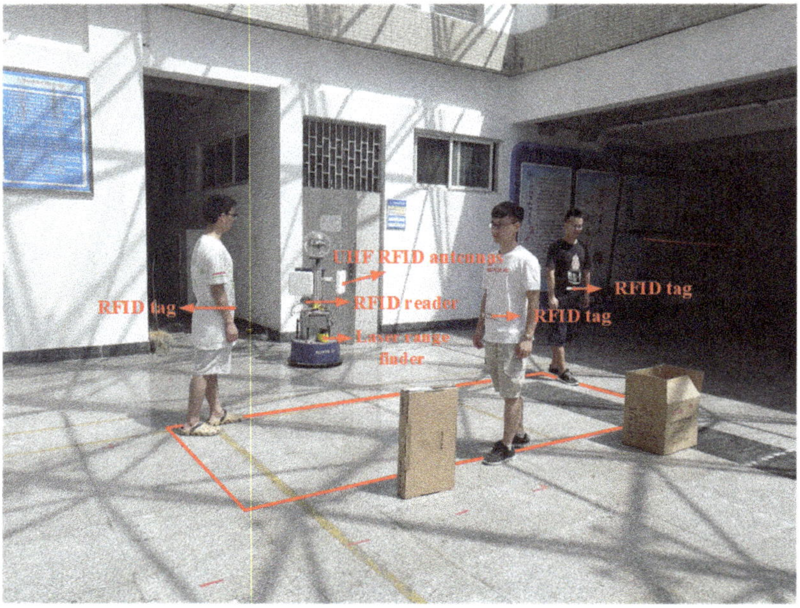

Figure 3. Experimental setup.

The experiments have been tested in a space that is a rectangular area of 4 m × 2 m, as shown in Figure 3. The robot had been placed one-meter perpendicular to the rectangle's long side. During the experiment, the position of the robot remained unchanged. The RFID tags worn by the three persons moved along the edge of the rectangle. Each of them walked three circles, and each person's speed was different (but each person's speed was kept constant while walking). We set a marker every one meter on the ground shown by the red line in the Figure 3 in order to calculate the positioning error. Each time the experimenter went across to a mark, the software records the time when he arrives and stores it in the database. We calculate the average velocity of the experimenter between the two adjacent markers, calculate the coordinates of their position at each moment, and treat this coordinate as the true position of the experimenter. Because the distance between adjacent marks is short, the ground truth error can be ignored. Subsequently, the position estimated by the algorithm is compared with the ground truth to measure the localization error, which is defined as the Euclidean distance between the estimated position and the true position. We think that, if the error is less than 0.8m, the cluster and tag ID are successfully matched. The matching rate is obtained by comparing the number of clusters successfully matched with the total number of clusters.

5.2. Impact of Different Methods on Experimental Results

In this section, we used three different methods to compare the final localization accuracy and matching rate. In the first method (BC), we only use Bray–Curtis similarity to match the radial velocities estimated by two sensors, respectively. For the second method (PF + BC), we only use particle filter to estimate the historical coordinates of each cluster. As for the third method (PCC + PF + BC), we added the Pearson correlation coefficient in the update stage of particle filter to further constrain the particles. We set $\rho = 0.1$, $\xi = 2$, $w = 25$, $N = 200$, $\varphi = 90°$, and the results of the three methods are compared in Table 2, and the estimated trajectories of moving objects are shown in Figure 4.

Table 2. Comparison of experimental results based on different approaches.

Methods	Localization Error (m)	Matching Rate (%)
BC	0.76	79.2
PF + BC	0.65	83.1
PCC + PF + BC	0.33	90.2

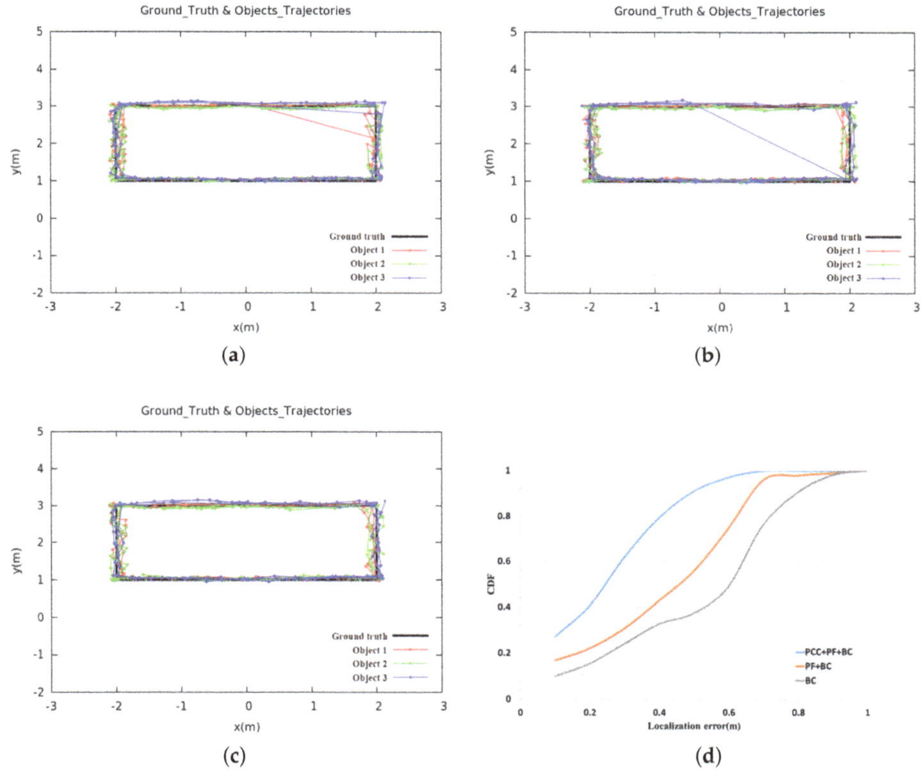

Figure 4. Comparision of the estimated trajectories based on different approaches. (**a**) BC; (**b**) PF + BC; (**c**) PCC + PF + BC; (**d**) CDFs.

If the obstacles are present in the certain environment, the BC-based approach cannot locate the moving objects very well, as shown in Table 2 and Figure 4. The localization error is 0.76 m, and the identification rate is 79.2%. From Figure 4a, it can be seen that the object's trajectory jumps greatly. That is because we assume that the moving distance of the same object at adjacent times should be the smallest, so we find the cluster $C_{t-1}^{(i')}$ with minimal distance at the previous time $t-1$ for each cluster $C_t^{(i)}$ at time t through Equation (11). When the distance between the object and the obstacle is very close, the algorithm may incorrectly obtain the historical information of the moving object, which causes the radial velocity matching to fail. Only after the object is far away from the obstacle by a certain distance, the algorithm can recover, trajectory jump of BC-based approach as shown in Figure 4a. Although the particle filter-based approach improves the accuracy and matching rate compared with the BC-based approach, the result is still not good enough: the localization error is 0.65 m and the matching rate is only 83.1%. We use the traditional particle filter method to update the particle's weight (Equation (6)), and then obtain the historical information of each object. Because particle filter mainly uses laser data to estimate the historical trajectory of the object, if human moves close to or away from the obstacles during the movement, the performance of the system will become very poor. This is because, in this case, the traditional particle filter algorithm cannot correctly reflect the matching degree between the particle and the object when processing the weight of the particle in the update stage. In the resampling stage, after many iterations, there are too few effective particles to approximate the true state of the object, which makes the historical position estimation of the object deviate, which further leads to the failure of radial velocity matching and trajectory jump. From Figure 4b, we can also clearly see the trajectory jump. The method of combining PCC and particle filter can better constrain the particles, the localization error is reduced from 0.65 m to 0.33 m, and the matching rate is increased from 83.1% to 90.2%. When compared with Figure 4a,b, the objects trajectories that are estimated by this method (Figure 4c) are smoother. Besides, if an object moves along the circular arc equidistance from the RFID antenna, there will be no difference in distance, and thus the radial velocity tends to zero, which is also one of the sources of error in this research.

5.3. *Impact of Different Parameters on Experimental Results*

5.3.1. The Influence of Antenna Settings on Experimental Results

We evaluated the localization error under various antenna combinations, as the detection range of RFID is directly confined by antennas on robot. We set $\rho = 0.1$, $\xi = 2$, $w = 25$, $N = 200$, $\varphi = 90°$, and the result is shown in Table 3. Due to the limitation of the coverage of the RFID antenna, the localization accuracy based only one antenna is low (for example, when only the right antenna is used, the localization error is 0.82 m). However, when we use two antennas, the measurement range of the antenna can completely cover the entire experimental environment, and the localization error is reduced from 0.82 m to 0.33 m. Therefore, two antennas are used in this paper for the localization and identification of the dynamic multi-objects.

Table 3. Comparison of localization accuracy under different antenna settings.

Antenna Combination	Methods	Localization Error (m)
Only Right Antenna	BC	0.88
	PF + BC	0.96
	PCC + PF + BC	0.82
Only Left Antenna	BC	0.68
	PF + BC	0.89
	PCC + PF + BC	0.61
Both	BC	0.76
	PF + BC	0.65
	PCC + PF + BC	0.33

5.3.2. The Influence of Phase Shift Threshold φ on Experimental Results

An appropriate setting of the RFID phase threshold is the key to the algorithm for eliminating π phase jump and accurately estimating the object's moving radial velocity. In this paper, it is assumed that the phase difference of the same object in the adjacent time should not exceed the threshold. In this section, we set $\rho = 0.1$, $\xi = 2$, $w = 25$, $N = 200$. Table 4 shows the localization accuracy under different phase shift thresholds.

It can be seen from Table 4 that when φ is set too small (e.g., $\varphi = 10°$), the algorithm will erroneously remove the originally correct phase information, resulting in the matching rate of only 81.6% and a localization error of 0.72 m. When φ is set too large (e.g., $\varphi = 180°$), the algorithm cannot correctly handle the phase where the jump occurs, resulting in the localization error up to 0.83 m and matching rate is only 66.8%. Therefore, only by setting an appropriate phase shift threshold can we better eliminate the effect of phase jumps while retaining normal phase information. We set $\varphi = 90°$, the matching rate can reach 90.2%, and the localization error is only 0.33 m, in order to obtain better experimental results.

Table 4. The influence of phase shift threshold φ on experimental results.

φ	Methods	Localization Error (m)	Matching Rate (%)
	BC	2.41	54.9
10°	PF + BC	1.98	64.0
	PCC + PF + BC	0.72	81.6
	BC	0.88	66.4
30°	PF + BC	0.72	82.1
	PCC + PF + BC	0.34	88.2
	BC	0.79	68.1
60°	PF + BC	0.63	85.2
	PCC + PF + BC	0.36	87.6
	BC	0.76	79.2
90°	PF + BC	0.65	83.1
	PCC + PF + BC	0.33	90.2
	BC	0.76	78.8
120°	PF + BC	0.68	81.4
	PCC + PF + BC	0.37	84.2
	BC	0.77	78.9
150°	PF + BC	0.67	81.8
	PCC + PF + BC	0.38	83.8
	BC	0.86	66.7
180°	PF + BC	0.81	67.1
	PCC + PF + BC	0.83	66.8

5.3.3. The Influence of The Number of Particles N on The Experimental Results

In the traditional particle filter algorithm, the number of particles will have an effect on the results. In this paper, Pearson correlation coefficient is added to constrain the particles in the update stage of particle filter, so we carry out experiments to analyze the impact of the number of particles on the localization error and matching rate in this case. We used CPU with core i5-7300 HQ, 2.50 GHz, and 8 GB ram in the experiment, and set other parameters exactly the same as before. Table 5 lists the results.

Table 5. The influence of the number of particles N on the experimental results.

Number of Particles N	Localization Error (m)	Time Consumption (ms)
5	1.41	2.96
20	0.44	3.99
50	0.37	5.16
100	0.35	5.82
200	0.33	6.44
400	0.32	8.30
1000	0.33	12.17

It can be seen in the above Table 5, a small N (such as $N = 5$) gives an increase in the localization error, since the small number of particles cannot effectively represent the probability density. The positioning accuracy gets improved when increasing the number of particles, similarly, performing filtering with the large number of particles also consumes more time. With a large N

(such as $N = 1000$), we almost get the same localization results. For considering the accuracy and the time-consuming of the algorithm, we choose $N = 200$ in our experiment.

5.3.4. The Influence of The Time Window Size w

The size of time window directly affects whether all moving objects can be found in each window. In the experiment, we set $\rho = 0.1$, $\xi = 2$, $\varphi = 90°$, $N = 200$, and the experimental results under different w sizes are compared in Table 6.

Table 6. The influence of the time window size w on the experimental results.

Window Size w (s)	Method	The Average Distance Traveled in Each Window (m)	Localization Error (m)	Matching Rate (%)	Time Consumption (ms)
	BC	0.93	1.14	66.9	28.16
5	PF + BC	0.93	0.88	72.5	34.42
	PCC + PF + BC	0.93	0.71	81.8	42.56
	BC	2.79	0.86	67.2	30.3
15	PF + BC	2.79	0.67	82.8	38.36
	PCC + PF + BC	2.79	0.37	87.0	46.42
	BC	4.65	0.76	79.2	33.28
25	PF + BC	4.65	0.65	83.1	43.1
	PCC + PF + BC	4.65	0.33	90.2	51.07
	BC	6.51	0.81	68.3	34.69
35	PF + BC	6.51	0.70	82.3	46.97
	PCC + PF + BC	6.51	0.33	87.7	54.81
	BC	9.18	0.83	67.6	38.41
50	PF + BC	9.18	0.64	83.8	54.69
	PCC + PF + BC	9.18	0.35	87.6	62.24

It can be seen from Table 6 that, as compared with the other two methods (BC and PF + BC), our proposed method (PCC + PF + BC) integrates Pearson correlation coefficient into the update stage of particle filter, and matches the laser cluster-based radial velocity and the phase-based radial velocity in each window has the least localization error and the highest matching rate. In addition, if the window is too small (e.g., $w = 5$), when the object is occluded during the experiment, the algorithm cannot find all objects in a period time, which will cause the failure of radial velocity matching. Consequently, we obtain the error up to 0.71 m, and the matching rate reduces to 81.8%. A suitable window size (e.g., $w = 25$) can ensure that all moving objects can be found in each time window, the matching rate can be 90.2%, and the localization error is only 0.33 m. If the window is too large (e.g., $w = 50$), the algorithm can always filter out the appropriate cluster to match with RFID tag in each window, which has little impact on the experimental results, but redundant information may occupy the computing resources of the system, and affect the real-time performance of localization and identification. Besides, as compared with $w = 25$, the average time-consuming is increased to 62.24 ms. Therefore, it is very important to choose an appropriate window size. This paper uses the window size $w = 25$ in order to ensure that the experiment has small positioning error and high matching rate. Figure 5a–c shows the localization error at different locations with $w = 25$.

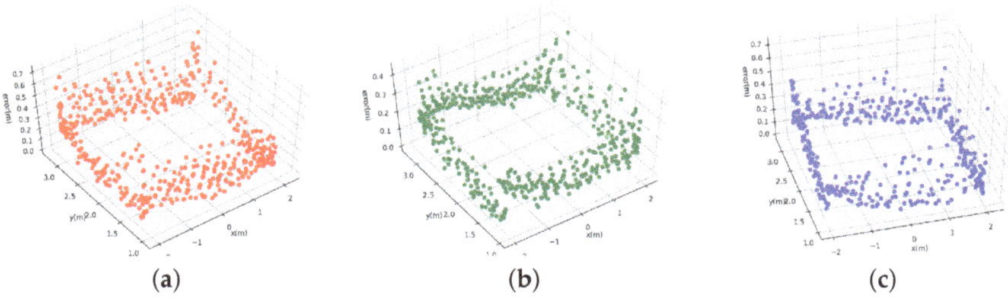

Figure 5. The positioning error at different locations with w=25. (**a**) The first object; (**b**) The second object; (**c**) The third object.

5.4. Evaluation of the Approach with a Complex Path and a Different Environment

We expand the original experimental scene to an area of 8 m × 4 m and move it to the area closer to the wall in order to verify the robustness of the whole system. The specific experimental scenario is shown in Figure 6. The experimental parameters are set to be the same as the original experimental parameters, and the result is shown in Figure 7. As can be seen from Figure 7c, as compared with Figure 7a,b, the object's estimated paths are basically consistent to the ground truth, and the average localization error of all humans is 0.52 m, which is only 0.19 m worse than the original experiment. The BC-based approach (BC) cannot locate the moving objects very well, the localization error is 0.90 m. Although the particle filter-based approach (PF + BC) improves the accuracy when compared to the BC-based approach, the result is still not good enough: the localization error is 0.76 m. The experimental results show that our approach is able to achieve identification and localization of multiple objects with good positioning accuracy.

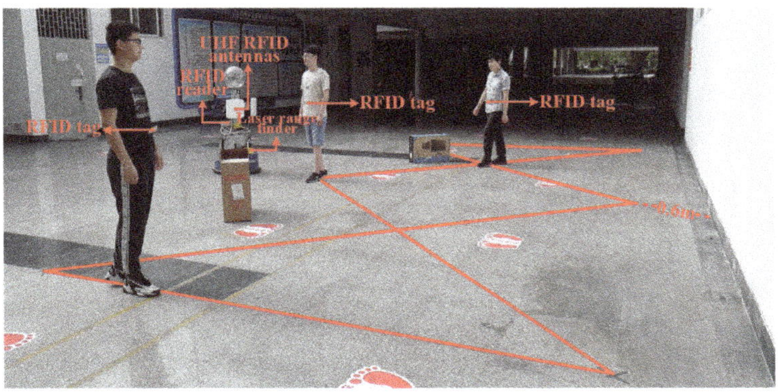

Figure 6. Complex paths experimental setup.

We conducted experiments in the lobby of our campus building in order to verify the actual use of our system in an indoor environment. In this experiment scene, the ceiling is about 2.3 m above the ground, and there are several walls in the environment. The experimental scenario is shown in Figure 8, the results are shown in Figure 9 and Table 7. As can be seen from Figure 9, the object's paths estimated by our method (PCC + PF + BC) are basically consistent to the ground truth, and the average localization error is 0.44 m, which is similar to our previous experiments. Similarly, we also compare the localization error of the other two methods (the BC-based approach (BC), and the particle filter-based approach (PF + BC)), as shown in Table 7.

The experimental results show that our approach is able to achieve identification and localization of multiple objects with a similar localization accuracy when compared to our previous experiments.

In practical applications, we can deploy this system in such an environment to identify and locate the passing pedestrians.

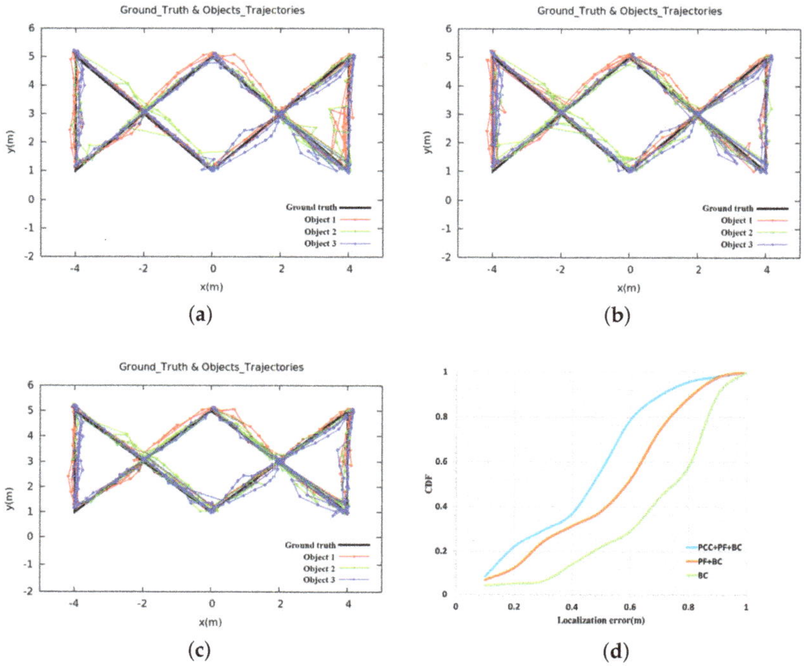

(a) (b) (c) (d)

Figure 7. Complex paths results. (**a**) BC; (**b**) PF + BC; (**c**) PCC + PF + BC; (**d**) CDFs.

Figure 8. Experimental setup of complex path in indoor environment.

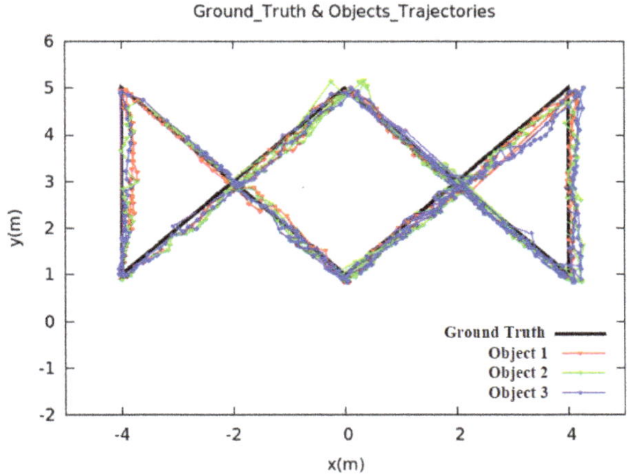

Figure 9. Experimental results of complex path in indoor environment.

Table 7. Comparison of experimental results of complex path in indoor environment based on different methods.

Methods	Localization Error (m)
BC	0.82
PF + BC	0.68
PCC + PF + BC	0.44

6. Conclusions

This paper proposed an approach for fusing the RFID and laser data in order to achieve dynamic multi-objects identification and localization by combining Pearson correlation coefficients and particle filter. The Pearson correlation coefficient and particle filter are combined to find out the historical path of each cluster, and then the radial velocity is estimated based on the cluster's position at adjacent times. At the same time, the radial velocity of the moving object is estimated using the phase difference between the adjacent moments of RFID, and those two are matched by the similarity algorithm based on the sliding time window to realize the identification and localization of multiple dynamic objects. The experiments show that the method that is proposed in this paper can achieve a matching rate of 90.2% in an environment with obstacles and a localization error of 0.33 m. In the future, we will overcome the problem of phase ambiguity, and further improve the matching rate and localization accuracy. Another research direction is visual sensors to overcome the problem of positioning failure of moving objects after long-term occlusion.

Author Contributions: W.F. analyzed the data and wrote the paper; R.L. proposed the idea and designed the experiments; R.A. revised the paper; Y.H. and Z.Q. conducted the experiments; H.W. and Z.C. reviewed the paper. All authors have read and agreed to the published version of the manuscript.

Funding: This research received no external funding.

Acknowledgments: This work is supported by Sichuan Science and Technology Program 2019YFH0161, 2019JDTD0019, National Natural Science Foundation 61601381, 61701421 and 61471306, and partially by China's 13th Five-Year Plan in the Development of Nuclear Energy under the grant number of 2016[1295].

Conflicts of Interest: The author declare no conflict of interest.

Abbreviations

The following abbreviations are used in this manuscript:

UHF	Ultra High Frequency
BC	Bray-Curtis
RFID	Radio Frequency Identification
PF	Particle Filter
PCC	Pearson Correlation coefficent

References

1. Zafari, F.; Gkelias, A.; Leung, K.K. A survey of indoor localization systems and technologies. *IEEE Commun. Surv. Tutor.* **2019**, 21, 2568–2599. [CrossRef]
2. Luo, J.; Zhang, Z.; Liu, C.; Luo, H. Reliable and cooperative target tracking based on WSN and WiFi in indoor Wireless Networks. *IEEE Access* **2018**, *6*, 24846–24855. [CrossRef]
3. Xu, D.; Han, L.; Tan, M.; Li, Y.F. Ceiling-based visual positioning for an indoor mobile robot with monocular vision. *IEEE Trans. Ind. Electron.* **2009**, *56*, 1617–1628.
4. Wang, M.M.; Liu, Y.; Su, D.; Liao, Y.F.; Shi, L.; Xu, J.H.; Miro, J.V. Accurate and real-time 3-D tracking for the following robots by fusing vision and ultrasonar information. *IEEE/ASME Trans. Mechatron.* **2018**, *23*, 997–1006. [CrossRef]
5. Liu, G.; Liu, S.; Muhammad, K.; Sangaiah, A.K.; Doctor, F. Object tracking in vary lighting conditions for fog based intelligent surveillance of public spaces. *IEEE Access* **2018**, *6*, 29283–29296. [CrossRef]
6. Xing, B.Y.; Zhu, Q.M.; Pan, F.; Feng, X.X. Marker-based multi-sensor fusion indoor localization system for micro air vehicles. *J. Sens.* **2018**, *18*, 1706. [CrossRef] [PubMed]
7. You, W.D.; Li, F.B.; Liao, L.Q.; Huang, M.L. Data fusion of UWB and IMU based on unscented Kalman filter for indoor localization of Quadrotor UAV. *IEEE Access* **2020**, *8*, 64971–64981. [CrossRef]
8. Li, C.D.; Yu, L.; Fei, S.M. Real-Time 3D motion tracking and reconstruction system using camera and IMU sensors. *IEEE Sens. J.* **2019**, *19*, 6460–6466. [CrossRef]
9. Peng, J.; Zhang, P.; Zheng, L.X.; Tan, J. UAV positioning based on multi-sensor fusion. *IEEE Access* **2020**, *8*, 34455–34467. [CrossRef]
10. Shi, F.D.; Liu, W.H.; Wang, X.; Ding, J.; Shi, Y.J. Design of indoor lidar navigation system. *Infrared Laser Eng.* **2015**, *44*, 3570–3575.
11. Zhao, X.M.; Sun, P.P.; Xu, Z.G.; Min, H.G.; Yu, H.K. Fusion of 3D LIDAR and camera data for object detection in autonomous vehicle applications. *IEEE Sens. J.* **2020**, *20*, 4901–4913. [CrossRef]
12. DiGiampaolo, E.; Martinelli, F. Mobile robot localization using the phase of passive UHF RFID signals. *IEEE Trans. Ind. Electron.* 2014, 61, 365–376. [CrossRef]
13. Tian, Q.L.; Wang, K.K.; Salcic, Z. A low-cost INS and UWB fusion pedestrian tracking system. *IEEE Sens. J.* **2019**, *19*, 3733–3740. [CrossRef]
14. Wu, J.; Yu, Z.J.; Zhuge, J.C.; Xue, B. Indoor positioning by using scanning infrared laser and ultrasonic technology. *Opt. Precis. Eng.* **2016**, *24*, 2417–2423.
15. Qu, H.K.; Zhang, Q.Y.; Ruan, Y.T. Laser radar based on scanning image tracking. *Chin. Opt.* **2012**, *5*, 242–247.
16. Tang, J.; Chen, Y.W.; Jaakkola, A.; Liu, J.B.; Hyyppa, J.; Hyyppa, H. An UGV indoor positioning system using laser scan matching for large-area real-time applications. *Sensors* **2014**, *14*, 11805–11824. [CrossRef]
17. Zhang, H.; Qiao, C.; Kuang, H.P. Object geo-location based on laser range finder for airborne electro-optical imaging systems. *Opt. Precisioneng.* **2019**, *27*, 8–16. [CrossRef]
18. Liu, R.; Liang, G.L.; Wang, H.; Fu, Y.L.; He, J.; Zhang, H. Dynamic object location with radio frequency identification and laser information. *J. Electron. Inf.* **2018**, *40*, 55–62.
19. Wang, D.Z.; Posner, I. Voting for voting in online point cloud object detection. *Robot. Sci. Syst.* **2015**, *1*, 1317.
20. Huang, R.L.; Liang, H.W.; Chen, J.J.; Zhao, P.; Du, M.B. Detection, tracking and identification of dynamic obstacles in driverless vehicles based on laser radar. *Robot* **2016**, *38*, 437–443.
21. Tong, X.Y.; Wu, R.; Yang, X.F.; Teng, S.H.; Zhuang, Y. Infrared and laser fusion object identification method. *Infrared Laser Eng.* **2018**, *47*, 167–174.

22. Liu, R.; Husikć, G.; Zell, A. On Tracking dynamic objects with long range passive UHF RFID using a mobile robot. *Int. J. Distrib. Sens. Netw.* **2015**, *11*, 781380. [CrossRef]
23. Fan, X.Y.; Wang, F.; Gong, W.; Zhang, L.; Liu, J.C. Multiple object activity identification using RFIDs: A multipath-aware deep learning solution. In Proceedings of the 2018 IEEE 38th International Conference on Distributed Computing Systems (ICDCS), Vienna, Austria, 2–6 July 2018; pp. 545–555.
24. Ni, L.M.; Liu, Y.H.; Lau, Y.C.; Patil, A.P. LANDMARC: Indoor location sensing using active RFID. *Wirel. Netw.* **2004**, *10*, 701–710. [CrossRef]
25. Hightower, J.; Borriello, G.; Wang, R. *SpotON: An Indoor 3D Location Sensing Technology Based on RF Signal Strength*; Technic Report 2000-02-02; Department of Computer Science and Engineering, University of Washington: Seattle, WA, USA, 2000.
26. Zhao, Y.Y.; Liu, Y.H.; Ni, L.M. VIRE: Active RFID-based localization using Virtual Reference Elimination. In Proceedings of the 2007 International Conference on Parallel Processing (ICPP 2007), Xi'an, China, 10–14 September 2007; pp. 56–56.
27. Alvarado Vasquez, B.P.E.; Gonzalez, R.; Matia, F.; De La Puente, P. Sensor fusion for tour-guide robot localization. *IEEE Access* **2018**, *6*, 78947–78964. [CrossRef]
28. Choi, B.S.; Lee, J.W.; Park, K.T. A hierarchical algorithm for indoor mobile robot localization using rfid sensor fusion. *IEEE Trans. Ind. Electron.* **2011**, *58*, 2226–2235. [CrossRef]
29. Suparyanto, A.; Fatimah, R.N.; Widyotriatmo, A.; Nazaruddin, Y.Y. Port container truck localization using sensor fusion technique. In Proceedings of the 2018 5th International Conference on Electric Vehicular Technology (ICEVT), Surakarta, Indonesia, 30–31 October 2018; pp. 72–77.
30. Wang, C.S.; Chen, C.L. RFID-based and Kinect-based indoor positioning system. In Proceedings of the 2014 4th International Conference on Wireless Communications, Vehicular Technology, Information Theory and Aerospace & Electronic Systems (VITAE), Aalborg, Denmark, 11–14 May 2014; pp. 1–4.
31. Li, X.Y.; Zhang, Y.Z.; Marsic, I.; Burd, R.S. Online people tracking and identification with RFID and Kinect. *arXiv* **2017**, arXiv:1702.03824.
32. Parr, A.; Miesen, R.; Kirsch, F.; Vossiek, M. A novel method for UHF RFID tag tracking based on acceleration data. In Proceedings of the 2012 IEEE International Conference on RFID (RFID), Orlando, FL, USA, 3–5 April 2012; pp. 110–115.
33. Faramondi, L.; Inderst, F.; Pascucci, F.; Setola, R.; Delprato, U. An enhanced indoor positioning system for first responders. In Proceedings of the International Conference on Indoor Positioning and Indoor Navigation, Montbeliard-Belfort, France, 28–31 October 2013; pp. 1–8.
34. Vorst, P.; Zell, A. A comparison of similarity measures for localization with passive rfid fingerprints. In Proceedings of the ISR 2010 (41st International Symposium on Robotics) and ROBOTIK 2010 (6th German Conference on Robotics), Munich, Germany, 7–9 June 2010; pp. 1–8.
35. Vorst, P.; Koch, A.; Zell, A. Efficient self-adjusting, similarity-based location fingerprinting with passive UHF RFID. In Proceedings of the 2011 IEEE International Conference on RFID-Technologies and Applications, Sitges, Spain, 15–16 September 2011; pp. 160–167.
36. Shyam, R.; Singh, Y.N. Face recognition using augmented local binary pattern and Bray Curtis dissimilarity metric. In Proceedings of the 2015 2nd International Conference on Signal Processing and Integrated Networks (SPIN), Noida, India, 19–20 February 2015; pp. 779–784.
37. Samantaray, A.K.; Rahulkar, A.D. Comparison of similarity measurement metrics on medical image data. In Proceedings of the 2019 10th International Conference on Computing, Communication and Networking Technologies (ICCCNT), Kanpur, India, 6–8 July 2019; pp. 1–5.
38. Li, X.; Zhang, Y.; Amin, M.G. Multifrequency-based range estimation of RFID Tags. In Proceedings of the 2009 IEEE International Conference on RFID, Orlando, FL, USA, 27–28 April 2009; pp. 147–154.
39. Li, S.S. An improved DBSCAN algorithm based on the neighbor similarity and fast nearest neighbor query. *IEEE Access* **2020**, *8*, 47468–47476. [CrossRef]
40. Fu, Y.L.; Liu, R.; Zhang, H.; Liang, G.L.; Shafiq, U.R.; Liu, L.X. Continuously tracking of moving object by a combination of ultra-high frequency radio-frequency identification and laser range finder. *Int. J. Distrib. Sens. Netw.* **2019**, *7*. [CrossRef]
41. Zhu, H.; You, X.; Liu, S. Multiple ant colony optimization based on Pearson Correlation Coefficient. *IEEE Access* **2019**, *7*, 61628–61638. [CrossRef]

42. Chen, G.; Liu, Q.; Wei, Y.K.; Yu, Q. An efficient indoor location system in WLAN based on database partition and euclidean distance-weighted Pearson Correlation Coefficient. In Proceedings of the 2016 2nd IEEE International Conference on Computer and Communications (ICCC), Chengdu, China, 14–17 October 2016; pp. 1736–1741.
43. Cardoso, A.D.V.; Nedjah, N.; Macedo Mourelle, L.D. Accelerating template matching for efficient object tracking. In Proceedings of the 2019 IEEE 10th Latin American Symposium on Circuits & Systems (LASCAS), Armenia, Colombia, 24–27 February 2019; pp. 141–144.

sensors

MDPI

Article

An Enhanced Indoor Positioning Algorithm Based on Fingerprint Using Fine-Grained CSI and RSSI Measurements of IEEE 802.11n WLAN

Jingjing Wang † and Joongoo Park *

School of Electronic and Electrical Engineering, Kyungpook National University, 80 Daehak-ro, Buk-gu, Daegu 41566, Korea; wjj0219@naver.com

* Correspondence: jgpark@knu.ac.kr; Tel.: +82-10-8560-6580

† Current address: Mobile Software and Navigation Laboratory, Kyungpook National University, Room 723, IT-1, 80 Daehak-ro, Bukgu, Deagu 41566, Korea.

Abstract: Received signal strength indication (RSSI) obtained by Medium Access Control (MAC) layer is widely used in range-based and fingerprint location systems due to its low cost and low complexity. However, RSS is affected by noise signals and multi-path, and its positioning performance is not stable. In recent years, many commercial WiFi devices support the acquisition of physical layer channel state information (CSI). CSI is an index that can characterize the signal characteristics with more fine granularity than RSS. Compared with RSS, CSI can avoid the effects of multi-path and noise by analyzing the characteristics of multi-channel sub-carriers. To improve the indoor location accuracy and algorithm efficiency, this paper proposes a hybrid fingerprint location technology based on RSS and CSI. In the off-line phase, to overcome the problems of low positioning accuracy and fingerprint drift caused by signal instability, a methodology based on the Kalman filter and a Gaussian function is proposed to preprocess the RSSI value and CSI amplitude value, and the improved CSI phase is incorporated after the linear transformation. The mutation and noisy data are then effectively eliminated, and the accurate and smoother outputs of the RSSI and CSI values can be achieved. Then, the accurate hybrid fingerprint database is established after dimensionality reduction of the obtained high-dimensional data values. The weighted k-nearest neighbor (WKNN) algorithm is applied to reduce the complexity of the algorithm during the online positioning stage, and the accurate indoor positioning algorithm is accomplished. Experimental results show that the proposed algorithm exhibits good performance on anti-noise ability, fusion positioning accuracy, and real-time filtering. Compared with CSI-MIMO, FIFS, and RSSI-based methods, the proposed fusion correction method has higher positioning accuracy and smaller positioning error.

Keywords: indoor fingerprinting localization; Gaussian filter; Kalman filter; received signal strength indicator; channel state information

Citation: Wang, J.; Park, J. An Enhanced Indoor Positioning Algorithm Based on Fingerprint Using Fine-Grained CSI and RSSI Measurements of IEEE 802.11n WLAN. *Sensors* **2021**, *21*, 2769. https://doi.org/10.3390/s21082769

Academic Editor: Simon Tomažič

Received: 17 March 2021
Accepted: 12 April 2021
Published: 14 April 2021

1. Introduction

With the development of wireless communication technology, the demand for location-based services (LBS) has increased greatly. The complex indoor environment makes GPS [1–4] signals that belong to the outdoor positioning system vulnerable to multi-path effects, which makes it impossible to apply in indoor positioning. The GPS [1–4] signal is blocked or reflected by the wall, so the satellite signal cannot be received in the indoor environment. Therefore, GPS cannot be used for positioning in an indoor environment. Common indoor positioning technologies include ultra-wideband (UWB), infrared, radio frequency identification (RFID), ultrasonic, LED visible light, ZigBee, Bluetooth, WiFi, geomagnetic, etc. There are many localization methods for wireless sensor networks. At present, these methods can be categorized into the followings: distance/angle-based positioning algorithm (Range-Based) and distance-independent positioning algorithm

(Range-Free). The representative range-based positioning algorithms are time of arrival (TOA) [5], time difference of arrival (TDOA) [6], angle of arrival (AOA) [7], and received signal strength indication (RSS) [8,9]. On the other hand, representatives of range-Free positioning algorithms are centroid algorithm, DV-Hop algorithm, Amorphous algorithm, APIT algorithm, etc. The performance evaluation criteria for a positioning algorithm mainly include positioning accuracy, node density, fault tolerance and adaptability, power consumption, cost, etc. Among them, the widespread deployment of indoor WiFi wireless access point (AP) has rendered the WiFi positioning technology feasible, and the location method based on location fingerprint is becoming the mainstream of indoor positioning due to its low cost and simple method. The traditional fingerprint location method usually includes two phases: offline training phase (offline phase) and online prediction phase (online phase). In the offline phase, the wireless signal characteristics of each reference point Reference Point (RP) are collected as fingerprints, and a database is established. In the online stage, by matching the wireless signal fingerprint characteristics of the user's location with the database information, the location corresponding to the fingerprint with the highest matching degree is obtained as the user's final location. Most of the existing fingerprint location systems use RSSI as coarse-grained fingerprint information, which is simple and does not need additional equipment configuration.

IEEE 802.11 protocol does not give specific processes or algorithms to generate RSSI. RSS refers to the MAC layer signal strength received by the client. Because RSS is coarse-grained information, it is often affected by multi-path effect and noise signal, and the location performance is not stable, so it can not meet the requirements of indoor precise location. In recent years, CSI [10,11] can be obtained through commercial Wi Fi devices. CSI is the channel state information that measures the channel condition. It belongs to PHY layer and comes from the sub-carriers decoded in Orthogonal Frequency Division Multiplexing (OFDM) [12] system. CSI is fine-grained physical information, which is more sensitive to the environment, so it is applied in the fields of action recognition, gesture recognition, tracking, and so on. The CSI based on the physical layer makes up for the shortcomings of the average superimposed amplitude processing method in the traditional RSSI. The phase information of each sub-carrier is added to provide more precise and stable signal characteristic information for indoor WiFi fingerprint positioning technology. CSI can minimize the effects of multi-path and noise by analyzing the transmission of different sub-channel signals. CSI has opened up a new space for the WiFi-based indoor positioning technology and the research effort has been directed to this field.

In Reference [13], CSI amplitude information is used for positioning and important fingerprint features. The phase information is not used. A positioning system Pinloc is proposed in Reference [14]. Pinloc estimates the distribution of a single sub-carrier in the complex plane through the probability density function, and then realizes the position estimation. However, the Pinloc system only uses the frequency diversity of CSI and does not consider the effect of spatial diversity. Reference [15] proposed a data processing method to reduce the phase deviation, but this paper did not consider the combined effect of random phase deviation and random delay deviation. Reference [16] also designed a fine-grained indoor fingerprint location system FIFS using CSI. Although FIFS utilizes both frequency diversity and spatial diversity, this method only aggregates the power of all sub-channels. The CSI-MIMO positioning system [17] only considers the amplitude and phase information of each sub-carrier. However, CSI-MIMO only considers spatial diversity and frequency diversity when generating location fingerprints, neglecting the characteristics of multiple receiving antennas, and the uniqueness of the location can not be reflected further. In recent years, the industry has proposed many fingerprint-based positioning methods, using different features (such as RSSI and CSI) as fingerprints to distinguish different locations. Although CSI-based methods generally have higher accuracy than RSSI-based methods, it is found that the positioning results of different methods usually compensate each other. By fusing different features, a more accurate positioning result can be obtained than a single feature. Following these backgrounds, this paper proposes a high-precision fingerprint positioning method based on RSSI and CSI.

Currently, most methods based on CSI fingerprints do not incorporate RSS information, thereby reducing the computing resources required for positioning. Therefore, this paper uses RSS and CSI, which are two kinds of information with different granularity, to realize regional location and precise location, respectively, and to make full use of the advantages of different granularity information. Compared with the distance estimation method, the complex indoor environment has less influence on fingerprint method. Based on the above reasons, this paper studies the indoor location method based on the hybrid fingerprint of RSS and CSI. Combining the respective characteristics of RSS information and CSI information, this paper proposes an improved fingerprint precise positioning algorithm based on RSSI and CSI. In the offline phase, the mobile terminal collects the RSSI signal and CSI signal from each AP in the known reference position. After denoising the original RSSI signal and CSI amplitude value by using the Kalman filter, the phase of CSI is preprocessed to build a more robust position fingerprint database. At the same time, to reduce the complexity of the fingerprint data and positioning error, we use the weighted k-nearest neighbor (WKNN) algorithm [18] to perform fingerprint matching and find the position coordinates in the online stage. Analyzing the experimental results shows that the improved fingerprint positioning algorithm has higher positioning accuracy in indoor environments.

The main contributions of this paper can be divided into the following:

1. This paper proposes a novel cross-layer approach including MAC layer information and physical layer information that enables fine-grained indoor fingerprint location algorithm in OFDM-MIMO WLANs.
2. The obtained RSSI value and CSI amplitude value are denoised, and CSI phase value is linearly transformed. The processed measurements information can express the difference of fingerprints between different locations.
3. The proposed algorithm reduces the dimension of the amplitude and phase values of CSI, and constructs a fingerprint database that can map the location feature data.
4. In this paper, an indoor fingerprint location method based on RSSI and CSI in high load AP environment is proposed. It improves the difficulty of getting RSS and CSI information of AP in high load WiFi channel due to beacon delay. The proposed method can be used in a high-load AP environment.
5. The positioning accuracy of the proposed method in two typical indoor environments is high. This method is higher than several traditional localization algorithms, and it is a more accurate WLAN Indoor fingerprint location algorithm.

This article includes the following chapters: Section 2 introduces RSSI, CSI, and WKNN. In Section 3, the proposed fingerprint location method is described. We process the original RSSI and CSI information, respectively, and obtain the effective fingerprint database in the offline stage. We use the WKNN algorithm to estimate the location through the fingerprint database established by the effective RSSI and CSI. Section 4 analyzes the experimental results and discusses the performance of the algorithm. Finally, Section 5 is conclusions and future work.

2. Related Work

2.1. Characteristics of RSSI

RSSI refers to the received signal strength received by the client, which belongs to the MAC layer and comes from each packet. The multi-path propagation of wireless signal refers to the reflection, diffraction, and scattering of electromagnetic waves on the propagation path. The signal received by the node does not come from a single path but from multiple paths. Due to the different path distances of electromagnetic waves, the arrival time of nodes is different, and there is phase difference. The positive and negative superposition of different phase differences will enhance or reduce the original signal, resulting in multi-path attenuation, which makes the indoor electromagnetic environment present regional and special. Therefore, the channel multi-path structure is unique for each location, which is called the RSSI position "fingerprint" [19,20].

WLAN vendors can privately define RSSI values. Through the description in the 802.11-2007 standard [21], we know that RSSI is measured by the receiving network card, and there is path attenuation in the middle, so the sender can not determine the specific receiving power of the receiver. The process or algorithm of RSSI generation is not given in the 802.11 protocol. According to the protocol, the RSSI value ranges from 0 to 255, and the RSSI value increases monotonically with the energy of PHY preamble. Therefore, only the PHY preamble measurement can be selected as the RSSI value. The instantaneous value of RSSI is obtained by integrating baseband IQ power. The theoretical calculation formula of RSSI is as follows:

$$RSSI_{Instan} = \sqrt{I^2 + Q^2} \tag{1}$$

The average value of RSSI within 1 s is obtained by averaging the instantaneous values of 8192 RSSIs. The average value of RSSI is calculated as follows:

$$RSSI_{Ave} = \frac{\sum_{i=1}^{8192} RSSI_{Instan}}{8192} \tag{2}$$

Power is sampled directly in time domain. Since most wireless signals are MW level, they are polarized and converted into dBm, which does not mean that the signal is negative.

2.2. *Channel State Information Amplitude and Phase*

CSI data belongs to the physical layer information of wireless communication protocol. The physical layer (PHY) of 802.11 protocol [17] is the interface between MAC and wireless media. CSI can be represented by the value of each element in the channel gain matrix H [22,23]. Channel impulse response (CIR) can describe the multi-path effect of wireless channel. It can be expressed in the following formula.

$$H_k = \|H_k\| e^{j\angle H_k} \tag{3}$$

where H_k is the CSI of the kth sub-carrier. $\|H_k\|$ and $\angle H_k$ represent amplitude and phase of kth sub-carriers, respectively. H appears in a complex form $a + bi$. CSI represents the coefficient of a wireless channel. We can get the modulus $\sqrt{(a^2 + b^2)}$ and argument $\theta = argtan\frac{b}{a}$ of the complex number, that is, the corresponding amplitude and phase.

Channel frequency response (CFR) [24] can be used to describe transmission by amplitude frequency and phase frequency. And because CFR and CIR are Fourier transforms for each other in the case of infinite bandwidth, CFR can be expressed as follows.

$$h(\tau) = \sum_{i=1}^{N} a_i e^{-j\theta_i} \delta(\tau - \tau_i) \tag{4}$$

where a_i is the amplitude attenuation of the ith path. θ_i is the phase offset of the ith path. τ_i is the time delay of the ith path. N is the total number of paths propagated. $\delta(\tau)$ is a Dirac impulse function.

Since CSI is the frequency response of multiple sub-carriers, it can accurately describe frequency selective channels. According to the Fourier transform of CIR and CFR, multiple propagation paths can be distinguished in time domain.

2.3. *Comparison of CSI and RSSI*

RSSI is the superposition of multiple path signals, which is very unstable. In a certain range, the probability of RSSI coincidence in different locations is relatively large, which makes it difficult for RSSI to complete high-precision indoor positioning, and it is only suitable for rough estimation of positioning range. CSI is not the superposition of all the sub-carrier information, it describes the signal of multiple paths, has more characteristics, and contains the channel state information of multiple sub-carriers. The amplitude and phase of CSI sub-carriers in different positions are crossed, so single sub-carrier cannot be distinguished. But if we use the diversity of CSI frequency and all the sub-carrier data, through a certain algorithm, we can distinguish different locations. Table 1 shows the differences between RSSI and CSI.

Table 1. Compare the difference between received signal strength indication (RSSI) and channel state information (CSI).

Category	RSSI	CSI
Layer	MAC layer	Physical layer
Granularity	Coarse-grained	Fine-grained
Time resolution	Packet	Multipath signal cluster
Frequency resolution	None	Subcarrier
Stability	Low	High
Dimension	One dimension	High dimension
Power consumption	Low	High
Mathematical value	Real number	Complex number
Universality	All Wi-Fi devices	Some Wi-Fi devices

2.4. Weighted K-Nearest Neighbor (WKNN) Algorithm

The K-nearest neighbor (KNN) is a deterministic algorithm. It obtains the fingerprint of the first K reference points which are closest to the location point in space and calculates the location estimation value of the K reference points according to the coordinates of the K reference points. The basic distance space can be calculated as Equation (5).

$$d_q = \left(\sum_{i=1}^{n} |s_i - S_i|^q \right)^{\frac{1}{q}}, \tag{5}$$

where s_i is the RSSI from the positioning phase. S_i is the RSSI from the fingerprint database. The variable q depends on the distance preferred by the algorithm. When $q = 2$, it represents Euclidean distance. Although it is not accurate compared with the probabilistic algorithm, the KNN is one of the most popular algorithms because of its low computational complexity.

$$(\hat{x}, \hat{y}) = \frac{1}{k} \sum_{i=1}^{k} p_i \quad p_i \in D_{1:k}. \tag{6}$$

The weighted k-nearest neighbor (WKNN) [25] algorithm was proposed to improve the KNN algorithm. KNN algorithm is to calculate the mean value of the coordinates of K reference points, while WKNN multiplies each reference point by a weighted coefficient, and then G weights and averages K reference points to obtain the location coordinates of the point to be located, as shown in Equation (7).

$$(\hat{x}, \hat{y}) = \sum_{i=1}^{k} \frac{\frac{1}{d_i + \varepsilon}}{\sum_{j=1}^{k} \frac{1}{d_j + \varepsilon}} \cdot (x_i, y_i), \tag{7}$$

where d_i represents the spatial distance between the fingerprint data of the ith reference point and the data measured in the online stage, ε is a positive number tending to 0, and the denominator in the formula is 0. (x_i, y_i) is the position coordinate of the ith reference point, and $(\hat{x}, \hat{y})$ is the position coordinate of the point to be located.

3. Proposed Indoor Fingerprint Localization Architecture and Methodology

To ensure location accuracy, this paper proposes an indoor hybrid fingerprint location algorithm based on RSSI and CSI. Because of the influence of multi-path effect, AP loss, heterogeneous network interference, and other factors, the collected fingerprint data can not be directly used, and this paper will carry out the relevant data processing work.

This part first introduces the structure of the indoor fingerprint positioning model. Secondly, it describes the process of RSSI and CSI data fusion after the establishment of the fingerprint database.

3.1. Indoor Fingerprint Localization Architecture

The indoor fingerprint location method proposed in this paper includes two parts: offline phase and online phase. In the offline phase, a modified device supporting OFDM-MIMO is used to collect RSSI and CSI measurement data at the reference point. Then, according to the proposed method, the abnormal packets which significantly deviate from the whole in the continuous RSSI and CSI packets are removed. According to the location fingerprint generation method designed in this paper, the position fingerprint containing RSSI, CSI amplitude, and CSI phase information is generated. The fingerprint uses the characteristics of frequency diversity and space diversity of the Orthogonal Frequency Division Multiplexing (MIMO) system to obtain CSI and RSSI values on different antennas. In each sub-carrier, RSSI and CSI are aggregated in the unit of receiving antenna. At the same time, the anti-jamming ability and stability of the location fingerprint are further enhanced and the uniqueness of the location is better reflected, which helps to improve the accuracy of the location. Next, the location fingerprints generated in each reference location are collected to construct the location fingerprint map and database. In the offline phase, the coordinates of the target location are obtained by matching the fingerprint database. Figure 1 shows the flow chart of the positioning system structure.

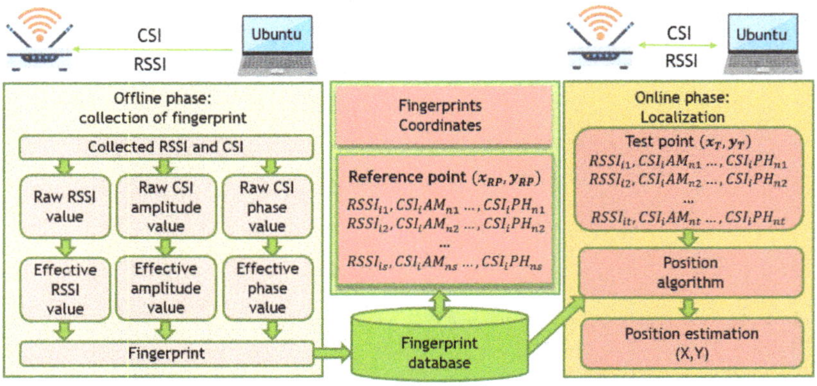

Figure 1. Indoor fingerprint localization architecture.

3.2. Proposed Indoor Fingerprint Localization Methodology

This section describes the fingerprint generation process of the fusion location method proposed in this paper. The features of the RSSI are its low characteristic dimension and easy obtainability, however, it also suffers from poor signal stability and low spatial resolution. On the other hand, the CSI retains the characteristics of fine signal granularity, high stability, high characteristic dimension, and high computational complexity. Based on the above characteristics, a fingerprint fusion localization method based on RSSI and CSI is proposed. Table 2 shows the available information about CSI [26]. It can be seen from Table 1 that the CSI obtained contains RSSI values on different antennas. To obtain more accurate positioning, we use MATLAB to obtain the RSSI value on the receiver antenna. According to the mathematical expressions of RSSI and CSI in time domain and frequency domain and their corresponding characteristics, RSSI and CSI are compared and preprocessed.

Table 2. Channel state information (CSI).

Data Information	Properties
Bfee-count	Number of Bfee count beamforming
Nrx	Number of receiver antennas
Ntx	Number of transmitter antennas
rssi-a,rssi-b,rssi-c	RSS of each receiving antenna
rate	Transmission rate of each data packet
noise	noise
CSI	CSI is a 3-dimensions array of Nrx $\times$ Ntx $\times$30

3.2.1. Processing of Raw RSSI Based on Gaussian-Kalman Filter

For indoor environment, the multi-path effect is obvious and RSS is more unstable. There may be some abnormal values in the measurement results of RSSI packets. If these abnormal values are used as the received signal strength, the positioning results will greatly deviate. The indoor location algorithm based on fingerprint similarity comparison needs to take the whole RSS vector as a whole. The fluctuation and noise of an RSS signal will have a great impact on the overall comparison results. We first process the RSSI value using Gaussian filtering [27]. The basic principle of Gaussian filtering is to establish a Gaussian distribution model for numerical values. Gaussian function is introduced in this method, and Gaussian function is discretized. The Gaussian function value at the discrete point is the weight value, that is, the RSSI values at the high probability occurrence regions are selected. The Gaussian filter can effectively filter the data significantly deviated from the true values and suppress the positioning error caused by signal mutation, but it is not effective in dealing with shadow effect, energy reflection, and other interference problems. Next, to smooth the output of RSSI data, this paper also uses the Kalman filter algorithm [28,29]. Kalman filtering can effectively filter indoor interference noise that obeys a normal distribution. After the Kalman filtering, the smoothed RSSI value of the reserved part is taken as the mean value and the final effective RSSI value is obtained. The above analysis shows that the proposed Gaussian-Kalman linear filter can obtain a more effective measurement value RSSI.

In this paper, the Dell notebook modified the wireless network card was used for data collection at Kyungpook National University (KNU), and the RSSI on different antennas was obtained. The detailed process of the experimental environment and data collection is described in detail in Section 4.1. Figures 2 and 3 show the raw RSSI values of AP1 obtained at the reference point (0,1) in the corridor of the IT1 building and at the reference point (0,1) in the lobby of the IT2 building, respectively. It can be seen from Figures 2 and 3 that the RSSI value obtained on antenna c is abnormal and cannot be used for positioning, so it needs to be filtered out. At the same time, we perform Gaussian filtering on the RSSI values obtained at antenna a and antenna b. Figures 4 and 5 show the Gaussian filtered RSSI value of AP1 on the effective antenna obtained at the reference point (0,1) in the corridor of the IT1 building and at the reference point (0,1) in the lobby of the IT2 building, respectively. To filter out the residuals of the Gaussian filtering, we further perform the Kalman filtering on the obtained smooth values. Figures 4 and 5 also show the Gaussian-Kalman filtered RSSI values obtained in the two environments.

The algorithm can effectively eliminate the mutation data and noise fluctuation in RSSI fingerprint data. We use the aforementioned data processing method to uniformly process the RSSI values obtained in other indoor environments. The algorithm realizes the accurate and smooth output of RSSI values and is used to establish an accurate fingerprint database, which makes the subsequent positioning more accurate.

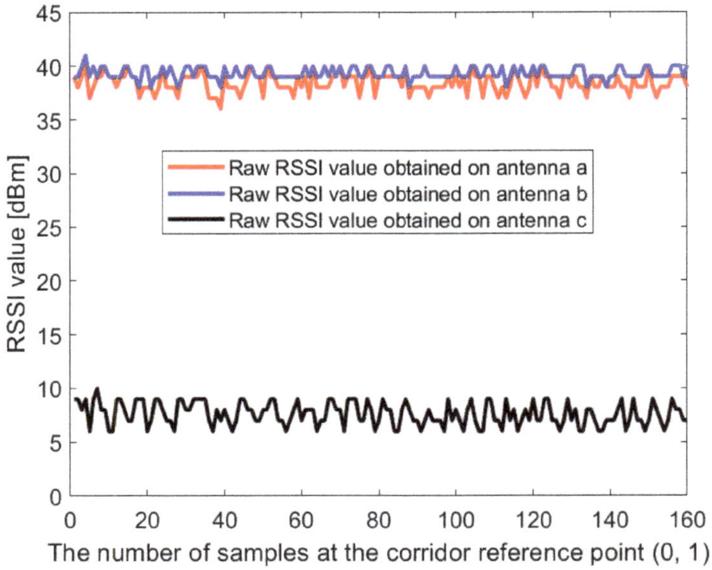

Figure 2. The raw RSSI value of access point AP1 obtained at the reference point (0,1) in the corridor of IT1 building.

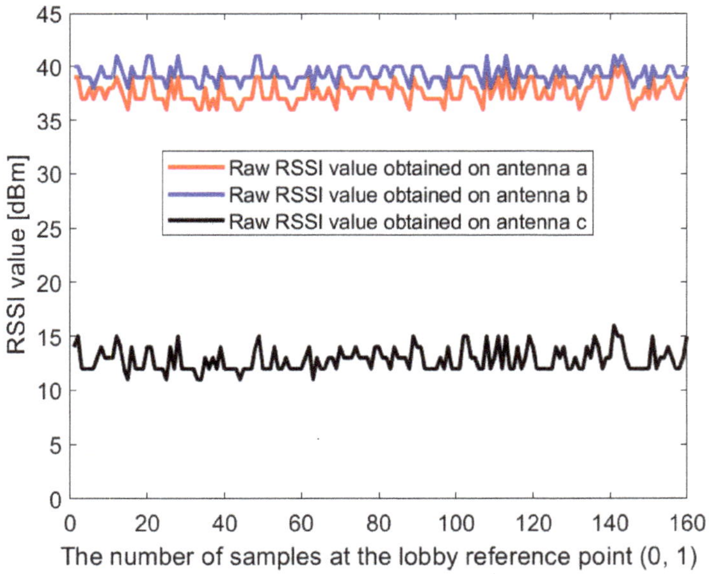

Figure 3. The raw RSSI value of AP1 obtained at the reference point (0,1) in the lobby of IT2 building.

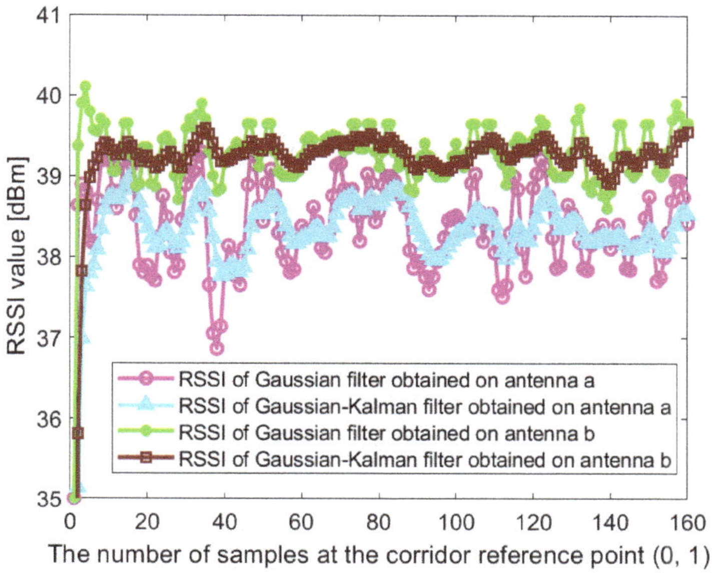

Figure 4. The Gaussian filtered RSSI value and the Gaussian-Kalman filtered RSSI value of AP1 on the effective antenna obtained at the reference point (0,1) in the corridor of IT1 building.

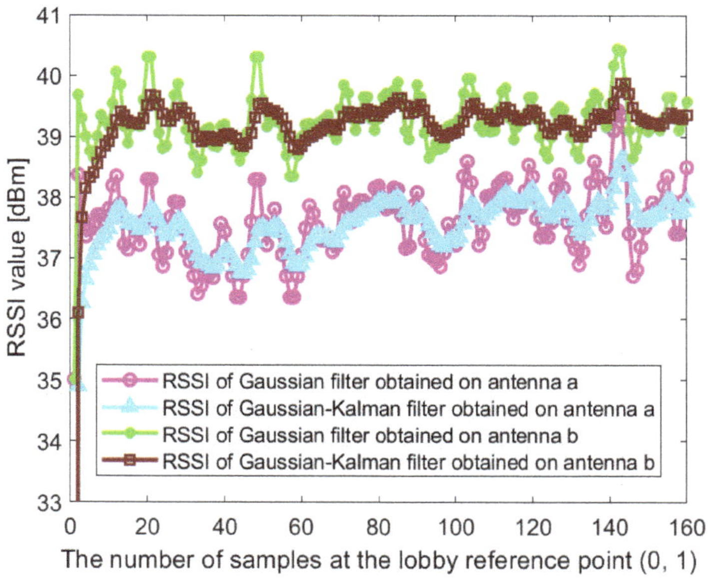

Figure 5. The Gaussian filtered RSSI value and the Gaussian-Kalman filtered RSSI value of AP1 on the effective antenna obtained at the reference point (0,1) in the lobby of IT2 building.

3.2.2. Kalman Filtering and Dimension Reduction Processing Based on CSI Amplitude Value

The Wi-Fi signal using the 802.11n transmission protocol can use the OFDM modulation method to modulate the signal to each sub-channel for transmission and to extract CSI on each channel. The CSI data of physical layer is mainly used for wireless network optimization and it is generally difficult to access these data. In recent years, Inter and Atheros network card suppliers have processed some of their network card firmware programs and some relevant organizations have opened corresponding software development packages. We can modify the open-source driver of network card in Linux and WIN system and use debugging mode to obtain CSI data of some wireless network cards. The Intel5300 network card provides sub-carrier level channel measurement for OFDM system, convert the measured value into more abundant multi-path information, and provide more stable measurement value and higher positioning accuracy.

As mentioned above, all software and scripts to read and parse channel measurements are performed by the MATLAB. Finally, a channel matrix with 30 sub-carrier groups is obtained, that is, the 802.11n channel state information is acquired. By giving full play to the advantages of the frequency diversity and space diversity of the MIMO system, while using the characteristics of multiple receiving antennas, the location fingerprint can better reflect the uniqueness of the location. We aggregate CSI in units of receiving antennas under each sub-carrier:

$$\begin{aligned} H_{csi-1} &= \sum_{m=1}^{p} h_{m1} \\ H_{csi-2} &= \sum_{m=1}^{p} h_{m2}, \\ &\cdots \\ H_{csi-q} &= \sum_{m=1}^{p} h_{mq} \end{aligned} \tag{8}$$

where p is the number of antennas at the transmitter; $h_{m1}, h_{m2}, \cdots, h_{mq}$ are sub-carriers received by receiving antennas $1, 2, \cdots q$, respectively. Figures 6 and 7 show the acquired unprocessed CSI on different receiving antennas in the two environments. It can be seen from Figures 6 and 7 that the CSI value obtained on antenna 3 cannot be used. Here, $q = 2$. Then, the amplitudes are calculated as

$$H_{am-i} = |H_{csi-i}|, \quad i = 1, 2. \tag{9}$$

For each sub-carrier, the mean value of the two amplitudes obtained by the receiving antenna is calculated, so that the dimension of the amplitude becomes 30 dimensions. Figures 8 and 9 show the amplitude of the sub-carriers for 160 measurements in two different indoor environments. The Kalman filtering was performed on the amplitudes of the 30 CSI sub-carriers obtained from different antennas. The obtained CSI value of the Kalman filter is represented by $\widehat{H_{am}}$. The ordinate is the magnitude of the amplitude and the abscissa is the time scale. The Kalman filter smooths the waveform further and the CSI amplitude value positioning using the filtering results is feasible. The amplitude of the two sub-carriers at both ends of the obtained sub-carrier are subtracted and dimension-reduced, the effective amplitude fingerprint from an antenna is reduced to 15 dimensions:

$$CSI_{AM_n} = \frac{1}{2}\left[\widehat{H_{am_{30}}} + \widehat{H_{am_1}}, \widehat{H_{am_{29}}} + \widehat{H_{am_2}}, \cdots, \widehat{H_{am_{k/2+1}}} + \widehat{H_{am_{k/2}}}\right], \tag{10}$$

where $k = 30, n = 1, 2, \cdots, 15$. Figures 10 and 11 show the amplitude of the effective CSI sub-carriers on antenna a at reference point (0,1) of AP1 in two different indoor environments.

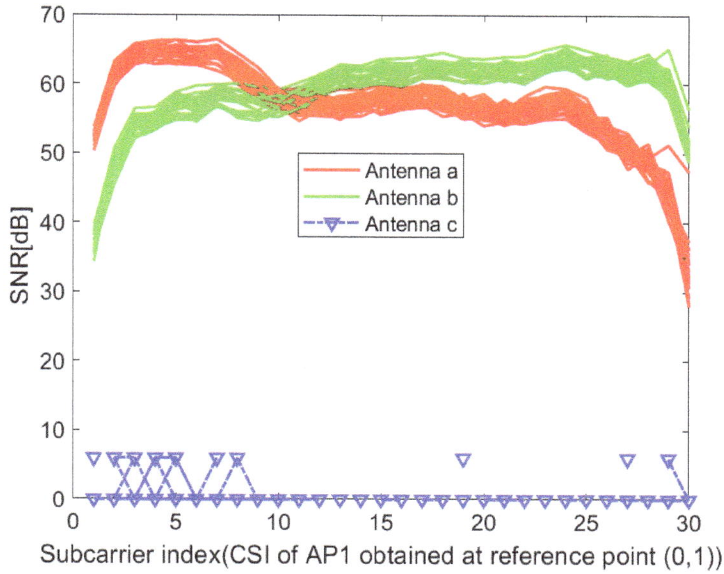

Figure 6. Raw CSI obtained on different receiving antennas at reference point (0,1) of AP1 in the corridor of IT1 building.

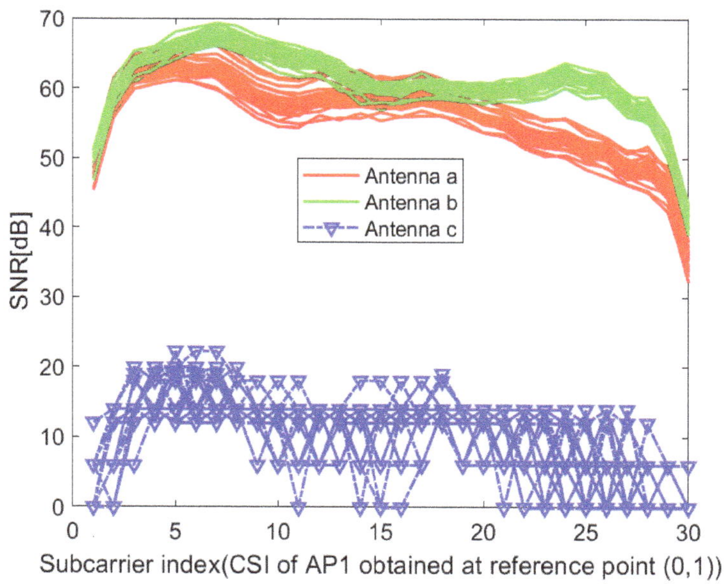

Figure 7. Raw CSI obtained on different receiving antennas at reference point (0,1) of AP1 in the lobby of IT2 building.

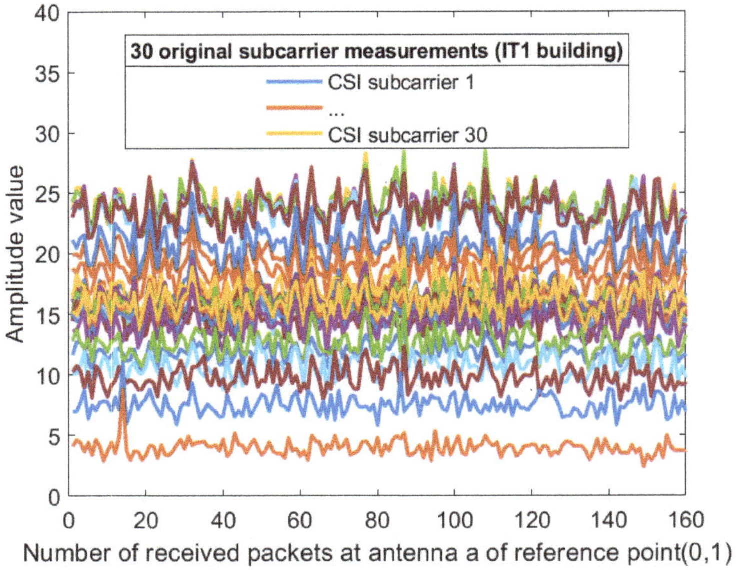

Figure 8. The amplitude of the 30 CSI sub-carriers on antenna a at reference point (0,1) of AP1 in the corridor of IT1 building.

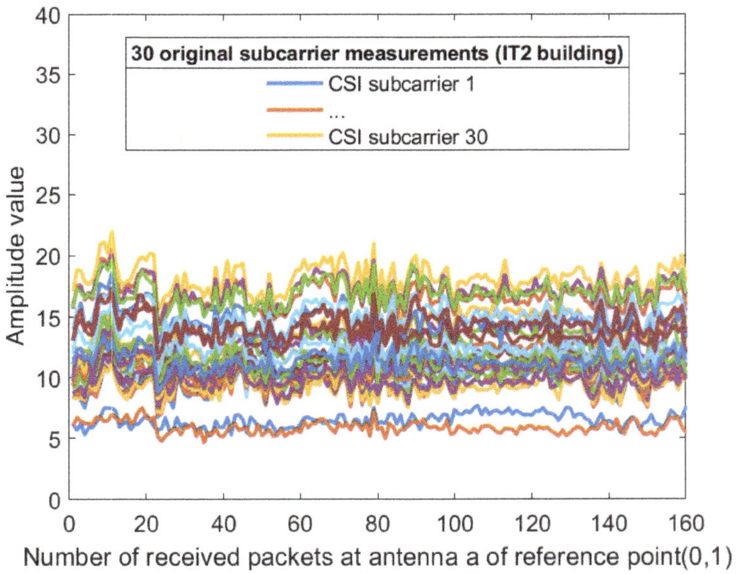

Figure 9. The amplitude of the 30 CSI sub-carriers on antenna a at reference point (0,1) of AP1 in the lobby of IT2 building.

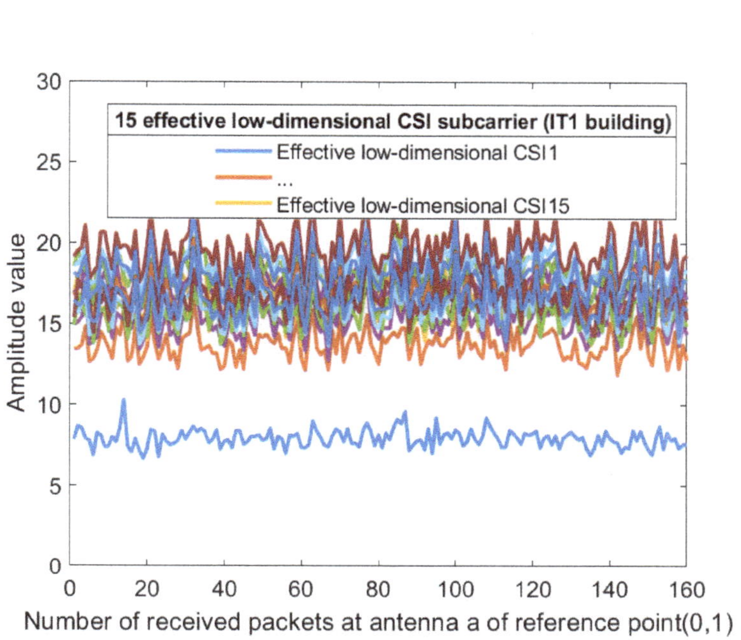

Figure 10. The amplitude of the effective CSI sub-carriers on antenna a at reference point (0,1) of AP1 in the corridor of IT1 building.

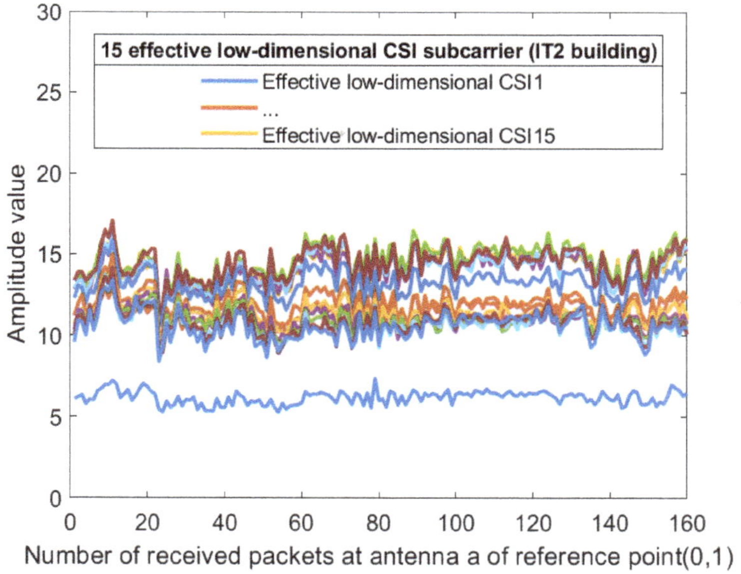

Figure 11. The amplitude of the effective CSI sub-carriers on antenna a at reference point (0,1) of AP1 in the lobby of IT2 building.

3.2.3. Linear Transformation and Dimension Reduction of CSI Phase Values

Figures 12 and 13 show the phase of each sub-carrier of the acquired CSI. If the center frequencies of the receiver and transmitter cannot be accurately synchronized (frequency offset), the carrier frequency offset (CFO) will occur. In addition, ADC also produces sampling frequency offset (SFO) [30]. The i-th sub-carrier phase can be expressed as:

$$\widehat{\varphi}_i = \varphi_i - 2\pi \frac{k_i}{N}\delta + \beta + w_Z. \tag{11}$$

Among them, φ_i is real phase information, δ is timing offset of receiver, β is unknown phase offset, and w_Z is phase measurement noise. k denotes the index of the i-th sub-carrier. N is the FFT size.

The existence of the error offset causes the original phase information to not be directly used. Assuming that the value of measurement noise w_Z is small, to eliminate the error of δ and β, linear transformation can be used as follows:

$$a = \frac{\widehat{\varphi}_i - \overline{\varphi_1}}{k_n - k_1} = \frac{\varphi_i - \varphi_1}{k_n - k_1} - \frac{2\pi}{N}\delta, \tag{12}$$

$$\mathrm{b} = \frac{1}{n}\sum_{1\le j\le n}\widehat{\varphi}_j = \frac{1}{n}\sum_{1\le j\le n}\varphi_j - \frac{2\pi\delta}{nN}\sum_{1\le j\le n}k_j + \beta, \tag{13}$$

where a is the slope of the received response phase. b is the offset. If the sub-carrier frequency is symmetric, then $\sum_{j=1}^{n} k_j = 0$, $\varphi_i - ak_i - b$ can eliminate the error introduced by δ and β. Because the real phase can not be obtained, only the relationship between the calibrated phase $\widetilde{\varphi}_i$ and the real phase can be obtained, and the difference is a constant multiple Ci related to frequency.

$$\sigma^2_{\widetilde{\varphi}_i} = c_i\sigma^2_{\varphi_i}, \tag{14}$$

$$c_i = 1 + 2\frac{k_i^2}{(k_n - k_1)^2} + \frac{1}{n}. \tag{15}$$

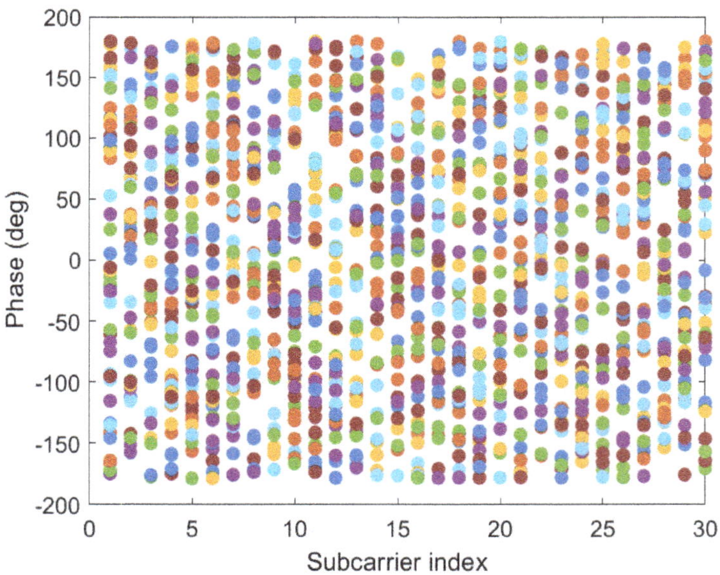

Figure 12. The phases of the 30 CSI sub-carriers on antenna a at reference point (0,1) of AP1 in the corridor of IT1 building.

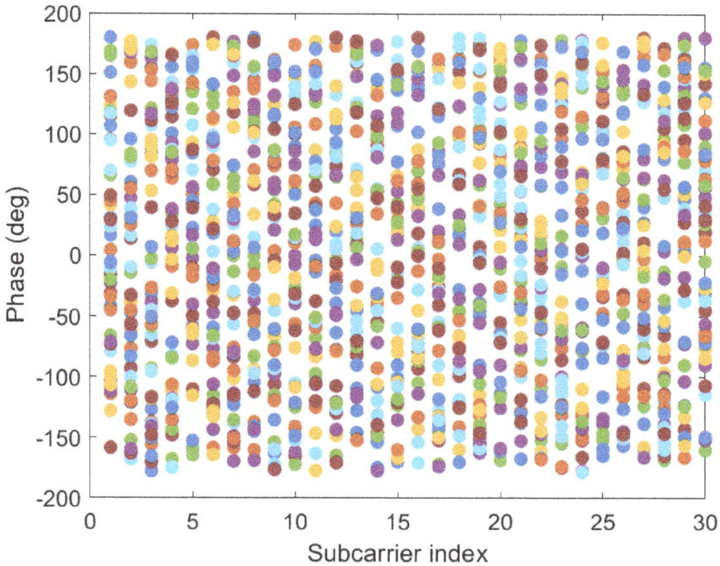

Figure 13. The phases of the 30 CSI sub-carriers on antenna a at reference point (0,1) of AP1 in the lobby of IT2 building.

Figures 14 and 15 give the phase information processed linearly, which is much more stable than the original phase in Figures 12 and 13. The phase received by the receiving end has random errors due to time delay and multi-path effects, which are the main factors affecting the accuracy of phase estimation, and they need to be eliminated to improve the accuracy. The CSI is processed by a linear transformation, which effectively eliminates the phase error and obtains a higher-precision estimated phase. It is possible to establish more accurate indoor fingerprints, and, for each sub-carrier, the three calibration phases are subtracted to make the fingerprint of phase information 30 dimensions:

$$H_{ph} = H_{ph-1} - H_{ph-2}, \tag{16}$$

where $H_{ph-i} = \widetilde{\varphi}_i, i = 1,2$.

After analyzing the trend of CSI phase, this paper proposes a data dimension reduction method.This method is easy to operate and reduces the calculation at the same time. Since the information contained in the adjacent sub-carriers is nearly similar, we remove the amplitude of the even-numbered adjacent sub-carriers according to this feature, and only preserve the phase information of the odd-numbered sub-carriers. Finally, the effective CSI obtained by dimension reduction can be used to establish offline CSI fingerprint database. Dimensionality reduction processing is carried out for the phase information of the two sub-carriers symmetric about the center sub-carrier, and the phase information from an antenna is reduced to 15 dimension:

$$CSI_{PH_n} = \left[\widehat{H_{ph_1}}, \widehat{H_{ph_3}}, \cdots, \widehat{H_{ph_{k/2}}}, \cdots, \widehat{H_{ph_{k-3}}}, \widehat{H_{ph_{k-1}}}\right], \quad k = 30, \quad n = 1,2,\cdots,15. \tag{17}$$

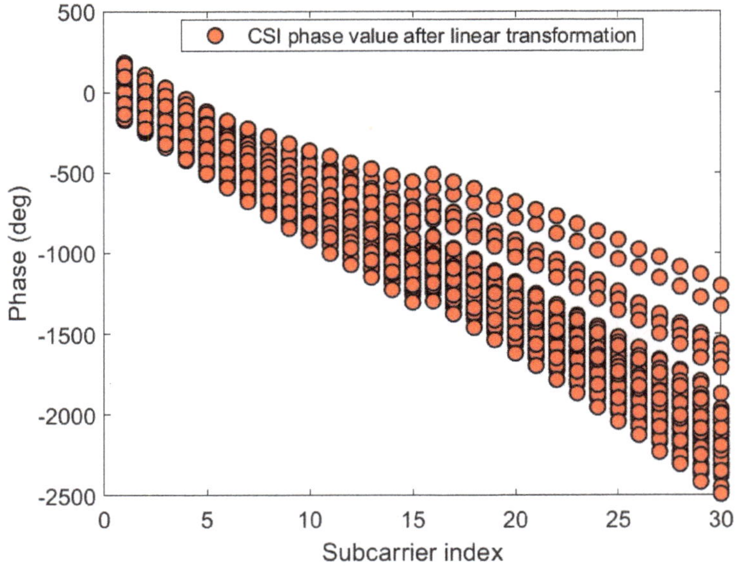

Figure 14. The phases after linear transformation of the 30 CSI sub-carriers on antenna a at reference point (0,1) of AP1 in the corridor of IT1 building.

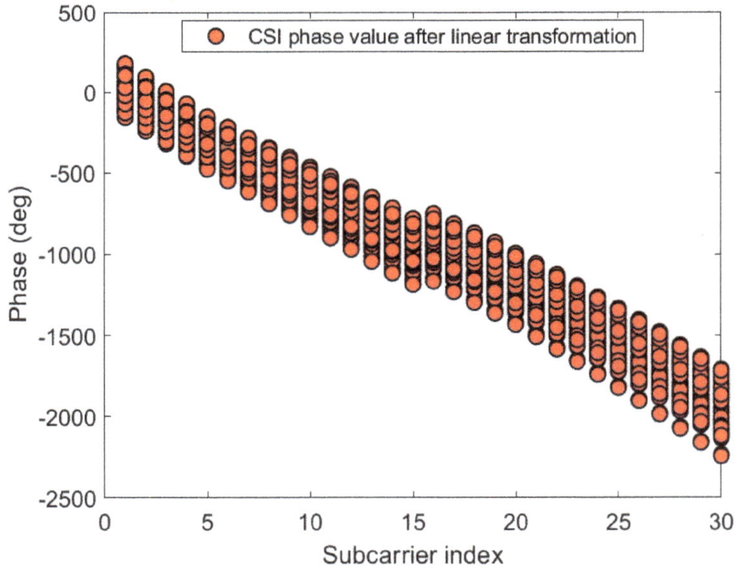

Figure 15. The phases after linear transformation of the 30 CSI sub-carriers on antenna a at reference point (0,1) of AP1 in the lobby of IT2 building.

Figures 16 and 17 show the variation trend of CSI phases in different dimensions under different indoor environments. It can be seen that, even though the phase dimension of the sub-carrier is 15, the phase change trend can be clearly presented.

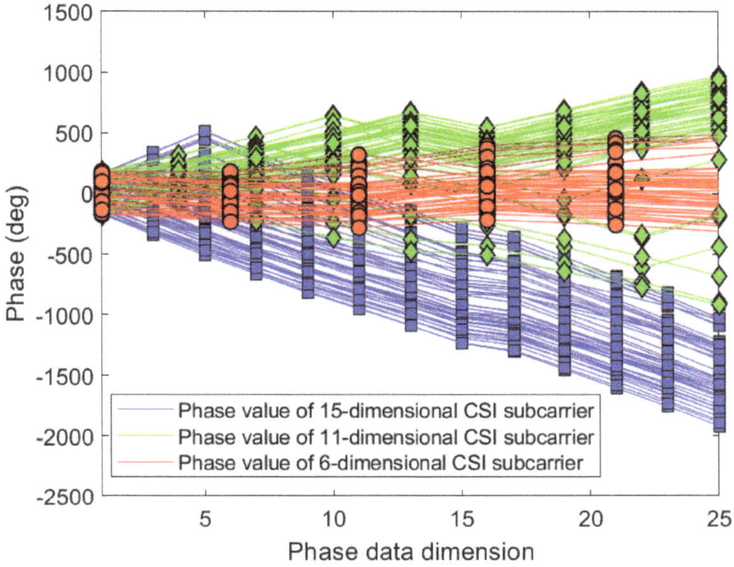

Figure 16. Compare the phases change trends with different dimensions on antenna a at reference point (0,1) of AP1 in the corridor of IT1 building.

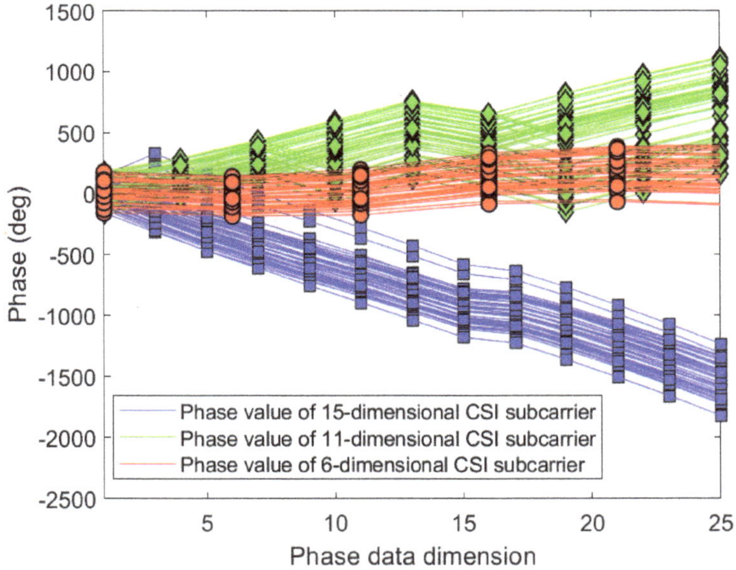

Figure 17. Compare the phases change trends with different dimensions on antenna a at reference point (0,1) of AP1 in the lobby of IT2 building.

3.2.4. Location Fingerprint Generation Based on Data Fusion

By the aforementioned processing of RSSI, CSI amplitude, and CSI phase information, we obtain the fingerprints containing 2×1 dimensional RSSIs, 2×15 dimensional CSI amplitude, and 2 × 15 dimensional CSI phase information, respectively. (x_m, y_m) can give the uniqueness of the position coordinates.

$$(x_m, y_m) = [RSSI_i \quad CSI_iAM_n \quad CSI_iPH_n], \tag{18}$$

where m represents the number of packets collected. i represents the number of antennas and $i = 1,2$. $n = 1,2,\cdots,15$. In addition, when there are multiple access points (APs) in the positioning system, they can be aggregated for the participate of location fingerprints.

$$(x_{RP}, y_{RP}) = \begin{bmatrix} RSSI_{i1}, CSI_iAM_{n1}, CSI_iPH_{n1} \\ RSSI_{i2}, CSI_iAM_{n2}, CSI_iPH_{n2} \\ \cdots \\ RSSI_{is}, CSI_iAM_{ns}, CSI_iPH_{ns} \end{bmatrix}, \tag{19}$$

where s represents the number of APs, and, in this paper, $s = 3$. $(x_{RP}, y_{RP})$is the position coordinates of the reference point stored in the fingerprint database during the offline phase. Online phase, the user's fingerprint information is compared with the fingerprint database to obtain several different areas. In this paper, we select t test points for fingerprint matching, and the coordinate value of the test point is (x_{Test}, y_{Test}). The WKNN algorithm is used to aggregate the above regions and to select the optimal estimation location.

$$(x_{Test}, y_{Test}) = \begin{bmatrix} RSSI_{i1}, CSI_iAM_{n1}, CSI_iPH_{n1} \\ RSSI_{i2}, CSI_iAM_{n2}, CSI_iPH_{n2} \\ \cdots \\ RSSI_{it}, CSI_iAM_{nt}, CSI_iPH_{nt} \end{bmatrix}. \tag{20}$$

4. Experimental Environment and Performance Evaluation

4.1. Experimental Environment

Figures 18 and 19 verify the performance of the proposed algorithm in the corridor on the 3rd floor of IT1 building and the hall on the 1st floor of IT2 building in KNU. The floor plan is shown in Figures 20 and 21.

Figure 18. Real corridor environment on the 3rd floor of Kyungpook National University (KNU) IT-1 building.

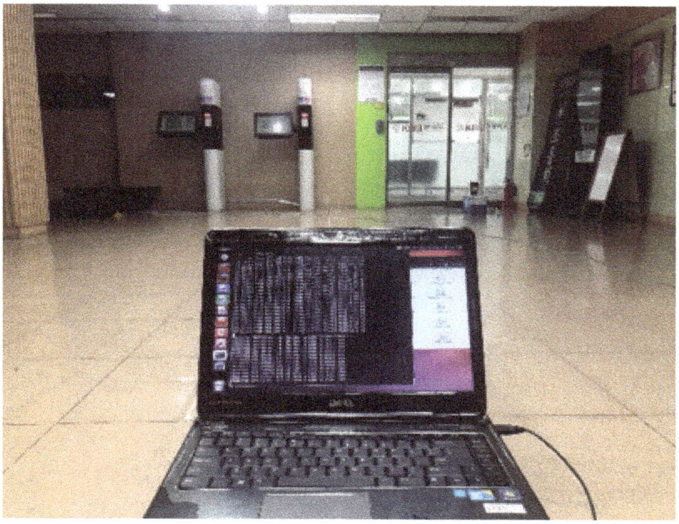

Figure 19. Real lobby environment on the 1st floor of KNU IT-2 building.

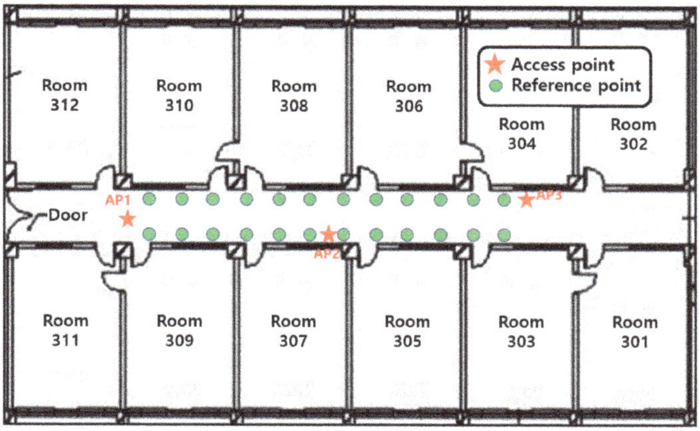

Figure 20. Floor plan of the corridor on the 3rd floor of KNU IT-1 building.

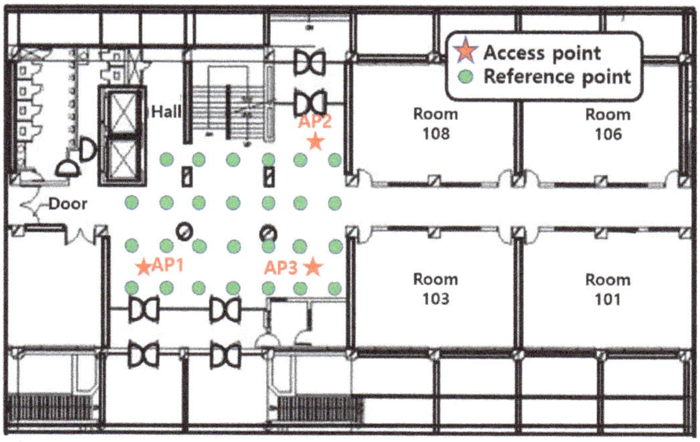

Figure 21. Floor plan of the lobby on the 1st floor of KNU IT-2 building.

The receiver of this paper is a laptop with ubuntu14.04.4 operating system and intel5300 wireless network card. The network card is attached with three dual-frequency external antennas of 2.4 G/5 GHz. The green dots in the figure indicate the positions of reference points. The accuracy of RSSI signal and CSI signal is the highest at a height of 1 m from the ground, so the distance between the AP antenna and the ground is 1 m. When collecting RSSI signal and CSI signal data, the tester holds a portable notebook device and stands at the reference point for data collection. The reference point receives signals from three APs. Each reference point collects RSSI data and CSI data for 30 min and saves them to the PC. The RSSI data and CSI data are extracted by MATLAB, and the original RSSI data and CSI data are filtered at the same time. Take the effective RSSI value and the effective CSI value to establish the database as the feature vector of the point. The experimental starts from the lower-left corner, and its coordinates are (0,0). The grid size of the fingerprint database is 1 m. In the corridor environment, fingerprint information of 24 locations are collected, and 27 locations are collected in the lobby environment. The performance of the algorithm is evaluated by the distance error between the estimated position of the reference point and the real position.

4.2. *Performance Evaluation*

In the indoor environment, there are many factors that affect the positioning results, such as the number of people in the positioning area and their status, static or moving, receiver location and antenna orientation, and the influence of other indoor equipment movements, etc. We consider these conditions that are difficult to control and evaluate the generality of the proposed algorithm. To weaken these external factors, the same set of location data is processed to compare the performance of the location algorithm.

4.2.1. Impact of the Number of Packets

During the construction of fingerprint database, the number of samples makes a great impact on the positioning performance. For the similarity comparison method of fingerprint database, the location accuracy largely depends on whether the fingerprint samples in the fingerprint database can accurately describe the distribution of the overall fingerprint or not. For this experiment, each unit area is sampled for 5 min, and nearly 1000–3000 CSI samples can be sampled for each AP. In the case of the same environment parameters, we selected different numbers of sample sets for training to verify the existing indoor location method. We also investigated the influence of the different sample set number in the proposed fingerprint location method on the positioning performance. The sampling frequency of sampled data is equal to 1500 packets/s.

Figures 22 and 23 show the impact on positioning performance when the number of samples in the process of establishing a fingerprint database is 100, 500, 1000, and 2000, respectively. In the two experimental scenarios, we select three APs to establish fingerprints and set $k = 3$ in the WKNN algorithm. The number of samples in the on-line phase is about 50, 250, 500, and 1000, respectively. When the sampling rate is increased from 100 to 500 packets, as shown in the figure. When it increased from 500 to 1000, the positioning accuracy showed a trend of improvement. When it is between 1000 and 2000, the positioning accuracy has not improved. Based on the above experimental results, the positioning accuracy of 1000 packets achieved the highest score under certain conditions. We also compared the average distance errors at reference point (1.1) and reference point (4,2), as shown in Figures 22 and 23. It can be seen from Figures 22 and 23 that selecting 1000 data packets at different reference points to build a database obtained more accurate positioning.

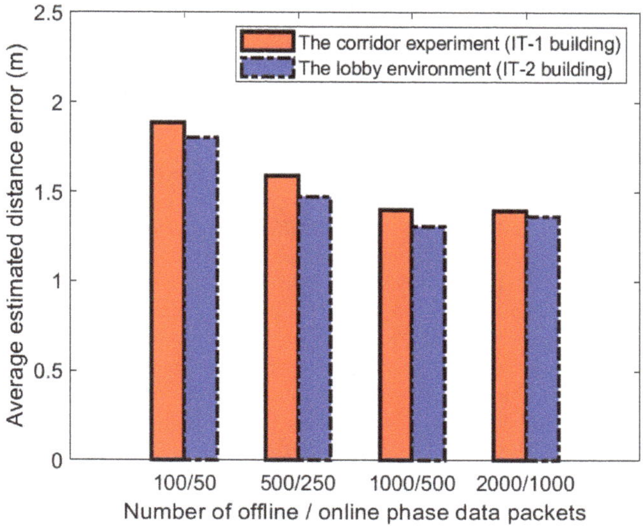

Figure 22. Distance estimation errors at reference points (1,1) in two different experimental environments.

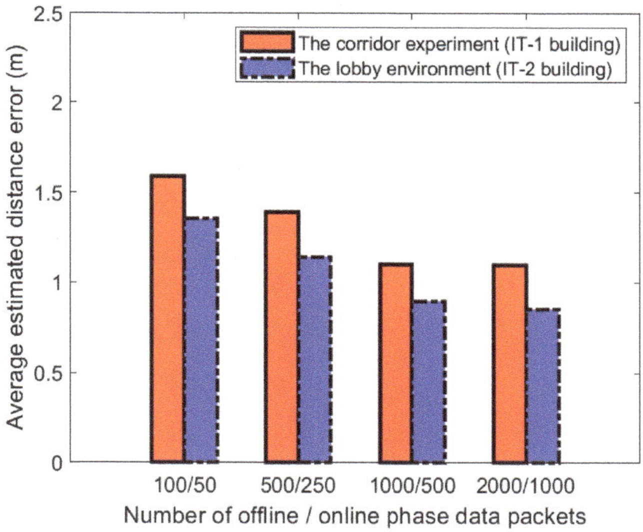

Figure 23. Distance estimation errors at reference points (4,2) in two different experimental environments.

4.2.2. Comparison with Existing Fingerprint Location Methods

In the positioning error analysis, cumulative distribution function (CDF), standard deviation, and average error were used to analyze the performance of the positioning method. In two indoor environments, the proposed method is compared with the RSSI-based fingerprint positioning system and two CSI-based fingerprint positioning systems (FIFS and CSI-MIMO). In the test process, WKNN is used as fingerprint matching algorithm and three APs are used to generate fingerprint. We use Euclidean distance to estimate the similarity measure between the reference point (RP) coordinates (X_{RP}, Y_{RP}) and the test point (TP) coordinates (X_{TP}, Y_{TP}). The formula is as follows:

$$\hat{d} = \sqrt{(X_{RP} - X_{TP})^2 - (Y_{RP} - Y_{TP})^2}. \tag{21}$$

The CDFs of location distance error in the two scenarios are shown in Figures 24 and 25. Figure 24 is the CDF figure obtained from the experiment in the corridor of IT-1 building. It can be seen from the figure that RSSI-based approach is vulnerable to environmental interference and is very unstable, and the error is the largest among the four positioning methods. FIFS and CSI-MIMO use channel state information as fingerprint eigenvalues. Due to the finer granularity of channel state information, the multi-path interference can be suppressed to some extent and the positioning accuracy of these two methods is higher than that of the RSSI-based approach. The proposed location method, based on FIFS and CSI-MIMO, and according to the characteristics of multiple antennas, improves the positioning accuracy by improving the fingerprint feature dimension. The red curve in Figure 24 shows that the probability of positioning error within 1.5 m reaches 60%, which is 33.33% higher than that of CSI-MIMO and 51.24% higher than that of FIFS. Figure 25 is the CDF diagram of the experiment in the lobby of IT-2 building. Compared with Figure 24, the accuracy rate is improved, which indicates that the environment affected the accuracy of the experiment. However, it can be seen from Figure 25 that the positioning accuracy of the proposed method is still higher than those of the FIFS and CSI-MIMO methods. The probability of positioning error within 1.5 m is 81.4%, which is 4.7% higher than that of CSI-MIMO and 41.4% higher than that of FIFS. Finally, the proposed method, FIFS method, and CSI-MIMO method are compared in two experimental environments by using the average positioning error, standard deviation. The results are shown in Tables 3 and 4.

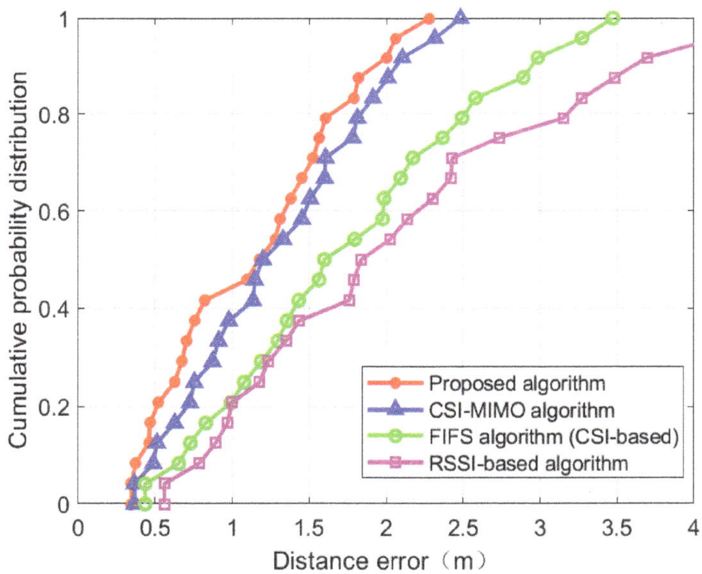

Figure 24. Comparing the cumulative distribution function (CDF) value of localization error of four algorithm in IT-1 building.

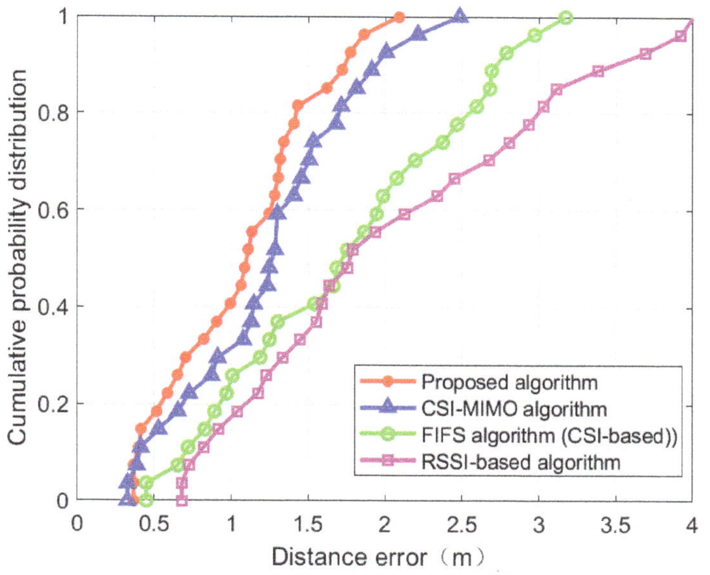

Figure 25. Comparing the CDF value of localization error of four algorithm in IT-2 building.

Table 3. Comparison of four indoor fingerprint location algorithms (corridor of IT-1 building).

Fingerprint Algorithm	Average Distance Error (m)	Standard Deviation (m)
RSSI-based algorithm	2.122 m	1.097 m
CSI-based algorithm (FIFS)	1.802 m	0.853 m
CSI-MIMO algorithm	1.319 m	0.605 m
Proposed algorithm	1.171 m	0.587 m

Table 4. Comparison of indoor fingerprint location algorithms (lobby of IT-2 building).

Fingerprint Algorithm	Average Distance Error (m)	Standard Deviation (m)
RSSI-based algorithm	2.078 m	1.007 m
CSI-based algorithm (FIFS)	1.767 m	0.781 m
CSI-MIMO algorithm	1.269 m	0.559 m
Proposed algorithm	1.094 m	0.488 m

It can be seen from the table, in terms of the average positioning error, the performance of the proposed method is about 1.171 m in the corridor environment, which is about 12.6% higher than CSI-MIMO and 36.9% higher than that of FIFS. The average positioning error of the RSSI-based method is about 2.122 m, it is lower than the positioning accuracy of the proposed method; In the lobby experimental scenario, the proposed method is about 1.094 m, which is 15.9% and 61.5% higher than CSI-MIMO and FIFS, respectively. The average positioning error of the RSSI-based method is about 2.078 m. Through the series of experiments, we confirmed that the method proposed in this paper was effective in improving the accuracy of positioning. The positioning method proposed in this paper can effectively improve the positioning accuracy by the following reasons. The reasons for the proposed algorithm to improve accuracy are as follows. Firstly, the fingerprints in the fingerprint database are integrated with RSS and CSI, which are not based on CSI or RSS alone; secondly, the fingerprint features are preprocessed effectively in the offline phase, and the integrity of the data is preserved; thirdly, WKNN is used to reduce the positioning error. In addition, since the proposed method performs the dimensional reduction for the fingerprint, its computational complexity is also improved over the CSI-MIMO and FIFS.

5. Conclusions and Future Work

In this paper, a hybrid fingerprint algorithm is proposed to improve the positioning accuracy of RSSI-based and CSI-based fingerprint location methods. In the offline phase, the amplitude values of RSSI and CSI are eliminated and filtered, and the phase value of CSI is introduced by linear transformation. This paper also proposes a method to reduce the fingerprint feature dimension of high-dimensional data CSI obtained in OFDM-MIMO system. Finally, combined with the effective RSSI, effective CSI amplitude value, and CSI phase value, a novel fingerprint database containing more abundant indoor location information was established. In the online phase, the fingerprint data is matched according to the WKNN algorithm, which effectively improves the positioning accuracy. This paper uses MATLAB to verify the effectiveness and superiority of the algorithm in two indoor environments. Compared with existing algorithms, the algorithm based on RSS and CSI hybrid fingerprints can improve positioning accuracy. At the same time, because additional hardware is not necessary for the proposed algorithm to perform positioning, its applicability is also high. Future research directions include the evaluation of the performance of the proposed algorithm in a three-dimensional indoor environment, an indoor laboratory environment with more obstacles and personnel flows.

Author Contributions: J.W. proposed the idea and implementation methodology, performed all experiments and wrote the paper, verified the experiment process and results; J.P. reviewed and edited paper, and supervised the work. All authors have read and agreed to the published version of the manuscript.

Funding: This study was supported by the BK21 FOUR project funded by the Ministry of Education, Korea (4199990113966). This work is supported by Smart City R&D project of the Korea Agency for Infrastructure Technology Advancement(KAIA) grant funded by the Ministry of Land, Infrastructure and Transport(or Ministry of Science and ICT) (Grant 21NSPS-B149843-04).

Institutional Review Board Statement: Not applicable.

Informed Consent Statement: Not applicable.

Data Availability Statement: Not applicable.

Conflicts of Interest: The authors declare no conflict of interest.

References

1. Rycroft, M.J. Understanding GPS. Principles and applications: Kaplan E. D. (ed.), 1996, 554 pp. Artech House, £75, hb, ISBN 0-89006-793-7. *J. Atmos. Sol. Terr. Phys.* **1997**, *59*, 598–599. [CrossRef]
2. Farrell, J.A.; Barth, M. *The Global Positioning System and Inertial Navigation*; McGraw-Hill Professional: New York, NY, USA, 1999; ISBN 007022045X.
3. Misra, P.; Enge, P. Global Positioning System: Signals, Measurements, and Performance—Revised Second Edition (2011). *Int. J. Wirel. Inf. Netw.* **2006**, *206*, 43.
4. *Global Positioning System: Theory and Applications*; American Institute of Aeronautics and Astronautics: Reston, VA, USA, 1996; Volume II. [CrossRef]
5. Wang, G.; So, A.M.C.; Li, Y. Robust Convex Approximation Methods for TDOA-Based Localization Under NLOS Conditions. *IEEE Trans. Signal Process.* **2016**, *64*, 3281–3296. [CrossRef]
6. Tiemann, J.; Wietfeld, C. Scalable and precise multi-UAV indoor navigation using TDOA-based UWB localization. In Proceedings of the 2017 International Conference on Indoor Positioning and Indoor Navigation, IPIN 2017, Sapporo, Japan, 18–21 September 2017.
7. Ma, Y.; Wang, B.; Pei, S.; Zhang, Y.; Zhang, S.; Yu, J. An Indoor Localization Method Based on AOA and PDOA Using Virtual Stations in Multipath and NLOS Environments for Passive UHF RFID. *IEEE Access* **2018**, *6*, 31772–31782. [CrossRef]
8. Sadowski, S.; Spachos, P. RSSI-Based Indoor Localization with the Internet of Things. *IEEE Access* **2018**, *6*, 30149–30161. [CrossRef]
9. Rusli, M.E.; Ali, M.; Jamil, N.; Din, M.M. An Improved Indoor Positioning Algorithm Based on RSSI-Trilateration Technique for Internet of Things (IOT). In Proceedings of the 6th International Conference on Computer and Communication Engineering: Innovative Technologies to Serve Humanity, ICCCE 2016, Kuala Lumpur, Malaysia, 26–27 July 2016.
10. Halperin, D.; Hu, W.; Sheth, A.; Wetherall, D. Predictable 802.11 packet delivery from wireless channel measurements. *ACM SIGCOMM Comput. Commun. Rev.* **2010**, *40*, 159–170. [CrossRef]
11. Yang, Z.; Zhou, Z.; Liu, Y. From RSSI to CSI: Indoor localization via channel response. *ACM Comput. Surv.* **2013**, *46*, 1–32. [CrossRef]

12. Han, S.; Li, Y.; Meng, W.; Li, C.; Liu, T.; Zhang, Y. Indoor localization with a single Wi-fi access point based on OFDM-MIMO. *IEEE Syst. J.* **2019**, *13*, 964–972. [CrossRef]
13. Wu, K.; Xiao, J.; Yi, Y.; Chen, D.; Luo, X.; Ni, L.M. CSI-based indoor localization. *IEEE Trans. Parallel Distrib. Syst.* **2013**, *24*, 1300–1309. [CrossRef]
14. Sen, S.; Radunovic, B.; Choudhury, R.R.; Minka, T. Spot Localization Using PHY Layer Information. In Proceedings of the ACM Mobisys, Lake District, UK, 25–29 June 2012.
15. Shi, S.; Sigg, S.; Chen, L.; Ji, Y. Accurate Location Tracking from CSI-Based Passive Device-Free Probabilistic Fingerprinting. *IEEE Trans. Veh. Technol.* **2018**, *67*, 5217–5230. [CrossRef]
16. Xiao, J.; Wu, K.; Yi, Y.; Ni, L.M. FIFS: Fine-grained indoor fingerprinting system. In Proceedings of the 2012 21st International Conference on Computer Communications and Networks, ICCCN 2012—Proceedings, Munich, Germany, 30 July–2 August 2012.
17. Chapre, Y.; Ignjatovic, A.; Seneviratne, A.; Jha, S. CSI-MIMO: An efficient Wi-Fi fingerprinting using Channel State Information with MIMO. *Pervasive Mob. Comput.* **2015**, *23*, 89–103. [CrossRef]
18. Halperin, D.; Hu, W.; Sheth, A.; Wetherall, D. Tool release: Gathering 802.11n traces with channel state information. *Comput. Commun. Rev.* **2011**, *41*, 53. [CrossRef]
19. Zegeye, W.K.; Amsalu, S.B.; Astatke, Y.; Moazzami, F. WiFi RSS fingerprinting indoor localization for mobile devices. In Proceedings of the 2016 IEEE 7th Annual Ubiquitous Computing, Electronics and Mobile Communication Conference, UEMCON 2016, New York, NY, USA, 20–22 October 2016.
20. Zegeye, W.K.; Amsalu, S.B.; Moazzami, F.; Dean, R.A.; Astatke, Y. Minimum euclidean distance algorithm for indoor WiFi Received Signal Strength (RSS) fingerprinting. In Proceedings of the International Telemetering Conference, Glendale, AZ, USA, 7–10 November 2016.
21. IEEE Computer Society LAN/MAN Standards Committee. IEEE Standard for Information technology-Telecommunications and information exchange between systems-Local and metropolitan area networks-Specific requirements Part 11: Wireless LAN Medium Access Control (MAC) and Physical Layer (PHY) Specifications. *IEEE Std 802.11* **2007**. [CrossRef]
22. Wang, J.; Park, J.G. A Novel Fingerprint Localization Algorithm Based on Modified Channel State Information Using Kalman Filter. *J. Electr. Eng. Technol.* **2020**, *15*, 1811–1819. [CrossRef]
23. Zhang, P.; Liu, J.; Shen, Y.; Jiang, X. Exploiting Channel Gain and Phase Noise for PHY-layer Authentication in Massive MIMO Systems. *IEEE Trans. Inf. Forensics Secur.* **2020**. [CrossRef]
24. Magsino, E.R.; Ho, I.W.H.; Situ, Z. The effects of dynamic environment on channel frequency response-based indoor positioning. In Proceedings of the IEEE International Symposium on Personal, Indoor and Mobile Radio Communications, PIMRC, Montreal, QC, Canada, 8–13 October 2017.
25. Xue, W.; Hua, X.; Li, Q.; Yu, K.; Qiu, W.; Zhou, B.; Cheng, K. A New Weighted Algorithm Based on the Uneven Spatial Resolution of RSSI for Indoor Localization. *IEEE Access* **2018**, *6*, 26588–26595. [CrossRef]
26. Wang, J.; Park, J.G. A novel indoor ranging algorithm based on amreceived signal strength indicator and channel state information using annextended kalman filter. *Appl. Sci.* **2020**, *10*, 3687. [CrossRef]
27. Yang, B.; Guo, L.; Guo, R.; Zhao, M.; Zhao, T. A Novel Trilateration Algorithm for RSSI-Based Indoor Localization. *IEEE Sens. J.* **2020**, *20*, 8164–8172. [CrossRef]
28. Zhou, C.; Yuan, J.; Liu, H.; Qiu, J. Bluetooth Indoor Positioning Based on RSSI and Kalman Filter. *Wirel. Pers. Commun.* **2017**, *96*, 4115–4130. [CrossRef]
29. Xue, W.; Qiu, W.; Hua, X.; Yu, K. Improved Wi-Fi RSSI Measurement for Indoor Localization. *IEEE Sens. J.* **2017**, *17*, 2224–2230. [CrossRef]
30. Shin, B.; Lee, J.H.; Lee, T.; Kim, H.S. Enhanced weighted K-nearest neighbor algorithm for indoor Wi-Fi positioning systems. In Proceedings of the 2012 8th International Conference on Computing Technology and Information Management, ICCM 2012, Seoul, Korea, 24–26 April 2012.

MDPI
St. Alban-Anlage 66
4052 Basel
Switzerland
Tel. +41 61 683 77 34
Fax +41 61 302 89 18
www.mdpi.com

Sensors Editorial Office
E-mail: sensors@mdpi.com
www.mdpi.com/journal/sensors

www.ingramcontent.com/pod-product-compliance
Lightning Source LLC
Chambersburg PA
CBHW041939260926
51850CB00018B/300
* 9 7 8 3 0 3 6 5 1 9 1 3 5 *